WATER

WATER

The Epic Struggle for Wealth, Power, and Civilization

STEVEN SOLOMON

NEW YORK • LONDON • TORONTO • SYDNEY • NEW DELHI • AUCKLAND

HARPER PERENNIAL

A hardcover edition of this book was published in 2010 by HarperCollins Publishers.

FIRST HARPER PERENNIAL EDITION PUBLISHED 2011.

Designed by William Ruoto

Maps by Brittany Watson

The Library of Congress has catalogued the hardcover edition as follows:
Solomon, Steven.
Water : the epic struggle for wealth, power, and civilization / Steven Solomon.–1st ed.
p. cm.
Summary: "A narrative account of how water has shaped human society from the ancient past to the present"–Provided by publisher.

ISBN 978-0-06-054831-5

1. Water and civilization. I. Title
CB482.S65 2010
553.7–dc22 2009027500

ISBN 978-0-06-054831-5 (pbk.)

12 13 14 15 OV/RRD 10 9 8 7 6 5

For Claudine Macé

Water mingles with every kind of natural phenomenon; and more than one might imagine, it has also mingled with the particular destiny of mankind.

—Fernand Braudel, *Memory and the Mediterranean*

When the well is dry, we learn the worth of water.

—Benjamin Franklin, *Poor Richard's Almanac*

CONTENTS

LIST OF MAPS

PROLOGUE

In 1763 a twenty-seven-year-old instrument maker named James Watt repaired a model of a Newcomen steam engine owned by the University of Glasgow. Britain was in the grip of a dire fuel famine resulting from the early deforestation of its countryside, and many of the primitive engines invented by Thomas Newcomen a half century earlier were working to pump floodwater from coal mines so that more coal could be excavated as a substitute fuel. While repairing the Newcomen machine, Watt had been startled by its inefficiency. Filled with the spirit of scientific inquiry then going on in the Scottish Enlightenment, he determined to try to improve its capacity to harness steam energy. Within two years he had a much-more-efficient design, and by 1776 was selling the world's first modern steam engine.

James Watt's improved steam engine was a turning point in history. It became the seminal invention of the Industrial Revolution. Within a matter of decades, it helped transform Britain into the world's dominant economy with a steam-and-iron navy that lorded over a colonial empire spanning a quarter of the globe. Britain's pioneering textile factories multiplied their productivity and output by shifting from waterwheel to steam power and relocating from rural riversides to new industrial towns. Steam-driven bellows heated coke furnaces to produce prodigious amounts of cast iron, the plastic of the early industrial age. Watt steam engines helped overcome Britain's fuel famine by pumping excess water out of coal shafts—and put the discharge to use by supplementing the water supply of the inland canals that had sprung up to expedite the growing shipments of coal from the collieries to the markets. Watt steam engines abetted the rise of urban metropolises, and improved the health and longevity of their residents, by pumping up freshwater from rivers for drinking, cooking, sanitation, and even firefighting. From Watt's steam engine, a new industrial society took hold that launched human civilization on an altogether new trajectory. World and domestic balances of power were recast, and mankind's material existence, population levels, and expectations increased more in just two centuries than they had in all the thousands of preceding years.

Yet as momentous as it was, Watt's innovation for exploiting steam power was but one of a long list of water breakthroughs that have been causally entwined with major turning points in history—much like the one unfolding before us today. Water has strongly influenced the rise and decline of great powers, foreign relations among states, the nature of prevailing political economic systems, and the essential conditions governing ordinary people's daily lives. The Industrial Revolution was akin to the Agricultural Revolution of about 5,000 years ago, when societies in ancient Egypt, Mesopotamia, the Indus Valley, and northern China separately began mastering the hydraulic arts of controlling water from large rivers for mass-scale irrigation, and in so doing unlocked the economic and political means for advanced civilization to begin. Ancient Rome rose as a powerful state when it gained dominance over the Mediterranean Sea, and developed its flourishing urban civilization at the heart of its empire on the flow of abundant, clean freshwater brought by its stupendous aqueducts. The takeoff event and vital artery of China's medieval golden age was the completion of its 1,100-mile-long Grand Canal, which created a transport highway uniting the resources of its wet, rice-growing, southerly Yangtze region with its fertile, semiarid Yellow River northlands. Islamic civilization's brilliance was sustained by the trading wealth that accompanied the opening of its once-impenetrable, waterless deserts by long-distance camel caravans that spanned from the Atlantic to the Indian oceans. Open oceanic sailing was the West's breakthrough route to world dominance, which it built upon through its leadership in steam, hydraulic turbines, hydroelectricity, and other water technologies of the industrial age. The sanitary and public health revolutions of the late nineteenth and early twentieth centuries that underpinned mankind's unprecedented demographic transformations sprung from efforts to provide freshwater free of filth and conditions inhospitable for disease-carrying organisms. America's historical rise, too, was explained in important part by its mastery and integration of its three diverse hydrological environments: exploitation of the industrial waterpower and transport potential of the year-round rivers in its temperate, rainy eastern half, highlighted by the catalytic Erie Canal; the naval domination of its two ocean frontiers, and ascendance to world leadership, following the epic building of the Panama Canal; and the triumph over its arid Far West by its pioneering innovation of giant, multipurpose dams inaugurated with its Depression-era Hoover Dam. The worldwide diffusion of giant dams, in turn, was a linchpin of the Green Revolution, and ultimately the emergence of today's global integrated economy.

That control and manipulation of water should be a pivotal axis of power and human achievement throughout history is hardly surprising. Water has always been man's most indispensable natural resource, and one endowed with special, seemingly magical powers of physical transformation derived from its unique molecular properties and extraordinary roles in Earth's geological and biological processes. Through the centuries, societies have struggled politically, militarily, and economically to control the world's water wealth: to erect cities around it, to transport goods upon it, to harness its latent energy in various forms, to utilize it as a vital input of agriculture and industry, and to extract political advantage from it. Today, there is hardly an accessible freshwater resource on the planet that is not being engineered, often monumentally, by man.

Whatever the era, preeminent societies have invariably exploited their water resources in ways that were more productive, and unleashed larger supplies, than slower-adapting ones. Although often overlooked, the advent of cheap, abundant freshwater was one of the great growth drivers of the industrial era: its usage grew more than twice as fast as world population, and its ninefold increase in the twentieth century rivaled the more celebrated thirteenfold growth in energy. By contrast, failure to maintain waterworks infrastructures or to overcome water obstacles and tap the hidden opportunities water always presents has been a telltale indicator of societal decline and stagnation.

Every era has been shaped by its response to the great water challenge of its time. And so it is unfolding—on an epic scale—today. An impending global crisis of freshwater scarcity is fast emerging as a defining fulcrum of world politics and human civilization. For the first time in history, modern society's unquenchable thirst, industrial technological capabilities, and sheer population growth from 6 to 9 billion is significantly outstripping the sustainable supply of fresh, clean water available from nature using current practices and technologies. Previously, man's impact on ecosystems had been localized and modest. Across heavily populated parts of the planet today, much of the rivers, lakes, and groundwater on which growing societies depend are becoming dangerously depleted by overuse and pollution. As a result, an explosive new political fault line is erupting across the global landscape of the twenty-first century between water Haves and water Have-Nots: internationally among regions and states, but just as significantly within nations among domestic interest groups that have long competed over available water resources. Simply, water is surpassing oil itself as the world's scarcest critical resource. Just as oil conflicts were central to twentieth-century history, the struggle over freshwater is set to shape a new turning point in the world order and the destiny of civilization.

Humanitarian crises, epidemic disease, destabilizing violence, and corrupt, failed states are already rife in the most water-deprived regions, where 20 percent of humanity lacks access to sufficient clean freshwater for drinking and cooking and 40 percent to adequate sanitation. Those who have predicted that the wars of the twenty-first century will be fought over water have foremost in mind the water-starved, combustible Middle East, where water looms omnipresently over every conflict and peace negotiation, and where those with oil are desperately trying to postpone their day of reckoning by burning it to pump dry aquifers and desalinate seawater in order to sustain farms and modern cities in the desert. Freshwater is an Achilles' heel of fast-growing giants China and India, which both face imminent tipping points from unsustainable water practices that will determine whether they lose their ability to feed themselves and cause their industrial expansions to prematurely sputter. The buffeting global impact will be especially far-reaching for the fates of water-distressed developing nations that are reliant on food imports to feed their swelling, restive populations. While the West, too, has some serious regional water shortages, its relatively modest population pressures and generally moist, temperate environments make it an overall water power possessing significant water resource advantages. If aggressively exploited, these advantages can help relaunch its economic dynamism and world leadership.

The lesson of history is that in the tumultuous adjustment that surely lies ahead, those societies that find the most innovative responses to the crisis are most likely to come out as winners, while the others will fall behind. Civilization will be shaped as well by water's inextricable, deep interdependencies with energy, food, and climate change. More broadly, the freshwater crisis is an early proxy of the twenty-first century's ultimate challenge of learning how to manage our crowded planet's resources in both an economically viable and an environmentally sustainable manner. By grasping the lessons of water's pivotal role on our destiny, we will be better prepared to cope with the crisis about to engulf us all.

PART I

Water in Ancient History

CHAPTER ONE

The Indispensable Resource

Earth has aptly been called the "water planet." It is, like ourselves, 70 percent water. Alone among the solar system's apparently lifeless planets and moons, it contains abundant surface water in all three of its natural states—solid ice, gaseous vapor, and, most important, flowing liquid. Water's pervasiveness and indispensable capability to transform and transport other substances played a paramount role in forging Earth's identity as a planet and the history of all life upon it. Its deceptively simple molecular architecture of one oxygen and two hydrogen atoms possesses a mighty range of powers and functions unique among Earth's substances. Water is the planet's universal solvent: its extraordinary capacity to saturate, dissolve, and mingle with other molecules to catalyze essential chemical reactions makes it Earth's most potent agent of change. It is water that conveys the life force of nutrients and minerals upward against gravity to crops, treetops, and the blood vessels of human beings. It is water that enabled the earliest forms of life to evolve and help create the planet's oxygen-rich atmosphere. Water's anomalous property of becoming less dense and more expansive as it freezes helps fracture rocks to promote geological change and fortuitously means that an insulating layer of ice forms first over the top of lakes and rivers, protecting the water-living creatures below.

Movements of liquid water and ice sheets over eons likewise carved many of Earth's geographic landscapes and defined the changing characteristics of its habitats and climates. It is water's exceptional capacity to absorb great amounts of heat before heating up itself that moderates seasonal surface temperatures and prevents the planet from becoming a perennially steamy hothouse like Venus or a frigid desert like Mars. It is the absence of water in the soil that causes deserts to suffer the extremes of heat in daytime and intense cold at night, while it is water's presence that maintains comfortable ranges in temperate zones. Moving water creates and constantly redistributes the planet's skin-thin layer of fertile topsoil, which when cultivated and magically watered in the right amounts yields civilized man's daily bread—primarily wheat in the Mideast and Europe, rice in south Asia, maize and potatoes in the Americas, and tubers in Africa.

Among water's most indispensable qualities is that it is Earth's only self-renewing vital resource. Evaporated water precipitates in a desalinated and cleansed form over the planet through Earth's continuous water cycle to restore natural ecosystems and make sustained human civilization possible. Although the remarkably constant total volume of accessible, self-renewing freshwater is infinitesimally tiny as a virtual few droplets of the planet's total water, it has sufficed to provide *all* of the water needed to support mankind throughout the entirety of human history—until today.

Water appeared on Earth early in the planet's infancy over 4 billion years ago, possibly through collisions with ice-bearing comets. Over time it assumed its familiar forms, such as oceans, ice sheets, lakes, rivers, streams, and wetlands on the surface, rainfall, snow, and water vapor in the air, and the invisible subsurface of shallow groundwater systems, soil moisture, and deep reservoirs of confined aquifers. Transmutations between water's three natural states help drive Earth's climate change cycles, prominently including the long fluctuations between cold, dry ice ages and warm, wet interludes like the present era.

Earth's last great ice age lasted some 90,000 years and reached its zenith about 18,000 years ago with ice covering one-third of the planet, compared to about one-tenth today. With so much water locked up in ice, global sea levels were about 390 feet lower. Now separate landmasses were traversable on foot. As the ice sheets melted and receded over the next several thousand years, they enriched the soil, filled underground aquifers, and created the contours of our present geography of lakes, rivers, and harbor-rich coastlines and filled in the shallow seas and channels—the English Channel, for instance, land-bridged England and continental Europe as recently as 9,000 years ago. Thick forests grew in the new temperate zones left behind by the glaciers, particularly in the Northern Hemisphere, where the glaciers had been concentrated. Then, about 10,000 years ago, the planet entered an anomalous interlude in which the climate became both warm and unusually stable. It was under these highly favorable climatic conditions that human civilization made its debut on Earth's stage.

Basic water conditions of aridity and moisture, seasonality and variable predictability patterns of precipitation, and river flow signatures and navigable lengths are defining elements of the planet's diverse range of habitats to which each occupying civilization tried to gainfully adapt during its few moments in history. Heat dispersal by ocean currents and the blanket of warm atmospheric water vapor help keep Earth habitable for humans from the equator to the subarctic latitudes. Within these boundaries are a half dozen main landscapes, each with a unique hydrological identity: Near the poles is the bitterly cold, low-rainfall, high-permafrost, and poorly drained tundra. The taiga, featuring large coniferous forests, lay south of the tundra in the Northern Hemisphere. Temperate forests, with good soil, ample rainfall, and rich flora, follow next, moving toward the equator. Then comes a belt of semiarid grasslands with less-fertile soil and erratic rainfall, such as in the barely cultivatable prairie of the North American Great Plains, Africa's savanna, and the steppes of central Asia. Interspersed among these regions are transition zones, notably one stretching from the Mediterranean to the Indus Valley and another in northern China, marked by drying, semiarid climates and several large rivers that flood over wide plains—the eventual cradle habitat of the ancient irrigation-farming-based civilizations. Between the 30-degree latitudes lie large deserts; around the equator are vast tropics with extreme rainfall, high temperatures, and rapid evaporation. Both are among the most water-fragile habitats on Earth—the former due to its dryness, and the latter due to its inundating, ever-soggy excess. Water also governs the crucial microclimates that exist within each basic zone. The seas play a dynamic role: it is the warm Atlantic Gulf Stream current that flows northeasterly from the Gulf of Mexico that keeps northern Europe wet and warm despite being at the same latitude as Canada's frigid Hudson Bay, just as the swift, northeasterly Kuroshio, or Japan Current, in the Pacific Ocean warms North America's coastal northwest. The Gulf Stream, in turn, also influences the prominent summer monsoons of Africa and Asia. Climatologists today postulate that the global conveyor belt circulation of deep and surface ocean currents acts as a key on-off switch of ice ages and is triggered by shifting mixtures of oceanic salinity and heat, particularly at the delicately balanced turnaround point in the North Atlantic. Similarly, the early signs of global warming are expected to express themselves in the form of more extreme precipitation events—more intense, frequent, and seasonally unpredictable storms, melts, and droughts. In short, every aspect of the past, present, and future of the planet and its inhabitants has been and will be powerfully influenced by water's pervasive impact.

Despite Earth's superabundance of total water, nature endowed to mankind a surprisingly minuscule amount of accessible fresh liquid water that is indispensable to planetary life and human civilization. Only 2.5 percent of Earth's water is fresh. But two-thirds of that is locked away from man's use in ice caps and glaciers. All but a few drops of the remaining one-third is also inaccessible, or prohibitively expensive to extract, because it lies in rocky, underground aquifers—in effect, isolated underground lakes—many a half mile or more deep inside Earth's bowels. Such aquifers hold up to an estimated 100 times more liquid freshwater than exists on the surface. In all, less than three-tenths of 1 percent of total freshwater is in liquid form on the surface. The remainder is in permafrost and soil moisture, in the body of plants and animals, and in the air as vapor.

One of the most striking facts about the world's freshwater is that the most widely accessed source by societies throughout history—rivers and streams—hold just six-thousandths of 1 percent of the total. Some societies have been built around the edges of lakes, which cumulatively hold some 40 times more than rivers. Yet lake water has been a far less useful direct resource to large civilizations because its accessible perimeters are so much smaller than riversides. Moreover, many are located in inhospitable frozen regions or mountain highlands, and three-fourths are concentrated in just three lake systems: Siberia's remote, deep Lake Baikal, North America's Great Lakes, and East Africa's mountainous rift lakes, chiefly Tanganyika and Nyasa. Throughout history, societies have also widely accessed shallow, slowly flowing groundwater, which is the underground counterpart of surface rivers and lakes.

The minuscule, less than 1 percent total stock of accessible freshwater, however, is not the actual amount available to mankind since rivers, lakes, and shallow groundwater are constantly being replenished through Earth's desalinating water cycle of evaporation and precipitation—at any given moment in time, four-hundredths of 1 percent of Earth's water is in the process of being recycled through the atmosphere. Most of the evaporated water comes from the oceans and falls back into them as rain or snow. But a small, net positive amount of desalted, cleansed ocean water precipitates over land to renew its freshwater ecosystems before running off to the sea. Of that amount, civilizations since the dawn of history have had practical access only to a fraction, since two-thirds was rapidly lost in floods, evaporation, and directly in soil absorption, while a lot of the rest ran off in regions like the tropics or frozen lands too remote from large populations to be captured and utilized. Indeed, the dispersion of available freshwater on Earth is strikingly uneven. Globally, one-third of all streamflow occurs in Brazil, Russia, Canada, and the United States, with a combined one-tenth of the world's population. Semiarid lands with one-third of world population, by contrast, get just 8 percent of renewable supply. Due to the extreme difficulty of managing such a heavy liquid—weighing 8.34 pounds per gallon, or over 20 percent more than oil—societies' fates throughout history have rested heavily on their capacity to increase supply and command over their local water resources.

Some societies developed in landscapes that offered relatively abundant, easily accessible, water resources with reliable availability and moderate variations; others have been hindered by more water-fragile and arduous habitats marked by dearth or excess and, worst of all, frequent, unpredictable shocks like extreme droughts and floods that overwhelmed otherwise sound hydraulic planning. Each unique environment imposed opportunities and constraints that helped shape that society's organizational patterns and history.

Adaptation is a constant necessity because water conditions are in flux. As historians Ariel and Will Durant have written, "Every day the sea encroaches somewhere upon the land, or the land upon the sea; cities disappear under the water . . . rivers swell and flood, or dry up, or change their course; valleys become deserts, and isthmuses become straits . . . Let rain become too rare and civilization disappears under sand . . . ; let it fall too furiously, and civilization will be choked with jungle." Natural secular climate change alters conditions slowly, but dramatically, over time. As recently as 5,000 years ago the Sahara Desert was verdant grassland with hippopotamuses, elephants, and cattle herders, whose water has since evaporated and seeped away into deep, fossil aquifers, while today's desiccating, windblown northern plain of the Yellow River was a watery swampland at the time it was a cradle of ancient Chinese civilization. Almost everywhere civilization has taken root, man-made deforestation, water diversion, and irrigation schemes have produced greater desiccation, soil erosion, and the ruination of Earth's natural fertility to sustain plant life.

How societies respond to the challenges presented by the changing hydraulic conditions of its environment using the technological and organizational tools of its times is, quite simply, one of the central motive forces of history. Repeatedly, leading civilizations have been those that transcended their natural water obstacles to unlock and leverage the often hidden benefits of the planet's most indispensable resource.

CHAPTER TWO

Water and the Start of Civilization

In *A Study of History,* British historian Arnold Toynbee influentially posited that the history of civilizations was centrally driven by a dynamic process of responses to environmental challenges. Difficult challenges provoked exceptional, civilizing responses in ascendant societies, while inadequate responses contributed to stagnancy, subordination, and collapse in declining ones. Prominent among the environmental challenges was water.

Throughout history, wherever water resources have been increased and made most manageable, navigable, and potable, societies have generally been robust and long enduring. Those that succeeded in significantly increasing their command and supply regularly were among the few that broke out of history's normative condition of changelessness and bare subsistence to enjoy spurts of prosperity, political vigor, and even momentary preeminence. Often major water innovations leveraged the economic, population, and territorial expansions that animated world history. Those unable to overcome the challenge of being farthest removed from access to the best water resources, by contrast, were invariably among history's poor.

Water's leading role in civilization was visible in the landscape of natural and man-improved waterways that ever have been history's directional arrows of exploration, trade, colonial settlement, military conquests, agricultural expansion, and industrial development. Wherever navigable waterways converged or where key river crossings or favorable harbors were established, influential urban centers arose at the center of civilizations. In every age, whoever gained control of the world's main sea-lanes or the watersheds of great rivers commanded the gateways of imperial power. If the advance of civilization could be charted on an electronic, time-sequenced map of the globe, it would show early city-states unifying up river valleys, along seacoasts, then spreading across nearby seas, and finally extending westward to link all the world's oceans and waterways in an ever-denser and faster-traveled web that has evolved into today's fitfully integrated global economy and world civilization.

It was also a common pattern of history that expansions driven by intensified use of water and other vital resources were followed by population increases that in turn so increased consumption that they ultimately depleted the further intensification capacity of the society's existing resource base and technologies. Such resource depletions thus presented each society with a moving target of new challenges requiring perpetually new innovative responses to sustain growth. This population-resource equation—the ever-shifting balance between each society's population size and the resources and know-how within its means to produce enough goods to sustain it—and its activating cycle of intensification and resource depletion, too, was one of the central dynamics of human and water history. History was littered with societies that declined simply because they could not overcome the deleterious local-resource depletions and population expansions accompanying their own initial success.

Signature water challenges evolved from era to era. Breakthrough responses that harnessed new water resources by novel means in one epoch sowed new conditions from which emerged the defining water challenges, and opportunities, of the next. At each turn of the cycle, the equation of water advantages recalibrated, altering the power balances among states and interest groups. Successful responses in every epoch, however, were invariably marked by intensified productivity in at least one of the five principal, interrelated ways water has been used throughout world history: (1) domestically for drinking, cooking, and sanitation; (2) economic production for agriculture, industry, and mining;

(3) power generation, such as through waterwheels, steam, hydroelectricity, and as coolant in thermal power plants; (4) for transportation and strategic advantage, militarily, commercially, and administratively; and (5) of growing prominence today, environmentally to sustain vital ecosystems against natural and man-made depletions and degradations. Whenever a major breakthrough occurred in any of these principal uses, such as Watt's improvement of the steam engine, it often had an outsized, transformational impact upon history by converting what had been a water impediment into a dynamic force for expansion. Time and again, too, an ascendant civilization's expansion involved the fusion of two or three diverse hydrological environments with its original habitats, such as the combining of a river's swampy delta with its upper river valley, a semiarid farming region of millet and wheat with another dominated by verdant monsoonal hillsides and rice farming, or a zone of wide deserts or temperate, rain-fed river and farming lowlands with the opportunities of open seas. Dynastic declines and the fall of expansive civilizations, when they came, also often occurred along the same hydrological fault lines.

Water tied man to Earth with a special bond. Fetuses gestated in water. Man and environment mutually exchanged water through the natural biological cycles of perspiration, exhalation and evacuation, and replenishment by drinking. A healthy, active person needed to consume at least two to three quarts of freshwater daily to stay alive—there was no substitute. Thirst came with only a 1 percent water deficiency; a 5 percent shortfall produced a fever; a 10 percent dearth caused immobility; death struck after about a week with a 12 to 15 percent water loss.

The special affinity between water and man was reflected in water's primary role in creation stories in diverse cultures throughout the world. "Almost every mythology sees the origins of life coming out of water," observed mythologist Joseph Campbell. "And, curiously, that's true. It's amusing that the origin of life out of water is in myths and then again, finally, in science, we find the same thing." Water was one of the Greeks' four primary terrestrial elements and one of five in ancient China and Mesopotamia. Water still plays a central role in common religious rituals of purification from Hinduism and Shinto to Islam, Christianity, and Judaism. Whether it was the rain dance of a tribal shaman, the ritual opening of an irrigation canal by an ancient king, or the dedication of a giant hydroelectric dam by a twentieth-century president, provision and control of sufficient water has conferred political legitimacy in all forms of human society.

Yet water's unique natural characteristics always simultaneously presented a double-edged challenge to civilized man: it was at once the necessary resource of survival that when brought under control yielded immeasurable benefits to society, but it also imposed one of the most formidable natural obstacles and limitations to growth. Water sustained life, but also, through the devastating shocks of drought, flood, and mudslide, could obliterate it on a terrifying scale, as witnessed in the quarter of a million deaths in the Indian Ocean tsunami of late 2004 and the traumatic 2005 inundation of New Orleans. While man needed water to live, drinking contaminated water and exposure to stagnant water infested with disease-carrying organisms was by far the leading cause of debilitating illness, infant mortality, and short life spans for all of history. Rivers, seas, and the waterless oceans of the deserts could be protective or constraining, in turn a separating defensive buffer between societies, or a bridge between peoples to open communication and trade, or a highway of invasion and conquest. Irrigation watered cropland, but also raised fertility-killing salts to the soil's surface. The secret of water's extraordinary potency to transform history was that whenever a society, in its constant struggle to wrench a surplus from nature, was able to innovate to make its water resources more manageable, abundant, potable, or navigable, it not only merely liberated itself from a major water-bound obstacle and constraint, but also unlocked and harnessed more of water's inherent, often hidden catalytic potential for growth.

A radical transformation in man's relationship to water played a pivotal role in the great transition to settled agriculture at the start of history. After eons as hunter-gatherers following their alimentary mainstay of herds of giant herbivores such as steppe bison, giant elk, and woolly mammoth from seasonal water hole to water hole, and gathering wild, edible plants along the way, some human tribes about 10,000 years ago began to adopt a settled economic life predicated upon the artificial transformation of nature through farming. As a hunter-gatherer, primitive man used water as he found it. As a settled farmer, managing water resources became essential to survival and growth.

Environmental change in climate and water conditions offered the most likely explanation of the mystery why hunter-gatherers suddenly traded their relatively undemanding and healthy lifestyle for the more-labor-intensive, less-healthy challenges of farming life. As the ice age glaciers retreated northward due to increased global warming and moisture at the start of the present warm period, tundra mosses and grasses also retreated and were gradually replaced by thick temperate forests. This forced the large herds ever northward after their food supply. An abrupt, 1,300-year-long mini ice age around 12,900 years ago may have accelerated the herds' disappearance. Some groups gave up following the herds to hunt smaller animals, fish, and gather the wild cereals and other edible plants that flourished on open landscapes. Experiments with settled farming and animal domestication ensued. Gradually, domesticated seed agriculture based on wild barley and emmer wheat grasses emerged in the Middle East's Fertile Crescent, which had transformed from prairie to semiarid landscapes as the climate changed. Farmers began to work the amply rain-fed, well-drained, and easier-to-work soils of wooded river valley hillsides with simple stone and wooden axes, hoes, and sickles. Slashing the bark killed enough trees for sunlight to nourish seeds planted in the loose leaf mold around the trunks. Scattered ash from burning the dead trees after two or three plantings temporarily revitalized the soil's depleted fertility for another few seasons. Finally, weed invasion forced such "slash-and-burn" method farmers to abandon the land and move on to clear new plots. Early walled, irrigated farming and trading settlements ultimately arose in a few favored locations. Jericho, possibly the world's oldest true city from about 7000 BC, with about 3,000 inhabitants, internal cisterns, grain storerooms, and a tower within its 10 acres, was situated at the lower slope of Mount Carmel near an ample "sweet," or freshwater, spring—in contrast to "bitter" water with traces of salt—known in the Bible as Elisha's Spring that irrigated the small, fertile, once-forested Jordan River valley and was later to lure the biblical Joshua and his Hebrew followers after the exodus from Egypt. Jericho's location also gave it access to the precious salt and trade routes of the Dead Sea. Salt had become vital to maintaining body fluid once the human diet had shifted to cereals.

Slash-and-burn farming on hillsides had one major drawback. It was always extremely vulnerable to erratic rainfall. The response to this environmental challenge produced one of history's most momentous innovations—the rise of large-scale, irrigated agriculture, and with it the birth of civilization. The earliest irrigated farming civilizations all developed along the semiarid plains of large, flooding, soil-bearing rivers where precipitation was too scant for rain-fed agriculture. In Mesopotamia, where civilization arose first, some hillside farmers moved down into the stoneless, muddy floodplains and swamps of the lower Tigris-Euphrates river valley in Sumeria near the mouth of the Persian Gulf. It would seem to be counterintuitive for farmers to relocate into a forbidding, miasmic habitat marked by scant rainfall and infestations of deadly waterborne diseases, and prone to violent floods and droughts. Yet the rivers possessed two prized resources that trumped all drawbacks—an ample, reliable, year-round supply of freshwater and a self-renewing source of fertile silt that spread across the cropland with the floods. If productively managed by the arduous building and maintenance of irrigation waterworks, the water supply and silt could produce bountiful yields many times greater than were possible on the rain-dependent hillsides. By specializing in the mass production on cultivated fields of one or two staples like wheat, barley, or millet, farm communities that mastered the techniques of irrigation ultimately produced grain surpluses that were stored as reserves for bad seasons when the floods were excessive or inadequate. These food surpluses in turn yielded rising populations, big cities, and all the early expressions of civilization—arts, writing, taxation, and the first large states—that were the precursors of modern society. The development of reed and wooden rafts, powered by oar and sail, also turned the rivers into high roads for trade, communication, and political integration. As political power concentrated and production of field seed grains expanded with better organization, these early, river-based irrigation civilizations became cradles of history's first great empires.

Irrigation farming societies also developed in other hydrological habitats based upon staple crops other than field agriculture of wheat and related grains. By the third millennium BC transplanted rice was being extensively cultivated in paddies along the naturally flooded fields of the monsoonal river valleys of Southeast Asia. This garden cultivation, too, required sophisticated, labor-intensive water management—storing the downpour, transplanting rice plants, and submerging and draining paddies at just the right levels and seasons, for instance, to support much more densely populated, civilized communities. Yet the monsoon-fed wet rice garden cultivation was not done on a scale as grand or politically centralized as in the semiarid wheat field agriculture, river-irrigation states. The seasonal rains delivered sufficient natural water supplies to support independent, smaller communities that did not need, and indeed could better resist, any centralized government command. In fact, the case of early wet rice societies supported the general observation that the way water resources presented themselves exerted a strong influence on the nature of the society's political system. By and large, where wealth-creating water resources were widely and easily accessible and not dominated by the existence of an arterial transport and irrigating waterway, there was a stronger tendency for smaller, more-decentralized, and less-authoritarian regimes to prevail.

The spread of civilization to cropland watered only by rainfall about a thousand years after the rise of large-scale irrigation societies represented one of history's most enduring, if slow-moving, expansive forces. In wheat-growing regions, the key development was the diffusion of the animal-powered wooden traction plow that facilitated the cultivation of large enough tracks to sustain fixed village communities. Yet nowhere did rain-fed farming ever produce the food surpluses, population densities, grand civilizations, and empires supported by irrigation; their heyday on the world stage, rather, depended upon the advent of other, later technological developments. Thus, for nearly all of history when wealth came from agriculture, one of the central dividing lines of human civilization lay between water-rich, irrigated agrarian states and water-challenged, sparsely populated, relatively poor, small, rain-fed farming communities.

Two other historical dividing lines were also marked by water usage. One was the gradual emergence of seafaring civilizations on the fringes of antiquity's irrigation empires in lands with marginal domestic agricultural fertility that earned their wealth principally by trading among neighboring states. Sea trade, which exploited the fast and cheap navigation potential of water's buoyancy, advanced in the Mediterranean Sea with the development of sail and oar-powered wooden cargo ships suitable for its relatively calm, enclosed waters and evolved very gradually into a significant historical force by the second millennium BC. In the Mediterranean, the Red Sea, and the Indian Ocean, sea traders facilitated cross-cultural and commercial exchange based upon market-set prices that over many centuries helped propagate a small but vibrant unregulated economic sphere in which the early beginnings of the modern market economy were nurtured.

The other great water demarcation line was the barbarian divide—the existential clash of societal organization and lifestyles between the nomadic pastoralist descendants of primitive hunter-gatherers and the expanding, civilized agricultural realm. The militarily skillful barbarian tribes of the central Asian steppes, the desert Bedouins of the Arabian Peninsula, and later the Nordic Vikings in their river longboats were far fewer in number and ranged Earth's more water-fragile and less-yielding landscapes, herding their animals between water holes, and trading with or, when strong enough, raiding or demanding tribute from civilized settlements. Periodically, they gathered enough strength under great warrior leaders to lead fearsome invasions that disrupted, overwhelmed, and ultimately reinvigorated civilized empires across the world. History recorded four great barbarian waves, starting with the Bronze Age charioteers of 1700–1400 BC and ending with the Turkish-Mongol invasions from the 700s through the fourteenth century, when the age of gunpowder and sheer manpower advantages decided matters for settled civilization. The slow, fitful expansion of civilized society around the globe was always synchronized with the breakthroughs and setbacks in irrigated and rain-fed cultivation, some of which were landmark events of world history. Wherever farming took hold, population advanced. In 8000 BC the planet was populated by about 4 million hunter-gatherers. After 5000 BC it began to double every 1,000 years. By 1000 BC world population had reached about 50 million. Then, under the prosperous shield of order-providing empires, it generally accelerated. By the late second century AD, some 200 million people inhabited the world. Even Watt's steam engine and the Industrial Revolution did not lessen civilization's reliance on agriculture. Instead, they provided new tools to enhance innovation production to feed the growing demand of a world inhabited by 6.5 billion by the early twenty-first century. Despite the dramatic expansion of total cropland, water volumes, and agricultural technologies, one thing that has not changed since ancient times is man's greater reliance on irrigation to feed himself. Today, two-fifths of the world's food is being grown on less than one-fifth of the planet's irrigated, arable land. All of human society today shares an irrigation legacy with the cradle civilizations of antiquity.

CHAPTER THREE

Rivers, Irrigation, and the Earliest Empires

One of the striking common features of ancient history was that all of mankind's four great cradle civilizations were wheat, barley, or millet field irrigation agricultural societies that arose in semiarid environments alongside large, flooding, and navigable rivers. For all their differences, Egypt around the Nile, Mesopotamia along the twin Tigris and Euphrates, the Indus civilization around the Indus, and China along the middle reaches of the Yellow River also shared similar political economic characteristics. They were hierarchical, centralized, authoritarian states ruled by hereditary despots claiming godly kinship or mandate in alliance with an elite class of priests and bureaucrats. All power was imposed top-down through control of water, which was the paramount factor of economic production, and managed through the marshaling of mass labor.

In his classic 1957 work, *Oriental Despotism,* Karl A. Wittfogel proposed a causal linkage between centralized authoritarian states and specialized, mass irrigation agriculture. The overriding challenge of so-called hydraulic society, he posited, was how to intensify exploitation of its silt-spreading, flooding river's potential water resources. The larger the river, the greater was the potential productive wealth, population density, and power of the ruling hydraulic state. Yet only centralized planning and authoritarian organization on an immense scale could exploit water resources to their productive maximum. Surplus yields depended critically upon delivery of adequate supplies of water at the right time to the right places as well as protection against catastrophic flooding. This required the forced, often brutal mobilization of hundreds of thousands, and sometimes millions, of peasant laborers during lulls in the farming season to construct and maintain irrigation and diversionary canals, sluices, water storage dams, protective dikes and levees, and other waterworks.

The bulky, inherently hard-to-manage physical property of liquid water itself, noted Wittfogel, "creates a technical task which is solved either by mass labor or not at all." Once conscripted and organized for waterworks, the workforces were readily mobilized by the state to construct its other celebrated grand monuments of hydraulic civilizations—pyramids, temples, palaces, elaborate walled cities, and other defensive fortifications like China's Great Wall. In further support of his hydraulic theory, Wittfogel observed that similarly organized theocratic, authoritarian, gigantic public-works-building agrarian societies, based on miraculously easy and fast-growing maize and potatoes, and responding to other labor-intensive water management challenges, were reinvented again much later in the New World, among the Olmec-Maya on cultivated swamp mounds of the tropical lowland habitats of Central America and on the bleak, terraced, and irrigation-channeled mountain plateaus of the Andes inhabited by the Incas and their predecessors.

Wittfogel's theory of hydraulic society fueled much debate over the decades, including whether the cooperative needs of irrigation created the large centralized state or vice versa. Yet such debate often sailed past the most salient point: the two social formations were complementary; they reinforced one another. Power and social organization in such societies depended absolutely upon regimented, concentrated control of the water supply. Whenever the water flow was interrupted, whether from natural or political causes, crop production fell, surpluses dissipated, dynasties and empires toppled, and starvation and anarchy threatened the entire social order. Ancient hydraulic societies tended to thrive where two prominent conditions existed: first and foremost, where the best available resources of water were highly concentrated in the state-controlled irrigation source; second, where the unifying presence of a dominating, navigable river gave the state command over regional communication, commerce, political administration, and military deployment.

For millennia authoritarian irrigation societies produced the most advanced civilizations in the world. Although the hydraulic model would be supplemented, and eventually superseded, by new social formations, it produced a recognizable prototype that has endured through history. Whatever the era, huge water projects requiring vast mobilization of resources tended to go hand in hand with large, centralized state activity. Vestiges of this hydraulic tendency were evident in the giant dams built in the twentieth century by centralizing liberal democratic, communist, and totalitarian states, often in the early stages of restoration periods.

Ancient Egypt was the prototype hydraulic civilization because its river, the Nile, was the consummate hydraulic waterway. The Greek historian Herodotus, who visited in 460 BC, famously described Egypt as the "Gift of the Nile." Indeed, Egypt's history was—and still is—almost entirely determined by what happened on and around the natural phenomena of its great river.

The Nile provided everything that was needed in virtually rainless Egypt. It was the only large source of irrigation water and its annual flood brought a thick, self-renewing layer of fertile black silt for its farmland. Unlike other great rivers, the annual flood season arrived and receded with clockwork predictability and in miraculous synchronization with the agricultural cycle of planting and harvesting. It was one of the easiest landscapes to manage for irrigation. Egyptian farmers needed merely to construct embankment breeches, sluice gates, extension channels, and some simple dikes to retain sufficient floodwater to soak the soil in the cultivated, low-lying basins beyond the river before releasing the excess to the next basin downstream. The Nile's steep gradient, furthermore, kept the river flowing steadily with good drainage that helped to flush out the soil-poisoning salts that afflicted artificial irrigation systems everywhere else. Indeed, the Nile was world history's only self-sustaining, major river irrigation system.

The Nile's natural beneficence also bestowed Egypt with a second great gift—it was a rare *two-way* navigable river. Its current and surface wind moved in opposite directions all year round, so it was possible to float downriver with the current and sail south upriver in simple, broad-bottomed vessels with square sails.

Finally, the wide, waterless desert beyond both banks provided a defensive barrier that helped insulate ancient Egyptian civilization against large-scale invasion for centuries. As a result of Egypt's total reliance on its single grand river, the flow of political power to the center in Egypt was simple, total, and unchanging. Throughout history, whoever controlled the Nile also controlled Egypt.

The Nile's bounty, however, depended upon one unpredictable variable beyond the Pharaoh's control—the extent of the river's annual flood. Excessive flooding inundated entire villages and wiped away cropland. Far worse were years of low flooding when insufficient water and silt resulted in famine, desperation, and chaos. To an astonishing degree, dynastic rises and declines throughout Egypt's long history correlated to cyclic variations in the Nile's floods. Good flood periods produced food surpluses, political unity between Upper Egypt's Nile Valley and Lower Egypt's marshy delta, waterworks expansions, Egyptian civilization's glorious temples and monuments, and dynastic restorations. Extended years of low flood, by contrast, were dark ages of privation, disunity, and dynastic collapses. Without Nile water, neither the wise nor the corrupt could rule effectively. Pharaoh's kingdom fractured between valley and delta and sometimes further into competing precincts ruled by warlords and menaced by bandits.

Ancient Egypt was marked by the rise of three great kingdoms—the Old Kingdom (circa 3150–2200 BC), the Middle Kingdom (2040–1674 BC) and the New Kingdom (1552–1069 BC)—and their respective dissolutions into the intervening First, Second, and Third Intermediate Periods. Nile flood levels were so important to determining tax revenue from the harvest and overall governance that they were assiduously monitored by priestly technocrats from Egypt's early beginnings by nilometers, which were depth gauges marked off on stones and originally situated at temples along the river. Nilometer records show that the fates of Egypt's subsequent occupiers were likewise driven by cyclic oscillations in the flood level of the Nile. In short, the rhythms of the Nile framed all the essential parameters of history and life in Egypt, including food production, population size, extent of dynastic reach, and conditions of peace or strife.

Nile flood levels, in turn, ultimately depended upon an occurrence far beyond Egypt's borders—the degree of the summer monsoonal rains that fell at the headwaters of the Blue Nile. The Blue Nile started in Ethiopia's Abyssinian plateau at over 6,000 feet at a spring venerated by the modern Ethiopian Orthodox Church. The southernmost source of the White Nile, the river's other main branch, was at a spring in Burundi in Africa's equatorial plateau lake region. The Blue and White Niles came together just north of Khartoum in the Nubian Desert before entering Egypt. By the time the Nile emptied into the Mediterranean Sea, its 4,168-mile journey made it the world's longest river. Yet by total water volume it was comparatively small—only 2 percent of the mighty Amazon, 12 percent of the Congo, 15 percent of the Yangtze, 30 percent of the Mississippi and 70 percent of Europe's Danube, Pakistan's Indus, or America's Columbia rivers. Virtually none of its net flow originated within rainless Egypt's own hot, arid borders. Since about half the White Nile's water evaporated in Sudan before reaching Egypt, some four-fifths of the river flow sustaining Egyptian civilization, and nearly all its precious silt, originated in the highlands and deep ravines of Ethiopia.

Every summer monsoonal rains swelled the Nile's tributaries in Ethiopia, triggering the downstream rush and the annual flood. The river normally rose in northern Sudan by May and by June reached the first cataract near Aswan in southern Egypt. By September the entire Egyptian Nile Valley floodplain was inundated under a turbid, reddish-brown lake, which then began to recede into the main river channel but left behind its thick, odorous residue of fertile black silt. By managing the overflow of water with simple irrigation works, Egyp-

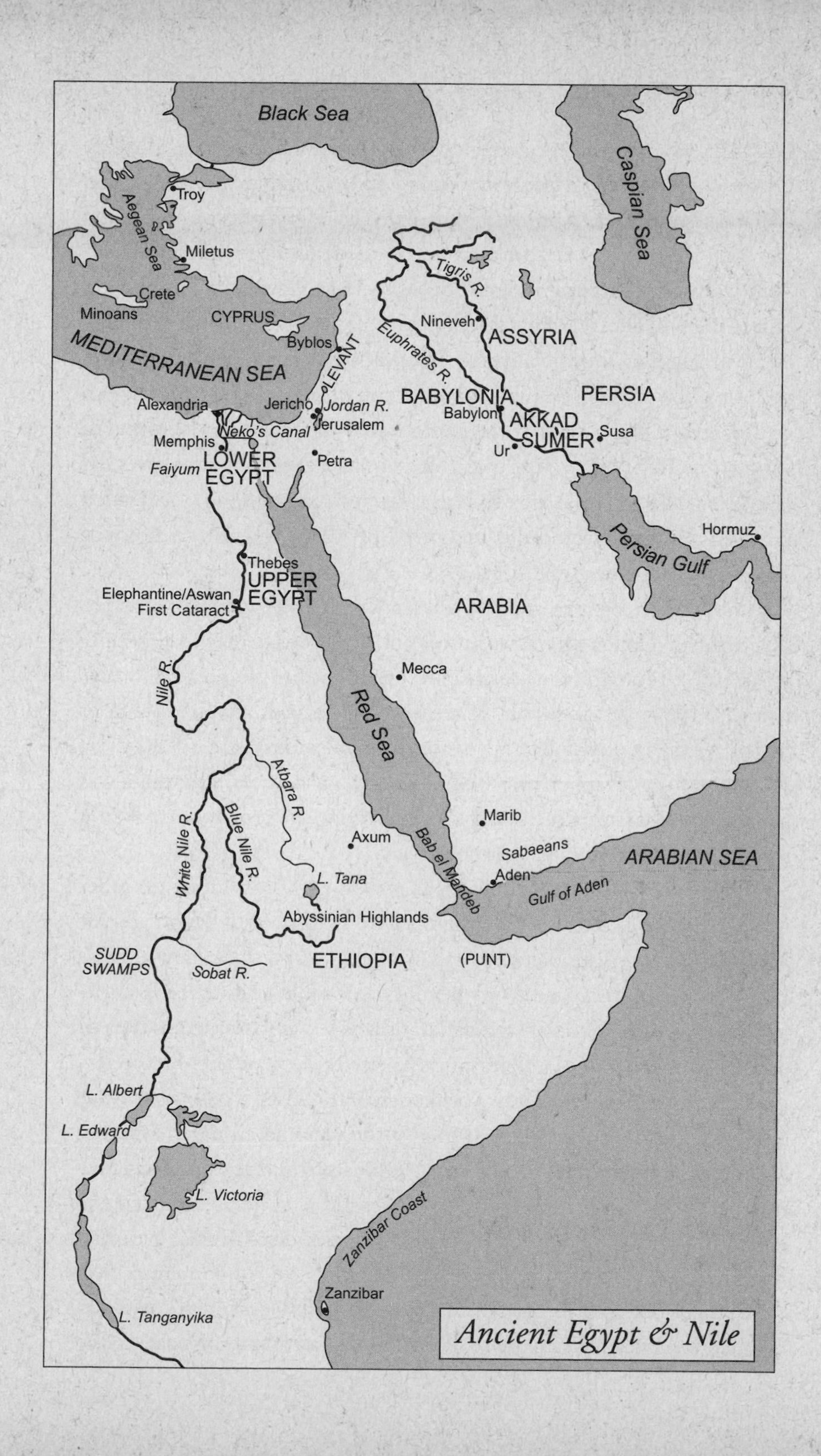

Ancient Egypt & Nile

tian farmers produced the ancient Mediterranean's richest breadbasket. Crops were planted in the waterlogged soil following the inundation and harvested in late April and May after the floods were gone; during the early summer the mud baked and cracked under the hot sun, aerating and reinvigorating the soil. Seeds were cast over the ground and buried by wooden scratch plows—a simple hoelike, wheelless implement dragged by a draft animal. Thousands of years of annual flood deposit built up 10-foot-high natural embankments ideal for human settlements on both sides of the river's nearly 600-mile length through the narrow Nile Valley. Just over the embankments were low-lying basins for farming, a total area less than modern Switzerland, into which farmers channeled the Nile's resources of water and silt to produce Egypt's emmer wheat and barley.

The Nile in Egypt consisted of two distinct hydrological and political zones. Upper Egypt was the Nile Valley from the first cataract at Aswan. Just north of modern Cairo began the fan-shaped, rich, labyrinthine 100-mile-long delta of reedy marshes and lagoons of Lower Egypt, whose topography and history were also partly molded by the fluctuating sea levels of the Mediterranean. When the kingdom was robust, one Pharaoh wore the double red and white crown symbolizing the unity of delta and valley, respectively.

The first to wear the double crown was Egypt's traditional founder Menes, the so-called King Scorpion, who as prince of Upper Egypt finally conquered the delta around 3150 BC and established Egypt's capital at Memphis at the delta's head. Power had consolidated over the previous centuries in both the delta and the valley from the battles of dozens of independent chieftains, themselves descendants of the nomadic hunter-gatherer bands who had settled closer to the river's water supply during the gradual drying of the regional climate. Whether Menes was a ceremonial title or an actual historical king, possibly identified with the early ruler Narmer, his legend accurately reflected the essential origins of Egyptian civilization, including his close personal identification with irrigation waterworks, and the fundamental duty of the ideal Pharaoh to control the flow of the Nile. Menes' royal ceremonial macehead, for instance, shows him as a conqueror wearing the white crown of the valley, dressed in a kilt and belted loincloth with a bull's tail, and using a hoe to dig an irrigation canal, while another figure removes the excavated dirt in a basket.

Menes' macehead paralleled records from other hydraulic societies showing sovereigns immersed in the daily functions of opening and closing floodgates, allocating irrigation water to peasants' fields, and directing waterworks construction. The older hydraulic civilization in Sumeria, with which Egypt had sea contact from the earliest times, was notably influential in the developmental trajectory of ancient Egypt's methods and tools. The hydraulic nature of ancient Egypt was also evidenced in the world's first recorded dam, a 49-foot masonry giant, supposedly built around 2900 BC to protect Menes' capital at Memphis from floods. Actual archaeological remains exist of another similar, masonry-faced, earthen reservoir dam 37 feet tall and 265 feet wide at the base, from between 2950 and 2700 BC about 20 miles south of modern Cairo. Far more common in ancient Egypt were simple, often short-lived earth-and-wood diversion dams to direct irrigation water during flood season.

The Nile's propitious characteristics and the simple, irrigated basin agriculture it supported visibly shaped all aspects of Egyptian culture, society, and daily life. At the apex of the hierarchical Egyptian state was the Pharaoh, the absolute sovereign who in the Old Kingdom was viewed as a living god who owned all the land and controlled the river. Supporting him was an administration of elite priest-managers, with such indicative titles as "inspector of the dikes," "chief of the canal workers," and "watcher of the nilometers." The priests' divine authority was validated by their command of such vital esoteric secrets as when the river would flood or recede, and when was the right moment to plant and to sow, and the technical engineering of durable waterworks. The state's totalitarian power was underpinned by the centralized collection, storage, and distribution of grain surpluses produced in good years. Labor on waterworks, and other state projects, was carried out by one of the oldest forms of manpower mobilization in world history—the compulsory, seasonal corvée. The peasant's duty to Pharaoh and state was so absolute that it continued into the afterlife; the peasant was often buried with clay statuettes that symbolically stood in for his perpetual work obligations after death. Control of water also helped foster the development of many of Egypt's early sciences and arts. The learned elites created calendars to facilitate farming, surveying tools to regrade and mark off land after inundations, and maintained written administrative records on parchment manufactured from the common papyrus reeds of the Nile delta. Papyrus, the oldest form of paper, represented one of history's earliest uses of water in manufacturing. It was fabricated by peeling away the outer covering of the reed, then cutting the stalks into thin strips, which were then mixed with water to activate its innate bonding properties. The thin strips were then layered, pressed, and dried.

A second linchpin of Pharaoh's power, repeated in societies throughout every age of history, was control of the region's key water transport highway. Control of shipping on the Nile allowed the Pharaoh to regulate all important transport of people and goods, and thus provided the means to exert effective rule over all Egypt. Barges laden with grain, oil jars, and other goods commonly plied Nile ports from Memphis to Thebes to Elephantine Island and, after 2150 BC, beyond to Nubia in modern Sudan when a canal was excavated through the solid granite at the Aswan water falls. The Nile artery, the richness of its valley and the delta, the predictable onset of its floods, and its protective surrounding desert also rendered Egypt one of world history's most inward-looking, changeless, rigidly ordered, and longest-enduring civilizations. Yet Egypt's simple basin agriculture was only a one-crop system with limited capacity to increase output beyond a certain ceiling. This capped Egypt's maximum population level and left Egyptians highly susceptible to famine and instability during prolonged periods of low Nile floods.

From about 2270 BC onward, the central authority and cultural grandeur of the Old Kingdom gradually disintegrated amid anarchic warfare among provincial chieftains, banditry, and famine. A climatic dry period in the Mediterranean region, which simultaneously disrupted civilization in Mesopotamia, contributed to a series of low floods on the Nile that undermined the agricultural economic basis of society.

Egyptian civilization's first dark age of disunity and competing fiefdoms lasted nearly two centuries. The return of abundant Nile floods resuscitated agricultural prosperity and facilitated the reunification into the Middle Kingdom following the military conquests and diplomacy of the rulers of Thebes in Upper Egypt in about 2040. The Middle Kingdom restoration was also associated with large new water projects and intensified food production, including the expansion of cropland into a large, swampy depression fed by high Nile floods called Faiyum. It was the prosperity of the Middle Kingdom that may have drawn the biblical family of Jacob to Egypt's delta during a period of drought and political upheaval in Palestine.

A series of droughts weakened the central state's power to resist Egypt's first full-fledged foreign invasion. In 1647 BC, the Hyksos, a Semitic-Asiatic group of Bronze Age charioteers who had penetrated Egypt through the increasingly porous Sinai Desert frontier, seized control of the delta almost without resistance. The Hyksos conquest traumatically altered Egyptian history by forcibly ending the cultural sense of fixed order and security that its isolated, predictable river environment had provided for so long.

When the hated Hyksos were finally expelled a century later and Egypt reunified under its New Kingdom, which would last half a millennium, Egypt assertively projected its cultural renewal outward through extensive foreign sea trade, military conquests of the Levant to the Euphrates and of Nubia in the south, and monuments of its native culture, such as the outsized temples at Luxor and Karnak, the modern Thebes. The New Kingdom renaissance coincided with three centuries of good Nile floods.

Harvests were increased by the intensive use of an ancient water lifting device, the shadoof. The shadoof, which likely originated in Mesopotamia centuries earlier and reached Egypt gradually, could raise 600 gallons of water per day. The device itself was a long pole on a fulcrum counterbalanced by a bucket at one end and a stone weight at the other. Two men operated it. One filled the bucket while the other leaned on the stone to lift the bucket and pour its contents into an irrigation channel that carried it to small plots. The shadoof allowed Egyptian farmers to irrigate a supplemental crop outside the main flood season.

In subsequent centuries, more powerful water-lifting devices were applied to the Nile. Archimedes' screw, invented by Greek polymath Archimedes and introduced by Hellenic rulers in the century following Alexander the Great's conquest in 332 BC, was a giant corkscrew encased in a long, watertight tube that raised river water through its grooves by hand cranking. Most important was the noria—a chain of pots attached to a large wheel turned by tethered oxen marching in a circle—which had arrived with the Persian conquerors in the sixth century BC. Noria pots filled up through dipping in the river and emptied their water into a pipe or channel as they descended from the apex of their arc. In first millennium BC Egypt, a single noria could raise water as much as 13 feet and irrigate 12 acres of a second crop in the off-season. It was also used to drain swamps for reclamation. By raising and redirecting heavy and bulky water, water-lifting technologies increased irrigated cropland by as much as 10 to 15 percent through the Greek and Roman occupations. The noria was so successful that it remained in continuous use until the twentieth century and the advent of electric and gasoline pumps. The noria was the water-lifting precursor of the seminal waterwheel, a transformative innovation that automatically harnessed power from flowing stream to ground flour and drove the first industries.

The New Kingdom also derived great wealth from Egypt's full embrace of foreign sea trade. Egyptians had been one of the first peoples to regularly sail the shorelines of the eastern Mediterranean. Old Kingdom vessels had regularly sailed to Byblos to secure precious, high-quality timber—so scarce in tree-poor, rainless Egypt—from the fabled cedar forests of Lebanon, to make boats, plows, and other vital tools of civilization. Pyramid illustrations show that as early as 2540 BC they ferried soldiers to Levantine ports in square-sailed ships that were simple adaptations of the river vessels that plied the gentle winds and currents of the Nile. Yet only with the reuniting of Egypt in the New Kingdom did sea trade become much more extensive. The spirit of the age was expressed by Queen Hatshepsut, one of the rare Egyptian female sovereigns and the first important queen of ancient history. Shortly after the start of her twenty-year reign in 1479 BC, inspired by an oracle of the god Amun, Hatshepsut launched a maritime trading expedition via the treacherous Red Sea to restore trade with Punt in the Horn of Africa, which was one of the two sources in the ancient world for prized, exotic luxuries like frankincense and myrrh, used for religious ceremonies and embalming mummies. The expedition returned with live frankincense trees that were transplanted in the queen's gardens and inaugurated three centuries of thriving shipping and trade extending from Punt to the eastern Mediterranean. Cedars from Lebanon, copper from Cyprus, silver from Asia Minor, and fine crafts and Asian textiles arrived from the north. Invariably, Egyptian military campaigns in the Levant accompanied the increasing overlapping of economic interests. Yet although Egypt profited handsomely from this sea trade, it never transcended its Nile-centric legacy to also become a true Mediterranean maritime civilization. Light, oversized Nile boats never were built to venture far beyond the safe, known shoreline routes. To transship larger cargoes across the open waters, the Egyptians instead relied upon abler vessels and seamen from Crete, where the Minoans presided over the Mediterranean's first great seafaring civilization.

The Mediterranean provided the highway for three waves of raiders that began to attack Egypt at the end of the thirteenth century BC. The Sea Peoples were an eclectic collection of seafarers put to flight as refugees from their homelands by the invasion of Iron Age barbarians from inland mountains who disrupted Bronze Age civilizations throughout the Middle East. After about a century, following a campaign that included one of history's early naval battles, the Egyptians finally repelled the last of the Sea Peoples, whose allies, the Philistines, settled in Palestine and would soon engage in battle with the Hebrews, who had earlier fled Egypt under Moses in search of new lands. Ultimately, however, the cyclic ebbing of good Nile floods marked a century of internal disorder and disintegration, while history's ever-expanding geographic overlap of regional empires brought new waves of foreign invasions that ushered in the longest and bleakest of Egypt's intermediate periods. For nearly four centuries, Egypt was subjugated by foreigners. Libyans, Nubians, and, briefly, the fierce, iron-weapon-armed Assyrians put their imprint on the Nile and its lands.

A brief resurgence of native Egyptian rule occurred in the late seventh century BC. Under the ambitious Pharaoh Neko II (610–595 BC), Egypt established a powerful navy, while its armies won battles and seized land in the war-torn Levant as far as the Euphrates River. Neko followed a policy of opening Egypt to the thriving Greek world and sea trade. He is most renowned in water history for excavating the first documented "Suez Canal," which he hoped would help Egypt compete in the Mediterranean. Neko's canal did not follow the same route from the Red Sea to the Mediterranean as the celebrated nineteenth-century canal. Instead, it connected the Red Sea to a branch of the Nile, and thus could unite Egypt's Mediterranean and Red Sea fleets of Greek-style galleys propelled by three banks of oars, known as triremes. It was dug wide enough to allow two vessels to pass each other. According to Herodotus, 120,000 died constructing it. Neko reportedly stopped work before completion when an oracle warned him that it would work to the advantage of his foreign enemies.

In fact, the canal was completed after the Persian conquest under King Darius I, who ruled from 521 to 486 BC, to facilitate shipping from Egypt to Persia. The new Hellenic dynasty of Ptolemies that assumed power after Alexander the Great's conquest hastened to dredge and extend the canal in the early third century BC. It was revitalized again in the early second century AD, under Emperor Trajan, during the height of the Roman Empire, but silted up under the Byzantines. The canal may have been opened and then later filled in by early Muslim rulers. In the early sixteenth century, the Venetians and the Egyptians discussed reopening the Red Sea to a Mediterranean connection in response to the Portuguese creation of the all-water spice trade route to India that broke the long, profitable Venice-Alexandria stranglehold on the region's trade with the East. But nothing came of it. Also fruitlessly, the Ottomans considered such a project during the sixteenth century; the sultan and the Christian king of Spain instead struck an armistice that allowed each side to focus its energies on fighting its own religion's heretics. Thereafter, the water link between the Red Sea and the Mediterranean remained closed until 1869, when the world-changing geopolitical and engineering marvel of the modern Suez Canal was opened to world commerce and naval forces.

Control of the Nile, and the caprices of its high and low floods, remained vital to the fortunes of all of Egypt's subsequent occupiers. The Greek and Roman overlords from Alexander's conquest in 332 BC to the fourth century AD, were blessed by good Nile floods and even rainfall, which facilitated an impressive expansion of cultivated land and intensified irrigation. For Rome, which took control in 30 BC, Egypt became the empire's export grain breadbasket, vital to maintaining its armies as well as the daily bread dole for Rome's legions of restive poor. The collapse of Byzantine rule to the Arab invaders in AD 640 followed a century of poor flooding. Three centuries of full Niles nourished Islam's heyday. Recurrent low Niles during the tenth to eleventh centuries, however, ultimately undermined the rule of the Fatimid founders of Cairo.

Due to its geostrategic centricity at a key crossroads of world commerce and politics, Egypt and its arterial river remained a pivotal theater in the struggle among the great powers through the centuries to modern times, including the nineteenth-century contest for global empire between England and France, and the twentieth-century Cold War between America and the Soviet Union over the Aswan Dam and regional influence in the oil-rich Arab Middle East.

Ancient Mesopotamia faced far more complex and adverse hydrological environmental challenges than those posed by the Nile. Yet even earlier than Egypt, it developed a hydraulic-model civilization that reflected the resources, cycles, and flow signatures of its flooding, silt-spreading twin rivers, the Tigris and the Euphrates, that coursed through the heart of the Fertile Crescent in modern Turkey, Syria, and Iraq. The rivers of Mesopotamia—which in Greek means "the land between the rivers"—helped create the ancient world's most precocious large civilization, featuring the wedge-shaped cuneiform that was mankind's first written language, the first large cities, sophisticated water-lifting and irrigation technologies, the wheel, towering spiral-shaped ziggurat temples, and dynamic, expansive empires.

Unlike the Nile, the twin rivers' distinguishing characteristic was that their floods spilled and receded unpredictably, often violently—and always out of rhythm with the needs of the farming cycle. When water was needed most, during autumn planting and plowing, the rivers were at their lowest. In the late spring, the nearly full-grown plants were imperiled with ruin from swollen rivers that flooded suddenly from torrential rainstorms of thunder and sheet lightning. Having two rivers with many branches complicated the hydraulics of the farming floodplain. Spillover from the higher yet slower-moving Euphrates, for instance, often drained easterly into the larger Tigris. Due to their shallow gradients, both rivers were prone to meander and, during big floods, to cut new courses to the sea, stranding existing cropland and entire communities of their life-giving water supply.

The key to civilization in Mesopotamia, therefore, rested upon skillful, year-round regulation of the twin rivers through extensive waterworks. Large reservoir dams stored water that was released in the growing season. Water was lifted to levels required to flood the furrowed fields. Strong protective dikes were needed to prevent flooding at the wrong time. A drainage network of sluices and diversion ditches was needed to combat waterlogging in the flat, poorly draining cropland. In short, if natural hydraulics made Egypt a gift of the Nile, Mesopotamia was an artificially contrived civilization whose success was achieved in defiance of the design of nature through the water-engineering ingenuity and willful organization of society.

In every way, Mesopotamia's material, social, and political existence was more volatile and uncertain than Egypt's. Instead of being protected behind a waterless desert barrier, its region was a natural crossroads of peoples, ideas, and goods, surrounded by potential raiders and rivals who lived in the rainy hills whose tributaries fed the twin rivers' plains. Extensive commerce and conflicts between city-states, and constant invasions by ever-larger empires, marked Mesopotamia's history. As the rivers' resources came under better control, moreover, political power tended to move upstream, where cropland was unspoiled and strategic command could be exerted over the rivers' unidirectional transport navigability and the region's water supply. "A society dependent on water coming from a canal that could be blocked several miles upstream from the fields it served was extremely vulnerable to warlike attack," writes historian William H. McNeill. "A position upstream was therefore always of supreme strategic importance in Mesopotamian politics and war, while downstream populations were always and inescapably at the mercy of whoever controlled the water supply." From Sumer in the fourth and third millennia BC to Akkad under Sargon I in 2334 BC, Babylon under Hammurabi in about 1792 BC, Assyria by 800 BC, and the Persians by 500 BC, the preeminent centers of Mesopotamian civilization moved generally upriver and its irrigation zone widened.

The Tigris and the Euphrates both rose in the Anatolian highlands of modern Turkey. The Euphrates started southwesterly through expanses of desert plateaus before veering sharply southeast to form the top of the funnel-shaped, flat floodplains with the southerly descending Tigris, which itself was replenished with snowmelt and runoff from tributaries flowing out of the Zagros Mountains of western Iran. The two rivers nearly merged in the region of modern Baghdad on the Tigris and ancient Babylon on the Euphrates, then ballooned gently to form the borders of the stoneless, easily farmed, fertile mudflats intersected with oft-shifting river channels of lower Mesopotamia, the biblical location of the Garden of Eden. South of the ancient Sumerian cities of Ur and Uruk, both rivers spilled into a full-fledged marshland rich in fowl, fish, and milk-bearing water buffalo but too swampy for farming. Finally, the rivers discharged their united flow into the Persian Gulf.

It was in hot, river-rich lower Mesopotamia, where precipitation was too meager for rain-fed farming, that permanent settlements based on large-scale irrigation agriculture first took hold after about 6000 BC. In the fourth millennium, many hundreds of years prior to Menes' founding of ancient Egypt, it flowered as Sumerian civilization. Everything depended upon mastery of the waters, whose secrets, according to Mesopotamian mythology, were revealed to man by Enki, the kind and wise fertilizing water god. Irrigation transformed Sumeria into a veritable garden, with abundant grains and nut and fruit trees, including the multiuseful date palm, close at hand.

The Sumerians' origins remain mysterious. It is suspected that they arrived by sea via the Persian Gulf. Their language was unlike

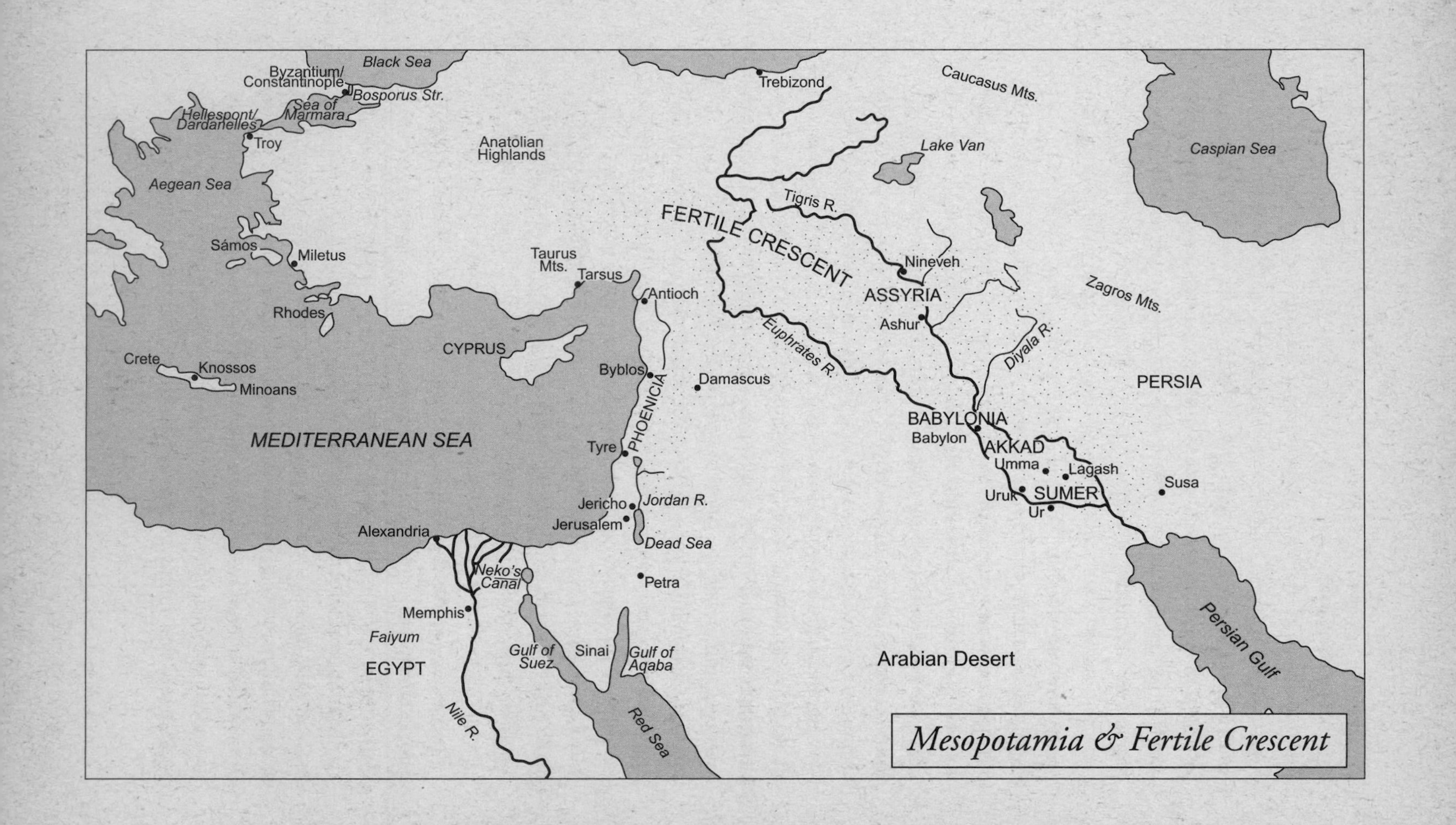

Mesopotamia & Fertile Crescent

any other known group, with a distinctive grammar and vocabulary. Sumeria was a civilization of walled city-states, each located about 20 miles apart, with their own stock of grain supplies used as payment in commercial transactions. About a dozen gradually rose to prominence. Uruk, just inland of marshes at the head of the Persian Gulf, was by 3400 BC the largest-known city on Earth, with more than two square miles within its walls. Ur, later the home of the biblical Abraham, was a port trading city on a since-vanished branch of the Euphrates with protective moats, canals, two harbors, a towering ziggurat temple in its center, and a population of 20,000 to 30,000.

In Sumeria begins the original urban revolution, and the civilizing influence of the city throughout history. In every age cities stimulated commerce and markets, the exchange of ideas, the arts, the division of labor, specialization, and the accumulation of surplus for investment that lay at the heart of economic expansion and the rise of great states. The great cities of history were integrally linked to man's uses of water and were, without fail, situated on rivers, lakes, oases, and seashores. City historian Lewis Mumford observes that "the first efficient means of mass transport, the waterway" was the most "dynamic component of the city, without which it could not have continued to increase in size and scope and productivity." For ancient Sumerian city-states, the waterway provided the economic lifeline that brought copper and tin to make bronze, stone, timber, and other vital raw materials absent in Mesopotamia. Sumerian vessels traded long-distance with Egypt via the Red Sea and plied the gulf and the Indian Ocean at least to the ancient Indus River civilization, known in written Sumerian records as Meluhha, from which they acquired beads of carnelian and lapis lazuli, timber, gold, and ivory.

The vital economic activity of the earliest Sumerian city-states, however, was irrigation agriculture. Each had its own farm work gangs comprised of many hundreds of farmers who worked large tracts of land that was owned, rented, or bequeathed by the gods. As in Egypt, coerced labor was done under schedules and regulations set by temple priests, who alone possessed the skills for calculating the changes of season, designing canals, and coordinating mass, collective effort. The religious provenance of the priesthood legitimized their taking large shares of the annual harvest surpluses for storage in the temple granaries.

Violent, unpredictable floods that destroyed waterworks and entire cities were an omnipresent, terrifying menace. Indeed, in Mesopotamian mythology the quasi-divine status of kings and the state's political legitimacy itself sprang from a purifying great flood sent by the gods to obliterate humanity and from whose watery chaos a new world order was born. The region's flood myth centered on a single, forewarned family that survived by building an ark—the progenitor of strikingly similar stories in Hindu mythology and the Noah story in Genesis. Mesopotamia's flood myth also reflected an acute awareness of water's precariously dual nature as both potential life-giver and great destroyer, as well as the king's obligation to avert floods while ensuring ample water for irrigation.

Farming in Sumeria started around the tributaries and main stem of the Euphrates, which was slower moving, easier to control, and richer in nutrient-bearing silt, and had a wider floodplain, than the Tigris. The higher-elevation Euphrates at first used the Tigris as an overflow drain for waters that fed through the network of primary and secondary canals that fed irrigation to the crops. Eventually, some of the wider irrigation canals were made large enough to carry ships and barges. Crops were grown on miles-long earthen embankments set amid the watery plain between the rivers and controlled by a matrix of dams, dikes, weirs, sluices, and ditches. One benefit of this arduous, artificial irrigation was that it permitted year-round, multicrop farming that yielded larger stockpiles than Egypt's single-crop basin system. Yet artificial irrigation also came with a terrible side effect that afflicted civilizations throughout history—salinization of the soil.

Contemplating the desolate, scrubby modern landscape of lower Mesopotamia, the twentieth-century excavator of Ur, British archaeologist Leonard Woolley, puzzled over what had happened to the former brilliant civilization: "Why . . . if Sumer was once a vast granary, has the population dwindled to nothing, the very soil lost its virtue?" The answer, determined Woolley's successors, was that increased salt accumulations in the poorly drained soil had depleted its fertility and the ecosystem foundation of Mesopotamian civilization. Over time, intensive irrigation farming had environmental side effects that undermined its sustainability. It tended to raise the level of groundwater to waterlog the soils, while water's capillary action drew deadly salt toward plant roots. Evaporation, which was especially rapid in hot, arid Mesopotamia, left the telltale crusted salt residue across the once-fertile surface—crop yields fell until finally little at all could grow. Mesopotamian tablets from 1800 BC duly record "black fields becoming white." To cope with salinization, the Sumerians shifted production from wheat to more-salt-resistant barley. In about 3500 BC, equal amounts of wheat and barley were being grown in Sumeria. A thousand years later, only 15 percent of the crop was wheat. By 1700 BC almost no wheat was being grown, and yields from both crops had declined by some 65 percent over seven centuries.

World history is replete with societal declines and collapses caused by soil salinization. Ancient Egypt was spared crippling soil salinization and waterlogging only because the Nile's propitious seasonal flooding and sloping valley drained away excess water and most salts in a timely manner. A second man-made environmental depletion also exacerbated Mesopotamia's agricultural crisis—deforestation. Wherever humans have settled on Earth, they have chopped down trees—for fuel, houses, boats, tools, and agricultural-land clearance—until their habitats were denuded. Many now-barren parts of Mesopotamia, as elsewhere in the neighboring Mediterranean rim, were once luxuriously verdant. Deforestation made landscapes drier and less fertile. It reduced rainfall as well as the capacity of the soil to retain what did fall. More of the fertile topsoil washed away in torrential downpours—a malevolent expression of water's power as history's greatest soil mover, surpassed only by modern industrial man himself.

In Sumeria the intersection of rising population and depleting agricultural resources finally created an unstable equilibrium wherein all easily irrigable cropland came under cultivation and the boundaries of city-states began to bump up against one another. Centuries of border conflicts over irrigation supplies and cropland resulted, including the world's first recorded water war, fought intermittently with several arbitrated temporary settlements from 2500 to 2350 BC, between neighboring city-states Umma and Lagash. The war seems to have been started by upriver Umma, first by seizing disputed land for cultivation and then by breeching irrigation canals from a branch of the Euphrates. It was ultimately won by Lagash, whose preserved tablets provide history's only account of the story. A decisive breakthrough for Lagash was its construction of an irrigation channel that gave it an independent water supply from the Tigris, and the ability to divert canal water supplying parts of Umma.

The first signs of disintegration of southern Mesopotamian civilization had become visible as early as 3000 to 2800 BC, following an abrupt, disastrous shift in the Euphrates—archaeologists aren't sure if it occurred naturally or inadvertently by man's waterworks. The river's course change displaced the water supply of prominent city-states, intensifying their life-and-death competition for water. Ultimate resolution of Sumerian city-states' contentious water and cropland challenges came from outside Sumeria—in the form of military conquest and unification by an upriver Semitic dynasty founded by a commoner who rose to power, Sargon of Akkad.

By legend, Sargon, who ruled from about 2334 BC, had a prestigious water pedigree: as an abandoned baby, he had been set to float in a basket along the river and found by a gardener—a mythic foundling-water motif later adopted in the Exodus story of Moses and in ancient Rome of the twins Romulus and Remus. He usurped power from a city-state king he had served meritoriously, raised an army, defeated the whole of Sumeria, and, like Egypt's Menes some eight centuries before him, created the region's first large, unified state. Its center was Akkad, a new city north of lower Mesopotamia whose location today is unknown, but may lie buried under modern Baghdad. Sargon's empire absorbed Sumerian high culture and invigorated it with a centralized military and political system in which former city-state rulers became governors loyal to him, now exalted as a semidivine king of kings. Cultivation of wheat and barley expanded to unspoiled lands upriver, and agricultural estates were granted to loyal allies. Irrigation works were a vital lever of his political authority. A new system of taxation—whose effectiveness was an unfailing measure of state power throughout history—was implemented by which local farming income flowed to the central bureaucracy to solidify the empire. To obtain metals, timber, and other vital resources not readily available in muddy Mesopotamia, he traded with distant societies and waged military action in the Levant. Excavated large Akkadian cities fit the classic, hydraulic state pattern: large granaries of barley and wheat delivered by farm carts along paved roads, central state distribution of carefully measured rations of staples like grain and oil according to jobs performed and to age, and a central acropolis linking the states to the deities.

Successful as it was, Sargon's Akkadian empire lasted only a century, collapsing about the same time as Egypt's Old Kingdom. While ancient legend attributed the collapse to "The Curse of Akkad" caused by an abomination one of Sargon's heirs committed against the preeminent god of air and storm, Enlil, modern science has identified another explanation: regional climate change—the prolonged arid and cold period that gripped the Mediterranean area at the time. Regional climate change would also help explain the collapse of Egypt's Old Kingdom during the same period. Archaeological soil evidence from the site of a major lost city in north Akkad has revealed that drought was so severe in one dirt layer corresponding to 2200 to 1900 BC that even the earthworms had perished.

After a chaotic interval, imperial restoration was briefly achieved by southerly Ur. Records show, however, that despite the expansion of waterworks, Ur's fragile revival was constantly menaced by poor barley harvests, floods, and neighboring enemies. Much later it was entirely abandoned to the desert sands, when the Euphrates changed course away from its walls and the gulf coast receded.

Reunification returned upriver to an enlarged area after two centuries under a powerful new dynasty centered in Babylon, on the Euphrates. The greatest of the Babylonian kings was Hammurabi, who ruled for forty-two years after inheriting the throne in 1792 BC. Befitting the expectations of kingship of the era, Hammurabi proclaimed himself the divine "provider of abundant waters for his people" who "heaps the granaries full of grain" and strived to validate his legitimacy by delivering both. In the early part of his reign, he dedicated himself to the smallest details of essential internal developments, such as digging irrigation canals and fortifying cities.

In the latter part of his reign, especially while establishing full dominance over Mesopotamia after 1766 BC, Hammurabi also used water as a political tool and a military weapon. To win loyalty and revive the old Sumerian heartland he had just conquered, for instance, he built a canal to serve its leading city-states. To subdue his great enemy city-state, Eshnunna, on a tributary of the Tigris near modern Baghdad, he dammed the river upstream and then released it in a devastating torrent. Withholding Euphrates River water was another weapon used with such deadly force by his son to subdue rebellious cities, including Nippur, Ur, and Larsa, that farmland in the far south turned to barren steppes and did not recover; the diversions proved difficult to reverse, causing many of the land's inhabitants to migrate north.

Hammurabi is most renowned in history for the world's first written public code of justice—pithily summarized as "an eye for an eye, a tooth for a tooth"—which was inscribed on a seven-foot-high stone stele erected in Babylon's main temple. The 282 laws of the Code of Hammurabi illuminated the conditions and primary concerns of ancient Babylonia. Prominent among them was water. Many laws dealt with the individual responsibilities for the operation of irrigation dams and canals, with penalties that reflected their critical importance to the order of Babylonian society. For example, Law 53 read: "If anyone be too lazy to keep his dam in proper condition, and does not keep it so; if then the dam break and all the fields be flooded, then shall he in whose dam the break occurred be sold for money and the money shall replace the corn which he has caused to be ruined." Laws 236, 237, and 238 dealt with the restitution due by a negligent boatman to the owners of lost cargo or a sunken vessel. Hammurabi's code also encompassed some political protections of the private property rights and contracts of merchants, whose trading activities brought needed goods to Mesopotamia, and a surprising many on the rights of women whose arranged marriages weren't working out.

Hammurabi's five successors ruled from central Mesopotamia for 155 years. The subsequent centuries, however, were filled with regional upheaval, headlined by two great waves of barbarian invasion. The first was associated with the Bronze Age charioteers; the second, which subsided after 1100 BC, with the Iron Age invaders. Iron—and its later, harder cousin, steel—was one of world history's most transforming innovations, comparable in impact to electricity or the computer silicon chip in modern times. As with these modern inventions and many other industrial technologies, iron production depended crucially upon skillful use of freshwater. Iron-making technologies that began in the Caucasus Mountains around 1500 BC were mastered in nearby northern Syria. Unlike the bronze alloy of copper and tin that was easily smelted at the temperatures of ordinary fire, iron ores required firing from much-hotter-burning charcoal. The absorption of carbon created "steeled iron," which became famously hard when quenched red hot in water. Yet the quenching had to be skillfully interrupted in order to prevent too-rapid cooling that would result in a uselessly brittle metal.

Hard iron weapons and tools dramatically changed the military and economic balances of power. Bronze Age empires everywhere were overthrown. States that mastered and applied iron became the great powers of the new age. Foremost among these was the Assyrian Empire, which, during its peak between 744 and 612 BC, consolidated power throughout the Fertile Crescent as far as Egypt. Following the upriver trajectory of Mesopotamian history, Assyria's heartland was in the north, on the Tigris near modern Mosul, then a wilderness region with ample rainfall in which lions still roamed. The Assyrians were infamous for their bloodthirsty and relentless military campaigns in which their armies "gleaming in purple and gold" came down upon their enemies "like a wolf on the fold," as poet Lord Byron put it. Yet their success also rested on an arsenal of masterful hydraulic projects executed with martial discipline, precision, and iron tools, including some innovations and achievements that make them among the greatest water movers in history. The Assyrians were prolific and expert builders of dams, which they used to increase water supply both for irrigation and the domestic needs of their large cities. This was best exemplified by the formidable hydraulic works undertaken by King Sennacherib, who ruled from 704 to 681 BC, to increase the water supply needed to expand his luxurious, double-walled capital city at Nineveh, with its 15 great gates, and its surrounding plantations of fruit trees and exotic flora, including the water-thirsty cotton plant, which the Assyrians called the "wool-bearing tree."

Nineveh stood on the Khosr River just above its confluence with the Tigris. The engineering challenge was that the Tigris lay too far below the city to raise enough water for the growing capital. Instead, between 703 and 690 BC, Sennacherib undertook three separate projects to obtain more water via the Khosr. First, he dammed that river 10 miles to the north and diverted it to Nineveh through an open-air canal. When that didn't provide enough water, he augmented its flow by damming and rerouting 18 small streams and springs from the hills 15 miles to the northeast. When that still failed to satisfy Nineveh's growing thirst, in 690 BC his engineers built a masonry dam at an oblique angle across a deep gorge to divert the water of another river through 36 miles of winding channels to feed the Khosr. At one juncture, a massive 1,000-foot-long, 40-foot-wide stone aqueduct with five arches was constructed to carry the canal across a valley toward Nineveh. Among other sophisticated hydraulic features of Nineveh's elaborate and integrated water system was the employment of a water-pressurized, U-shaped, inverted siphon pipe to carry water across and then up a topographical depression.

The Assyrians also institutionalized one of world history's landmark breakthroughs in obtaining clean urban drinking water—the qanat. Originating in the hilly region of what is today eastern Turkey and northwestern Iran, qanats were long, deep, slightly inclined tunnels hewn through subterranean rock face into underground mountain aquifers and ran by gravity to lower-lying population centers. Being underground, they lost little water to evaporation—a major problem in hot environments. Their construction depended on iron tools and required sophisticated mining and engineering capabilities, including precise gradients and the cutting of vertical shafts for maintenance access and ventilation. The testament to the qanat's great success was its ubiquity throughout central Asia and the Mediterranean rim, from southern Spain and Morocco to the west to northern India to the east. Romans built them when they occupied the region. They were mainstays throughout the realm of Islamic civilization. Spanish colonists introduced them, much later, in Mexico. Qanats even remained actively used into the twentieth century and provided much of Tehran's water supply until the 1930s.

Since qanats required large quantities of water to be drawn from deep wells in a confined space, they encouraged the Assyrians to innovate improvements in water-lifting techniques based upon wheel-based pulleys. The spread of qanats overlapped the sixth century BC development of the earliest Greek aqueducts. Between them, ancient Middle East and Greco-Roman engineering tried almost every water supply technique aside from river water pumping used by civilization until the nineteenth century.

Sennacherib was famous in the Bible for his long siege of Jerusalem in 701 BC during the reign of King Hezekiah in response to a rebellion across Palestine against Assyrian hegemony. Jerusalem's historical greatness in antiquity owed as much to its water supply as its strategic trade crossroads location. The city's main water source was the Gihon Spring just outside its walls. Its pre-Hebrew occupants, the Jebusites, had connected the spring to the city by a 1,200-foot-long secret underground water tunnel to protect themselves against siege. Yet the tunnel became their undoing in about 1000 BC when King David discovered its whereabouts and Hebrew soldiers stole through it to take the city by surprise.

David's successor, Solomon, promptly solidified the new kingdom by expanding the city's water supply with three large external reservoirs to feed the city's internal network of cisterns and rain-collecting water tanks. Keenly aware of this history as he fortified his defenses in anticipation of Sennacherib's siege three centuries later, King Hezekiah ordered the digging of a new secret water tunnel underneath Jerusalem to transport water from the source of the Gihon Spring to a reservoir inside the city walls. Cut through sheer bedrock with a precise gradient, the 1,800-foot S-shaped tunnel has carried water almost continuously for 2,700 years. In the event, all the rebel strongholds except Jerusalem fell to Sennacherib's soldiers. Failing to find the hidden Gihon Spring or the secret water tunnel, the Assyrians decided to withdraw after Hezekiah agreed to pay a heavy tribute as reparation.

One rebellious city that did not escape Sennacherib's vengeance was Hammurabi's fabled Babylon. In 689 BC he overran the city after a fifteen-month siege, looted its treasures, massacred or deported the population, and reduced its main buildings to rubble. He prepared to seal its doom by flooding it with waters diverted through channels dug from the Euphrates. At the last moment, however, Sennacherib's son rescinded his father's plan in deference to the city's storied past; as king, he later rebuilt the city in an effort to wed Babylonians agreeably to Assyria. His leniency proved to be a grave political mistake. Within less than a century, Babylon had risen again, leading the overthrow of Assyria's empire and sacking many of its great cities.

Babylon's revival reached its zenith under the reign of King Nebuchadrezzar II from 605 to 562 BC. Nebuchadrezzar rebuilt the fabled city with the concept that it was the organizing, renewing center of the chaotic universe, with resplendent ornaments within and without its immense, 10-mile walled perimeter and majestic gates. These included the spiraling ziggurat known in the Bible as the Tower of Babel and one of the ancient world's seven wonders, the Hanging Gardens. The mechanically watered gardens were built by Nebuchadrezzar to please his Medean wife, who longed for the forested hillsides of her youthful home in what is now Iran. They are believed to have consisted of a series of terraced roof gardens of overhanging trees and plants rising on a mountainlike palace of stone balconies. Irrigation water was lifted in pots from the Euphrates by a tall noria waterwheel powered by man or animals, and flowed down from terrace to terrace. The stones were waterproofed, as were the walls of Babylon itself, against seepage by the use of viscous, tarlike bitumen.

Babylon's imperial revival did not long endure. The city finally met its doom on October 12, 539 BC, when it was overrun by the region's rising new superpower, the Persian Empire of Cyrus the Great, in the aftermath of Cyrus's major victory over the Babylonian army at the confluence of the Tigris and Diyala rivers—one of the many battles in history to be fought at strategic riversides. In an uncorroborated history of Herodotus, the city's denouement came after a long, fruitless siege when Cyrus tried one last stratagem—manipulating the Euphrates River, which ran straight through the center of Babylon and presented the only soft point in its formidable defensive walls. He stationed troops near the river's entrance and egress from the city. Upriver, other soldiers excavated a large diversion channel that redirected the river's flow away from the city. As the river's level to Babylon fell to "only deep enough to reach about the middle of a man's thigh," the Persian army waded across and successfully breached the city's floodgates before the Babylonian defenders inside realized what was happening.

Cyrus and his successors, including Darius and Xerxes, built the world's largest empire, stretching from the sands of Libya to the Jaxartes (modern Syr Darya) and Indus rivers in Asia, an area about the size of the continental United States. Its center was in the high Iranian plateau at Susa, east of Mesopotamia. They adopted and advanced hydraulic methods throughout their domain. They revitalized Mesopotamia by widening the irrigated cropland with a new grid pattern of canals, many of which were navigable by barges. An army of slaves dredged the waterways of silt. Salinization and waterlogging problems were mitigated by planting weeds when land was fallow to lower water tables and by trying not to overirrigate. As in classic hydraulic societies, the Persian sovereign visibly supervised big hydraulic operations, including allocation of irrigation water, which was distributed, in principle, to those who needed it most. It was they who introduced the water-lifting noria to Egypt, and who made the first systematic attempts to dam the main channels of the Euphrates and Tigris by constructing artificial cataracts as impediments to naval invasion from upriver; Alexander the Great, in his conquest of Persia two centuries after Cyrus, systematically removed many of them.

Herodotus also reports that wherever the mighty Persian king and his armies traveled within his empire, he was careful to drink only water, properly boiled, from one single river near Susa. "No Persian king ever drinks the water of any other stream," Herodotus wrote, "and a supply of it . . . is brought along in silver jars carried in a long train of four-wheeled mule wagons wherever the king goes." Whether apocryphal or no, Herodotus's assertion highlights the very real dangers posed by drinking water from any unknown source, as well as ancient beliefs in the mystical powers of regeneration and purification ascribed to special water sources.

Until its fall to Alexander, the mainly land-based Persian Empire was the unrivaled power of the age. The chain of events leading to its ultimate undoing, however, started with its failure a century and half earlier to vanquish the upstart naval power of small Greek city-state Athens.

In the Indus River valley of modern Pakistan, and later along the Yellow River in China, advanced ancient irrigated agrarian civilizations developed with familiar hydraulic patterns along flooding, silt-rich, navigable rivers in semiarid landscapes where precipitation was too sparse and unreliable for large, rain-fed farming. Until the 1920s, the advanced ancient Bronze Age civilization that thrived along the Indus from about 2600 to 1700 BC was lost to history. Its very existence had been discovered only accidentally when British railroad builders in colonial India unearthed some ancient bricks. Excavations revealed an enormous city of 30,000 to 50,000 inhabitants buried under centuries of Indus mud. Mohenjo Daro, on the lower Indus, was as large as any Mesopotamian city of its day and laid out in a carefully planned, rectangular grid with an elevated defensive sanctuary and a lower level. Later archaeologists uncovered scores of settlements and cities clustered along the Indus and the Arabian Sea shoreline of an entire lost civilization. They found a second, nearly identically designed giant city, Harappa, on a dried-up Indus tributary upriver in the Punjab ("the land of five rivers"), as well as a large port city linked to the sea by a mile-long canal. In all the Indus civilization occupied an area larger than its Mesopotamian or Egyptian contemporaries.

The character of this civilization remains enigmatic. Its pictorial right-to-left written language is undeciphered. But it far antedates, and

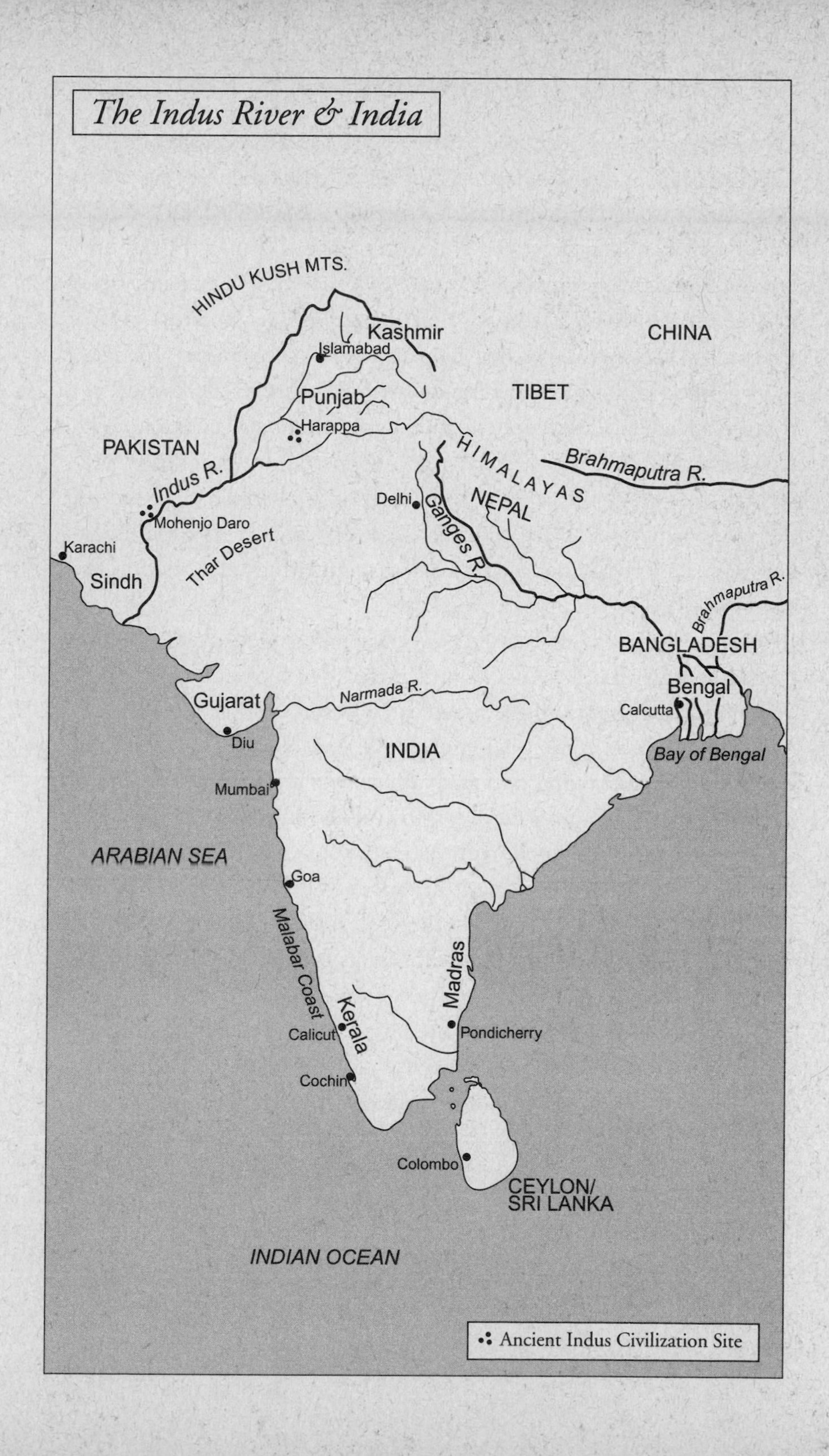

The Indus River & India
HINDU KUSH MTS.
Kashmir
CHINA
Islamabad
Punjab
TIBET
Harappa
PAKISTAN
HIMALAYAS
Brahmaputra R.
Indus R.
Delhi
Ganges R.
NEPAL
Mohenjo Daro
Karachi
Thar Desert
Sindh
Brahmaputra R.
BANGLADESH
Bengal
Gujarat
Narmada R.
Calcutta
Diu
INDIA
Bay of Bengal
Mumbai
ARABIAN SEA
Goa
Malabar Coast
Madras
Kerala
Calicut
Pondicherry
Cochin
Colombo
CEYLON/
SRI LANKA
INDIAN OCEAN
Ancient Indus Civilization Site

has no linguistic link with, the Sanskrit of the ancient Vedas and subsequent Hindu Indian civilization that inherited its domain. By every indication it was a classic hydraulic society. Its centralized, redistributive organization was suggested by the existence within its brick-built cities of capacious granaries for wheat and barley. The endemic malaria detected in the unearthed skeletal remains was a signature of the disease-bearing mosquitoes that bred in the standing water of irrigation channels and afflicted hydraulic societies everywhere. Typical of monsoonal habitats, irrigation methods apparently involved the storage of water during the wet season for release in the dry months. Trade artifacts show that the Harappan civilization probably had extensive sea trade contacts with Mesopotamia from very early on, and it is likely that Sumerian civilization had a similar stimulating impact upon its rapid growth as it had upon Egypt.

Among the Indus civilization's most intriguing features was its advanced urban hydraulics, which anticipated developments in ancient Rome by 2,000 years and the nineteenth-century sanitary awakening by 4,000 years. Its communal Great Bath at Mohenjo Daro, located in a building's inner courtyard and sunken into a platform with entry stairs at either end, was a deep, large tank about the size of an average modern swimming pool with its own water supply and drainage channels, and waterproofed with bitumen. Whether it was used for ritual purifications as in later Hindu rites, hygiene, or social gatherings as in Roman baths is unknown. But its linkage to the extensive, underground municipal sewer network, indoor toilets, and water wells in its many two-story houses reflected a precocious understanding of sanitary water supply and waste removal that would be later rediscovered elsewhere as a necessary cornerstone of urban civilization.

Perhaps the Indus civilization's greatest mystery was why it suddenly disappeared from history around 1700 BC. It used to be thought that its abrupt demise was due to its being overrun by the invasion from the northwest of light-skinned, fair-haired, Indo-European Aryan horsemen and charioteers, cousins of the Germanic, Celtic, and Hellenic warriors, whose descendants eventually established the Vedic Hindu civilization both in the Ganges and the Indus river valleys. Yet by the time of the Aryan invasion, the Indus civilization seems to have been in severe decline. The main culprit, instead, was likely its unpredictable, fragile hydrological environment.

The Indus valley was fed by two main water sources: snowmelt from the surrounding Himalayas and Hindu Kush Mountains to the north and west, and the intense deluge of the seasonal, highly variable monsoon. Rapid silt buildup in the flat floodplain made the Indus region highly prone to violent inundation. Like the Euphrates, the river's notoriously errant tributaries frequently abandoned their channels to carve new routes to the sea. Increasing regional desiccation from climate change, with steady encroachment from the Thar Desert in the east, added to the hydrological fragility. From about 2000 BC, the Indus region appears to have been ravaged by many massive, destructive floods. Many dry riverbeds, including once-great tributaries of the Indus—and possibly a vanished twin river—provided widespread evidence of rivers that had radically changed course, forcing large towns and farms to be abandoned. Mohenjo Daro itself was rebuilt at least three times. In the end, unpredictable flooding, droughts, soil salinization from irrigation, and rising water tables likely undermined its sustainable prosperity and caused its population to decline and emigrate.

The Indus civilization's fall fit a common historical pattern. The vanished, 50-mile-long irrigation canals along the Moche and Chicama rivers in pre-Inca Peru, the large spiderlike canal systems built by the Hohokam, or "gone people," native Americans in modern Arizona between AD 300 and 900, and the lost Pueblo societies that succeeded them there testified to the similar fate of many irrigation societies located in water-fragile habitats when subjected to water shocks, including long or intense periods of climate change. Ancient Petra, the rock-carved city in modern Jordan, based on seasonal wadi agriculture and caravan trade, collapsed in AD 363 when an earthquake destroyed its elaborate, cistern-based water system.

The Maya in the rain-fed, poor soils of the seasonal tropical forests of the Yucatán Peninsula ingeniously built an advanced corn-and-beans-based civilization between AD 250 and 800 upon fragile water foundations marked by pronounced unpredictability of winter aridity and little perennial surface water. They did so initially by slashing and burning fast-regenerating vegetation and then, with greater productivity and population size, draining and dredging an array of irrigation canals and farming on raised earthen mounds and hillside terraces. They also carved deep, underground cisterns in the porous limestone bedrock to store fast-collecting, seasonal groundwater runoff for year-round domestic needs. The Mayan civilization's rapid collapse and 90 percent population decline in three stages after AD 800 was likely propelled by several interrelated depletions that undermined its water resource engineering: deforestation from hillside farming as population pressures grew triggered soil erosion that cluttered its jungle canals and farming mounds with poorer soils and intensified regional aridity during dry seasons; as farm productivity suffered, internecine warfare for food increased among neighboring communities; the final blow was probably the onset of the worst long-term drought cycle in 7,000 years. The geographic pattern of collapses across the Yucatán Peninsula closely tracked the diminishing availability of accessible stored groundwater.

When high civilization was reborn in India about a millennium later following the arrival of iron, it was centered first in the Ganges River valley, an altogether different habitat characterized by luxuriant forests, heavy, monsoonal rainfall, and a river system that carried several times more flow than the Indus. The iron ax was the key innovation that cleared the jungle, followed by heavy plows pulled by eight oxen or more to turn the fertile soil for planting. From about 800 BC, monarchies with hydraulic-state attributes based increasingly on large-scale rice cultivation that could sustain denser populations began to take hold along the Ganges from valley to delta. Increasingly powerful, centralized authorities directed the skilled and labor-intensive chores of timely field flooding and draining, water storage, diking, and building and maintaining canals and embankments.

In time India's two monsoons produced two harvests, which doubled rice's intensified productivity. Over the centuries Indians developed esoteric arts for trying to read cloud patterns and ocean signals to anticipate the onset of the monsoons, which was critical for the timing of planting and the feeding of India's population. To the present day, no satisfactory solution has ever been found to the unpredictability of the monsoon's start date and the extreme variability of its volume, which remain the largest single variables in India's economic growth.

Ultimately, all of the northern half of India, from the semiarid floodplains of the Indus to the moist Ganges valley to the soggy deltas where the giant Ganges and Brahmaputra rivers spill into the Bay of Bengal, was united through the conquests of the "Indian Julius Caesar," Chandragupta, founder of the Mauryan dynasty that established India's first golden age, from 320 to 200 BC, in the aftermath of Alexander the Great's retrenchment from the Indus. In a pattern of declines and restorations reprised throughout history, a second Gupta golden age, likewise hallmarked by large-scale, centralized waterworks, flourished from about AD 300 to 500 in the reunification of these distinctive, hydrological environments. To cope with the seasonal extremes of the monsoons, Indians from this period onward, particularly in the west, began to build hundreds of distinctive, elaborately carved, templelike stepwells, three to seven stories deep, that women and children descended to retrieve water stored from diluvial periods.

Despite its superficial appearance of geographic unity, however, the Indian subcontinent was really a checkerboard of disparate hydrological and topographical environments that fostered local economic and cultural autonomy and defied easy political unification. No arterial waterway linked India's independent regions into a coherent, politically unified society. Until the coming of the British colonial overlords in the nineteenth century's age of steamships and railroads, no one managed to rule over all of India. Yet even Britain's unification of the subcontinent would be short-lived. Following its independence after World War II, the three distinctive hydrological regions of the northern India heartland fractured asunder into three separate nations: Pakistan, along the spine of the Indus; India, along the main Ganges valley; and Bangladesh, in the swampy Ganges-Brahmaputra delta.

In the center and south of the subcontinent there were other distinctive Indias, too. Coastal India, with its face to the sea, flourished as a key link in the ancient world's expanding Indian Ocean trade network. Goods had moved by sea among India, Mesopotamia, Egypt, and Southeast Asia from the early rise of civilized societies. India traded by sea in the first millennium BC with the Sabaeans from the Arabian Peninsula in modern Yemen. The Sabaeans carried cargo, including precious frankincense and myrrh grown on trees only there and in Punt in Africa's Horn, journeying overland by caravan to the Levant and Egypt.

This east-west sea trade intensified in the first century BC when Mediterranean-world sailors in the Red Sea made the historic breakthrough of mastering the Indian Ocean's two-way, seasonal monsoons to reach southern India. Shortly thereafter, sea trade routes were extended to the Malayan peninsula and the Spice Islands of modern Indonesia. In the waters of Southeast Asia travelers from the West exchanged goods with Chinese ships, creating a permanent sea link across the Old World, from China to the Mediterranean. In Muslim and colonial European times, the long-distance Indian Ocean trade route would become history's single greatest highway to world power and empire. It would be paralleled by the overland central Asian Silk Roads that also connected China, India, and the Levant. Together, the sea and land trade routes from the Far East to the Mediterranean became the axis of a market-driven international economy, with India as both a central hub and a coveted object of subjugation in the ongoing power struggle for wealth and world leadership that accompanied the rise of a new maritime civilization based principally upon sea trade.

CHAPTER FOUR

Seafaring, Trade, and the Making of the Mediterranean World

On the dry, hilly, island-flecked eastern Mediterranean fringes of antiquity's irrigated empires arose a cluster of seafaring societies that secured their wealth and defense primarily through maritime commerce and naval power. Over time these small states spawned a different kind of civilization that contrasted starkly with that of the centralized, authoritarian hydraulic societies. Its distinctive characteristics were a private-sector market economy, individual property and legal rights, and a representative democracy for those who qualified as citizens. Taking form first in ancient Greece, its traditions spread through the Mediterranean world with the Hellenist diaspora, the reach of the Roman Empire, and later through the activities of the small seafaring republics Venice and Genoa. With the eventual breakthrough of global ocean sailing in the sixteenth century, it rose to world prominence through the influence of leading Western liberal market democracies, the Dutch United Provinces, the British Empire, and the United States. Today, its imprint informs many prevailing norms of the integrated global economy.

The small, ancient maritime trading states of the eastern Mediterranean were forced to international seafaring not by preferred choice, but because of their domestic agricultural and water resource limitations. Their scant rainfall, hilly terrains, small strips of arable land, and short rivers unsuitable both for long inland navigation and mass-irrigated farming simply were inadequate to produce enough food to sustain large, prosperous populations. Yet the challenges of their harsh geography did present one opportune route to increase economic surplus. The sea itself, for those able to master the art of navigating its waters, offered a ready and cheap highway to link societies along its length that were willing to exchange their grain and other basic resources for the specialized goods produced indigenously around the Aegean—above all, its prized olive oil and wine. The region's jagged coastlines also presented good harbors conducive to maritime trade and fishing. The formidable natural sea barrier itself, furthermore, helped defend the independence of small states against the superior armies of the land-based hydraulic empires nearby.

For ancient Aegean civilizations, transport by sea came to play a role analogous to that of river transportation in Egypt and Mesopotamia. Despite its 2,500-mile length and stormy dangers, the tideless, shallow, and relatively tranquil Mediterranean (Latin for "sea between the land") was one of Earth's most welcoming seascapes for sailors. Save for the narrow eight-mile-wide passage at the Strait of Gibraltar—known in antiquity as the Pillars of Hercules—opening to the Atlantic Ocean, it was virtually an immense, enclosed lake. At its northeast extremity a pair of twin straits, the Dardanelles (known as the Hellespont in antiquity) and the Bosporus, separated Europe from Asia and led to the huge inland Black Sea and the resources of central Asia. In the southeast, between Egypt's Nile delta and the Sinai Peninsula, only a small neck of land separated it from the Red Sea and beyond to the Indian Ocean. On its eastern rim the wealthy ports of the Levant offered access to goods arriving by overland caravans from all over the Near East and beyond. Within its civilized circumference arrived all the food, raw materials, manufactured goods, and luxuries necessary to support an accomplished maritime trading civilization.

In antiquity three main sea trade routes traversed the length of the Mediterranean: one sailed along the shoreline ports of southern Europe; a parallel southern route tracked the harbors of North Africa; the third, central route sailed the open waters between major islands such as

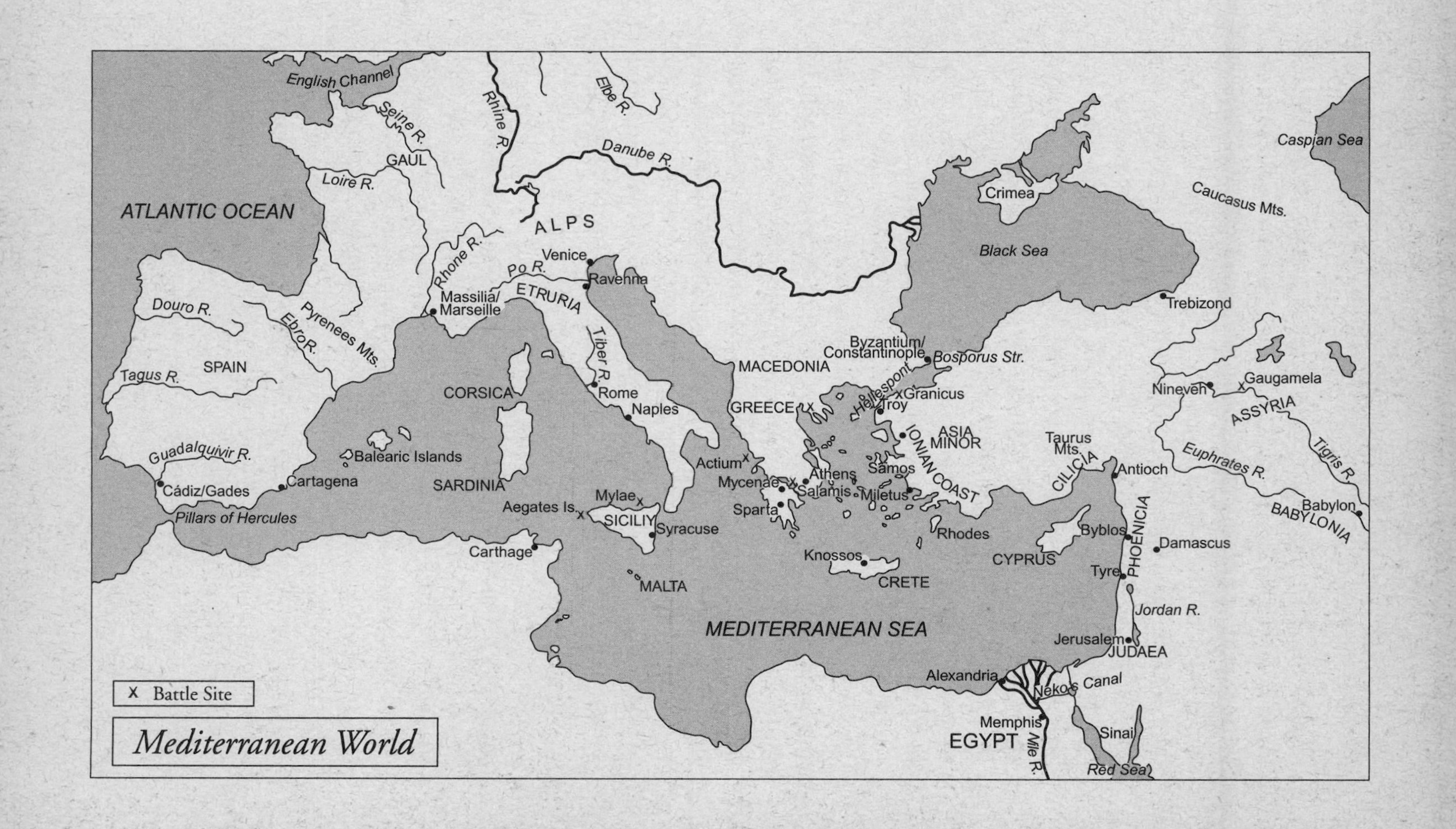
English Channel
Seine R.
GAUL
Loire R.
ATLANTIC OCEAN
Rhine R.
Elbe R.
Danube R.
ALPS
Rhone R.
Venice
Po R.
Ravenna
Massilia/
Marseille
ETRURIA
Douro R.
Pyrenees Mts.
Ebro R.
SPAIN
Tagus R.
Tiber R.
Rome
Naples
CORSICA
SARDINIA
Balearic Islands
Guadalquivir R.
Cádiz/Gades
Cartagena
Pillars of Hercules
Aegates Is.
Mylae
SICILIY
Syracuse
Carthage
MALTA
MACEDONIA
GREECE
Actium
Mycenae
Athens
Salamis
Sparta
Byzantium/
Constantinople
Bosporus Str.
Hellespont
Granicus
Troy
ASIA
MINOR
IONIAN COAST
Samos
Miletus
Rhodes
Knossos
CRETE
MEDITERRANEAN SEA
Crimea
Black Sea
Trebizond
Caspian Sea
Caucasus Mts.
Nineveh
Gaugamela
ASSYRIA
Tigris R.
Euphrates R.
Babylon
BABYLONIA
Taurus
Mts.
CILICIA
Antioch
CYPRUS
Byblos
PHOENICIA
Damascus
Tyre
Jordan R.
Jerusalem
JUDAEA
Alexandria
Nekos Canal
Memphis
EGYPT
Nile R.
Sinai
Red Sea
x Battle Site
Mediterranean World

Cyprus, Rhodes, Crete, Malta, Sicily, Sardinia, and the Balearics. Each could be navigated simply by following a series of visual landmarks without need of a compass or sextant. The gravest peril was frequent winter gales, which changed direction rapidly and created treacherous crosscurrents. Thus, for most of ancient history, the main sailing season was confined to April to October. Due to the unidirectional, west-to-east direction of the winds, easterly sailing toward Levantine entrepôts was speedy by the travel-time measures of the ancient world. Distances in the opposite direction, by contrast, seemed immense and required laborious and skillful effort of oar and sail—it commonly took up to sixty days to navigate from the Levant to the Mediterranean's midway point, where the sea was nearly land-bridged by the large island of Sicily. Indeed, the Sicilian bar effectively created two Mediterranean basins, a western and an eastern, that developed as more-or-less self-contained worlds. In the western Mediterranean in the third and second millennia BC, a vanished, ancient seafaring people laid thousands of puzzling religious stone megaliths on the arc of islands and shorelines stretching from Malta, Sardinia, Spain, and Morocco, up the North Atlantic coast to Brittany, Ireland, and Stonehenge, and into the northern seas as far as Scandinavia. In the eastern Mediterranean basin an enduring, robust seafaring civilization developed that shaped the course of history.

Sea trade in the Mediterranean advanced during the fourth millennium BC with the development of large cargo vessels constructed of planks and powered by sails as well as traditional oars. By combining the force of the winds with the water's properties of low friction and buoyancy, sails enabled cargoes to be shipped efficiently over long distances in an age when overland transport was slow, perilous, and often impossible. Thus was established the large cost advantages of sea transport over land that has endured to the present day, and with it an international marketplace that gave societies a chance to increase wealth through economic specialization, including those specialized in facilitating trade. A further advance came after 2200 BC when the rudder was introduced to supplement the steering oar.

The distinction of being history's first true great Mediterranean maritime civilization belonged not to the precocious but timidly shore-hugging Egyptians, but instead to an island people born on the sea—the Minoans of Crete. It was the Minoans who blazed many of the early Mediterranean trade routes. By 2000 BC, Crete was the trade crossroads of the region and for over half a millennium exerted a powerful economic, cultural, and naval influence throughout the Aegean and the eastern Mediterranean. The slender, 160-mile-long island's most important natural asset was its strategic location between the lucrative markets of the Levant, Asia Minor, and Egypt and the raw materials of the western Mediterranean. The Minoans were especially well positioned to profit from the Bronze Age because from Crete they could easily bring together both bronze's constituent metals—copper from the traditional deposits of neighboring Cyprus and Cilicia on the southern Anatolian coast, and tin from the mines of Etruria (Italy), Spain, and overland from distant Gaul and Cornwall, England. Bronze had first appeared in Mesopotamia around 2800 BC, and in Egypt by 2000 BC. Minoan workshops added further wealth by producing outstanding metalwork, weapons, tools, and pottery that were desired throughout the Old World. Minoan sea power rested on two kinds of vessels—a commodious, rounded, slow-sailing merchantman for commerce and a sleek and nimble long ship for raiding and defense that cruised under a single sail until battle, when its lone bank of oarsmen maneuvered its pointed ram into its enemy's hull.

As their wealth accumulated, the Minoans constructed lavish, multistoried palaces and large cities, and devoted themselves to the arts of civilization. One striking feature of their greatest city, Knossos, was that in an age of heavy fortification, it remained unwalled. This was one of history's earliest testimonies to the major defensive advantage provided by the open sea throughout the age of sail, as well as to the supremacy of the Minoan navy. Their domestic waterworks were sophisticated. Water tanks at the palaces of King Minos—probably a title, like Pharaoh, that applied to every ruler—flushed away human waste from indoor lavatories, while its cities were underlain with terra-cotta drainpipes and sewers. Agriculturally, terraces and dams maximized

Minoans' potential to grow olives and grapevines on their island's semiarid, mountainous terrain. As the Minoans settled the Aegean, they passed much of their civilization, including an early form of Greek writing, on to the ancient Greeks who followed them.

Minoan life was cataclysmically disrupted around 1470 BC by a huge volcano 70 miles to the north that vaporized most of the island of Thíra (Santorin). The explosion rocked Crete with earthquakes, clouds of ash that buried some of its cities, and a huge tidal wave that decimated the harbors along its northern coast. Greatly weakened, the Minoans survived only a century more before succumbing to a rising culture on mainland Greece that they had helped nurture, the Mycenaens.

Until about 1200 BC, the Greek-speaking Mycenaeans flourished on the trade routes and predatory naval power acquired from the Minoans. Mycenae itself was the city-state, according to Homer's *Iliad,* ruled by King Agamemnon, who led the Greek city-states in their concerted naval and armed forces campaign from about 1184 BC against the walled fortress city of Troy to retrieve the beautiful wife of his brother, King Menelaus of Sparta. But it wasn't Helen's face alone that launched the Greeks' celebrated thousand ships. It was the lure of booty. Perched on a hill in northwest Asia Minor overlooking the mouth of the Hellespont, Troy possessed great wealth from tolls on passage through that strategic strait, its silver mines, and payment of tribute from its weaker neighbors. Nor was it the artifice of the wooden Trojan Horse that was chiefly responsible for Troy's ultimate defeat. It was the Mycenaeans' incomparable fleet, which gave them uncontested control of the sea supply lines and the ability to sustain the ten-year-long siege that ended with the sack of Troy.

Yet even by the time of the Trojan War, Bronze Age Mycenaeans and other Aegean mainlanders were being displaced from their homes by invaders from the north armed with superior iron weapons. Many took to the seas and became the Sea Peoples who raided Egypt at the end of the New Kingdom. Many Mycenaean refugees ultimately resettled on the Aegean Islands and the rugged Ionian coast of Asia Minor and endured the dark age that disrupted civilized life throughout the Mediterranean and the Near East for three centuries.

When advanced civilization reemerged, three great powers competed for control of the sea-lanes across the entire length of the Mediterranean: the Phoenicians of the Levant; the Etruscans, who emerged in Italy and provided the first kings of Rome, but whose provenance remains mysterious to the present day; and the classic Greek city-states, which ultimately became the cradle of modern Western civilization. Of these, the earliest to rise were the Semitic Phoenicians, who left the West their modern alphabet.

The Phoenicians had two advantageous assets to launch their Mediterranean seafaring career: good harbors at Tyre, Sidon, and Byblos and abundant cedar and other wood resources for shipbuilding and export from Levantine forests that were greatly coveted by great powers in Mesopotamia and Egypt. From 1000 to 800 BC, Phoenician traders had the Mediterranean virtually to themselves. Phoenician ports were crowded with large cargo ships. They created one of the most daring seafaring trade societies in history. On their long sea voyages, Phoenicians even sailed by night and ventured far out of sight from the coasts. They established great colonies across the Mediterranean, including at Carthage at the southern gateway to the western Mediterranean near modern Tunis. They sailed through the Pillars of Hercules into the Atlantic Ocean to set up a settlement at the fine harbor at Gades, modern Cádiz in Spain. Thanks to their blockading grip over the strategic Pillars of Hercules from the late sixth century BC, Phoenician vessels for four centuries exploited a virtual trade monopoly in the raw material resources available along Europe's Atlantic coast and northern seas. Phoenician vessels sailing under commission by Egyptian Pharaoh Neko may have attempted in about 600 BC to circumnavigate Africa by sailing south through the Red Sea, and a century later Phoenicians of Carthage successfully colonized Africa's west coast—all two millennia before the Portuguese changed world history by accomplishing the circumnavigation feat. For a long while, the Phoenicians' domestic assets and adventurously earned sea-trading wealth were sufficient to offset their major geostrategic liability of being located adjacent to the powerful land empires of the Near East. After the eighth century BC, however, the Phoenician homeland was overrun by the armies of Assyria and the enduring heart of Punic civilization migrated westward to Carthage.

Where the Phoenicians were first and foremost traders, the loose conglomeration of sovereign Greek city-states that arose by the eighth century BC along the Ionian coasts and neighboring Aegean islands were great colonists. Between 750 and 550 BC, they founded some 250 colonies, including at Syracuse on wheat-growing Sicily and in 658 BC at Byzantium, the future Constantinople and Istanbul, on the Bosporus Strait gateway to the Black Sea and the golden wheat fields of the Crimea on its northern coast. Their economy was initially based on trading their homeland's highly prized wine and olive oil for basic grains and raw materials, although their comparative advantage gradually diminished as competing colonists transplanted vineyards and olive trees throughout the Mediterranean.

The leading Greek Ionian city-state was Miletus. In the eighth century BC, mariners from Miletus discovered how to reliably navigate during the warm sailing season to the Black Sea through the Hellespont and Bosporus straits, which for much of the year were barred by sudden, racing currents, difficult eddies, and strong northeasterly headwinds. Although the city's famous harbor has long since silted up, in the seventh and sixth centuries BC, Miletus was the leading trader of goods between the Mediterranean and Black seas. The latter was almost a Milesian lake. Tons of wheat and fish were transshipped from the Black Sea on Milesian vessels, financed by Milesian bankers and traded at great profit to the grain-poor city-states of the Aegean.

The wealth and urbanity of Miletus produced some of the formative thinkers of early Greek civilization, including the renowned father of Greek philosophy, Thales. One of the Seven Wise Men of ancient Greece, Thales was also a mathematician, a statesman, and an astronomer; he famously predicted the total eclipse of the Sun of May 25, 585 BC. Thales philosophized that there was a single primal substance of all things—water. Observing water's protean forms—ice, liquid, and gas—and natural processes, such as evaporation, in which water seemed to turn into air, rainfall where water appeared out of air, the ongoing silting up of river mouths, and the existence of freshwater springs that bubbled up out of the ground, he reasoned that everything on Earth was a manifestation of water in some transformed aspect. Later

Greek philosophers such as Aristotle downgraded water to only one of four primary elements—water, air, fire, and earth. Thales' hypothesis of water's primacy, which had a parallel in early Babylonian cosmology's placement of water as the first element of creation, left a profound imprint on Greek thinking in which reason and scientific observation became an exalted means to knowledge.

Miletus also played an instigating political role in the events that culminated in the rise of Athens and the most celebrated flowering of Greek civilization in the fifth century BC. By the mid-sixth century BC, the land-based Persian Empire founded by Cyrus the Great had gained hegemony over much of Asia Minor, including Ionia. In 499 BC, following a series of intrigues, Miletus led a rebellion of Ionian Greek cities against Persian overlordship. It enlisted the alliance of Athens, which sent a few ships across the Aegean to assist. Within five years King Darius's Persian army crushed the rebellion and ransacked troublemaking Miletus. In 490 BC Darius sent a modest fleet of Persian soldiers to punish tiny upstart Athens for its complicity. But the plan, which depended upon traitors opening Athens's gates from inside, failed when the Persians lost a battle upon landing at the plain of Marathon and speedy runners carried the news to Athens some 26 miles away before the regrouped Persian fleet arrived—the genesis of the modern "marathon" race.

When the Persians returned a decade later to exact a crushing revenge on Athens and its allies, Darius's son, King Xerxes, amassed an overwhelming 180,000- to 360,000-man military force and a 700- to 800-ship naval fleet, most of which it had commandeered from its subject states Egypt, Phoenicia, and Ionian Greece. In the spring of 480 BC, Xerxes' army crossed the Hellespont on two bridges constructed of boats that they lashed together. To avoid a march through rugged mountains, the Persian soldiers painstakingly dug a canal through the Athos peninsula. As it marched, the army was sustained by seaborne supplies previously stocked at depots along the northern coast of Greece and on its accompanying fleet.

All but some 20 Greek cities surrendered without a fight to the advancing Persians. Panicked Athenians consulted the oracle at Delphi, which in its usual enigmatic manner advised putting their faith in their wooden walls. No one knew whether this meant the traditional fortifications at the Acropolis or the hulls of the naval fleet it had built in preparation for the Persian invasion at the clever politicking of its brilliant young leader, Themistocles. Three years earlier, the forty-four-year-old Themistocles had persuaded Athens's democratic assembly to invest the windfall from a recent discovery at a state-owned silver mine in a modern naval fleet that could attack Persia's weak point—its extended naval supply lines—rather than relying upon the traditional army. When a small, valiant force of Spartans and Athenians were finally defeated by the advancing Persians at the mountain pass at Thermopylae to open the road to Athens, Themistocles made his decision. He ordered the evacuation of Athens, which the Persians proceeded to ransack and burn, while he retreated with his navy. Xerxes pursued, intent on destroying it, lest it retain the potential to disrupt Persian supply lines and command of the Aegean. At a narrow channel between Salamis Island and the mainland just west of Athens, the Persian fleet caught up to the Greeks.

On the morning of September 23, 480 BC, Xerxes ascended a hillside and, from a majestic throne, sat back to watch history's first recorded major sea battle, which he confidently expected to culminate with the utter demolition of the Greek navy. In the three-mile-long and one-mile-wide sound below him he could see the entire Greek fleet of 370 ships bottled up at either end by his own fleet, which had twice as many warships. Yet unbeknownst to Xerxes, Themistocles had consciously *enticed* the Persians to fight the showdown battle inside the confined waters at Salamis.

Although his ships were fewer in number, Themistocles' navy consisted of newly designed triremes, one of the great warships of history. With three banks of oars manned by 170 oarsmen who rowed with their backs to the direction they traveled, the trireme lay low in the water and had vastly more power than the traditional, two-banked, 50-oared, 100-foot-long Aegean penteconter. Under expert seamanship, it was fast and nimble, able to sprint into battle at nine knots and turn around in one minute in an arc of two and a half ship lengths.

It had two main weapons: a new, more deadly and readily extricable bronze-tipped ram to punch a hole in the hull of its enemies; and armored marines, who launched projectiles at the enemy ship as the ram closed in and, as necessary, grappled alongside to board for hand-to-hand combat.

Themistocles knew that Greeks could not win a battle in the open seas, where the enemy could deploy its entire fleet of much heavier, slower-moving war galleys. So with great cajoling, he maneuvered his reluctant allied Greek commanders to allow themselves to be entrapped in the strait at Salamis in the hopes of inducing Xerxes' navy to fight in constricted waters, where their numerical advantage would be diminished and the Athenian-designed triremes' tactical strengths most exploitable. On the morning of the battle, Themistocles lured the Persian fleet deeper into the narrow strait with a retreating maneuver, then ordered a sudden about-face. His rams charged the bewildered Persian fleet. Many Persian ships foundered, while trailing Persian vessels pressed in upon them from the rear. Almost half the fleet sank. Only about 40 Greek triremes were lost.

With his naval supply lines interdicted, and his land forces suddenly vulnerable, Xerxes expeditiously withdrew much of his army to Asia Minor. Fearful that the Greek fleets might cut off their escape route by destroying the bridge of boats at the Hellespont, the Persians beat a starving, dysentery-racked retreat, in which, by Herodotus's account, "they gathered the grass that grew in the fields, and stripped the trees, whether cultivated or wild, alike of their bark and of their leaves, and so fed themselves."

Not only was Greece saved, but Athens's supremacy on the sea was also established for generations to come. Salamis also was one of the earliest dramatic examples of the asymmetrical advantages of naval power in enabling small, less-populous states in the age of sail to offset the balance-of-power advantage of much-larger, predominantly land-based rivals. At once it enlisted the natural impediment of the sea as a stalwart defensive ally, while conferring formidable, tactical advantages in controlling supply lines, projecting offensive military force, and disruptively blockading enemy ports. Economically, it provided key advantages in transporting and profiting from trade in international goods.

Throughout history, from ancient Greece to the British Empire and the modern era of nuclear-powered navies, naval superiority has always been a key axis of power. Although great battles like Salamis were relatively rare events, by determining control over the seas they often were associated with decisive turning points. In the Mediterranean world, given its absence of strategically dominant inland waterway routes, it has been especially true that empires repeatedly rose and fell on naval sea power. Subsequent sea battles from Rome's victory over Carthage off Sicily in the First Punic War to the defeat of the Spanish Armada by England in 1588, the Napoleonic War battles of the Nile and Trafalgar in 1798 and 1805 and World War II's decisive ocean battles involving the *Bismarck* in the Atlantic and Midway in the Pacific reiterated the early lesson of Salamis.

The Battle of Salamis catapulted Athens, like Minoan Crete before it, into the role of eastern Mediterranean naval and commercial superpower. The rebuilt city-state soon became the wealthy epicenter of a glorious burgeoning of arts, philosophy, rhetoric, politics, history, mathematics, and scientific inquiry that created the foundations of Western civilization. Immediately after Salamis, Athens and the Greek world experienced its classical golden age.

Athens's turn to naval power at Salamis under Themistocles spurred a democratizing influence as well. It elevated the voice—eventually institutionalized in voting rights—of the large number of poor oarsmen required to man the galleys and diminished the relative influence of the traditional army, which was drawn more heavily from the aristocracy.

Reinforced by Athens's international free-trade policy, an articulated market economy with safeguarded private property rights evolved. The new Athens did not prosper merely by trading local olive oil and wine to obtain wheat and other vital goods. Its port city, Piraeus, became a thriving international clearinghouse market for goods destined for ports throughout the region. As in Venice, Amsterdam, London, and New York and other great shipping entrepôts of future centuries, a complex of private warehouses, shippers, bankers, wholesalers, and other commercial service suppliers grew up around its docks. A rudimentary commodities market for grain developed, in which a benchmark grain price for the entire Mediterranean was established from the demand and supplies arriving from the Black Sea, Sicily, and Egypt. The government treasury bulged with revenues from the 2 percent toll levied on all cargo using its port. Athens's democratic citizens' assembly, in turn, encouraged these burgeoning private markets by improving the port with breakwaters, docks, dredging, and other public services to accommodate more and larger ships. To protect its maritime commerce, Athens also provided naval escorts for the grain freighter convoys making the slow journey to Piraeus from the Crimea through the Bosporus and the Hellespont. As the unspoken marriage of convenience between government and private markets yielded rising prosperity and power for both, the democratic base of the Athenian polis became more representative and pluralistic.

Seafaring culture itself further nurtured the evolution of a new model of society based on representative, liberal market democracy for vested citizens. In contrast to the centralized river irrigation and land-oriented hydraulic states in which the populace had few practical economic alternatives other than complying with the policy commands and heavy taxes of the central government, private sea merchants had the natural freedom to trade in harbors where taxes were lower for the services offered and their rights better safeguarded. Thus it was no coincidence that many of history's leading seafaring trading states were also its leading representative market democracies and shared lineage with the political economic traditions born in Athens. With the rise of Athens, civilization's great dichotomous tension was joined between the main competing forms of mobilizing economic goods and manpower—by authoritarian government command, on the one hand, and market price signals and private profit incentives on the other—that have competed for supremacy in numerous forms through the ages into the twenty-first century.

Athens's glory age came to an end when its political ambitions overreached its naval power. It lost the Peloponnesian War when Sparta matched its sea power sufficiently to be able to impose a blockade on its ports to starve it into submission. In 338 BC it submitted to the rising, neighboring northern inland kingdom of Philip of Macedon, who plowed the wealth earned from his introduction of irrigated agriculture in Macedon's fertile central plain into building an expansive military force capable of controlling the gateway city on the Bosporus, Byzantium, and with it gained a strategic stranglehold over the Greek bread supply coming from the Black Sea. Yet in one of history's many unforeseeable twists, Athens's fall to Macedon became the instrument of implanting its Hellenic civilization and Greek language far and wide throughout the Mediterranean and Eurasia. The agent of this brilliant diffusion was Philip's son, Alexander the Great.

Tutored in Greek civilization as a teenager by no less than Aristotle himself, Alexander came to power as a twenty-year-old in 336 BC upon Philip's assassination. From the time he led his army of 43,000 soldiers and 6,000 cavalry across the Hellespont two years later to launch his conquest of the Persian Empire, Alexander never lost a single military engagement in a triumphal 15,000-mile, eight-year military march. When he died in 323 BC in Nebuchadrezzar's palace at Babylon at only thirty-two years of age, he reigned over the expanse of the Old World from the Nile to the Indus. His conquests were a dividing line of ancient history, inspiring later great conquerors from Caesar to Napoléon and instilling a millennium-long flourishing of Greek cultural influence in the western half of the Old World that was ultimately subsumed into Islamic and Christian European civilization.

In creating his empire, Alexander displayed a versatile mastery of the military and civil arts of water management. Although at the outset he had no navy, he understood the paramount strategic importance of controlling the sea-lanes, and he devoted many of his earliest military campaigns to neutralizing his rivals' naval advantages through unconventional land assaults from the rear that closed all enemy Mediterranean ports throughout Syria, Phoenicia, and Egypt. Turning inland, he hastened to cross the Tigris in 331 BC before his Persian enemy could employ that river barrier in its defensive stand, and then won a decisive victory at Gaugamela, not far from the ancient Assyrian city of Nineveh. Beyond Persia into central Asia and India, Alexander not only overcame armies but also diverse hydrological habitats, including the mountain snows of the Hindu Kush, the turbulent Oxus River, and central Asia's parched steppes. After crossing the Indus River in the spring of 326, he penetrated victoriously into the Punjab by attacking during the torrential monsoons, a time when Indian troops normally took a hiatus from fighting and conditions rendered their formidable charioteers, archers, and elephant corps less effective.

Already lord of the largest terrestrial empire in history, Alexander now quested to reach "Ocean"—the vast water body Aristotle and other Greek scholars believed surrounded Earth—and gain the enlightenment of knowledge. But when he urged his exhausted soldiers to press forward into the unknown, dense Ganges forests in search of Ocean, they refused to go on. Only this brought Alexander's stunning conquests to their end. But he determined to use his return to explore the coast of the uncharted Persian Sea and the harsh Gedrosian desert, which no army ever had crossed. A fleet of ships was built and outfitted. Alexander himself barely survived the six-month trip down the Indus River to the sea when an enemy arrow pierced his lung. To convoy his swollen entourage of up to 85,000 troops plus noncombatant camp followers across the Gedrosian desert, the army dug wells to provide water for itself and the ships, which stocked a four-month food supply for the army. As water supplies dwindled after mountain barriers forced Alexander to turn inland, however, the journey became a march of desperation. Alexander turned it into an opportunity for inspirational leadership. Like a common soldier he marched on foot. When a soldier found a small water source and brought the first drink in his helmet to his king, Alexander first asked whether there was enough for all the troops; upon being told "no," he dramatically poured out the water and announced he would wait until all his men could slake their thirst before drinking. Up to 25,000 are believed to have perished on the march.

Less than two years later, in June 323 BC, Alexander himself died of a fever at the end of a long night of banqueting at Nebuchadrezzar's old palace in Babylon. With him died his master plan to rebuild the famous city as the capital of his new Hellenic empire. Yet his legacy flourished through the Greek civilization that took root in the vigorous rebuilding he and his successors undertook wherever they had conquered. Declining irrigation systems in Egypt and Mesopotamia were rejuvenated and expanded with Greek hydraulic engineering, resulting in blossoming production, wealth, and the civilized arts. Ports and harbors were upgraded, and shipbuilding expanded in the Levant. Wherever Alexander passed he founded new cities—many named Alexandria. His most enduring legacy was Egypt's Alexandria, a splendid seaport and capital city for the 1,000 years of Hellenic and Roman rule. Within a century of its founding, Alexandria became the Mediterranean's most vibrant entrepôt and the heart of a Hellenic renaissance that was transmitted, through Rome and Islam, to Western civilization. Alexander personally selected the site because of its good anchorage and ample freshwater from a nearby lake, and designed the master plan for the city. To illuminate the way for toll-paying ships into its famous deep, double harbor, his successors erected one of the Seven Wonders of the World, the towering lighthouse at Pharos—probably taller than the Statue of Liberty in New York harbor—whose bronze mirror reflected sunlight by day and fire by night some 35 miles distant. Alexandria's world-famous library, built by copying all manuscripts that came into its busy harbor, became the central repository for much of the ancient world's literature and knowledge. In its heyday from 306 BC to the fire of 47 BC, it held as many as 700,000 items.

Greek science, mathematics, and medicine, a research institute, and an observatory flourished anew in Alexandria. Archimedes of Syracuse, the great mathematician, inventor, and father of hydrostatics, studied in Alexandria. In addition to the achievements of Euclid, Plotinus, Ptolemy, and Eratosthenes, Ctesibius, a contemporary of Archimedes, invented a floating mechanism of reliable regularity to calibrate the important water clock, or *klepsydra* ("time thief"), as well as a hydraulic organ. In the first century AD, Hero of Alexandria invented, as an amusement, a working miniature model of a steam engine—had he been motivated to build a full-scale working version, the world might have had the steam engine some 17 centuries before James Watt applied the same scientific principles to the machines that launched the Industrial Revolution.

It was also from Alexandria in the late second century BC that Greek sailors operating in the Red Sea made the breakthrough discovery of how to navigate the two-way monsoonal ocean winds to sail directly between the Gulf of Aden and southern India and thus projected themselves more prominently into the growing long-distance Indian Ocean trade between Orient and Occident that played such a dynamic role in early world history. In Roman times, Alexandria became the major harbor for exporting Egypt's indispensable grain surpluses to the imperial Italian capital. To Islam, it bequeathed a seafaring culture. In the Middle Ages, the Alexandria-Venice nexus reigned as the Mediterranean's premier commercial hub and a vibrant interface between Islamic and Western civilizations.

For several centuries, Alexander's own embalmed and perfumed body lay in state in Alexandria in a resplendent sarcophagus with transparent cover. No less than Julius Caesar came to pay homage. It survived for two centuries after suzerainty of Egypt passed to Rome in 30 BC.

Rome was the first great power to dominate the entire Mediterranean. Located at the sea's midpoint, it was strategically well positioned to enrich itself both from the natural resources of the western basin and the vibrant markets and know-how of its advanced, civilized eastern half. For several centuries it derived wealth and power by ruling over its sea routes with an authority reminiscent of the control hydraulic-irrigation societies had exerted over their great rivers.

Although deservedly famous for its great armies, Rome's rise as a superpower actually began in the third century BC when it seized command of the western Mediterranean's sea-lanes. Its unique genius as a civilization, indeed, resided in combining its military power with its pragmatic, well-organized, and large-scale applications of engineering technologies—the control and use of water prominent among them. Through water engineering, Rome mastered shipbuilding and seafaring infrastructures for its navy, drainage for the imperial highways used by its army, and construction of massive aqueducts and urban water systems to create something new in civilization—the giant metropolis.

By legend, the city's founders were the semidivine twins Romulus and Remus, who, like their predecessors Sargon and Moses, had been set afloat to the fates in the river. They had been nursed by a she-wolf (still the city's emblem) who happened upon them at the Tiber River shoreline, reared by a shepherd, and eventually set up an original settlement on the Palatine Hill near the river. Historical Rome, indeed, commenced with separate tribal settlements atop its famous seven spring-fed hills near where an ancient salt trade route crossed a shallow fording point in the Tiber near the Tiber Island. By the eighth century BC, the city came under the rule of the Etruscans, from whom Rome inherited many of its advanced hydraulic arts of drainage and irrigation. By draining large swamps from Tuscany to Naples and embanking the silt-rich Po River against uncontrolled flooding, the Etruscans made Italy's limited agricultural resources yield enough food to sustain a pre–Athenian age civilization sufficiently prosperous to challenge Carthage and the Greek colonies of the mid-Mediterranean.

It was under Etruscan rule that Rome's first large engineering work was completed in the sixth century BC. The Cloaca Maxima, or great sewer, drained the swampy, malarial valley between the city's Seven Hills to create what became the center of ancient Rome's civic and commercial life, its Forum. So well constructed that it still functions today, its egress point into the Tiber can be seen along the embankment from the Tiber Island bridge and its malodorous effluvia smelled through the air vents in the Forum's ruins.

The Romans cast off their Etruscan kings in 509 BC. They set up, like their ancient Athenian contemporaries, an aristocratic republic. Governed by two annually elected consuls, a Senate, and landed patrician families, the Roman Republic would endure in form and as an ideal for centuries. It protected private property and other rights with written law and exalted the virtues of the simple, independent citizen-farmer who was expected to put down his hoe and pick up his arms when war necessitated—ideals extolled in the late eighteenth century by America's founding fathers. Without the existence of a central, arterial river to provide transport and large-scale irrigation through the Italian peninsula, Rome's economic and political power consolidated slowly, facilitated by its construction of a network of well-drained major roads that fanned out from the capital in all directions, starting with the southeasterly Appian Way in 312 BC. Rome's rise was abetted indirectly by the sea power of the Greeks of Syracuse, who early in the fifth century destroyed Etruscan naval power. By 270 BC Rome controlled the entire Italian peninsula. Across the Strait of Messina, on the northeastern tip of the rich, grain-growing island of Sicily, Rome's expanding ambitions collided with the great Mediterranean naval empire founded by the Phoenicians centuries earlier—Carthage.

The turning point of Rome's rise in history as a great power was its three Punic Wars, through which it won command of the Mediterranean. The First Punic War began in 264 BC and lasted twenty-three years. Rome's initial objective was simply to remove Carthaginian garrisons from eastern Sicily. But its besieging armies were frustrated by the resupplies Carthage delivered from its western Sicily strongholds, which in turn were supplied by sea from its capital city in North Africa. Thus, to neutralize eastern Sicily, Rome needed to control the sea-lanes around the island.

Yet Rome at this point was exclusively a land power. To achieve its objectives, it would have to become one of history's rare land-based civilizations that successfully transformed itself into a dominant sea power as well. In 260 BC the Senate authorized the construction of 20 triremes and 100 quinqueremes, five-bankers manned by 300 oarsmen. Since Romans didn't know how to design a war galley, they relied upon the know-how of Greeks from the cities of southern Italy and Sicily. In a remarkably few months, with rowing crews training on dry land all the while, the neophyte fleet, with more than 30,000 men, was ready to set sail out of Rome's harbor at Ostia at the mouth of the Tiber to confront Carthage's larger, experienced, and redoubtable navy. In effect, the battle for Sicily was to be a proxy contest with Carthage for dominance of the entire western Mediterranean. Rome's warships did not try to match Carthage's light, fast fleet designed for rapid maneuvers and ramming executed by skilled seamen. Instead, they were designed pragmatically to mobilize the advantage of Rome's infantry strength by making the sea battle more like a land fight. They were heavier, slower, and steadier in bad weather, with large decks to hold more marines. They were designed to pull alongside with grapples and board the enemy for hand-to-hand combat. A brilliant stroke was added when the fleet was readying for battle in Syracuse—some say at the suggestion of its ingenious resident, Archimedes—to attach an upright 36-foot-long gangplank with a heavy spike at the outboard that could swing down over the bow and embed sturdily into the nearing enemy vessel both to frustrate its ram and to permit swift boarding by Roman soldiers.

Against improbable odds, Rome's navy triumphed in the war's first major sea battle in August 260 BC off Sicily's northern shore near Mylae. What followed was nineteen years of naval-warfare attrition, both from losses inflicted by the enemy and even larger losses wrought by sea storms. One Sicilian storm in 255 BC cost hundreds of Roman ships and more than 100,000 lives; another two years later off southern Italy sank most of the navy's rebuilt fleet. In the end, Rome won the First Punic War chiefly by its relentless perseverance in rebuilding its fleet and its tolerance for sustaining heavy losses in vessels and manpower, all the while improving its seamen's skills. The final battle was at the western tip of Sicily near the Aegates Islands on March 10, 241 BC.

The strategic advantage won by Rome in the First Punic War—of being the most potent naval force in the western Mediterranean—proved to be decisive in the outcome of the Second Punic War, from 218 to 201 BC. It was during the second war that Carthage's brilliant general Hannibal famously led his army and a contingent of elephants from his base in Spain across the Ebro River into Gaul, over the Alps, and into Italy, where he marauded victoriously throughout the Italian countryside for over a decade in what proved to be a fruitless bid to trigger local insurrection against Roman rule. In the end Hannibal was unable to sustain his supply lines without naval replenishment and was ultimately defeated on the banks of the Metaurus River in 207 BC, when reinforcement troops traveling overland under his brother Hasdrubal failed to arrive in time. Indeed, it was Rome's naval superiority that had compelled Hannibal to attempt the treacherous overland invasion of Italy in the first place. Had Hannibal been able to go "by the sea, he would not have lost thirty-three thousand out of the sixty thousand veteran soldiers with whom he started," concluded Captain A. T. Mahan in his classic *The Influence of Sea Power upon History*. Sea-power superiority also delivered Rome the means to counterattack Carthage. Through the naval supply buildup of a formidable army base in northern Spain, it besieged and finally took Carthage's Spanish stronghold at Cartagena ("New Carthage"), and then forced Carthage's surrender in 202 BC through attacks on its North African homeland across the Mediterranean.

The first two Punic Wars transformed the trajectory of Roman history. The conversion of the entire western Mediterranean basin into an unchallenged Roman lake brought Rome its first taste of the fruits of ruling a provincial empire and propelled its rise as one of history's great powers. All the grain wealth of Sicily, the mineral deposits of southern Spain, the tin, silver, and other resources that moved from the Atlantic through the Pillars of Hercules, and slave manpower from defeated populations came into Roman hands. Gradually, Rome ceased striving for basic self-sufficiency from its low-yielding home soils and began to rely on shipments of imported grain for its daily bread. Large estate owners abandoned drainage projects to reclaim marginal cropland in favor of producing higher-value-added, tradable luxuries like olives, wine, and livestock, often with slave labor. Class tensions polarized as wealth became concentrated in fewer hands, while individual military commanders compensated by advancing the interests of free commoners, who increasingly served as their professional troops.

Initially, Rome took uneasily to its changing political cultural identity as a hegemonic maritime power. Only gradually during the course of the second century BC did it accept the inexorable demands of its success to extend its dominance over the eastern Mediterranean as well. Nevertheless, whenever possible, it exerted its weighty influence indirectly through the soft power of financing trade and being the largest import market, while leaving naval patrol duties in the east to maritime allies like Rhodes and Pergamum. As late as 100 BC, Rome had scaled back its fleet in the eastern Mediterranean to skeletal size.

All that changed dramatically during the first century BC when pirates began to exploit Rome's minimal naval presence. The largest group of buccaneers, headquartered in Cilicia on Asia Minor's rugged southern coast, possessed more than 1,000 ships and a formidable arsenal, and was ruled by a well-organized, hierarchical command. By 70 BC they had become an intolerable nuisance by interfering with vital grain shipments to Rome and by brazenly raiding coastal highways as far away as Italy and kidnapping prominent Roman citizens for ransom. One famous hostage was the young Julius Caesar, who was seized while on a ship bound from Rome to Rhodes, where he was to study law. During his captivity Caesar amicably suggested to his captors that due to his importance, they should double their initial ransom demand—which they readily did—and promised, with equal geniality, that after his release he would return to crucify each and every one of them. In fact, as soon as he was freed, he raised a fleet in Miletus and killed as many as he was able to catch, though as a reward for their decent treatment of him he allowed their throats to be slit before nailing them to the cross.

Gripped by a sense of national crisis at the threat to its food supply, Rome's Senate finally acted. In 67 BC it commissioned General Pompey to rid the Mediterranean of the pirate menace and invested him with almost unlimited power to accomplish it. In one of the most spectacularly successful naval operations in history, Pompey amassed a force of 500 ships and 120,000 marines and launched a methodical, sector-by-sector sweep of pirate enclaves eastward from Gibraltar. In less than three months, all the pirates were defeated and the buccaneer capital in Cilicia was besieged into submission.

Pompey did not stop there, however. Without authorization from the Senate, he sailed his formidable fleet to the Near East, where he brought Syria, Judaea, and the cities of Antioch and Jerusalem under Roman rule. He returned to Rome in 62 BC as a conquering hero with fearful power, and entered a ruling triumvirate with Caesar and Crassus. Pompey's naval operations revived Roman sea power and organized it into a permanent naval force. Thereafter, it always would be a crucial component of Rome's ability to wage war and enforce its will on others. At first, however, it was turned inward upon itself in two decades of bloody civil wars that were ignited on January 11, 49 BC, when Caesar and his army marched across the muddy little Rubicon in northern Italy, which violated the republic's forbidden boundary line, and amounted to an attempted coup d'état. The ensuing civil war between Caesar and Pompey was fought across the breadth of the entire Mediterranean from Spain to Egypt, with Caesar's breakout from Pompey's blockade in the Adriatic playing a major role in his ultimate triumph prior to Pompey's assassination in Egypt. When Caesar, now dictator for life, himself was murdered at the Senate in Rome on March 15, 44 BC, civil war erupted anew.

Fittingly, the final, decisive battle that ended the civil wars and inaugurated the imperial era was fought at sea, in 31 BC, off the Actium promontory near the Gulf of Corinth in Greece. On one side was the allied force of Caesar's leading general, Mark Antony, and his lover, Egyptian queen Cleopatra. On the other side was Octavian, later honored by the Senate with the supreme title Augustus Caesar, the young grandnephew and adopted son of Caesar. In command of Octavian's fleet was his brilliant military commander, lifelong right-hand man, and civic colossus of the Roman Empire in his own right, Marcus Agrippa.

To try to offset his inferiority at sea, Octavian had raised a new navy of 370 ships. Recognizing that his enemy's more-expert crews and nimbler, more-lethal vessels rendered futile any attacks based on the conventional tactic of ramming, Agrippa, in a stroke of genius reminiscent of Rome's design innovation of the spiked gangplank in the First Punic War, armed the ships with a new weapon he conceived: a catapult that fired arrows leashed to a rope and tipped with an iron-clawed grapnel that enabled his marines to clutch onto enemy galleys from much farther range than the conventional, hand-thrown grapnel, and pull themselves in by windlasses for hand-to-hand combat. With the help of the catapult grapnel, Agrippa's fleet won decisive battles off Sicily in 36 BC that reversed Octavian's waning fortunes and went on to win control of the sea war on the Mediterranean. By the time of Actium, he held enough strategic bases to interdict Egypt's grain supply freighters, and thus slowly starve Antony's huge military forces, including its Actium fleet, into submission. At the battle itself, Agrippa enjoyed a numerical warship advantage of 400 to 230. Before the day had ended, Cleopatra and Antony had fled for Egypt, where, a year later, they committed suicide, while Octavian's Rome seized direct possession of the Mediterranean's last nominally independent great state and with it the prize of the rich Nile granary.

Octavian acquired the title of Emperor Augustus and prudently consolidated his power by, among other actions, establishing a well-organized, permanent professional navy to police the Mediterranean. Over the next 200 years of the Pax Romana, Rome's empire was extended from the Atlantic to the Persian Gulf, from North Africa to the northern British Isles, and from central Europe through the Balkans. To secure the frontier against barbarian tribes, naval squadrons controlled some 1,250 miles of natural defensive water barriers, including the Rhine, the Danube, and the Black Sea.

One of Julius Caesar's unfulfilled visions had been to join the Rhine and Danube rivers by a canal and thus create a navigable, arterial water route through continental Europe's heart, a Nile of Europe. In the event, the Rhine-Danube boundary remained the defensive frontier between Roman civilization and the barbarian world—an equivalent of China's Great Wall—and never became the central transport waterway unifying northern and central Europe. In the Middle Ages, the old Rhine-Danube frontier again shaped history as the rough, axial dividing line between Catholic and Protestant Europe. Ultimately, it took 2,000 years, until 1992, for political conditions to be conducive for the completion of the 106-mile-long Rhine-Main-Danube canal linking the North Sea and the Black Sea and helping to integrate Europe into a single economic community.

Augustus famously boasted of his legacy that he had found Rome a city of brick and left it a city of marble. Indeed, under the order established by the Roman Empire, wealth and commerce soared. Goods were sucked in by inexorable political and economic gravity from the empire's provinces along its rivers and bordering seas—North, Baltic, Black, Red, and Atlantic—toward its ravenous mouth and stomach in the central Mediterranean. In an era where it was difficult to move any large quantities by land, river and sea transport were Rome's vital lifelines.

At the empire's height, staples and luxuries poured into its bustling ports from distant foreign civilizations spanning the Old World. Grain that became the daily bread dole for commoners came from Egypt, North Africa, and the Black Sea; the rich and powerful enjoyed wools from Miletus, Egyptian linens, silks from China, Greek honey, peppers, pearls, and gems from India, Syrian glass, marble from Asia Minor, and aromatics from the Horn of Africa and the Arabian Peninsula. The mutual attractive force of trade between Rome and its counterpart empire in the Far East, Han China, increasingly found shipping routes through the narrow Strait of Malacca between the Malaya Peninsula and Sumatra to stimulate a vibrant exchange across the long-distance Indian Ocean highway to give critical mass to the nascent global market economy that took hold in this era. Over a hundred trading ships per year sailed the monsoons for India through the Red Sea, parts of which were patrolled for pirates by the Roman navy. Throughout the Mediterranean, the infrastructures of shipping and trade were improved and expanded. In order for large cargo ships to arrive directly at Rome instead of being transshipped in smaller boats from the natural, deep port near Naples, for example, Emperor Claudius in AD 42 constructed a man-made harbor from the dredged marshes north of Rome that was linked to the Tiber by an artificial canal and towpath; inside the harbor, called simply Portus, was a large lighthouse modeled on Alexandria's Pharos lighthouse.

Rome earned its economic surplus both from being the center of sea trade and from imperial exploitation of the rich provinces around the Mediterranean rim whose own political economies were increasingly molded to the necessities and pulse beat of the giant Roman metropolis. With some 1 million inhabitants at its height, Rome was far and away the largest city in Western history and would remain so for nearly 2,000 years. Such a size was far more than it could support on local Italian agriculture and industry. Therefore, as Rome grew rich upon the provincial resources on its periphery, it also grew increasingly dependent upon them for its internal stability. During Rome's zenith, chronically high urban unemployment resulted in a welfare state with up to one-fifth of the often restive population receiving subsidized bread from public storehouses and entertainment at public spectacles—gladiatorial contests, ship races and various games, in venues like the Colosseum and the Circus Maximus. Rome's basic food security required the reliable importation of about 300,000 tons of grain per year. Two-thirds came from destinations within several days' sailing. But one-third came from the Nile Valley in Egypt, which was a difficult and dangerous thirty- to sixty-day voyage into the prevailing westerlies. Emperors from Augustus onward thus placed high state priority on protecting the fleet of huge grain ships that crossed the open waters from Alexandria to Rome. Each cargo carrier was up to 180 feet long and 44 feet deep—larger than any ship that crossed the Atlantic until the early nineteenth century. One famous grain cargo ship passenger who voyaged to Rome in AD 62 was the prisoner St. Paul. The Nile became so important as a breadbasket that Egypt was forbidden by edicts to export its grain anywhere else. Egypt's irrigation was intensified and its cultivated acreage expanded under the Romans, facilitated by a long period of good Nile floods, and even rainfall.

Rome vigorously exploited another of history's seminal water technologies to help it produce the daily bread for so many hungry soldiers and citizens—waterpower. To grind grain into flour to make bread, Rome built vast numbers of waterwheel-powered gristmills on streams and artificial conduits fed by aqueducts that transmitted the energy captured from the flowing current to turn the wheel and the millstone attached to it. As early as the first century BC, Roman engineers had made the ingenious breakthrough of moving the traditional horizontal waterwheel to a position vertical to the water, and to multiply the power it generated through the use of gearing. Many of the water mills built by Rome to feed army garrisons and cities were impressively large and powerful. The famous fourth century AD Roman water mills at Barbegal near Arles, France, used water forced along a six-mile-long aqueduct to drive eight pairs of wheels. It could grind 10 tons of grain daily. It was in imperial Rome that water-powered mills were transformed from small, household and local community devices into tools of large-scale, centralized bread production. As such, they became key instruments of state power.

Why the Romans never fully exploited the enormous work potential of their own advanced waterwheel techniques beyond grinding bread flour is one of the vexing questions of its history. They possessed sufficient know-how to apply waterwheels to industrial uses, such as driving mechanical saws, fullers' beaters, tilt hammers, or bellows to heat iron furnaces. But what they may have lacked, given their surplus of expendable slave labor, was the economic incentive to invest in labor-saving mechanization.

One new water engineering technology that Romans did profitably employ was hydraulicking for mining. Hydraulicking used powerful jets of water that were far more productive than manual digging in the hills of Spain to extract the gold used for its coinage and financial system. Roman engineers released water from large tanks erected 400 to 800 feet over the mining site to generate waterpower sufficient to shear away hillsides and break up rock formations that exposed the valuable gold veins. In the mid-nineteenth century, hydraulicking would have its most famous, intensive modern application at the height of the California gold rush.

Although not famed for their technological originality, Romans did use water to make one transformational innovation—concrete—around 200 BC that helped galvanize their rise as a great power. Light, strong, and waterproof, concrete was derived from a process that exploited water's catalytic properties at several stages by adding it to highly heated limestone. When skillfully produced, the end process yielded a putty adhesive strong enough to bind sand, stone chips, brick dust, and volcanic ash. Before hardening, inexpensive concrete could be poured into molds to produce Rome's hallmark giant construction projects. One peerless application was the extensive network of aqueducts that enabled Rome to access, convey, and manage prodigious supplies of wholesome freshwater for drinking, bathing, cleaning, and sanitation on a scale exceeding anything realized before in history and without which its giant metropolis would not have been possible. That it amply served the poor as well as the rich was likewise a notable development in the history of civic society. Throughout its empire, Rome's aqueducts supported the robust health of towns and frontier garrisons whose soldiers' fitness for battle was a critical element of its army's superiority. Its mobilization of public water that served all classes established a landmark civic standard embraced later by industrial democratic Western societies.

Freshwater conduits had been in use for centuries before censor Appius Claudius built Rome's first aqueduct, the 10-mile-long subterranean Aqua Appia, beneath its first major paved roadway, the Appian Way, in 312 BC. Some four hundred years earlier the Assyrians had built their aqueducts augmenting Nineveh's water supply and Hezekiah had excavated Jerusalem's secret water tunnel. In 530 BC the Greek island of Samos likewise cut a water tunnel two-thirds of a mile long, while classical Athens had several aqueducts. The technical high point of Hellenist water engineering was the Ionian city of Pergamum's early second century BC 25-mile-long aqueduct with double and triple terra-cotta piping and a pressurized section that enabled water to cross a low valley and then rise again on the other side against the natural force of gravity.

What distinguished Rome's public water supply infrastructure was not its originality, but rather its precision, organizational complexity, and grand scale. Spectacular ruins of the famous three-tiered, 160-foot-high arches of southern France's Pont du Gard, the still partly functioning, narrow-arched aqueduct bridge at Segovia, Spain, and the celebrated Roman baths at Bath in England offer glimpses of Rome's widespread hydraulic accomplishments. Roman water systems underpinned the empire in southwestern Europe, Germany, North Africa, and Asia Minor, including at Constantinople, the "New Rome" established by Emperor Constantine at Byzantium on the Bosporus in AD 330.

Yet nowhere was Rome's public water system more influential than in Rome itself. Indeed, Rome's rapid growth to a grand, astonishingly clean imperial metropolis corresponded closely with its building its 11 aqueducts over five centuries to AD 226, extending 306 miles in total length and delivering a continuous, abundant flow of fresh countryside water from as far away as 57 miles. The aqueducts funneled their mostly spring-fed water through purifying settling and distribution tanks to sustain an urban water network that included 1,352 fountains and basins for drinking, cooking and cleaning, 11 huge imperial baths, 856 free or inexpensive public baths plus numerous, variously priced private ones, and ultimately to underground sewers that constantly flushed the wastewater into the Tiber.

As in all ages from antiquity to the present, the pattern of water distribution read like a map of the society's underlying power and class structures. Nearly one-fifth of total aqueduct water during the empire's heyday went to meet the watering needs of patricians' suburban villas and farms. Inside the city walls, paying private consumers and industries and those granted water rights by the emperor were water-Haves who received another two-fifths of Rome's freshwater. Public basins and fountains used freely by ordinary people, by contrast, received only 10 percent of total aqueduct water. Nevertheless, like the bread dole, provision of a minimum amount of free water was an essential pillar of the state's political legitimacy that Roman officials were careful to maintain. The remainder of aqueduct water was allocated to the emperor's ever-growing demands for public monuments, baths, nautical spectacles, and sundry other public purposes. Rome's patrician families enjoyed hot and cold indoor running water, sanitary bathrooms, and water closets that were unsurpassed in comfort until modern times. Unlike today's highly pressurized, enclosed pipe systems, Rome's aqueducts flowed from their source by natural gravity through precisely sloping gradients maintained over long distances; only in the city was pressurized plumbing employed to raise water to elevated locations. Most of the aqueducts were subterranean. But about 15 percent of the system was above ground and ran along its famous arched structures to maintain its gradient over uneven terrain.

Sustaining and housing a population of 1 million may not seem like much of an accomplishment from the vantage point of the twenty-first century with its megacities. Yet for most of human history cities were unsanitary human death traps of inadequate sewerage and fetid water that bred germs and disease-carrying insects. Athens at its peak was only about one-fifth the size of Rome, and heaped with filth and refuse at its perimeter. In 1800, only six cities in the world had more than half a million people—London, Paris, Beijing, Tokyo, Istanbul, Canton. Despite Rome's hygienic shortcomings—incomplete urban waste disposal, overcrowded and unsanitary tenements, malaria-infested, surrounding lowlands—the city's provision of copious amounts of fresh, clean public water washed away so much filth and disease as to constitute an urban sanitary breakthrough unsurpassed until the nineteenth century's great sanitary awakening in the industrialized West.

Although there are no precise figures in ancient records on how much freshwater was delivered daily, it is widely believed that Roman water availability was stunning by ancient standards and even compared favorably with leading urban centers until modern times—perhaps as much as an average of 150 to 200 gallons per day for each Roman. Moreover, the high quality of the water—the Roman countryside offered some of the best water quality in all Europe, and still does so today—was an easily overlooked historical factor in explaining Rome's rise and endurance.

Yet it was a universal testimony to water's perennial economic and human value that even in conditions of relative plenty man constantly desired to have more of it. In an amusing reminder of unchanging human nature, Senator Julius Frontinus, who became Rome's Water Commissioner in AD 97, in his famous short treatise *On the Water Supply of the City of Rome* urged harsh punishment for the many water thieves who "have laid hands upon the conduits themselves by penetrating the side walls."

Frontinus modeled himself, almost reverentially, upon the single most illustrious creator of Rome's public waterworks—Augustus's loyal military commander, schoolmate, and virtual coemperor for much of his reign, Marcus Agrippa. In AD 33 Agrippa, acceding to Augustus's request, assumed the office of aedile and with it responsibility for Rome's municipal works and services. Actium was two years in the future and Augustus—still known as Octavian—faced waning public support at home with the outcome of the civil war with Mark Antony very much in doubt. A famously self-effacing, plebeian-born protégé of Julius Caesar, Agrippa enjoyed wide popularity with commoners that Augustus lacked. His year-long aedileship would become the most lauded and influential in Roman history. At its start, Rome's public infrastructure, following years of civil discord and war, lay in a crumbling, neglected state. It ended with revolutionary improvements—on a scale often associated with historic dynastic restorations and renewals of civilization—that not only resurrected Rome's municipal infrastructure and services but also Augustus's popularity and much of the political support he needed to overcome Antony, then far removed in Egypt with Cleopatra.

Waterworks were the centerpiece of Agrippa's urban renewal program. In only one year, largely at his own personal expense, he repaired three old aqueducts, built a new one, and greatly expanded the capacity and distribution reach of the entire system. Some 700 cisterns, 500 fountains, and 130 ornately decorated distribution tanks were also constructed, and 170 free public baths were opened for both men and women. He cleaned out the sewers, famously rowing through the Etruscan-built Cloaca Maxima on an inspection tour. In addition, he put on splendid games, distributed a dole of oil and salt, and on festive occasions offered free barbers.

Rome's municipal water system became Agrippa's lifelong passion. In the years after his aedileship, even as he ruled over the eastern half of Augustus's growing empire, led important military campaigns and was considered a leading successor to the emperor when Augustus fell gravely ill, he acted as the city's unofficial, permanent water commissioner and spent lavishly from his own funds for the purpose. In 19 BC he built a sixth new voluminous aqueduct, the Virgo, whose water was acclaimed for its purity and coldness, which he used partly to supply Rome's first large public bath near today's Pantheon.

The Virgo aqueduct, much of which lay underground, had the historical distinction of being the only line never to completely stop flowing through Rome's subsequent dark centuries; today, Virgo water flows in Bernini's famous Quattro Fiumi (Four Rivers) fountain in Piazza Navona and terminates at the Trevi Fountain, where the relief in the left panel shows Agrippa himself supervising the construction of the Virgo with the design plans unscrolled before him. Upon his death in 12 BC, at age fifty-one, he bequeathed his slaves to Rome's water system maintenance crew. His master water system plan was adopted as the basis of the official imperial water administration created a year after his death by Augustus. It guided Rome's water management thereafter, including the major new aqueducts built until the early second century. Longer term, Agrippa's civil works set a standard and concept of public municipal service for all classes, a democratic legitimacy, and tool of exercising political power that is influential in modern liberal Western democracies.

Agrippa's innovation of the first large public bath—soon magnified in scale, vanity, and variety of activities by the 11 monumental imperial baths erected by succeeding emperors—became the model central institution of social and cultural life in ancient Rome. The traditional Republican era bath was transformed from a simple "sheltered place where the sweaty farmer made himself clean" into a multifaceted, sometimes luxurious "community center and a daily ritual that defined what it meant to be Roman," writes historian Lewis Mumford. "The Roman bath compares with the modern American shopping center." A typical Roman's day at the baths started after a day's work and lasted several hours. A large facility consisted of a cluster of activity rooms and big bathing chambers surrounding an open, central garden. The richest baths were adorned with statues, floor mosaics, and marble or stucco reliefs on the walls. Bathers generally would first get an oil rubdown in the *unctuarium* before exercising in one of the gymnasiums. Bathing started in the hot *caldarium* and steam room or *sudatorium,* much like a modern Turkish bath, heated by furnaces from below; Romans didn't use soap, but instead scraped the dirt off their perspiring skin with a curved metal instrument, the *strigil.* Next, they lounged at length with friends in the *tepidarium* or warm baths, often conversing and carousing together. Then they dipped in the cold bath, or *frigidarium,* and swam in the pool. Lastly came a rubdown with oils and perfumes. Along the way they ate snacks and sipped wine served to them by attendants, read books from the bath's library, got massages, relieved themselves in multiseated latrines along the walls of the baths, and sometimes indulged in drunken carousing and love-making. A full range of bathhouses, from free to costly, were available for all classes. At some, men and women bathed naked together, a practice whose repeated banning by emperors testified to its persistence. From one outpost of the empire to the other, Romans reinforced their Roman identity by practicing the daily social and hygienic rituals of the bath.

Just as the high and low floods of the Nile tracked the prosperous and low periods of civilization in Egypt, Rome's great eras of achievement and population growth corresponded to its periods of aqueduct building and expanding water supply. The early aqueducts were built during Rome's Italian peninsular expansion as rising population levels overtaxed the city's resources of local freshwater springs and wells and potable Tiber River water. The transformative victory in the Second Punic War in AD 201 was followed by an intensive burst of Republican era aqueduct building featuring Rome's third aqueduct, the voluminous, 57-mile-long Aqua Marcia in 144 BC, which for the first time distributed ample good water across the social spectrum. Agrippa's aqueduct constructions sufficed until the mid-first century AD, when Emperor Claudius increased the water supply by about 60 percent with two new aqueducts, and Trajan added a third in AD 103 to keep pace with the doubling of Rome's population in the early imperial period. The end of aqueduct construction in the early third century, by contrast, reflected the plague-ridden fall in the city's population and the early decline of the Western Roman Empire; indeed, the last aqueduct was built in AD 226 mainly to serve the decadent luxury of refurbishing the emperor's baths rather than the needs of the citizenry.

Other water-related depredations also marked Rome's decline. Rome's heavily fortified European frontier river barriers were breached by Germanic barbarians: In AD 251 the Goths crossed the Danube; in AD 256 the Franks broke through the Rhine. Both pillaged deep into the empire. In the same period Rome began to lose control of the seas and the security of its food and raw material lifelines to its provinces came under steady assault by pirates, Goths, and other barbarian tribes. Rome's underfinanced and diminished navy increasingly retreated. Hyperinflation, heavy taxation, recession, and severe pestilential disease debilitated the empire's economy from within. Without, new defensive walls were erected around the capital by Emperor Aurelian in AD 271.

Although the empire earned a temporary, century-long reprieve through the administrative reforms, reassertion of military power, and authoritarian economic command of several resourceful soldier-emperors, notably among them Diocletian and Constantine, its command of the sea-lanes and defenses that underpinned its control of vital supplies from its provinces was irreparably breaking down. This was importantly illustrated in its Egyptian breadbasket. In response to onerous taxation payable in grain—tax "rates" were calculated by Roman governors according to the Nile's annual flood level—cultivated cropland in Egypt by the third century had shrunk by half as its farmers grew weary of working for overlords and abandoned their fields. Draconian new grain taxes imposed in AD 313 worsened the long-term situation. Finally, in AD 330 Emperor Constantine transferred the capital itself to a new, more defensible and economically strategic location at the ancient Greek city of Byzantium overlooking the Bosporus Strait gateway to the Black Sea. The regular Rome-bound Egyptian grain shuttles were redirected to the "New Rome," renamed Constantinople. Rome's remaining population was left to fend for itself. As often in history, the change in main water transport routes signaled the shift in destinies among leading powers and civilizations.

The Western Roman Empire's final demise accelerated in the late fourth century AD. The proximate cause was a new wave of incursions by Gothic and other barbarian tribes put to flight by the invasion into eastern Europe from the central Asian steppes of a fearsome, nomadic tribe, the Huns. The Huns, who eventually settled in the Danube valley, themselves had been propelled into motion by their own ejection from their Asian homelands by an even more warlike group, the Mongolian Juan-juan war confederacy, which also constantly menaced China. By AD 410, when traitors opened Rome's gates to Alaric, the Visigoth leader, for the traditional three days of sacking, the imperial Western seat of its government already had fled to safety at Ravenna, where the mucky coastal marshlands afforded better natural defenses from cavalry and barbarian armies.

Rome's aqueducts also figured prominently in the city's subsequent history and its ultimate renaissance as a center of world civilization. In the mid-sixth century, Byzantine Roman emperor Justinian made a major effort at revival when he tried to retake Italy from the Goths. The Eastern Empire had prospered from its seat at Constantinople and even had launched a formidable new navy. Justinian assigned the exceptionally talented general Belisarius to undertake the recovery of Italy. In 536 and 537, Belisarius successfully conquered northward from Sicily, took Naples by sending 400 soldiers undetected through an aqueduct he had drained during the siege, and then entered Rome without a fight when the Goths evacuated it as indefensible. Destroying Rome's aqueducts was one of the first targets of the Goths in their countersiege against Belisarius. Water ceased to flow almost everywhere; baths, drinking fountains, basins, and sewers went dry. Citizens were forced to crowd into the low-lying areas closer to the Tiber and rely upon the river and wells for their freshwater. The cutoff of the aqueduct flow also shut down the big waterwheel-powered gristmills on the Janiculum, the steep hill near the modern Vatican where much of the city's daily bread was produced. The ever-resourceful Belisarius responded by constructing floating water mills, moored between two rows of boats, under the Tiber bridges where the currents are artificially accelerated—such floating water mills later became commonplace under medieval European city bridges. The Goths tried to jam or break the waterwheels by throwing the bodies of slain Roman soldiers and tree trunks into the Tiber, but Belisarius thwarted them by laying a protective chain across the river to catch the debris.

The Goths also secretly probed the empty aqueduct channels in the hopes of gaining surprise entry into the city. They might have succeeded had not a Roman sentry at the Pincio Hill gate glimpsed the flickering torch light of Goth soldiers as they passed a shaft rising to the surface from the subterranean channel of the Aqua Virgo. The sentry concluded he had seen the gleaming eyes of an errant wolf. Belisarius, however, insisted on an investigation that exposed the Goths' incursion. He ordered the sealing up of all the aqueduct channels. After defending Rome, Belisarius moved north. In AD 540 he took back Ravenna, which the Goths had made their capital. His success and growing popularity, however, made Justinian uneasy about his ambitions. He was recalled. In the end, Justinian's dreams of imperial restoration scarcely outlived him. A new wave of barbarian invasions, this time led by the Germanic Lombards, soon overwhelmed Italy.

By the end of the sixth century, with most of its aqueducts and sewers in ruins and its buildings crumbling, as Rome biographer Christopher Hibbert describes it, "Rome's decay was pitiable . . . the Tiber carried along in its swollen yellow waters dead cattle and snakes; people were dying of starvation in hundreds and the whole population went about in dread of infection . . . The surrounding fields, undrained, had degenerated into swamps" infested with malaria-bearing mosquitoes. The city's population had shrunk to only 30,000. An anti-Lombard alliance between the Papal States and the Frankish Carolingians—reaching its apogee in AD 800 with the St. Peter's Christmas Day coronation of Charlemagne as Holy Roman Emperor—and a papal effort to marshal peasant labor to repair some of the aqueducts failed to endure. By 846 Muslim pirate vessels traveled up the Tiber and plundered St. Peter's.

The Mediterranean West's free-market seafaring and republican democratic traditions, however, were not totally extinguished on the Italian peninsula. Instead, they were transplanted after AD 400 to a cluster of islands in a very shallow 200-square-mile saltwater lagoon intersected by a few deep channels at the head of the Adriatic Sea to which prosperous Roman citizens from the countryside had fled for safety from the barbarian marauders. Venice was destined to become the most precocious of the early Italian city-states, the preeminent sea trading and naval power in the Mediterranean, a progenitor of the modern market economy and the longest lived democratic republic in world history. By AD 466 the dozen tiny island communities began to elect representative tribunes to coordinate affairs among themselves. The first doge, or duke, was elected ruler in AD 697 in what would be an unbroken line of democratically chosen successors until the Venetian republic was finally overrun 1,100 years later in 1797 by French conqueror Napoléon Bonaparte.

Venice lent its nascent naval power in the Adriatic to assist Belisarius and the Byzantine Empire, in what would become a long, complex, competitive alliance between the two greatest sea powers of Christian civilization in a Mediterranean soon threatened by the ascendant commercial and military forces of Islam. It was through Venice that the historic bridge of continuity was established between the early republican seafaring trading traditions born in the ancient Mediterranean and the sea-oriented, liberal market democracies that later rose to world preeminence in post-Renaissance western Europe.

The revival of the city of Rome itself began in 1417 with the end of the Great Schism and the return of the reunified papacy to Rome in the person of Martin V. For want of drinking water, much of Rome's population at the time was still clustered in ramshackle houses close to the filthy Tiber. One of Martin's earliest acts upon returning to Rome was to repair the still partly functioning Virgo Aqueduct that had eluded total destruction by the Goths. Over the next two centuries, several of Martin's successors, notably including Nicholas V, Gregory XIII, Sixtus V, and Paul V—known collectively to historians as the "Water Popes"—dedicated themselves to rebuilding Rome's water system and adorning it with the high Renaissance fountains still admired today. As the water returned, so did Rome's population and the city's grandeur. Rome's population doubled to 80,000 by 1563, reached 150,000 in 1709, and rose to 200,000 by the time of the birth of the Italian state around 1870. Fittingly, the last pope before Rome's integration into democratic Italy completed the redesign of the Republican-era Marcia aqueduct, which became Rome's first to operate under modern-era pressure with pumps.

CHAPTER FIVE

The Grand Canal and the Flourishing of Chinese Civilization

Although intensive irrigation society developed latest in China among the river-born, cradle civilizations of antiquity, its water management achievements surpassed all the others. China's inventive, adaptable, and wide-ranging water engineering responses to its diverse environments was the foundation of what became the most precocious, preindustrial civilization in world history. "The Chinese people have been outstanding among the nations of the world in their control and use of water," observed Joseph Needham in his classic *Science and Civilisation in China.*

China's ancient civilization arose in a landscape markedly different from other hydraulic societies. It began in the semiarid inland north where the Yellow River in its middle reaches exited the barren steppe highlands of Mongolia and carved a large bend through plateaus covered with deep deposits of soft, flakey, yellowish rich soil, called loess, left by the receding ice age. The climate on these stark plateaus, larger than the size of California, was harsh: frigid in winter, scorching hot in summer, prone to droughts, dusty whirlwinds, and occasional summer downpours that eroded the soft cliffs and washed its loess soil into the Yellow, choking it with the thick silt that gave the river its name and enriching the north China floodplain into which it spilled. Yet the plateau's combination of ample river water, easily farmed and drained soils, and military defensibility provided fertile conditions for a single season of intensive, field grain agriculture. The best adapted crop was millet, a tough grain capable of surviving prolonged dry periods. Gradually, farming was extended throughout the large, loess-enriched, northern floodplains. Most extraordinarily, however, China's civilization achieved the rare accomplishment of hurdling its geographic origins over time to transplant itself far beyond its mother river region to a radically different habitat south of the 33rd parallel dominated by the voluminous Yangtze River. In contrast to the semiarid north, the Yangtze region was rainy, humid, verdant, mostly hilly, heavily monsoonal, and civilized by the intensive cultivation of an entirely different crop, wet rice.

The outstanding, transformational event that catapulted Chinese civilization above all its contemporaries, and marked one of water history's turning points, was the completion in the early seventh century AD of the Grand Canal—still mankind's longest artificial waterway, extending over a distance equal to that between New York and Florida. The south-north-running canal linked China's two disparate, giant river systems and habitats to create the world's largest inland waterway transportation network. Just as the Nile had unified Upper and Lower Egypt, China became integrated into a militarily defensible nation-state with a strong, centralized government that commanded an expansive diversity of highly productive economic resources. The Grand Canal played a catalytic role not only in China's becoming the world's most precocious civilization during the Middle Ages but also in the country's fateful fifteenth-century decision to turn its back on the rest of the world that ultimately led to its prolonged, slow decline.

The Grand Canal was so successful because it bridged China's underlying hydrological fault line: north China's chronic insufficiency of accessible freshwater resources to fully irrigate its superabundance of rich soil to achieve its maximum food-growing potential, and south China's opposite profile of having more water than could be productively employed on its less fertile soils. Managing this north-south water and land resource mismatch has been a recurring, central technical and political challenge of Chinese governance in every era since imperial times.

Both the 3,400-mile-long Yellow and the 3,915-mile-long Yangtze originated in the Tibetan plateau in the Himalayas. Beyond that their signature flows and environmental characteristics diverged sharply. The Yellow was shallow and by far the world's siltiest river—30 times siltier than the Nile and nearly three times more than the famously muddy Colorado River. A dipperful of its water was commonly said to contain 70 percent mud. It was the rapid buildup of eroded silt from the loess plateaus that caused the Yellow to frequently overflow its banks in unpredictable, devastating floods across its lower plains. So many millions perished and lost their livelihoods in these fearsome floods over the centuries that the river became known as "China's Sorrow." Its greatest floods—some carving new paths as far as 500 miles away to the Yellow Sea—repeatedly fomented political and economic upheavals throughout Chinese history. Building tens of thousands of miles of levees to try to contain the Yellow within its banks, and rebuilding them after the inevitable failures, was thus always a top political priority of every Chinese dynasty.

The huge Yangtze, by contrast, carried some 15 times more water than the Yellow, with deep navigable channels and many large tributaries that made it an ideal transport highway for large vessels once its waters had descended the mountains and wound its way through its deep canyons and gorges to enter its enormous lower basin and swampy delta. The Yangtze's seasonal monsoon floods regularly inundated the region; every half century or so, however, the combined rush of descending water and the engorged flow from its tributaries created giant waves that overwhelmed all man-made flood control infrastructure and resulted in devastating floods. When China's climate was moister in ancient times, the central section of the Yangtze had been a gigantic swamp, far too wet to sustain large-scale civilized human settlement. Gradual desiccation, and Chinese advances in water redirection, terracing, drainage, and other wet rice irrigation techniques gradually transformed the region into prosperous farmland. By medieval times it was producing the greater part of China's food, with rice surpluses distributed along its extensive tributary network and to the Yellow River region in the north via the Grand Canal and coastal sea routes. Politi-

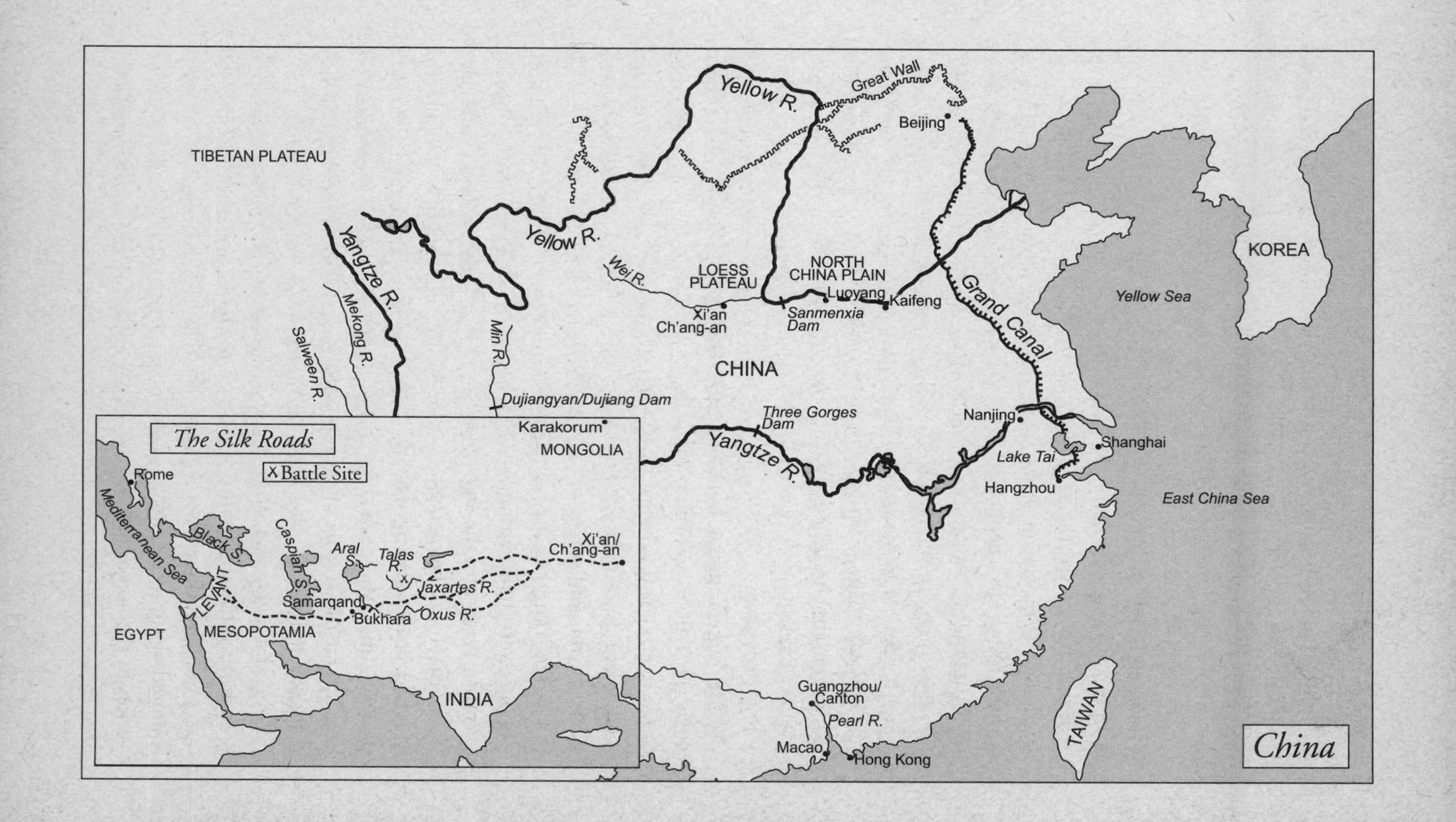

China
TIBETAN PLATEAU
Yellow R.
Yellow R.
Great Wall
Beijing
KOREA
Yellow Sea
Yangtze R.
Mekong R.
Salween R.
Min R.
Wei R.
LOESS PLATEAU
NORTH CHINA PLAIN
Luoyang
Kaifeng
Xi'an
Ch'ang-an
Sanmenxia Dam
Grand Canal
CHINA
Dujiangyan/Dujiang Dam
Three Gorges Dam
Yangtze R.
Nanjing
Shanghai
Lake Tai
Hangzhou
East China Sea
Guangzhou/ Canton
Pearl R.
Macao
Hong Kong
TAIWAN
The Silk Roads
X Battle Site
Karakorum
MONGOLIA
Rome
Mediterranean Sea
Black S.
Caspian S.
Aral S.
Talas R.
Xi'an/ Ch'ang-an
Jaxartes R.
Samarqand
Bukhara
Oxus R.
LEVANT
EGYPT
MESOPOTAMIA
INDIA

cal control of the "golden waterway" of the Yangtze thus joined flood control as a vital linchpin of Chinese power. So closely correlated was river management and governing power that the very Chinese character for "politics" is derived from root words meaning flood control.

The traditional founding father of China's Yellow River civilization was Yu the Great. A water engineer, Yu rose to power on the merits of his accomplishment as the tamer of the great floods that ravaged settled life in the Yellow basin before recorded history. By having "mastered the waters and caused them to flow in great channels," he made the world habitable for human society. In honor, the tribal confederation elevated him to leadership. He went on to found the Bronze Age Xia dynasty from about 2200 to 1750 BC, and he became venerated as the lord of the harvest in association with the river's early irrigation works.

Yu's legend reflected the paramount importance of water control in Chinese history. At birth it was said he emerged, fully formed, straight from the dead body of his father, who had previously tried and failed to control the floods by damming and diking the river's flow, and had been put to death for stealing magic soil from heaven in order to build a dam. After careful study and surveys, Yu took up his father's task by the different approach of laboriously dredging river channels and digging ditches and canals, including one bored through a mountain, in order to divert excess floodwaters to the sea. He labored selflessly alongside the workers, and after many years, finally succeeded in bringing the Yellow River and its floodplain under control. Confucius hailed him as the ideal of the humble, qualified government official who used his power for the public good, and thus the aspiring role model for China's technocratic elite who governed in support of its emperor.

Water management helped frame the historic Chinese philosophical debate about the right principles for man's governance of himself and his relations to the natural order. The sixth century BC Taoists argued that humble water's yielding, yet relentless flow that wore down all hard and strong obstacles expressed the essence of nature and provided an exemplary model for human conduct. Taoist engineers designed waterworks to allow water to flow away as easily as possible, exploiting the dynamics of the natural ecosystem, just as they urged Chinese leaders to gradually win support for their goals through persuasive dialogue. Their main rivals, the Confucians, on the other hand, advocated a more forceful manipulation of both nature and human society to achieve the public good. They believed that rivers had to be forced, through dikes, dams and other obstructive constructions, to do man's bidding as defined by rulers and technocrats. Although the Confucian view prevailed as the guiding tendency of Chinese hydrology from the Han Empire in the late third century BC to the twenty-first century postcommunist state, the underlying principles framed a fundamental engineering debate which has reemerged on the global level today as the world seeks environmentally sustainable solutions to the water scarcity crisis.

After nearly half a millennium, Yu's Xia dynasty was displaced as the predominant power by the Shang and later the Zhou dynasties. Each was centered along a different, but overlapping part of the inland Yellow basin and flourished on indigenous irrigation agriculture without significant river or seaborne trade with other regions. The Shang was a Bronze Age tribe that with the help of the chariot imposed an aristocratic rule from about 1750 to 1040 BC over an area centered in the fertile north China plain and within reach of the Yellow River region's tin and copper deposits. Although they were one of the earliest literate cultures east of Mesopotamia, the Shang's many primitive customs included ancestor worship, human sacrifice, and various ritual consultations with the spirits by priestly diviners. The excavations at their city of Anyang yielded tens of thousands of "oracle bones" that were consulted by priests to reveal answers to the vital questions of life and death such as whether it would rain or when the barbarians from the north would attack.

North China's climate was still much warmer and moister than today, and large-scale irrigation depended on extensive reclamation of cropland by draining fens and marshes by mass organized manpower. Their highly stratified social organization and large public works, including extensive walled cities, fit the model of the hydraulic civilization. Startling confirmation that millet was one of their staple crops was made in 2005 when archaeologists exploring the remains of an ancient village buried by an earthquake and flood discovered a bowl containing a well-preserved 4,000-year-old millet noodle 20 inches long.

With the conquest and amalgamation of the Shang culture by its former vassals, the Zhou dynasty, centered on its western border along a tributary of the middle Yellow River, more of the distinctive character of the emerging Chinese state took shape. While retaining the older dynasty's use of kinship as the basis of political organization, the Zhou introduced the enduring political concept that the emperor's ruling legitimacy stemmed not solely from divine right of birth but from a "Mandate from Heaven" based upon moral performance. Water control was a key test of the mandate. A good emperor was expected through magic and ritual to be able to deliver vital things like rain, peace and good harvests; droughts and floods, on the other hand, were events that signaled heaven's disapproval. One legitimizing boon to Zhou crop irrigation was the innovation during their reign in the sixth century BC of productivity-enhancing iron tools. The advent of iron, however, also stimulated the deployment of new weaponry. From about 400 BC, what would emerge as the unified Chinese state was forged over nearly two centuries of incessant warfare between seven competing regional powers. During this period, the flight of northern farmers from the war zones accelerated the migration of Chinese civilization to the cultivation of rice paddies in the warm, wet south. Although China's rice farming would not achieve its full critical mass until the seventh century AD, by the time the Ch'in dynasty consolidated its victory over its rivals in 221 BC—giving China its modern name—its domains extended throughout the Yangtze basin to the eastern seaboard.

For a dynasty whose own rule lasted only fifteen years, the Chin's legacy was remarkable. Their new political structure of all-powerful emperor with a centralized bureaucracy replaced the old feudal system. Uprooted aristocrats were compelled to move to the emperor's capital, while their local estates were superseded by a system of provinces and counties ruled by governors loyal to the emperor. Standardization was applied to weights and measures, writing systems and currencies, census-taking was begun, and taxation ruthlessly levied.

Like many great founding or restoring dynasties, the Ch'in were prodigious builders. Their accomplishments included building a vast road network and early segments of the Great Wall against marauding northern nomads. Of critical importance to their rise and legacy was the construction of large-scale, sophisticated irrigation and transport waterworks. Three in particular stood out. In their home state near the middle Yellow River they completed the Cheng-kuo Canal in 246 BC. By diverting water from two tributaries of the Yellow, it irrigated vast acreage in the Wei river valley north of its capital of Xi'an, site of the famous life-sized terra-cotta army of 8,000 soldiers, horses, and chariots that guarded the original Ch'in emperor's tomb. Although heavy silting limited the irrigation canal's productive life to a century and a half, the great increase in food and population it yielded played a vital role in providing the wealth, weaponry, and manpower the Ch'in needed to complete their conquest of China during the Warring States period.

Even grander and more impressive were the irrigation works of western Sichuan, north of the upper Yangtze, undertaken by a water engineer so accomplished he seems almost to have been an avatar of Yu the Great. Li Bing had been appointed provincial governor in 272 BC, nearly half a century after the region's conquest by Ch'in generals. To enrich the province and win loyalty from the local population, he embarked upon an ambitious hydraulic engineering scheme intended to provide at once flood protection and reliable irrigation from the rapidly flowing, unpredictable waters of the Min River to the surrounding farming floodplain. Li Bing's celebrated waterworks—still flowing today—were constructed chiefly along Taoist precepts. Rather than directly block the river's forceful flow with a dam, a series of diversion weirs were built from flexible bamboo cages filled with rocks that were situated at a juncture where the natural contours of the river facilitated its division into an outer and an inner channel. The weirs could be adjusted to direct more of the water to one channel or the other depending upon conditions—to the outer channel to divert water against flooding or to the inner channel when irrigation water was needed. Li

Bing emplaced three upright stone figures in the water as signal gauges. When their feet grew visible, the weir's gates were to be opened to water the fields; when their shoulders became covered, the gates were closed. To complete the irrigation diversion so it could reach the farmland in the Chengdu plain below, Li Bing's workers laboriously cut a channel through the mountainside by heating the rock by bonfire, then dousing it with water until it cracked and could be chipped away. Li Bing's waterworks transformed the plains of eastern Sichuan into one of China's most affluent irrigated farming zones. Covering some 2,000 square miles it sustained a population of 5 million—the maximum supported by the Egyptian Nile from ancient times until the nineteenth century. The outer channel also provided navigability. Later, in medieval times, Li Bing's flowing Min River canals in the plains found additional employment turning thousands of waterwheels to hull and grind rice, and to power textile spinning and weaving machinery.

Li Bing also improved Sichuan's production of precious salt by drilling early brine wells, some more than 300 feet deep, that drew salt directly from its underground sources rather than relying upon traditional salt harvesting from briny pools that had seeped up from the earth. His successors learned to use long, bamboo tubes with leather flap valves to create suction to draw the saltiest water from the deepest recesses. Bamboo plumbing became the mainstay not only of salt works, but eventually was applied ubiquitously throughout south China's rice paddies by farmers as conduits for pump-lifted and relocated water and also in cities as rudimentary water mains.

The Ch'in's third extraordinary water project was the Ling Chu, or Magic Canal, the world's first transport contour canal, which was dug by following the natural topography of the surrounding landscape to avert complex tunneling and water-level management problems. By controlling and joining two rivers that flowed near each other in opposite directions, the 20-mile-long Magic Canal created a waterway link through the mountain ranges dividing northern and southern China. Built on the orders of the Ch'in emperor to support the conquering armies he had sent south in 219 BC, the Magic Canal made it possible to travel by boat through natural waterways and earlier channel cuts all the way from the lower Yellow River, south to the Yangtze, and beyond to the port of Canton—an astonishing distance covering 1,250 miles. Nothing like it had ever existed before in history.

The greatest beneficiary of this unprecedented precursor of the Grand Canal was not the Ch'in, however, but their immediate successors, the Han. Under the four centuries of Han rule, from 206 BC to AD 220, China's powerful centralized state and high civilization flourished as one of the two greatest on Earth. Historians frequently have noted the many historical parallels between the Han and Roman empires. Their periods of greatest power, wealth, and influence were contemporaneous, their empires were of comparable geographic size, they flourished at the extreme edges of the civilized world at the time, and the proximate causes of the demise of each were barbarians attacking from the northern frontiers. Of course their political economies, cultures, and hydrological underpinnings were quite dissimilar. Rome did little intensive irrigation, relied for its wealth upon its Mediterranean sea-linked network of colonies, encompassed many cultures, and honored individualism. The Han Empire, by contrast, was the epitome of a hydraulic state: inward-looking and land-oriented, based upon intensively irrigated agriculture, and governed top-down by a despotic emperor and a cadre of expert technocrats overseeing mass peasant laborers.

The Han wasted little time in marshaling forced labor to add and improve canal segments to the great transport waterway they'd inherited, along with so many other remarkable, nation-building achievements, from the Ch'in. Much of their success was also owed to the extensive construction of irrigation and flood control canals, dams, and dikes, including some forty major water projects to control the Yellow River. Under the Han's centralized administration, the patchwork of cropland in the Yellow River valley was organized into a single, intensively irrigated continuum that created China's classic landscape and served as the economic and political heartland of the empire. Treadle chain pumps, a simple but extraordinarily useful small-scale technology for lifting water operated by the simple stepping motion of as few as one or two individuals, invented in the first century AD, was widely applied across China for drainage and irrigation and to supply drinking water. Eventually all water planning was centralized in a national office, establishing a tradition that has endured to the present day.

By 100 BC the Han state had become the largest landowner, with government monopolies also instituted over vital goods like iron, salt, and wine. Private merchants and the nascent profit-driven market system that had begun to develop under the Ch'in but conflicted with Confucian precepts of governance were suppressed by regulation. Sovereign taxing power was used to weaken disfavored classes and accrue authority to the Han state. In time, all urban markets became government controlled, with officially set commodity prices and taxation on commerce that filled the treasury.

The Han's bid for state domination over economic life was made easier by the fact that wealth creation remained predominantly based on intensively irrigated agriculture at inland locations along mostly navigable, relatively easily governed arterial rivers. Despite China's long coastline, sea trade—always problematic for sovereign states to control—remained underdeveloped due to the geographical fact that there were simply few enticing, easily reached Far Eastern civilizations with whom to profitably trade. Although some unregistered itinerant merchants survived and even flourished trading between cities and at the peripheries of society, mainstream Chinese civilization developed a strongly inward-looking orientation that tended to accrue great power to the central state.

The Han emperors encouraged the expansion of industry, some using water as a vital input. Most importantly this included its precocious iron casting industry. A process for casting iron into molds, one of history's key inventions, had been discovered in China as early as the fifth century BC, nearly 1,800 years before cast iron became widespread in Europe. In the third century BC, the Chinese iron masters discovered a heating and cooling process that produced a malleable cast iron with the strength and solidity that rendered it nearly as good as steel. The Han employed it in important applications, such as making cast-iron plowshares for agriculture and pans in which brine could be evaporated for the mass production of salt. Within two years of nationalizing all cast-iron manufacturing in 119 BC, Han leaders had established 48 state foundries that employed thousands of workers. To achieve the high temperatures necessary for casting iron, the Chinese employed efficient bellows to stoke the blast furnaces. An early innovation that greatly increased cast-iron production was the application of waterwheel power to the bellows. In AD 31, noted Chinese engineer Tu Shih invented a powerful, widely imitated, water-driven bellows used to produce cast-iron agricultural implements.

Unlike the Romans, who used waterwheels chiefly to grind grain and for mining, the Chinese also pioneered the large-scale application of waterpower for industrial production. Indeed, for well over a millennium, China was the human civilization's leader in harnessing water as energy to do useful work. Powerful vertical waterwheels, with gearing to turn several shafts, were used to operate trip-hammers to pound iron into shapes, hull rice, and crush metallic ores, as well as for other applications, by the AD 200s and 300s, many centuries before they appeared in Europe. By AD 530 Buddhist monasteries in the northeastern city of Loyang were even operating waterwheel-powered flour-sifting and -shaking machines based on the same essential design as that used by the steam engine—albeit with the crucial absence of steam power itself—that would galvanize England's eighteenth-century Industrial Revolution.

Not surprisingly, the Chinese would become world pioneers in ensuing centuries in applying waterpower to the ancient art of silk making—one of trade history's great monopolies that enriched imperial China for centuries. The art of producing silk filament from the cocoons of the mulberry silkworm and weaving it into textiles was first discovered as far back as the Stone Age. Hot water played a critical role in the silk-making process, in what was perhaps history's earliest example of water use in industrial production. To net one pound of raw silk, silkworms had to eat 100 pounds of mulberry leaves and produce about 15 pounds of cocoons. Great skill was required to unwind the silk filament from the delicate cocoons. It was accomplished by first soaking the cocoons in boiling water to kill the chrysalis; the strands were drawn out, then joined together and finally woven to produce the soft fabrics desired worldwide.

The Romans first encountered silk in 53 BC when fighting the Parthians in modern Iran. By the first century AD, Roman demand for popular Chinese silk became such a burden on Rome's balance of trade that Emperor Tiberius tried to forbid the importation of silk garments. China's monopoly advantage in the silk trade with the Roman world endured for another 500 years. It was finally broken by a famous case of industrial espionage when two Christian Byzantine monks traveling in China stashed silkworm cocoons in the hollow of their staffs and returned with them to Constantinople, which promptly established its own lucrative silk industry by the end of the sixth century.

The Han began shipping large quantities of silk to Persia and the Levant on taxed and protected camel caravans across the famed 4,000-mile-long Silk Roads of arid central Asia in 106 BC shortly after discovering, to their surprise, the existence of a fairly advanced civilization in the far West—Rome. The availability of freshwater sources dictated the trade routes and the scale of the camel trains. The several Silk Roads started at the huge city of Chang'an (the former and later X'ian) in the Yellow River valley, skirted inside the Great Wall to the Jade Gate and then went beyond China's borders to follow a string of oases across the harsh wind- and sand-blown high deserts of central Asia at the foot the Himalayan, Altai, and Tien Shan mountains. Oases formed wherever mountain streams rushed down, sometimes flooded by snowmelt or violent storms. A northern and a southern route came together between the Jaxartes and Oxus rivers and crossed through Samarkand and Bukkara in modern Uzbekistan, before following various roads through Persia and Mesopotamia to Roman Syria on the Mediterranean; another branch route headed south to India.

The entire trek was made possible only by the astonishing strength and water-storing capacity of the two-humped Bactrian camel, which unlike its one-humped Arabian cousin was able to tolerate the freezing temperatures of the high Asian deserts. Caravans of sturdy, woolly Bactrian camels plodded 30 miles per day carrying 400 pounds of goods on their backs. Although larger trains offered greater safety, most of the caravans were no larger than 50 men and their beasts since that was all the scarce water resources en route could support at any one time.

Many of the oasis outposts thrived as important entrepôts of civilization, where both high-value luxury goods and new ideas were exchanged free from government control. In addition to silk, China shipped iron goods, ceramics, jade, and lacquer; the West sent back gold, ivory, precious stones, coins, glass, Persian sesame seeds and nuts, and, from India, spices and perfumes. Buddhism entered China and the Far East from India along the Silk Roads, its teachings spread by two Buddhist monks in the first century AD, overlapping the simultaneous early spread of Christianity in the Roman Empire.

Trade on the Silk Roads reached its apex in the seventh and eighth centuries. Then, suddenly, after the annihilation of a Chinese expeditionary force by a Muslim army at the Talas River near Samarkand in 751 helped trigger the collapse of Chinese power across central Asia, the Silk Roads closed for over four centuries. It was one of history's obscure skirmishes that in retrospect had outsized consequences. The silk trade was redirected to the Indian Ocean spice routes that were increasingly dominated by Muslim shipping, accelerating the rapid rise and global reach of Islamic civilization. The Silk Roads were finally reopened for travel by the Mongol Empire, whose thirteenth-century conquests under Genghis Khan and his heirs spanned from China to Persia. It was at the end of the thirteenth century that Venetian Marco Polo, a jewel merchant, made his famous trading expeditions from Venice to the Mongol overlords of Cathay (China). But by that time the Muslim-controlled, two-way seasonal monsoonal Indian Ocean sea route had established itself as the most reliable and economical way to transport goods between East and West and retained the predominant share of the traffic.

With the fall of the Mongols, the glory days of the Silk Roads ended forever. But the combination of these two overland and sea routes succeeded in establishing an enduring, unregulated, long-distance, Old World trading network based on market economic exchange that competed with and would eventually supplant the traditional authoritarian command organization of centralized states. International trade had reached a peak at the contemporaneous high point of the Roman and Han Empires in the early Christian era. It then eroded with their declines. But gradually, with the restoration of civilized order in the Orient and Occident it built up again. By about AD 1000 the web of international exchange had achieved a critical mass of sufficient density and volume to propel its expansion throughout the second millennium and ultimately evolve into the integrated global market economy of the twenty-first century.

The Han empire finally collapsed in AD 220, like Rome, under pressure from barbarian raiders from its northern frontiers and internal depopulation and weakening from exposure to unfamiliar infectious diseases inadvertently imported on trading ships and Silk Road caravans. Indeed, the Han imperial state had never fully recovered from the combination of a short-lived usurpation and a disastrous course change of hundreds of miles in the flooded Yellow River in AD 11. The delay of several decades in repairing the river's damaged irrigation and protective infrastructure led to inadequate food supplies, famine, disorder, and massive emigration along China's vital northern defensive frontier. Manpower shortages, in turn, weakened the size of the Han's military presence. But the essential underlying reason for the Han's eroding northern strength was its inadequate supply and control of irrigation water to grow enough food to maintain a sufficiently large army there—north China's relative shortage of water resources. A third-century Chinese report highlighted that "there was insufficiency of water for the fullest use to be made of the productive power of the soil." Nor was there yet in place any efficient network to transport compensatory southern food supplies to the northern borders.

The historical parallels between Rome and China endured into the sixth century AD with native efforts to reconstitute the fallen empires in each region. The initiative of Constantinople's Byzantine Roman emperor, Justinian, to reunify the Latin West ultimately failed, leaving Rome itself as a shrunken ruin of its former glory until the Renaissance. In China, by contrast, reunification succeeded. Restoration under the Sui from 589 to 617 AD and the successor T'ang dynasty until

AD 906 paved the way for China's medieval economic revolution and its golden age when Chinese society soared to the zenith among world civilizations.

Why did China's reunification succeed, yet Rome's fail? One outstanding distinction was China's building of the imperial Grand Canal linking the Yangtze and the Yellow rivers. Completed at a breakneck pace in only six years in 610 by the Sui—who, like the Ch'in, were prodigious and ruthless infrastructure builders—using some 5 million conscripted male and female laborers, the nearly 1,100-mile-long, elongated S-shaped Grand Canal linked up local canal segments that had been built episodically since the fifth century BC and added new segments. In all, it created a stupendous 30,000-mile-long national inland waterway system that enabled a united China to ship vital rice supplies grown on the terraced hillside paddies of south China to the large population centers and army troops located on the Yellow River to defend against the continuing threat from bellicose nomadic horsemen from the Asian steppe.

Not only did the Grand Canal overcome the vulnerability that had vanquished the Han. By bridging China's north-south hydrological fault line, it synergized the natural and human resources of the two diverse geographical zones to help launch China's brilliant, medieval golden age. All China was invigorated with fresh economic and cultural energy. In contrast to the Han, the Sui and T'ang dynasties presided over a more robust double base, one in the traditional Yellow River valley in the north and an even more productive southerly one that had been steadily growing for centuries around the Yangtze.

The old Roman Empire and Europe, by contrast, lacked the unifying impetus of any such inland waterway. Its major arterial river system, the Danube-Rhine, was ill-suited to that purpose because it flowed away from the early hubs of Mediterranean European civilization, through difficult, rain-fed agricultural soils, and along an unstable frontier boundary besieged by warlike tribes. The open waters of the Mediterranean were far less effectively controlled and thus less conducive to playing a uniting role than large irrigation rivers. From this point on, the histories of China and Europe diverged widely.

While the European territory of the old Roman Empire remained a fragmented jigsaw of competing states and endured the stagnation of the long dark ages, the Grand Canal served as the electrifying fulcrum of China's medieval economic revolutions in transport, agriculture, and industry.

China's Grand Canal was one of mankind's stellar engineering achievements. It was the largest artificial transport waterway ever built and required a larger labor force than the one that erected the Great Wall, most working with their bare hands and shovels and uncounted hundreds of thousands giving their life to the project. It ultimately spanned eastern China for 1,100 miles from the port city of Hangzhou south of Shanghai to Beijing on the northern frontier with a gigantic channel 10 to 30 feet deep and up to 100 feet wide. It featured 60 bridges and 24 locks to manage elevation variations and summit level water flows. Its waters teemed with commercial vessels of all shapes and sizes, powered by sail, oars, and paddle wheels, transforming the world's most densely populated trading area into a single national economic market. Rice-laden barges came and went from huge granaries that the government maintained at key junctures along its route to furnish the food lifeline for China's national security. The easy movement facilitated centralized governance by grain tax collectors, bureaucrats, and soldiers en route to army garrisons. Due to its surpassing importance, the Grand Canal became a key political barometer, and driver, of Chinese history. Whenever the canal was threatened, cut, or left in a state of disrepair, China was generally in the throes of a crisis or immersed in prolonged decline or political lassitude. A robust Grand Canal system, on the other hand, spurred internal growth and security, rendered superfluous the pirate-harassed sea transport links between the southern food supply and the northern defensive garrisons, and generally encouraged China's inward-looking, autarkic impulses.

As a result of the canal, water transport became many times less costly, at least one-third less than shipping by land. Government policymakers made ongoing improvements a top priority. One key advancement was the world's first double canal pound lock at an opening onto central China's Huai River. It was built in AD 984 at the order of Ch'ia Wei-yo, a Sung government assistant transport commissioner, who was seeking ways to minimize theft and damage to ships and cargo caused by existing methods for moving vessels between differing water levels. At the time, cargo ships were manually hoisted out of the water with ropes by large labor crews, dragged up a slipway ramp cut into the bank, and then relaunched from a second slipway into the water at the level of the second waterway. Pound locks, which came into widespread use on the Grand Canal network during the eleventh century, impounded water between two lock gates; vessels were either lifted or lowered simply by adding or withdrawing water within the impounded portions. Boats could easily be lifted up to five feet in a pound lock. A graduated series of pound locks enabled canals to lift boats to unprecedented elevations—the summit level on the Grand Canal, for instance, was 138 feet above sea level. Moreover, the pound lock conserved precious water, allowing canals that often went dry in summer to operate more days of the year. One double lock built where a branch of the Grand Canal joined the Yangtze enabled the passage of ships five times larger than possible under the double slipway system.

River cargo volumes soared from early T'ang times. In the eighth century, combined tonnage of the government's 2,000 ship Yangtze salt and iron fleet alone reached a third of the total transported on all British commercial vessels in the mid-eighteenth century. During the Sung dynasty from 960 to 1275 government officials improved their communication by using the waterway to distribute the world's first national newspaper, an official government gazette. As shipping traffic increased, private shipping brokers became increasingly active, matching and administering contracts between buyers and sellers, storing goods at their warehouses and serving as clearinghouses of market prices and conditions. Where the commerce of inland waterways merged with the sea, great port cities with market activity arose to trade in spices, silk, and other luxuries with the rest of the world through shipping networks that extended across the Far East, India, Arabia, and the Mediterranean.

During the T'ang era, the Chinese were content to sail the seacoasts with the two-way monsoons, south in winter and north when the winds reversed in summer, and to rely upon Arab, Persian, and other foreign shippers for long-distance sea trading. By the Sung dynasty, Chinese nautical mastery was world class, featuring massive ships built with iron nails from its foundries, watertight compartments unknown in the rest of the world, a huge sternpost rudder for steering, buoyancy chambers, and distinctive, narrow fanlike sections of canvas sails stretched between bamboo masts that looked much like a Venetian blind. Navigation was facilitated by the invention in 1119 of the mariners' compass, one of many important Chinese innovations to migrate westward. As a result, Chinese seafaring gradually grew more ambitious. Yet seafaring in China's golden age never approached the awesome scale of its inland river shipping, which so impressed Marco Polo, native of Europe's greatest seaport, Venice, then with a comparatively tiny 50,000 inhabitants. Describing the Yangtze at the smallish port city of I-ching, he related that "the amount of shipping it carries and the total volume and value of its traffic . . . exceeds all the rivers of the Christians put together and their seas into the bargain . . . I have seen in this city fully five thousand ships at once . . . and there are on (the Yangtze's) banks more than two hundred cities, all having more ships than this."

The national Grand Canal waterway transport network also gave powerful impetus to China's rice-farming revolution (eighth to twelfth centuries), one of the decisive events in Far Eastern history. Originally a dry crop, wet rice had been extensively cultivated in naturally flooded fields alongside the monsoonal rivers of Southeast Asia by small communities since the third millennium BC. It first arrived in China from India around 2000 BC. After about 500 BC more intensive new irrigation methods that allowed cultivation on a larger scale and over a wider range of landscapes began to spread.

Rice irrigation demanded solving formidable water challenges. First, there was the problem of converting nature's hydrological excess of rainfall and floods from an insuperable obstacle into a productive irrigation resource. The transplanted young roots had to be kept submerged in shallow water for several months, after which the water was drained off. Farmers had to prepare, level, and wall the terraced paddies of south China's hillsides, drain and refill paddies at timely moments, and keep the entire system constantly flowing with muddy water in order to provide sufficient oxygen to the rice plants and to suppress infestations of malaria-carrying mosquitoes. A technical array of dams, sluice gates, water-lifting norias, simple treadle pumps, and a network of bamboo pipes enabled the process. The labor was immensely intensive. But so were the rewards. Inundation transformed poor, quickly depleted soils into perennial rice paddies that never had to lay fallow so that yields could sustain population densities far higher than those that could be fed by achievable yields of wheat or maize—thus providing the demographic profile that distinguished Asia's history. In China, it completed a population and dietary transformation that had been evolving gradually over many centuries as many Chinese farmers moved south away from the millet and wheat fields of the Yellow River basin.

China's rice farming revolution reached its apogee in the early eleventh century with the government's importation from Champa in central Vietnam of a variety of rice that matured in only sixty days. The fast-growing Champa rice also required less water than domestic varieties, so it could be grown on drier hillside paddies that soon were irrigated for the first time. In 1012, on the orders of Sung emperor Chen-tseung, samples of the seeds of Champa rice were distributed to farmers in a conscious effort to expand food production. The effects were astonishing. Suddenly, two and three crops could be grown on an expanded area of irrigated land. Rice production soared. China's population promptly followed. By the end of the twelfth century it reached 120 million—double the peak achieved by the Han in AD 2 and the early T'ang in the 700s. Some 75 million lived in the south, inverting the historical population balance favoring the north, and transforming China permanently into a densely populated nation primarily of rice eaters—a profile that defined its economic and social structure until the twenty-first century. Abundant food shipments along the Grand Canal drove the rise of Earth's largest urban centers of the times, cities such as Hangzhou, Kaifeng, Louyang, and Beijing, that also pulsed with resurgent private market commerce and industry. During the Sung dynasty, China's greatest age, these urban centers became hubs of a remarkable scientific renaissance, entrepreneurship, and a protoindustrial revolution six to seven centuries before Europe's. Many great inventions were developed, some which later migrated westward through the Indian Ocean and Silk Road trade routes to stimulate Islamic civilization and later still Europe's rise.

By 1100, China was by far the world's indisputable technological leader. Techniques were being used in smelting iron with coke, canal transportation, bridge design, water-powered textile manufacturing, and producing iron farm tools that were not paralleled in Europe for some 600 years. In addition, China was the first civilization to discover that mixing saltpeter (potassium nitrate) with carbon and sulfur produced a volatile substance that exploded when heated—gunpowder. It also pioneered firearms, scientific instruments for measuring the heavens and navigation, hydraulic clockwork, printed books, moveable type, paper money, and the first toilet paper.

One of the great northern Sung cities, and probably the most important place in the world at the time, was the capital, Kaifeng. In 1100, it was larger than ancient Rome with a registered population plus army of about 1.4 million. Its strategic location near the junction of the Grand Canal and the Yellow River not only put it within easy supply of the rice barges from the south, but also along the water transport routes that brought abundant coal and iron from north China's mines. Kaifeng rose as an important industrial center. Spurred by deforestation which overtook the region by about 1000, Chinese iron smelters had made the breakthrough discovery of how to use coal in place of charcoal to cast iron from coke-burning blast furnaces. They also invented a decarbonization method to produce large quantities of hard steel from cast iron. Iron production soared. By 1078, China was producing 114,000 tons of pig iron, double England's total output 700 years later.

The story was similar in textile manufacturing. Before 1300, China's precocious textile artisans were operating water-powered spinning machines that could draw several silk filaments simultaneously from boiling water-immersed cocoons—some 400 years before England's first water-powered spinners began producing silk stockings in Derby to help launch the industrial factory system. Mechanical clocks, which China invented at least two centuries before Europeans, with sophisticated gearing, precision, and self-regulatory mechanisms, helped government administrators accurately keep the all-important official calendar. The famous 30-foot-tall noria-type water clock, with quarter-hourly gongs and bells, erected in 1090 in a Kaifeng pagoda was even used to calculate favorable heavenly synchronizations of the Chinese emperor's procreation schedule among his 121 wives and concubines.

Yet China's technological leadership did not make it invulnerable to its age-old menace of nomadic invasions from the north. In 1126 the barbarian Ch'in Tartars, armed with iron-weapon-making know-how imported from China and with its traditional, potent cavalry, overran Kaifeng and northern China. They, in turn, were expelled by the Mongols in 1234. The surviving Southern Song dynasty established a new capital at Hangzhou. Utilizing the natural protection afforded by the wide Yangtze River, it built its primary defense around a naval fleet of hundreds of newly designed, armored river- and canal-fighting vessels propelled by paddle wheels and treadmills and armed with onboard projectile-throwing machines, crossbowmen, and pikemen. Behind the Yangtze lay the second line of defense—the muddy rice paddies in which the Mongols' fearsome mounted cavalry bogged down. These defenses helped them survive the onslaught of Genghis Khan's heirs until 1279.

The ferocious Mongol conquests from 1206 led by Genghis Khan created the largest land empire in world history. The trademark of this tribal confederation of nomadic, mounted archers from the arid steppe was the merciless slaughter of defeated populations and their domesticated animals, wholesale pilfering, and razing of cities, irrigation works, and other vital infrastructures of civilized life. When Genghis died, undefeated, in 1227 the Mongol empire spanned the entire central Asia steppe from the Volga River in the west to the Amur River in the east. His successors expanded into eastern Europe as far as Poland and Hungary by 1241. Much of the Islamic Middle East capitulated to Mongol warriors, who in 1258 savagely destroyed Baghdad and its caliphate. Mongol armies reached the Adriatic by 1258 and the edge of Africa by 1260. By 1279 all of China was in Mongol hands, the first time in history that China was ruled completely by foreigners. The zenith of the Mongol empire was reached under Genghis's able grandson, Khubilai Khan, who ruled from 1260 to 1294. He set up China's Yuan dynasty, which had its seat of government in Beijing. It was Khubilai's Mongol-led China that Marco Polo served and famously described to the transcriber of his *Travels,* Rustichello of Pisa, while the two languished in a Genoa prison after the Venetian galley on which Marco had been traveling was captured during a September 1298 battle between Mediterranean trade rivals Venice and Genoa.

The Mongol conquests were to be world history's last great wave of invasions by nomadic, pastoralist warriors that had disrupted and challenged the settled, civilized lifestyle since the Bronze Age. Before the Mongols, the water-fragile central Asian steppes had produced invasions from bellicose confederations of Hsiung-nu, Juan-juan, and the Mongol's close cousins, Turkmen; the latter allied with China in defeating the Juan-juan in the mid-sixth century, and subsequently divided into two groups and eventually infiltrated and rose to political prominence in Islamic society. The nomads' ignorance of civilization's complex technologies, including sophisticated water management, proved to be one of their grave weaknesses in trying to govern the societies they conquered. The rule of the Mongol's Tartar predecessors, for example, had been undermined by a decline in canal traffic capacity and related diminution of iron and agricultural production, resulting from their failure to undertake timely restorations when the flooding Yellow River burst its restraining dikes in 1194 and cut a new path to the sea.

Well into the thirteenth century the Mongols themselves still crossed rivers using primitive, inflated skins and rafts. Their conquest of southern China succeeded after forty-five years only when a former Sung commander, Liu Cheng, defected to Khubilai and built a river naval fleet capable of challenging Sung dominance of the Yangtze River. The decisive confrontation was a five-year river-and-land siege of Hsiang-yang, a key river fortress that controlled the main access route to the

Yangtze heartland. When success came in 1273, south China finally lay open to the Mongols. With its cavalry bogged down in muddy rice fields, its infantry unfamiliarly led the invasion. The final remnants of Sung resistance were extinguished in early 1279 following their large naval defeat off the coast of Canton, when a loyal minister jumped in the ocean, drowning himself along with the last imperial prince.

Even then, the Mongol Yuan dynasty had very limited success exploiting the rich potential of China's water resources. Despite building an impressive naval fleet, Khubilai was unable to extend the Mongol's military land prowess to sea power, as exemplified by the unsuccessful large naval invasions against Japan in 1281 and Java in 1293. He failed as well to wrest ocean shipping from the Muslims. Like his predecessors, Khubilai devoted high priority to enhancing the Grand Canal. He restored the connection to the Yellow River, disrupted by the course change of 1194, straightened the canal, and added a northern extension to his new capital, Beijing, on the extreme northeastern frontier. Food security and prosperity increased. Yet his engineers failed to make the crucial canal innovation necessary to supply sufficient water to enable food transport ships to readily pass year-round over the summit level of the hills leading to Beijing. This contributed to the Yuan dynasty's downfall. Sea convoys temporarily mitigated the vulnerability, but finally piracy and rebellions in the south disrupted reliable food deliveries. The breakthrough "Heaven Well Lock" of the Grand Canal would be made during the restoration by the Ming dynasty, which finally toppled the despised, plague-weakened Mongolian Yuan after 1368.

Like many nativist restorations, the early Ming era was marked by a revival of old traditions, renewed economic and creative vigor, and xenophobia. Water engineering advances played a prominent role, above all in shipping, reconstruction of iron chain suspension bridges, and in major improvements of the Grand Canal. The Ming's superior command of water resources, in fact, had played a decisive role in driving the Mongols back into the northern steppe. In 1371 the Ming navy, armed with iron prows and firearms, broke through the chains and bridge of boats defending the gorge at Chü-tang, the key to controlling Sichuan.

Once in power, Ming seagoing vessels, meanwhile, reopened a 500-mile-long transport supply line of food, clothing, and weaponry that enabled the reconquest of southern Manchuria. Once victory was secured, this sea convoy from the south, manned by some 80,000 men, rapidly became an indispensable lifeline to supplying rice to Beijing and the Ming's northern defense lines. When the Ming relocated its own capital seat to Beijing in 1403, it simultaneously launched an enormous, state-run shipbuilding program to secure its control over the vital sea-lanes. Between 1403 and 1419, the shipyards near Nanking alone turned out 2,000 ships. The Ming fleet featured 3,800 ships by 1420, including 250 giant, long-distance "treasure ships," some up to 440 feet long and 180 feet wide, with square linen sails on four to nine masts towering up to 90 feet high, capable of carrying 450 to 500 sailors and displacing up to 3,000 tons apiece—ten times more than the flagship Vasco da Gama sailed in his historic voyage around the Cape of Africa into the Indian Ocean at the end of the century. The ships incorporated all China's advanced innovations, making it the supreme naval power of its time.

The Ming soon exercised their new sea power with a series of spectacular maritime expeditions that revealed China's clear naval superiority in the great age of sail that was just dawning around the world. The most famous of these were the seven expeditions between 1405 and 1433 commanded by Admiral Cheng Ho, a Muslim and court eunuch devoted to the emperor. Cheng Ho's first fleet had more than twice as many vessels as the Spanish Armada 150 years later, and included 62 giant treasure ships. It far outclassed the Arab dhows and Indian vessels it met in the Indian Ocean. In his seven voyages, Cheng Ho's 27,000 man fleet easily established control in the Indian Ocean, over the Malacca Strait, Ceylon, and Calicut, India, and became an influential force at Hormuz at the mouth of the Persian Gulf. Cheng Ho also sailed up the Red Sea, where some Muslim crewmembers disembarked to undertake a pilgrimage to Mecca, and south along the East African coast as far as Malindi in modern Kenya, where he obtained a giraffe as a novelty present for the emperor in Beijing. In contrast with the European voyages in the Indian Ocean during the following century which were undertaken to secure treasure, profitable trade routes, and eventual military dominance, Cheng Ho's primary mission was instead to win homage for the glory and power of the Ming rule. Few dared resist his demands of honor for the "son of heaven" in Beijing when his warships appeared off shore. Gifts were diplomatically bestowed upon those who acquiesced. Resisters were militarily disciplined—but not massacred, as they were by Europeans three-quarters of a century later. When a ruler in Ceylon showed reticence, for instance, Cheng Ho had him seized and shipped back to China's Imperial Court for proper disciplining.

Then, in 1433, all the expeditions abruptly ended. Edicts from the emperor strictly limited Chinese seafaring and contacts with foreigners, the construction of oceangoing ships, and even the very existence of ships with more than two masts. Cheng Ho's great warships were left to rot. Naval personnel were redeployed to smaller ships that plied the Grand Canal. Eschewing its power, China turned inward, away from the world.

It was a remarkable moment in history when a great power, possessing all the means to dominate all worlds it encountered and with vessels seaworthy enough to cross all the open oceans, including the Pacific to the New World, suddenly decided not to press its advantage. Historians have mused how world history would likely have been radically different had the Portuguese encountered a powerful Chinese empire controlling the key ports and sea-lanes of the Indian Ocean when they rounded the southern African cape in 1498 to establish the world-changing direct ocean link between Europe and the East. Indeed, one cannot help but further wonder whether Europe itself might have been subordinated and colonized if, instead of cutting off from the world, China had applied its maritime and industrial superiority to press southward around Africa, to master the Atlantic Ocean wind and current systems, and announced itself to Europe and the Americas before Columbus and da Gama ever hoisted sail.

Why did China suddenly turn inward? Xenophobia and angst about the revival of Mongol power in the north, where the modern Great Wall was being built, were motivating factors. But the world-history-shaping about-face in China's geopolitical strategy was made possible, and also driven, by the successful completion in 1411 of the greatest of all Ming water engineering triumphs—the New Grand Canal. The dredging, repair, and expansion of the entire Grand Canal had become a top priority once the Ming government moved China's capital back to Beijing in 1403. By providing the means to supply the northern frontier's fortresses with food and munitions, the Grand Canal became the vital defensive artery for the entire country. The existing sea transport system was not reliable enough to provide the needed food supplies for the northern frontier due to piracy and the inherent natural uncertainties of sea travel. To move supplies along the inland Grand Canal extension to Beijing, however, the Ming had to devise one canal innovation that had stymied the Yuan engineers—how to supply enough water to enable perennial passage, even in the dry season, over the highest point in the hills. Often large cargo ships were sidelined for up to six months until water levels refilled with the seasonal rains. The breakthrough Heaven Well Lock was made in 1411. The new lock split the combined flow of two rivers and allowed managers to regulate seasonal water flows through a network of 15 locks. Heaven Well Locks were introduced along the length of the Grand Canal, which at a stroke became a reliable, all-season inland waterway and all-important supply line of the Ming dynasty. With the government employing 15,000 boats and some 160,000 transport workers, food supplies to the north rapidly quadrupled.

The sea transport supply route became redundant and was shut down. "With the re-construction of the Grand Canal to Peking (Beijing) in 1411, and the abolition of the main sea transport in 1415," China historian Mark Elvin observes, "the navy became for the first time a luxury rather than a necessity." After 1415, shipbuilding resources were diverted to the building of canal boats; after 1419 all ocean shipbuilding ceased. The decision to end Cheng Ho's expeditions after 1433 and rely exclusively on China's internal resources, therefore, was but another sequential step in the same inward political direction.

The completion of the New Grand Canal proved to be the decisive turning point that enabled China to make its history-changing policy

U-turn and cut off from the rest of the world. Moreover, by artificially creating a more self-contained, command-controlled, hydraulic environment, the New Grand Canal also enhanced the centralized authority of the Ming state. The emperor and his conservative neo-Confucian mandarins, in alliance with the landed agricultural interests, used this power to suppress the surviving private merchant class that had been such a vibrant component of the Sung golden age. This contrasted starkly with contemporaneous developments in Europe, where the absence of a unifying inland waterway system and the focus on transport by sea helped produce smaller states, whose competition led to the expansion of unregulated trade and free-market enterprise.

Although economic growth continued after the mid-fourteenth century, China's inner dynamism and creative inventiveness gradually declined. This also helped illuminate the second historical enigma why it was that industrially advanced medieval China, possessing virtually all the requisite scientific know-how, did not make the next advances to create modern industrialism hundreds of a years before the decisive breakthroughs were finally achieved in the West. A key part of the answer, simply put, was that the reassertion of a strong, isolationist, centralized state inhibited the emergence of a market-driven economic engine that in eighteenth century England ultimately combined the profit motive with innovations in technology to make the breakthroughs that fueled the Industrial Revolution. Another part of China's failure to achieve early industrial takeoff also stemmed from the chronic surfeit of cheap manpower resulting from the dense populations produced by its rice farming society. This diminished both the political and economic incentives to develop labor-saving technologies, such as the steam engine, whose catalytic synergies with iron were to drive the early industrial age.

China's isolation lasted almost four centuries. Yet by trying to preserve its ways without engaging the innovative ferment of the outside world, it made itself vulnerable, once again, to external incursions. Just how far China had fallen behind technologically was stunningly demonstrated by mobile British steam gunboats during the first Opium War of 1839 to 1842, which forcibly reopened the helpless empire to the world. China's foreign trade contacts at the time were restricted to a single port, Canton. Seeking a way to balance a lopsided trade pattern favoring Chinese exports to the West of tea and other luxuries, the British had gradually cultivated a Chinese market for opium grown in its Bengal, India, colony. As Chinese opium addiction, and with it opium imports, mounted, Chinese officials in 1839 resolved to interdict importation of the drug. At first they appealed to England to halt its opium exports. With unimpeachable logic, they noted in a letter to Queen Victoria that opium was banned in England and that the same principle should apply to China. To the British, however, moral or legal consistency was subordinate to its mercantile and colonial interests. The Chinese implorations were rebuffed. In an act reminiscent of America's Boston Tea Party, Chinese officials seized some 30,000 chests of drugs from British and other European merchants and dumped many into the river. Britain's response was to dispatch a fleet of cannon-armed, paddle-wheel, steam gunboats to the mouth of the Canton River in June 1840. To the amazement of the Chinese, the steamboats seemed to have the power to fly across the water, regardless of wind or current. It took only half a dozen skirmishes for Britain to win the Opium War. British steamers sailed up Chinese rivers, entering the Yangtze River to take Shanghai and then the strategic choke point where the Grand Canal met the great river. When Nanking was threatened in August 1842, China capitulated to Britain's unequal and humiliating treaty terms. In addition to indemnities for merchant losses, the Chinese were compelled to cede in perpetuity to Britain the barren island of Hong Kong and to open five port cities to the free trade of low-cost British merchandise, which Britain reasonably expected would help enrich its world-class manufacturers. France and the United States soon demanded and received similar rights; the second Opium War in the late 1850s ended with Anglo-French forces occupying Beijing, more port openings, and the right of foreigners to travel inside China, including having diplomatic representatives dispatched to the emperor's Forbidden City at Beijing.

The humiliating defeat in the Opium Wars rendered publically visible the extent of the demise of China's 2,000-year-old empire. It added insult to the widespread disaffection with the ineffectual government and helped stir the rebellions that ultimately toppled it. A telltale sign and fomenter of this internal decay was once again waterworks deterioration. Millions died in three major dike breaks in the Yellow River between 1841 and 1843. In 1849, the worst flood in a century ravaged the lower Yangtze. The major shift of the Yellow River toward its present northern course in the 1850s caused major breeches in the Grand Canal. Northern sections of the canal were left unrepaired and the critical channel supplying Beijing was abandoned altogether following the Taiping Rebellion and other major uprisings in the 1850s and 1860s. Floods worsened in the late nineteenth century due to inadequate diking and waterworks maintenance, hastening the final days of the ruling Manchu dynasty and the long Chinese empire in the 1911–1912 revolution. The revival of the Grand Canal and other major water infrastructure, reminiscent of the restoration of new dynasties, began with the coming to power after a Japanese occupation and long civil war of the post–World War II Mao Zedong–led communist regime.

CHAPTER SIX

Islam, Deserts, and the Destiny of History's Most Water-Fragile Civilization

Golden age China overlapped and exchanged goods with a young, trading-based civilization that had emerged improbably out of the sparsely populated, parched desert of the Arabian Peninsula under the inspirational organizing banner of a new religion, Islam. During Islam's brilliant flowering from the ninth through twelfth centuries, its civilization held sway over an extensive domain stretching from Spain in the west, across North Africa, south from Egypt along the East African coast to the Zambezi River near modern Mozambique, east from the Levant to the Indus River, and northeast in central Asia beyond the Oxus River to the western borders of the fabled Silk Roads. The riches underpinning its illustrious civilization came from its control of the Old World hub of long-distance land and sea trading routes linking the civilizations of the Far East, the Near East, the Mediterranean, and sub-Saharan Africa.

From its stunningly rapid rise to its puzzlingly abrupt fall from history's center stage, the signature characteristics and historical destiny of Islamic civilization were overwhelmingly dictated by the challenges and responses to its scarce natural patrimony of freshwater. Islam's core habitat was a desert surrounded by two saltwater frontiers, the

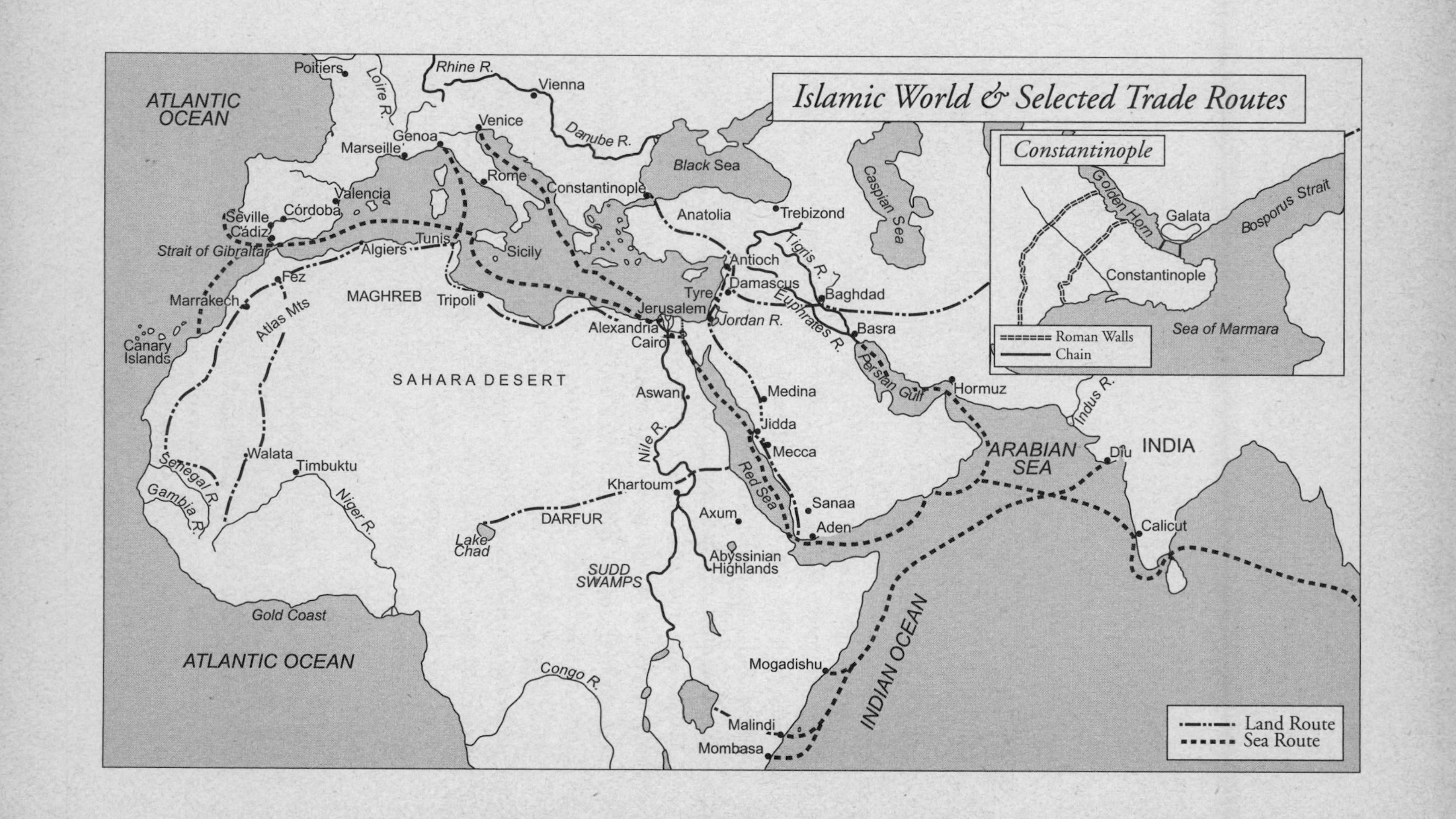

Islamic World & Selected Trade Routes
Constantinople
Golden Horn
Galata
Bosporus Strait
Constantinople
Sea of Marmara
Roman Walls
Chain
ATLANTIC OCEAN
Poitiers
Loire R.
Rhine R.
Vienna
Danube R.
Venice
Genoa
Marseille
Rome
Constantinople
Black Sea
Caspian Sea
Valencia
Córdoba
Seville
Cádiz
Strait of Gibraltar
Tunis
Algiers
Sicily
Anatolia
Trebizond
Antioch
Tigris R.
Damascus
Baghdad
Tyre
Euphrates R.
Jerusalem
Jordan R.
Basra
Fez
MAGHREB
Tripoli
Marrakech
Atlas Mts
Alexandria
Cairo
Canary Islands
Persian Gulf
Hormuz
Indus R.
SAHARA DESERT
Aswan
Medina
Jidda
Mecca
ARABIAN SEA
Diu
INDIA
Nile R.
Walata
Timbuktu
Senegal R.
Gambia R.
Niger R.
Khartoum
Red Sea
DARFUR
Sanaa
Axum
Aden
Calicut
Lake Chad
Abyssinian Highlands
SUDD SWAMPS
Gold Coast
ATLANTIC OCEAN
INDIAN OCEAN
Congo R.
Mogadishu
Malindi
Mombasa
Land Route
Sea Route

Mediterranean Sea and the Indian Ocean. Precious few fresh hydraulic resources watered its interior. Its deserts contained scattered date-palm-shaded oases, underground springs and wells, and some seasonal wadis. Only a few large rivers—such as the Nile, the Tigris-Euphrates and to a much lesser degree the Jordan—were capable of sustaining intensive irrigated agriculture and the civilized, urban life that clustered around it. No navigable river or artificial waterway like China's Grand Canal spanned the long distances of arid emptiness between water sources to unify and centralize the Islamic world's political, economic, and social centers. Its noted dearth of small, perennial rivers—its so-called stream deficit—additionally made freshwater an omnipresent natural resource challenge for drinking, irrigation, transport, and waterpower that put great stress on the population-resource balances of Islamic society in all but a few privileged locations.

Freshwater scarcity, in short, effectively rendered Islam a water-fragile civilization, extremely vulnerable to changes in natural and engineered hydrological conditions. As a result, its periods of abundance were temporary and its sufficiency rarely enduring. For centuries, the dearth of freshwater in its original Arabian habitat had been the primary obstacle confining its inhabitants to bare subsistence lifestyles. The Arab genius in transforming the obstacle of the hot, dry deserts, and subsequently the salty sea frontiers, into near-monopoly highways of trade was the key catalyst that launched Islam's hallmark rise to greatness as a civilization controlling the long-distance movement and transit between East and West. Its precarious hydrological foundations also ultimately helped explain why its preeminence unraveled so quickly after the twelfth century.

Islamic civilization started with Muhammad, founding prophet of its monotheistic religion and revealer of the Koran, its holy book. Arabians at the time were polytheistic animists with strong tribal social structures. Many were still nomadic pastoralists, raising camels and raiding trade caravans. Settled life at the sporadic oases supported only very small populations. One important settlement, Mecca, was built around a spring with "bitter," or salty, tasting water, and had only about 20,000 to 25,000 inhabitants. Mecca was located at an important restocking juncture for water and other supplies along the historic camel caravan trade route that carried frankincense, myrrh, and other luxuries between Yemen and the Mediterranean ports of the Levant. It was also specially advantaged because it was a regular destination of Arab pilgrims who came to venerate a black meteorite that had fallen nearby in antiquity and was regarded as divine.

Legend and Muhammad identified the origin of the Semitic Arabs as descendants of Ishmael, Abraham's son by his maid-concubine, Hagar. From the beginning, water was always highly esteemed in both desert Arab and Islamic society. By tradition, no man or beast can be denied access to drink from a man's well; the very transliteration of shari'aa, or Islam's governing religious law, means "the way" or "path to the watering place." Muhammad himself was born around 570 into a reputable but weaker clan of Mecca's leading Quraysh tribe. Many Quraysh were merchants who had leveraged the power from the tribe's control of water rights for the pilgrimage into lucrative participation in the camel caravan trade. Orphaned at a young age, the uneducated Muhammad grew up in the caravan trading business of his uncle and clan elder, Abu-Talib. Historians believe he traveled outside Arabia along the trade routes, where he encountered many new ideas and religions. At age twenty-five, he married a rich older widow with caravan business interests.

Everything was quite unexceptional about the life of Muhammad until about the age of forty. Then one night, in 610, while sleeping in a cave outside of Mecca, he had a supernatural experience. He had a vision of the Archangel Gabriel summoning him to be God's chosen emissary and to begin reciting the first part of His revelation. For much of the next decade, Muhammad preached to a small group of followers, asserting that he was the final prophet in a line of divinely inspired Jewish and Christian messengers from Abraham, Moses, and Jesus. Islam meant simply a "submission" to God in all facets of life. As his following grew, the leading families of the Quraysh tried to suppress him. Muhammad's position in Mecca became untenable when his uncle and tribal protector died in 619. Some of his adherents fled for Christian Ethiopia. In 622, Muhammad and a group of followers left Mecca for a settlement 200 miles north at the crowded, sweet water oasis of Yathrib, later renamed Medina, or "city of the prophet," where he'd been invited to arbitrate disputes between local tribes.

From Medina, Muhammad's power base grew rapidly. He expelled Medina's Jewish tribes when they refused to acknowledge him as the true prophet. To supplement the limited agricultural resources of the oasis, he led his followers to raid camel caravans from Mecca in an expanding alliance with converted Bedouins, who shared profitably in the stolen booty. Before long, Muhammad was engaged in armed struggle with the Quraysh, possibly over control of trade routes. Several victories reinforced the religious fervor of the Muslim faithful that God was on their side, and gradually convinced Meccan leaders to negotiate the peaceful submission of Mecca to Islam by 630. As Mecca's new leader, Muhammad abolished all blood and property privileges except custodianship of the cube-shaped Ka'bah shrine housing the black meteorite. Mecca replaced Jerusalem as the holy focal point of Muslim prayers.

Through control of the oases, marketplaces, and key caravan and trade routes, plus diplomacy backed by several military offensives, Muhammad rapidly united most of the tribes of the Arabian Peninsula under the banner of Islam. Yet when Muhammad died in 632 many of the tribal chiefs considered their oaths to Islam no longer binding and rebelled against the financial tribute Medina exacted from them. The first caliph, or "successor" to Muhammad, Abu Bakr, responded by organizing a regular army to quell the rebellions. The momentum of these military successes launched a growing Islamic fighting force of fierce nomadic tribesmen who soon reached the frontiers of Arabia's great neighboring empires, Byzantine Rome and Sassanian Persia.

Under the ambitious and strong-willed second caliph, Omar, Arab armies surged across these frontiers and unleashed one of world history's astonishing military juggernauts. Long-standing borders were swept away with stunning speed and world history's cultural map was permanently transformed by seeding Islam throughout the conquered territory. One of the earliest and greatest victories was the August 636 battle of the Yarmuk River, a tributary of the Jordan River at the modern border of Syria, Jordan, and Israel. Aided by a dust storm that concealed their approach and fired by the zeal of religion and the lavish booty of imperial conquest, a large Arab army decimated a huge Byzantine force that became trapped with its back to the river, which soon ran bloody with its dead. By 642 Islamic armies controlled all of Syria and Palestine as well as Egypt's Nile Valley—thus severing Byzantine Constantinople from two of its richest provinces. Other Arab armies meanwhile thrust eastward, seizing Mesopotamia and the wealth of its twin rivers by 641. By 651 the entire Sassanian Persian Empire had succumbed with astonishing ease. The historic border between Rome and Near East empires, stable for some 700 years, was obliterated in just fifteen years.

Historians have offered various explanations for the spectacular and improbable success of the small, modestly equipped Islamic armies against the huge Persian and Byzantine empires. Although both maintained the façade of imperial power, the two old empires had become internally enfeebled by wars, disease, political struggles, barbarian invasions, and economic corrosion from their failure to maintain agricultural water management infrastructure. In Persia, internecine political quarrels had weakened the central administration, which also failed to maintain the river-fed irrigation systems on the Tigris and Euphrates that had supported the original rise of its power. The resulting fall in crop yields undermined the society's cohesion. The Byzantine grip on Egypt had been weakened by a century of low Nile floods during which land under cultivation had shrunk by half. The consequent famine, and an overlapping plague, had diminished Egypt's population by the time of the Arab invasions in 639 to merely 2.5 million—half its Pharaohic height. The highly organized, religiously inspired Arab armies also created their own advantages, notably by the use of camel transport which helped them attack effectively over wide areas. In a typical battle, camels provided the supply trains until preparations were ready for horses, mounted by sword-brandishing cavalry, to make the final charge.

Islam's military expansion continued, albeit at a less prodigious pace, following a power struggle and civil war that ended with the assassination in 661 of the fourth caliph, Ali. This was a seismic event in Islamic history. Islam's ruling caliphate moved from Medina to Damascus, under the hereditary control until AD 750 of the Quraysh's powerful Umayya clan. Moreover, Ali had been Muhammad's cousin and husband of his daughter, Fatima. His demise ignited the bloody schism between establishment Sunni and dissident Sh'ia, who believe that legitimate leaders of Islam should descend only from the prophet's direct household.

Under the Umayyads, North Africa was slowly brought into the Islamic fold. Aided by its new Berber allies—and in ships loaned by the Christian Byzantine Empire—Islamic soldiers crossed the Strait of Gibraltar to easily overthrow in 711 the Catholic Visigothic kingdom in Spain. The western Mediterranean, dominated by Rome in its heyday, was transformed into a Muslim lake. Arab fleets also became a force to be reckoned with in the waters east of Sicily and Malta. On land, raiders skirmished with Europeans deep into northern France over the next quarter century. In the east, Muslim armies crossed the Hindu Kush Mountains and stormed the Indus Valley between 708 and 711. The Caucasus mountains and the rich Oxus valley became the northeast borders of Islam's empire following a Muslim defeat at the hands of Turkish steppe tribe warriors—many of whom were later converted to Islam—and a Muslim victory over a T'ang Chinese army at the Talas River in 751, an event which effectively closed the overland Silk Roads and diverted its trade to the Indian Ocean. Islamic armies also marched south along the African coast. By ejecting the Abyssinian Christians from the narrow Strait of Aden (the modern Bab-el Mandab), they took control of its tolls and opened up the entire Indian Ocean to Arab shipping. Large Arabian dhows were soon sailing the Indian Ocean's two-way, seasonal monsoons and currents as far as Malacca and China and back again, and displacing Hindu shipping throughout the Old World's richest long-distance-trading ocean.

By AD 750 Islam's empire effectively had attained its largest geographical reach. It was a far-flung and decentralized empire with several competing regional centers and political interests loosely unified by a common religion, a common Arabic language, and enormous wealth derived from an extensive land and sea trading market economy. By one estimate the caliphate's revenue was no less than five times greater than the Byzantine Empire's by 820.

It was Islamic civilization's meager freshwater patrimony for farming that compelled it to pursue its livelihood through trade and commerce by exploiting its occupancy of the lands at the crossroads of the civilized Old World. Its agriculture was confined to three main types of cultivation and habitats. Along the sandy coastlines where annual rainfall exceeded seven inches the olive tree provided nourishment, cooking oil, and lighting fuel. Around the scorching desert oases with temperatures of at least 61° Fahrenheit flowered the remarkably useful date palm with its eatable fruit, fibrous leaves for weaving, and trunk for scarce wood. Only in the few irrigable river valleys, or on plains where more than 16 inches of rain fell each year, could the basic grains for Islam's daily bread be cultivated. In between the expanses separating these agricultural pockets roamed small clans of nomadic, seasonal groundwater and grass-seeking desert pastoralists who bred camels and other animals that provided the milk, meat, clothing, and tent skin mainstays of their simple, subsistence lifestyles.

Freshwater scarcity thus profoundly shaped the nature, institutions, and history of Islamic society. Water imposed constraints on food production and set limits on the maximum size of Islam's sustainable population. For instance, in its halcyon days, Islam could support only 30 to 50 million people; at the time, China's population was triple that number and world population was ten times greater. As a result, Islam was a civilization that chronically lacked manpower and was forced to expand through religious conversion and conquest. Islam's religious universalism and Arab leaders' eventual acceptance of non-Arab converts were likewise shaped by this demographic shortfall. So was the unusual degree of tolerance with which conquered peoples, mercenaries, and even its large slave population were absorbed into its society.

Freshwater scarcity also forced Islam's population to be highly concentrated around each region's few good water sources. Overcrowded towns of exceptional size and a few cities of dazzling, world-class accomplishments, such as Baghdad, Cairo, and Cordoba, were characteristic of Islamic society. Typically, a grand mosque surrounded by commercial markets lay at each town center, which was encircled by a web of twisty, narrow, and unsanitary streets built on slopes from which the rare rains would wash away the refuse.

At the height of its glory, three disparate, rival regional power centers arose—Spain-Magrib, Egypt-Levant, and Mesopotamia-Persia—reflecting and magnifying the religious and tribal divisions within Islam. In such decentralized circumstances, economic organization by command was impossible. Instead, it was the invisible hand of market forces that governed the signature transit and trade that held together Islam's economy and helped stimulate the breakthroughs underpinning its civilization's rise. "Not being well endowed by nature," observes historian Fernand Braudel, "Islam would have counted for little without the roads across its desert: they held it together and gave it life. Trade-routes were its wealth, its raison d'être, its civilization. For centuries, they gave it a dominant position."

Water scarcity presented the primary obstacle standing between Islam and its historic rise to greatness through trade. First and foremost, it needed a way to cross the long expanse of its own hot, waterless interior deserts. Its first triumphant innovation, which at a stroke transformed the barren desert barrier into an insulated, exclusive Islamic trade highway, came by its disciplined organization of the hardy camel, with its prodigious water-storing capacity, into long trade caravans and military supply transports. A caravan of 5,000 to 6,000 camels could carry as much cargo as a very large European merchant sailing ship or a fleet of barges on China's Grand Canal. Islam's quasi-monopoly over this powerful pack animal provided it with the mobility to cross and exit its desert homelands—and to make its mark on world history.

The one-humped Saharan dromedary was specially adapted for the hot deserts. It could go without drinking water for a week or more, while plodding some 35 miles per day across the desert sands with a

200-pound load on its back. Water was stored in its bloodstream—its fatty hump, which grew flaccid during long journeys without nourishment, functioned as a food reserve—and it maximized water retention by recapturing some exhaled water through its nose. Once at a water source the camel speedily rehydrated by consuming up to 25 gallons in only ten minutes. It even could tolerate briney water. It possessed an uncanny memory for the location of water holes. Moreover, it could eat the thorny plants and dry grasses that grew on arid lands and were indigestible by most other animals. During a trip, camels could lose one-quarter their body weight, twice the amount fatal to most other mammals. The camel's extraordinary physical attributes made it possible for caravans to make the two-month, trans-Sahara trip from Morocco to Walata at the frontiers of the Mali Empire in Africa, which included one notorious stage of ten waterless days.

Like seas, deserts have played a distinctive role in history as expansive, empty spaces between distant civilizations. Initially, both imposed formidable geographic barriers of separation. But when traversed by some transport innovation, they were rapidly transformed into history's great highways of invasion, expansion, and cultural exchange that often abruptly realigned regional and world orders. Camels took Arab merchants and soldiers everywhere. Ultimately they reached Islam's other great water challenge—the frontier of its seashores. Islam's second water breakthrough was to extend its overland desert trade franchise to mercantile mastery of the Old World's great sea waterways, the Indian Ocean and much of the Mediterranean Sea. Its large dhows, their hulls made of planks tied together with date palm or coconut tree fibers and propelled by triangular lateen sails, which were highly maneuverable against headwinds, and steered with nimble stern rudders, became the caravans of the seas that carried Sinbad the Sailor on the adventures described in the classic literary cycle *The Thousand and One Nights.*

The long-distance trade route from the Moluccas or Spice Islands of Indonesia across the Indian Ocean to India and the West became in Muslim times the single greatest highway to world power and empire. At a time before Europeans had unlocked the secret arts of sailing across the open oceans and discovered the riches of the New World, and when the Silk Roads were closed, Arab dhows carried the lion's share of the world's most desirable goods and spread Islamic civilization throughout the richest coastal ports of call in the world.

Thanks to the Indian Ocean's unique, seasonally reversing wind system, Arab seamen could set out with a full load of goods between April and June by following the southwesterly monsoons, arrive within two months, do their trading, and fill up their cargo holds with profitable Oriental luxuries in time to catch the reliable fair winds from the northwest that would carry them home when cooling meteorological conditions reversed the monsoon's direction in winter. Arab vessels also plied the Mediterranean Sea from Spain to Alexandria and the Levant. Compared to the Indian Ocean, however, the Mediterranean's seaports offered far less alluring wealth while its unidirectional west-to-east winds made sailing more difficult.

By integrating its command over the resources of two disparate water environments—the waterless desert and the salty sea—Islam's influence soared. Camels and dhows defined its seamless land and sea caravan network that could transport goods and people between the four corners of the Old World. Disassembled dhows were transported by camel across the Sahara Desert for assembly and launch, camels and all, across the Red Sea. Once on the Arabian Peninsula, the ships were again disassembled and portaged for the long, landward rest of the journey along the wadis and oases to the ports of the Arabian Sea that led to the Indian Ocean. The preference for this laborious overland route was that for centuries the rocks and coral reefs, unpredictable winds, irregular currents, and pirate-infested waters of the deep, salty Red Sea were more perilous to navigate than the great deserts along its coasts. Many of the seaways and coastlines leading to the Indian Ocean's fabulously rich sea-lanes also were inhospitable and dangerous for seafaring. Arabia's absence of navigable rivers and its scant number of good harbors with sufficient freshwater made supplying ships extremely difficult; the lack of wood resources in the arid landscape was a second, water-related impediment. Adding to the navigational problems for seamen, the Arabian coast was notoriously stormy.

Islamic merchants nevertheless overcame these water obstacles. In

Mesopotamia goods went by river to Baghdad, then overland west to Syria and Egypt, north to Constantinople and Trebizond on the Black Sea, and east through northeastern Iran and thence to central Asia and China. Gold and slaves from Sudan, Oriental silk, peppers, spices and pearls, and much of everything else transited through Islamic lands by Arab traders. After about 1000, European vessels from the Republic of Venice and other rising small sea states increasingly handled the final transshipments from Alexandria and other Arab ports throughout the Mediterranean in commercial alliances that often transcended religious rivalries.

Islam's expansive economic power made it a great military force that encroached upon and threatened neighboring civilizations. The native sub-Saharan civilizations of the Niger River fell under domination by Muslim states following the conquest of Ghana in 1076. Much of East Africa, with the notable exception of the Abyssinian highlands of modern Ethiopia, also succumbed. In India, Hindu civilization was in retreat from Islamic conquests over hundreds of years through the seventeenth century. Europe, too, barely survived the onslaught of Islam's initial military juggernaut from 632 to 718, and remained at peril for several centuries in the heated clash of civilizations that continued in earnest across the Mediterranean throughout the sixteenth century.

Christianity, and all that later flowered into Western civilization, came closest to possible extinction in the year from AD August 717 to August 718. During those 12 months, a huge Muslim naval and army force of over 2,000 ships and 200,000 men laid siege to Constantinople, seat of the Byzantine Empire, inheritor of Rome's civilization, and Christendom's greatest city. Had the imperial city, located on the strategic triangular promontory overlooking the junction of the Bosporus Strait and the Sea of Marmara that controlled the narrow 225-mile waterway linking the Mediterranean and Black Sea trade routes and divided Europe from Asia, fallen under Islam's flag, the entire Mediterranean Sea likely would have become a Muslim lake. Europe's interior, via the river Danube and toward the Rhine, would have been wide open for an easy Muslim march of conquest. Europe, and the entire Western world, today might be Muslim. In the event, the siege of Constantinople would be an epic turning point in the clash of civilizations between Islam and the West. It was also a dramatic illustration of the geostrategic advantage of a strong water defense.

In the early eighth century, the Christian world outside Constantinople was sparsely scattered and doctrinally divided among Greek, Latin, Syriac, and Coptic churches. Rome, laid waste by multiple sackings and its aqueduct water supply system in ruins, was a shrunken shadow of its former self and under Byzantine protection. Latin Church missionaries were struggling to convert the barbarian European princes who ruled in the power vacuum left behind by the fallen Roman Empire. The conquests of Charlemagne and his crowning in 800 as the first Holy Roman Emperor still lay many decades in the future. Islam, by contrast, was still at the height of its explosive expansion following Muhammad's death.

The only serious setback experienced by the Arab conquerors in the seventh century had been their previous failure in 674–679 to conquer Constantinople. Their overland attack had faltered when the Saharan camel proved unable to tolerate the cold of the Anatolian highlands of Turkey. Their sea attack had hinged on the success of its large siege engines and catapults against Constantinople's double walls. It, too, failed when the Byzantines counterattacked by unleashing their terrifying, newly invented, secret chemical sea-warfare weapon—"Greek fire"—upon Arab ships. The chief characteristic of Greek fire was that it combusted spontaneously and fiercely upon contact with air and was inextinguishable even in water. The secret of its precise composition was lost during the Middle Ages and remains unknown today. It was a crude oil-based substance traditionally laced with sulfur, evergreen tree pitch or quicklime; by adding the right amount of saltpeter, the mixture became ferociously self-igniting. Only sand, vinegar, and urine were believed to dampen its smoky flames. Greek fire was usually blown by an air pump through long, bronze-lined tubes toward enemy ships, where it burst into flames; alternatively, it was catapulted at attacking ships in clay jars or fired in a hail of arrows that had been saturated in it. So many of the wooden hulls of the caliph's vessels were set ablaze and so much terror inflicted upon Arab sailors by this unnerving weapon that in 679 the Muslims withdrew, and even agreed to pay an annual tribute to Constantinople. Greek fire not only saved the Byzantines, but its secret gave them a long-lasting military advantage in sea warfare that endured for a long time.

Yet the hard experience of 674–679 also meant that when the Arabs returned in 717 for their revenge assault, they came better prepared and in much greater force. Once again Constantinople's defense hinged upon its supreme strategic location and sea power. The city's position made it at once easily supplied through either of the two long, narrow straits—to the east by the Bosporus, 18 miles long and less than half a mile wide in some places, or by the 40-mile-long and one- to five-mile-wide Dardanelles on the west—connecting the Black and Mediterranean seas. On the northeast side of Constantinople's peninsula, abutting the entrance to the Bosporus, was a wonderful, deep, five-mile-long harbor, the Golden Horn, which offered the only well-sheltered port in a turbulent stretch of sea. These natural geographical defensive advantages were reinforced by a great, half-mile-long chain across the harbor's mouth that the Byzantines could raise to block the entrance. The city's peninsular location meant that major fortification of walls and moat was needed only on its landward side. Its single defensive flaw was that it had only one good stream flowing into the Golden Horn to provide freshwater. To mitigate this vulnerability, Byzantine Roman water engineers had borrowed on the waterworks expertise of mother city Rome to build dams, a long-distance aqueduct, and giant underground cisterns within its walls to supply enough freshwater to withstand a siege.

The site, occupied since 658 BC by the prosperous Greek trading city of Byzantium, had been chosen, and renamed, by the Roman emperor Constantine I to replace beleaguered Rome as capital of the Roman Empire for its commanding strategic defensive and trading positions on the Black Sea. This "New Rome"—like the original it had seven hills, a bread dole for the poor, and a new Senate to entice the nobility to emigrate—was founded on AD May 11, 330. Moving to Constantinople had been one of Constantine's two historic decisions. The second, inspired by his vision of a heavenly cross heralding his power-consolidating victory in the battle of the Milvean Bridge (AD 312) on the Tiber on the outskirts of Rome, was to adopt Christianity as the favored religion of the Roman Empire. In the life-and-death struggle against Islam (717–718), the fate of Constantine's second great decision depended much on the strategic foresight of his first.

The fortunes of Constantinople and Christianity were improved in their hour of crisis by the seizure of the imperial throne a few months earlier by a gifted general, crowned as Emperor Leo III. The Muslim military strategy was to assault the city's double walls from the landward side with a massive army, while two fleets bottled up the Dardanelles and the Bosporus to deny any supply relief from Mediterranean or Black Sea ports. The initial land attack, however, failed. So the Muslims settled in for a long siege to be waged, as in the late 670s, primarily on the water. This time they succeeded in sealing up the Dardanelles. The Bosporus proved more difficult. When the Muslim fleet approached Constantinople, its lead ships got caught up in swift, unfamiliar currents; Leo III promptly lowered the chain across the Golden Horn and struck the disoriented Muslim ships with Greek fire, destroying and capturing many of them.

Nature then assaulted the Muslim besiegers in their outdoor tents with an abnormally bitter winter. Their resupply was delayed. Famine and disease entered the camps, compelling the besiegers to eat their animals and even dead men's flesh. As so often is the case in the history of warfare, noncombat causes claimed more lives than enemy weapons. The besiegers suffered the added indignity of having to dump many of their dead into the sea because snow froze the ground for many weeks to prevent burial.

When the warmth returned in the spring of 718, the Muslims' luck turned. Reinforcements of 400 ships and 50,000 men arrived from Egypt. One night they succeeded in sneaking past the Golden Horn to complete the blockade that would doom the city and the Byzantine Empire. However, many of the Arabs' Coptic Christian crew chose that moment to abandon their ships and desert to the Byzantines. Informed by their gift of priceless military intelligence, Leo III in June mustered a surprise counterattack with Greek fire that routed the blockading fleet. As Coptic Christian desertions mounted, Leo followed up with an unexpected land attack on the Asian side of the waterway. Caught off guard, thousands of Muslims were slaughtered. When, at Leo's connivance, the neighboring Bulgars began to attack the Muslim forces, and rumors flew that the Frankish army was en route to join in, the caliph lifted the siege on August 15, 718, and retreated. By all accounts only 30,000 of the 210,000-man Islamic force, and only five out of its over 2,000 ships made it back home.

Constantinople was saved. That the city's impregnability endured for another 500 years alongside a far wealthier and more vibrant Islamic civilization was testimony to the disproportionate military advantages of sea power and control of geostrategically important waterways. The city was finally sacked and effectively subjugated only in 1204—not by Muslims, but by fellow Christians diverted from their intended march to the Holy Land on the Fourth Crusade by the intrigues of the mercantile-minded sea power Venice and its redoubtable, blind, octogenarian doge, Enrico Dandolo. Venice thereafter exercised commercial hegemony over the straits, controlling the lucrative routes to the Black Sea. Constantinople finally fell to the Islamic Turks only in 1453.

The enormous consequences of Constantinople's victory in 718 rippled through history for centuries. The first major effect was the survival of Christian Europe as a significant cultural and geographic rival to Islam. In 732 an expeditionary Muslim force from Spain was defeated by Frankish leader Charles Martel, grandfather of Charlemagne, on a battlefield near Poitiers, France, in what would later be regarded by Christian historians as heralding the turning point in ending the Arab Muslim land expansion in Europe. By 1097, Christian Europe was strong enough for its knights to cross the Bosporus from Constantinople and launch the successful counterattack to retake the Holy Land from Islamic control—the First Crusade.

Christian Europe's major gains against Islam were won principally through sea power. Constantinople's triumph had ensured that the eastern Mediterranean, unlike its western half, never succumbed to Islamic dominance. Between 800 and 1000, both Muslim and Christian vessels vied for supremacy over the riches of the eastern Mediterranean, plundering where possible and trading when necessary. By 1000, the city-state Republic of Venice finally gained the upper hand as the great sea power and transshipper from the central Mediterranean to the rich ports of Alexandria and the Levant. Three centuries later, Genoese merchants broke the Islamic chokehold on the Strait of Gibraltar, opening the Atlantic sea-lanes to unify the Christian Mediterranean with the emergent world of northern Europe. From about the eleventh to sixteenth centuries, Christians increasingly controlled the Mediterranean while the Muslims reigned in the Indian Ocean. Thereafter, the "Voyages of Discovery," motivated partly by the Portuguese's and other Atlantic sea powers' covetous desire to break the Italian and Muslim monopoly on trade with the East, culminated in a breakthrough all-sea route around Africa to India and momentously transformed the power relationships of world history with Europe at its center.

Islam's gradual ejection from the Mediterranean following the defeat at Constantinople not only saved Christianity. It also had far-reaching effects within Islam itself. It set off a period of upheaval and renewal that reinvigorated Arab Islam by amalgamating it with older Near Eastern civilizations to help launch what proved to be its golden age. The defeat at Constantinople signaled the end of its juggernaut military expansion, which in turn upset the internal dynamics that had held together the growing fissures within the Islamic community. Previously, victory on the battlefield had yielded ample booty and tribute from defeated populations for distribution that had smoothed over internal rivalries among Arab tribes. With diminishing bounty to share, the tribal political system of Arab privilege run by the ruling Damascus-based Ummayad caliphate also began to fuel discontent among the growing number of non-Arab Muslim converts who increasingly supplied Islam's manpower but often felt unwelcomed with second-class status.

In 750, the Ummayads were toppled in a civil war by a coalition led by a rival family descended from Muhammad's uncle, Abbas. The Abbasids' new caliphate was based on inclusion of non-Arab Muslims, governance that was comparatively professional and efficient rather than run on the basis of tribal patronage and nepotism, and religious universalism that encouraged converts with equal rights and opportunities. The new caliphate's heartland was the productive, irrigated farmland of ancient Mesopotamia, where Arab conquerors had installed themselves as large landowners. The Abbasids' commercial orientation shifted toward the east and the Indian Ocean. To celebrate their rise, they founded a new city—Baghdad—strategically positioned in a place where the Tigris and Euphrates rivers flowed near one another. This location gave the city convenient access to abundant irrigated food from the muddy floodplains and intersected major trade routes to Persia and the East. It was in Abbasid Baghdad that a great Islamic civilization first began to flourish. From 762 to 1258, when it was destroyed by the Mongols, Baghdad was the largest and grandest city anywhere outside China.

While Islamic civilization was not notably innovative in water engineering, in its ascendant period it vigorously applied known Middle Eastern technologies to get the most from its freshwater-scarce habitats. Water management thus played a key role in sustaining the power and splendor of the caliphate. Old waterworks were restored and new ones constructed. Muslim irrigation had its greatest success around Baghdad, where five dam-fed, cross country canals running from the Euphrates to the Tigris watered extensive, productive cropland. East of the Tigris, Abbasid engineers expanded the Nahrwan Canal that had been started by the Sassanian Persians in the second century AD. Water released from a famous masonry dam on Iran's Kur River that was rebuilt in about 960 irrigated large fields of sugar, rice, and cotton.

The diffusion of Middle Eastern water technologies and crops supported the spread of high Islamic civilization across the Muslim world. Underground qanats increased domestic water supplies, while water-lifting norias and shadoofs supplemented field irrigation across North Africa and in Spain. Low-level diversion dams were widespread in

Muslim Spain and became an important acquisition of the Christian kings when they later expelled the Muslims from Spain.

Grand cities arose that competed with Abbasid Baghdad culturally and politically. Cordoba, lying inland on the river Guadalquiver, became the seat of a brilliant humanistic Islamic civilization in Spain presided over for a long period by the lone dynastic Ummayad family survivor of the Abbasid purge that followed the civil war. The river irrigated the surrounding plains and provided the means for transporting food and goods to Cordoba's marketplace. The tenth century saw the rise of a dazzling new city, Cairo, from which the Shiite Fatimids asserted their claim to the Islamic caliphate. The economic basis of Fatimid power was the fertile farmlands of the Nile and the extensive sea trade and camel routes through the Levant and Red Sea. By the early 1300s, Ibn Battutah, the renowned fourteenth-century Muslim traveler and diarist sometimes called "Islam's Marco Polo," marveled that because of its great size "in Cairo there are twelve thousand water-carriers who transport water on camels" throughout its sprawling network of streets and markets.

The grand Muslim cities like Cordoba, Cairo, Baghdad, and Granada, situated in hot, dry lands, displayed Muslim splendor and power by building sumptuous palaces, surrounded by shaded gardens with fountains and running water suggestive of paradise, and public baths as in ancient Rome. Wherever practicable in Islam's stream-poor landscape, Muslim engineers exploited waterpower to grind flour in traditional mills as well as to produce new products and goods. Floating water mills operated day and night on the Tigris River to produce Baghdad's daily bread while at the port city of Basra in southern Mesopotamia tidal-flow-powered mills did the same. At Basra water-powered mills also processed sugarcane, first crushing the cane and extracting its juice, which was then boiled down to produce refined, crystalline sugar. Other waterwheels powered big trip-hammers used by fullers to prepare woolen cloth and to pound vegetable fibers in water until it formed a pulp from which paper could be manufactured.

Paper production methods had come to the Islamic world serendipitously through the capture of Chinese prisoners skilled in papermaking during the victorious 751 battle of the Talas River in central Asia. These prisoners set up a workshop in Samarkand. From there papermaking technology later was transferred to Baghdad. In China the bark of the mulberry tree had long been used as the basic raw material. Lacking mulberries, rags, especially linen, were substituted in the Islamic world. The original manual production process was in two steps, with water playing a key role in both. First, torn-up rags were soaked, shredded, and beaten in vats with spiked clubs to produce a pulp—a manual process subsequently automated by pulp beaters powered by water. Next, the pulp was put in a vat of warm water, stirred, and strained through a molded wire latticework to produce rectangular sheets. The sheets were squeezed and hung dry, then rubbed as smooth as possible with stones, and finally immersed in a vat of gelatin and alum for stiffening. Baghdad's water-powered paper pulp mill process spread west to Spain, and from there to Christian Europe a century later.

Paper manufacturing played a catalytic role in diffusing knowledge rapidly through the wide availability of books. Baghdad, for instance, had over a hundred bookshops by 900. Books helped usher in a glorious era of humanist enlightenment in the sciences, arts, philosophy, and mathematics, along with economic prosperity, and relative tolerance and peace. Greek, Persian, and Sanskrit manuscripts systematically were translated into Arabic from the early ninth century at Baghdad's "House of Wisdom" created by Caliph al-Mamun. Ultimately, it would be through Islamic scholars centered in Cordoba—not the long-lived though decaying civilization of Constantinople and the Byzantine Empire—that Christian Europe became reacquainted with the works of Aristotle and its own classical Greek intellectual heritage. This rediscovery later flowered as the European Renaissance to help give birth to postmedieval Western civilization. Muslim scholars made many original discoveries that also migrated to Europe. Algebra, trigonometry's sine and tangent, the astrolabe and other navigational and geographical measuring instruments, the distillation of alcohol, and numerous medical treatments were among the most notable. Islamic alchemy contributed greatly to the development of the West's scientific knowledge and methods. Islamic instrument makers even were working on elaborate gear trains for water clocks to be driven by waterwheels in the same period that China was employing this technology. A distinguished tradition of thinkers, among the best known of whom were Avicenna and Averroës, influenced mainstream Western philosophical development.

Yet sometime toward the end of the twelfth century—some historians use the death of Averroës in 1198 as the benchmark date—Islam's most glorious era abruptly began to stagnate. Why the intellectual vitality and material growth suddenly drained away, and why its culture soon was eclipsed by more vigorous civilizations, remains one of the puzzling questions of history.

The most traumatic symbol of Islamic civilization's decline was the devastating Mongol sack of Baghdad on February 20, 1258. Mounted Mongol warriors, using gunpowder-fired weapons in their relentless surge of conquest across the Eurasian steppes from China to the Near East to the doorstep of central Europe, stormed the once-illustrious city to loot, burn, pillage, and slaughter. In customary Mongol fashion, hundreds of thousands of residents were massacred. The last caliph, in a calculated symbolic act of contempt, was trampled to death under the hoof of a Mongol horse. The obliteration of the Abbasid capital was completed by the destruction of many surrounding irrigation dikes and waterworks to render impossible any agricultural resurrection. It was the first time that non-Muslim invaders had been able to impose infidel rule in the Islamic heartland. Christian Europe was spared a similar agonizing fate as that experienced at the hand of the Mongols by both Islam and China only due to a fluke of history. News of the death of Genghis's son and successor, Ogadei, had reached the banks of the Elbe River during the 1241 conquests when Europe lay prone for the taking. Mongol commanders, uncertain how the power vacuum in Karakorum would be filled, voluntarily pulled back their forces into Russia. Eventually they invaded other regions, and looked beyond the relatively meager wealth of medieval Europe for richer prizes.

Yet Islamic civilization had been in critical decline long before the arrival of the vanquishing Mongol cavalry. Like the Persian and Byzantine empires overrun by the first Arab armies of the seventh century, the foundation of its economic prosperity had grown internally stagnant. A principal cause was faltering water management and its inability to keep technologically ahead of its inherent scarcity of freshwater resources. The agricultural productivity of Mesopotamia, for instance, deteriorated markedly with the rising political influence of Islam's nomadic converts who increasingly supplied the Arab caliphate's military manpower. Most notable of these were the Turks, who held effective power in Baghdad after 1055 under the nominal leadership of the Abbasids. Dependency on the Turks was a consequence of the limitation water scarcity had imposed on the size of the ruling native Arab population. While the Abbasid dynasty's founders had arduously rebuilt and maintained irrigation waterworks on the Tigris-Euphrates and Nahrwan canal system and had expanded cultivated cropland to its largest extent into the eleventh century, the recently nomadic Turks were steeped in the traditions of steppe herders who followed their sheep and horses between water holes and seasonal grasslands. Under Turkish influence, centralized political authority waned and Mesopotamia's irrigation system eroded. Inadequate maintenance caused irrigation and drainage canals to clog with silt. Soils became waterlogged and deadly salt rose to the surface of the floodplains between the twin rivers. As in ancient times, salt-whitened fields produced falling agricultural yields and declining population levels.

Deteriorating irrigation maintenance also helped cause both the Euphrates and the Tigris to make major disruptive course shifts around the year 1200. The Tigris's return to its former, more easterly channel north of Baghdad was a twin disaster because not only did this realignment dry up a large tract of irrigated cropland, but it also destroyed part of the 400-foot-wide Nahrwan transport and irrigation canal and the agricultural network it supported downstream. The agricultural decline in Mesopotamia coincided with a parallel shrinkage and collapse by the twelfth century of irrigation in Egypt. Thus both of the Islamic world's great breadbaskets fell into crisis at the same time. As always, the level of Nile floods was the key determinant of Egyptian prosperity and the political system that depended upon it. Ample Nile floods had buttressed the first three centuries of Arab rule. A period of low Nile floods between 945 and 977, however, eroded the amount of land under cultivation and paved the way for the conquest of Egypt by the Shiite Fatimids in 969. Fatimid rule was eventually undermined by two generations of low Nile floods that produced cannibalism, plague, and decaying waterworks. In 1200 one-third of Cairo's population perished from severe famine when disastrous low floods returned after a long period of normality. This catastrophe fueled the enduring Egyptian suspicion that the upriver emperors of Ethiopia somehow had made good on their threat to divert the Nile's waters. By the time the Mamluks, white Muslim slave soldiers of ethnic Turkish origins, seized power in Egypt in 1252, irrigated agriculture had fallen into such desuetude that the Nile breadbasket was able to support no greater population than the one Arab conquerors had inherited from the Byzantines in the seventh century. The revival of Nile irrigation awaited the water engineering projects of the Turkish and British overlords in the nineteenth and twentieth centuries.

In Muslim Spain the problem was less one of waterworks deterioration than of a failure to innovate to find more efficient ways to exploit their existing water resources. When the Christian Europeans reconquered Spain, they inherited an extensive irrigation network with highly developed social and administrative processes—including the famous water court at Valencia, the oldest democratic institution in Europe, whose elected judges have adjudicated irrigation disputes in public for over a millennium. But it was entirely based on Middle Eastern traditions of small-scale river diversion dams for irrigation, waterpower, and water supply. Muslim engineers had ample familiarity with large impoundment dams and aqueducts used in Spain by the ancient Romans. But they never experimented with them in order to improve their water use productivity. Their Christian successors did. Their successful innovations helped Spain flourish after 1492 when the armies of King Ferdinand and Queen Isabella expelled the last Moors from the Iberian Peninsula.

Lifting, damming, and channeling water has often made the difference between agricultural surplus and famine. Water-lifting irrigation devices, such as the workhorse Middle Eastern noria or chain of pots (above) and Archimedes' screw (left), have remained in continuous use since antiquity. Hand-built, water-storage earthen dams, like the one being reinforced by rural Kenyan villagers in 2004 (below), were mainstays of Egyptian and Mesopotamian hydraulic civilizations five millennia ago.

Wet rice farming requires intensive and sophisticated water management. Above, nineteenth-century Chinese farmers sow rice and move water on wooden treadle pumps according to traditional methods. Below, a highland Ethiopian farmer in 2008 plows his field in a manner virtually unchanged since the start of civilization.

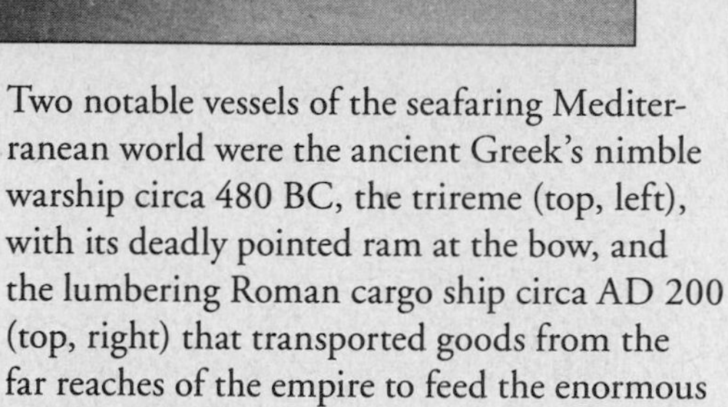

Two notable vessels of the seafaring Mediterranean world were the ancient Greek's nimble warship circa 480 BC, the trireme (top, left), with its deadly pointed ram at the bow, and the lumbering Roman cargo ship circa AD 200 (top, right) that transported goods from the far reaches of the empire to feed the enormous imperial capital.

Marcus Agrippa (right), Augustus' lifelong right-hand man, transformed the empire's health, military robustness, and civic society by institutionalizing the public supply of abundant, clean water through extensive aqueducts, such as the remnant at the Pont du Gard in southern France (below), to urban fountains, baths, and functioning sewers.

Li Bing (left) built sophisticated waterworks, like the still-functioning Min River diversion weirs in Sichuan, that spread prosperity and helped the Ch'in dynasty consolidate power throughout China in the third century BC.

The completion of the 1,100 mile Grand Canal (left) in the seventh century AD united the resources of the south's Yangtze and the north's Yellow River and catalyzed China's spectacularly advanced medieval civilization. The New Grand Canal in the early fifteenth century signaled the nation's fateful turn inward and voluntary withdrawal of its indomitable fleets from the high seas. A traditional seagoing Chinese junk (right).

Islam's glorious age, from the eighth to twelfth centuries, was built upon a vast trading network of desert-crossing camel caravans (top) and lateen sail-rigged, cargo-carrying dhows (middle), which spanned from Atlantic Spain across the Indian Ocean and south along Africa's coasts and interior river civilizations. Yet Islam's arid homeland's shortage of small rivers limited its use of waterwheels for power and irrigation, like the one functioning on the Orontes River at Hama, Syria (below), in the early twentieth century, and contributed to its rapid decline from world prominence after the twelfth century.

The heavy moldboard plow (top), widespread in northwestern Europe by the tenth century, was one of the seminal innovations of the region's agricultural revolution and belated economic rise.

Northern Europe's many navigable and fast-running rivers became arteries of commerce and production. Waterwheel-powered bread flour gristmills, like this old wooden waterwheel mill in northwestern France (right), were ubiquitous.

Water power was applied most notably to medieval industrial production, including rolling iron, as shown at this 1734 Swedish mill, and to drive huge leather bellows to heat furnaces for high-volume iron casting.

Europe's world dominance began after 1500 with the advent of transoceanic sailing and long-range naval cannonry, which followed the Voyages of Discovery championed by Portugal's Prince Henry the Navigator. Vasco da Gama (above, left) sailed around Africa's cape to India, and Columbus crossed the Atlantic to the New World, with the help of the discovery ship par excellence, the caravel (above, right).

The apogee of naval power in the age of sail was achieved by England, whose two greatest admirals, Francis Drake and Horatio Nelson (right), helped defeat the Spanish Armada and Napoléon, respectively. A 120-cannon French warship (left) from the Nelson-Napoléonic era.

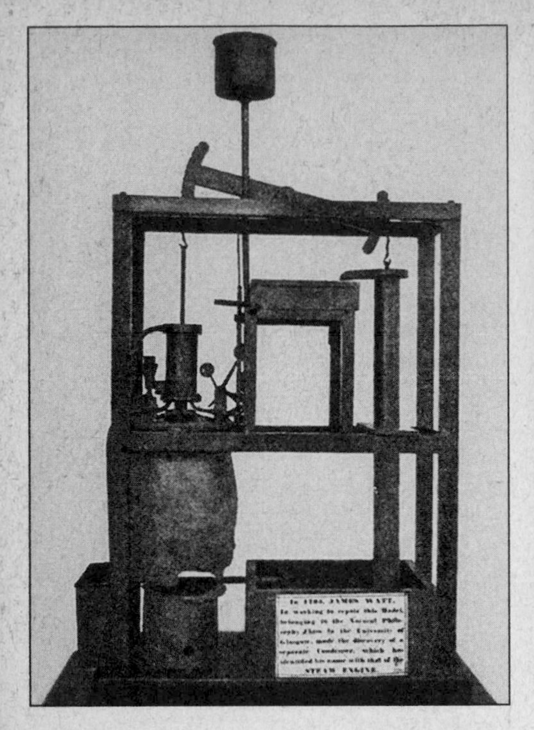

In 1763, James Watt (above, right) began repairing this model of Newcomen's early steam engine (above, left). The result was the modern steam engine, the seminal invention of the Industrial Revolution.

Powerful steam engines, like Watt's rotary motion 1797 model (below), superseded water-wheels to drive the world's great early automated factories and iron mills that underpinned the world dominance of the nineteenth-century British empire.

Robert Fulton's (below, left) *Clermont* (above) on New York's Hudson River ushered in the age of river steamboats. Fulton's vigorous advocacy for American canals also helped spur the development of the Erie Canal.

Forsaken in love, the Duke of Bridgewater (right) focused his energies on pioneering the building of a canal in 1761 from his coal mine to Manchester. His success ignited a national canal-building boom that transformed England's economy.

America's visionary canal builder was New York's De Witt Clinton (above), who celebrated the 1825 completion of the Erie Canal with a ceremonial wedding of the Lake Erie's water with the Atlantic Ocean at the Hudson River's mouth. By providing an economical east-west route across the Appalachian Mountains, the 363-mile-long Erie Canal (below) transformed America's destiny by linking New York and the eastern seaboard to the Mississippi Valley and the vast resources of the continental interior.

The arrival of abundant, clean freshwater from a new aqueduct network in upstate Croton in 1842 relieved celebratory New Yorkers of their chronic water scarcity and affliction by waterborne diseases.

English leader Benjamin Disraeli championed the sanitary reforms that triggered the industrial world's public health revolution. Disraeli later seized the opportunity that enabled England to gain influence over the vitally strategic Suez Canal, which linked the Mediterranean to the Red Sea and the Indian Ocean.

The entrepreneurial genius behind the 1869 Suez Canal was French Viscount Ferdinand de Lesseps. De Lesseps' later effort to build an interoceanic canal between the Atlantic and Pacific Oceans failed, but galvanized events that ultimately led to the creation of the Panama Canal.

The driving force behind the Panama Canal was America's greatest water president, Teddy Roosevelt (middle), who got behind the controls of huge steam shovel (top) during his heralded visit to the canal zone to make good on his promise to make "the dirt fly." The first steamer sailed through the canal's famous Culebra Cut in August 1914 (below).

Dust storms, like the one of April 14, 1935, approaching Rollo, Kansas, and created partly by man's mismanagement of a fragile water environment, ravaged the plains during the Great Depression. Pumping water accumulated over eons in huge, deep underground aquifers soon transformed the Great Plains into one of world history's great breadbaskets, but raised doubts about its long-term sustainability.

"I came, I saw, I was conquered . . ." President Franklin Roosevelt's September 30, 1935, dedication (left) of the Boulder (later renamed Hoover) Dam (below) inaugurated one of the great eras of water history. Giant, multipurpose dams transformed America's arid Far West, helped the country win World War II, and spread the worldwide Green Revolution.

Egyptian President Gamal Abdel Nasser was adulated by Arabs throughout the Middle East. He triggered a crisis with the great Western powers when he nationalized the Suez Canal in 1956 and contracted with the Soviet Union to construct the multipurpose high dam on the Nile at Aswan, a project he likened to the great pyramids.

Like great dynastic founders throughout Chinese history, Mao Zedong launched massive development waterworks to reengineer China. Today, the nation has nearly half the world's 45,000 large dams, including the controversial super giant at Three Gorges on the Yangtze, and a continental-scale south-to-north river water diversion scheme.

Water development has come much more slowly for two poor rural Kenyan villagers, who in 2004 laid a two-mile-long water pipe that finally brought clean, fresh water to their village from a rural well after a thirty-year wait.

Rachel Carson's (left) *Silent Spring*, highlighting the extensive toxic pollution of waterways, was one of the seminal birth moments of the modern environmental movement. Congress was finally galvanized to enact comprehensive clean water regulations after Cleveland's filthy Cuyahoga River, seen burning in 1952 with industrial pollution (middle), again burst into spectacular flames in June 1969.

Climate change, energy, and food issues are intimately interconnected with water. Retreating mountain glaciers from global warming, such as those that sustain Asia's great rivers and a fourth of humanity, threaten catastrophic droughts and hydroelectricity shortages in the dry season and devastating floods during the monsoons.

One fifth of humanity still lacks access to enough clean, fresh water for basic domestic needs and two-fifths for adequate sanitation, compelling hundreds of millions, especially children (left) and women, to forgo education and productive work to walk several miles each day to fetch water for daily survival. Billions in poor regions with inadequate water infrastructure, like Ethiopia's highlands near Lake Tana (middle) at the source of the Blue Nile, try to endure water's natural destructive excesses, such as floods, mudslides, and droughts.

With freshwater use increasing twice as fast as soaring world population growth, a perilous new planetary era of water scarcity is dividing global society and politics between freshwater Haves and Have-Nots, and establishing a fifth, vital historical use for water—to sustain the earth's life-giving water ecosystems.

Another major water fragility that undermined Islamic civilization was its shortage of small rivers. Islam's "stream deficit" not only inhibited its development of a speedy, safe, and extensive internal transport network. It also handicapped Islam in exploiting one of the rising sources of potential competitive advantage during the Middle Ages—waterpower. Although Muslim hydraulic engineering knowledge was more advanced than in Europe, the waterwheel never played as important a role simply because of its natural shortage of fast-flowing streams. At a time when rival Europe was learning to apply the abundant waterpower and transport potential of its many small rivers to the development of the early industry that helped drive its historical ascendancy, Islamic Spain continued to use the waterwheel almost exclusively for grinding grain and lifting water. The use of energy derived from waterpower was a seminal factor in the development of early industry. By the mid-twelfth century, Europeans had already attained parity with Islam in waterpower.

Islam's inability to maintain its early command of the high seas was also another key factor in its rapid decline after the twelfth century. In hindsight, the first mortal blow was its failure to defeat Constantinople in 718 and thus monopolize the Mediterranean as an Islamic Lake. This left the door open for European maritime states to build up their sea power. By the late eleventh century, they began to take over the key trade routes. Gradual expulsion from Mediterranean sea trade eliminated a major source of wealth and forced Islamic civilization to rely more extensively on its water-scarce, desert resources and propelled the abruptness of Islam's reversal of fortune.

Yet Islam also failed to fully consolidate its greatest opportunity of all—its control of the rich, long-distance Indian Ocean trade. In the Indian Ocean, the Muslims proved to be timid, shore-hugging sailors. They ventured into the open seas only when absolutely necessary. Nor were they intrepid explorers of the unknown. In Africa they traveled south as far as the treacherous Mozambique Channel between the mainland and the large island of Madagascar, but no farther. Ironically, the channel became known in Arab history as "the passage of the Franks" through which the Europeans—whom the Muslims called

"Franks"—sailed when they transformed history by rounding the cape of south Africa and bursting into the Indian Ocean at the end of the fifteenth century. Why Muslim seamen already preeminent in those waters did not try to push on around the African cape into the Atlantic long before Europeans made the breakthrough voyage in the opposite direction may appear in hindsight to be one of world history's great missed strategic opportunities. Yet, in fact, it was simple enough to understand. They had little economic incentive to do so—they already controlled the most profitable trade routes in the world.

Islam's decline at sea more broadly was due to its failure to transform itself into a true maritime civilization. While it occupied the second frontier of its seas, it never genuinely absorbed it into a dynamic new synthesis with its original desert-borne civilization. Alexandria, despite all the advantages of its wonderful large harbor and its central location at the trade interstices of the Mediterranean and routes east, never became a Muslim Venice. Islam coped with its deficit of streams, good harbors, and dangerous coastlines, but it did not truly overcome it. Culturally, Islam remained fundamentally land-oriented. It thus left itself vulnerable to being outflanked when Christian infidels mounted their great challenge to Islam at sea.

One recurrent lesson of history is that societies that passively live too long off old water engineering accomplishments are routinely overtaken by states and civilizations that find innovative ways to exploit water's ever-evolving balance of challenges and opportunities. Thus Muslims failed to meet the challenges posed, first by Chinese junks and then, after China's voluntary withdrawal from the seas, by the spectacular entrance in early 1498 of the Portuguese explorer Vasco da Gama into the Indian Ocean. As often is the case in hindsight, one civilization seems to surpass another, as Muslims did in displacing the Byzantines and Persians, with startling abruptness because advantages have been quietly building up for some time and then express themselves with full force all at once. Advances in navigation, shipbuilding, and sea weaponry had been steadily expanding the opportunities for cheaper, quicker, and safer sea transport and trade. But the ascendancy of sea power was not awesomely displayed until the moment da Gama's fleet rounded Africa's cape and crossed the Indian Ocean and pulled into the port at Calicut, India. In little more than a decade, sturdy Portuguese ocean vessels, armed with cannons that had an effective range of 200 yards, seized control of the Muslims' richest sea routes across the Indian Ocean to the Spice Islands. Portugal's opening of the all-water spice route to India also broke the long-standing Venice-Alexandria stranglehold on the trade of oriental goods throughout the Mediterranean. Venetian overtures to Egypt's rulers to reopen Pharaoh Neko's old Red-Sea-to-Nile "Suez" canal route as a countermeasure came to nought. As a result, the traditional overland camel and sea caravan routes that had yielded so much of Islam's wealth went into accelerated, lasting decline.

Sea power also emerged as the weakest link of Islam's militant revival under the Turks from the fifteenth century. The Turks were originally Far Eastern nomadic steppe cousins of the Mongols. Between the ninth and eleventh centuries, many Turkish tribes entered Islamic territories in the Middle East and converted to Islam, often serving as mercenary soldiers. Over time they became the military backbone, and then the political masters, of Islam. When the era of Mongol hegemony ended, the Ottoman Turks embarked on a military expansion into the Anatolian highlands of modern Turkey.

In 1453, Turks under the command of the young Mehmet II, thereafter "the Conqueror," sent a shudder through all European Christendom by finally taking Constantinople and making it their new capital. The final assault against the historic city was won with the help of a gigantic cannon built by a Hungarian engineer and by Mehmet's own master stroke against the nearly impregnable Golden Horn. His forces dragged 70 galleys overland and launched them behind the Byzantine imperial squadron guarding the Horn's entrance. For the next 200 years, the Turks spearheaded a new Islamic jihad against Christian Europe. In the sixteenth century formidable Turkish Islamic armies stormed through Greece, the Balkans, and Hungary and in 1529 laid siege to Vienna on the Danube in central Europe. During the peak of their power under Sultan Süleyman the Magnificent, who ruled from 1520 to 1566, even Rome itself felt threatened. As late as 1683 Turkish armies were capable of besieging Vienna a second time.

Europe had reacted to the first Arab-led Islamic expansion by launching the Crusades to retake the Holy Land; its response to the Turkish-led second clash of civilizations included a series of sea battles for control of the Mediterranean. Although the Turks' new fleets had reestablished Muslim sea power and had captured the strategic eastern Mediterranean island of Cyprus in 1570–1571, the Turkmen were no match for European vessels, sailing skills, and naval tactics developed in rough gales and currents of the Atlantic Ocean. In 1541, Lufti Pasha, a former grand vizier to Süleyman, became concerned that while the Ottomans were powerful on land, they were vulnerable to their Christian enemies at sea. His assessment was proven correct in the climactic sea battle between the combined fleets of Christendom and the Turks on October 7, 1571, off the Greek coast at Lepanto, not far from the site of the Battle of Actium that had ended Rome's civil wars. The bloody, four hour Battle of Lepanto is celebrated not just for signaling Christian Europe's decisive triumph over resurgent Islam at sea, but also for marking a turning point in the history of naval warfare—it was the first major sea battle in which gunpowder was important. The Turkmen fought mainly the way most major sea battles had been fought since antiquity—they tried to re-create the conditions of a land battle aboard ship. Their soldiers were armed mainly with bows and arrows and swords for close combat; their pilots and oarsmen tried to ram their galleys into enemy vessels or maneuver them alongside close enough for grappling hooks to draw them together to permit boarding and hand-to-hand fighting. The Christian fleets, by contrast, fought with weapons that presaged the dawning new era of naval warfare. Their galleys had cannons mounted on their bows and their soldiers were armed with muskets or arquebuses that could fire at and hit the enemy from a distance. The Venetians also unveiled an entirely new class of warship, a galleass, that was much bigger than traditional galleys, with 50-foot oars manned by half a dozen men, and large swivel guns. History's next great sea battle, the British defeat of the Spanish Armada seventeen years later in 1588, would complete the transformation to modern artillery-based sea warfare from a distance. At Lepanto, a total of 30,000 men died in the Christians' bloody victory over Islam. Among the Christian wounded was Miguel Cervantes, the author of the novel *Don Quixote,* who through his life proudly displayed his maimed left hand as proof of his role in the battle. Lepanto crippled the expansionist ambitions of the Turkish Empire by curtailing its mobility at sea and its access to the vital resources that moved along the world's sea-lanes.

The sea battle with Europe illustrated that Islam's demise from international preeminence was only partly due to its own absolute response to its internal water resource fragilities. What its neighbors were doing with their water resources determined outcomes too. Civilizations' responses to their water challenges throughout history were variable and always in flux. Some civilizations rose sooner because water conditions in their habitat were more favorable to being exploited by available technologies and forms of organization. Hydraulic civilizations, for instance, arose earliest because semiarid, flooding river valleys offered opportunities for irrigation they had the ready means to exploit. The development of Islam's trade, using camels to carry goods through its harsher desert habitat, took much longer. Still other regions with even more inauspicious water resource endowments faced challenges so daunting as to all but relegate their native societies to subordinated starting positions in the ongoing competition among societies.

Such was the destiny of much of sub-Saharan Africa, where geography presented formidable hurdles. Its equatorial rain forest regions, like tropical lowlands everywhere, were ecologically precarious habitats particularly inimical to the development of large, advanced civilizations. Its soils were permanently saturated sponges, extremely difficult to clear for farming and unhealthy for human settlement. Travel was hard, except by river. Nonetheless, impressive civilizations did develop in the surrounding drier, more hospitable tropical forests and transitional savanna lands, such as the successive empires that flourished around the Niger River and the headwaters of the Senegal and Gambia rivers. But for a long time these civilizations developed in isolation behind the barrier of large deserts and the impenetrable ocean that limited their ability to engage on equal terms with other societies in the cultural and economic exchange that has stimulated civilization in every age. It was not surprising that the external barriers of sub-Saharan African empires were breached first by neighboring civilizations propelled by superior competitive advantages in water technology—camel caravans by Arab traders and later by the oceangoing vessels of the Europeans. Following international history's inexorable progression from trade to raid and domination, the Muslims and Europeans pressed their advantages by imposing exploitative relationships upon Africans through trade, conquest, and colonization. The ultimate symbol of this inequity was the large slave trade in black human beings. For centuries this was a monopoly of the Arabs. But when European ships appeared on the Atlantic coast of Africa and offered cheaper and safer sea routes, as well as new markets in the New World, domination of the slave trade shifted from the Arabs to the Europeans.

The Europeans on the ocean-bounded, cold, and wet northwestern edge of the civilized Old World had also inherited water resources that were extremely challenging to tap and harness. For millennia non-Mediterranean northern Europe remained an impoverished backwater. But when its inhabitants finally broke out of the ocean-bound confines of their peninsula-shaped continent with the innovation of open sea sailing, they gained command over one of the most dynamic water advantages in all world history. For much of previous history, sea power mainly had helped small states survive defensively against much larger land-based states with powerful armies; naval prowess equalized the balance of power by enlisting the formidable difficulty of navigating the sea itself into the battlefield, and by stretching and harassing enemy supply lines. But with the advent of open sea sailing, control of the oceanic highways of the entire world suddenly was transformed into an overwhelming offensive advantage. With China's voluntary turn inward and propelled by their leadership at sea, Europeans were able to exert an extraordinary, global dominance that was to last half a millennium.

PART II

Water and the Ascendancy of the West

CHAPTER SEVEN

Waterwheel, Plow, Cargo Ship, and the Awakening of Europe

While the glory of Chinese and Islamic civilizations waned during late medieval times, another civilization was starting to flourish on the European edge of the Old World. Invigorated culturally by its synthesis of Christian religion with its rediscovered Hellenic-Roman roots, and economically by the fusion of the disparate resources of its precocious, semiarid Mediterranean south and its slower rising, colder, temperate north, Western civilization would consolidate an unprecedented 500-year supremacy over the world's wealth and political order. An ongoing series of water challenges and responses marked the path of Europe's historical ascent. Time and again, water's inherent, latent potency as a transforming agent was unleashed by the conversion of a formidable water obstacle into a vehicle of productive expansion. Most dramatically, the West was propelled by two key turning points in history—the advent of transoceanic sailing with long-range cannonry in the European Voyages of Discovery in the late fifteenth and early sixteenth centuries, and the gradual harnessing of waterpower for industry, first with waterwheels and, in the late eighteenth century, with the invention of the modern steam engine. Also driving the West's rise was the distinctive political economic order that arose in its most dynamic centers, featuring self-expansive, flourishing free markets and representative liberal democracies that sprung from the seeds originally planted by ancient Greece's seafaring city-states.

The European continent's geographic shape as a peninsula bounded on three sides by high seas—the warm, lakelike Mediterranean Sea to the south; the cold, rough, semienclosed North and Baltic seas in the lonely north; and in the west, the vast, stormy, tide-tossed Atlantic Ocean, for most of history at once the West's great impenetrable frontier and its protective barrier—fostered a natural maritime orientation central to its history. The continent's absence of a unifying, arterial inland waterway like Egypt's Nile or China's Grand Canal further pushed Europe's inhabitants toward its seascapes to communicate and trade among themselves. The Danube and Rhine rivers, which might have served as some part of that unifying backbone, flowed respectively east to the Black Sea and north toward the North Sea, both away from the main direction of early civilized European society in the Mediterranean; in fact the two great rivers had provided Rome's primary defensive barrier against incursions from the nomadic barbarians who lived beyond it in the northeast—they were Rome's Great Wall of China. Indeed, just as centralized, large hydraulic civilizations emerged along the arterial, irrigable rivers of some of antiquity's semiarid habitats, Europe's greater reliance on wide seas, rain-fed farming, and many small, navigable rivers helped foster its own distinctive political history of small, competing states linked by markets and friendly to the gradual development of liberal democracies.

Backward northern Europe originally sprang to life during the so-called Dark Ages from about AD 600 to 1000 from a sparsely populated, barbarian hinterland of the old Roman Empire into a settled, autonomously growing region of Christian civilization with critical impetus from a combination of water engineering including new plow technology, land drainage that expanded its rain-fed cropland, and exploitation of its small rivers for navigation and waterpower. Following the Byzantine emperor Justinian's failed bid in the sixth century to reconquer the Roman heartland, northern Europe on both sides

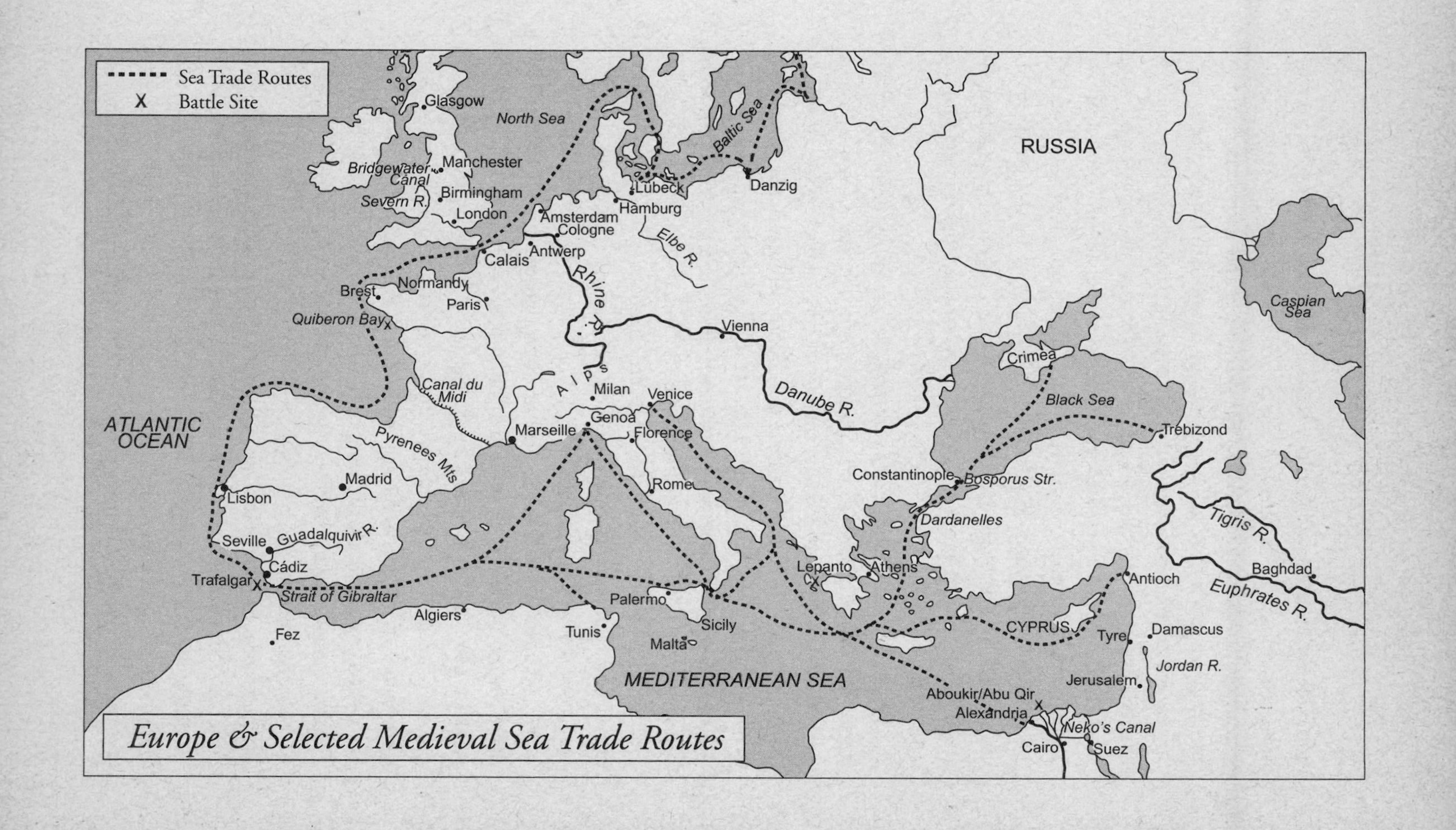

Europe & Selected Medieval Sea Trade Routes

of the Rhine and Danube underwent tumultuous centuries of power struggles among barbarians and settled societies from which ultimately emerged a decentralized feudal political system and manorial economy linked by independent walled towns and unregulated trade. The most important barbarian kingdom was that of the Franks, whose conversion to Christianity at the end of the fifth century and political alliance with the Roman papacy was crucial to the survival and spread of the Latin church. At their zenith under Charlemagne, who was crowned Holy Roman Emperor on Christmas Day 800 by Pope Leo III at St. Peter's, the Rhine Valley–centered Frankish kingdoms controlled almost all of modern France and modern Germany, the upper Danube, and northern Italy.

Yet in the ninth and tenth centuries, the stability of the Frankish and other consolidated administrations eroded under pressure from a new wave of barbarian raiders, including the terrifying Scandinavian Norsemen, or Vikings, who earned a living sailing up and down Europe's rivers and coasts in their long, shallow ships by raiding and exacting tribute from settlements situated along the shores. Ultimately repelled by walled citadels and castles defended by mounted, armor-clad knights, these barbarians, too, settled down to civilized life, adopted Christianity and, like converted barbarians throughout history, invigorated their new religion with fresh zeal. The Norsemen who settled in Normandy became the Normans that conquered England in 1066 and soon thereafter provided the knights who seized Sicily from Muslim control and then led the first crusaders from 1096 to 1099 in taking Jerusalem and the Holy Land for Christendom. By about AD 1000 most of northern Europe had been Christianized, and market forces had gained enough momentum to help launch the early stages of the commercial revolution from 950 to 1350 that propelled the West's early economic takeoff.

Northern Europe had always possessed promising physical attributes for development. Thanks to the blessing of the warm, Atlantic Ocean Gulf Stream current from the Caribbean, its northwest was a temperate climate zone suitable for nearly year-round farming despite its subarctic latitudes. It was rich in freshwater and other natural resources, had copious rainfall, and almost endless, indented sea coastlines with many good, natural harbors for shipping and trade. It had the potential backbone of an extensive waterway transport network in its many long, navigable rivers, mostly flowing north, that reached much farther inland than the rivers of Mediterranean Europe.

Yet throughout Roman times the region faced one insuperable water obstacle to agricultural expansion—excessive precipitation and poor natural drainage of its heavy, clay soils. Thick forest and swamps covered most of its flat, often-waterlogged plains. Tillage methods applied in the lighter, drier soils of the Mediterranean and Middle East, notably including the simple wooden, shallow, scratch plow pulled by an ox or pair of asses, were all but useless in the northern terrain. As a result, northern Europe's rain-fed cropland was limited to the slash and burn methods of hillside patches where drainage was adequate and in the few other locations where soils were naturally more permeable and could be worked by laborious, small-scale farming methods. With agriculture perennially hovering around near-starvation levels, population in the region was low and life spans short.

The seminal breakthrough of the agricultural revolution that stirred northern Europe's economic awakening came with the heavy wheeled moldboard plow. Pulled by a team of four to eight oxen, the moldboard plow had deep, curved iron or iron-covered blades that turned over deep furrows and produced high dirt ridges unlocking the fertility of the heavy soils over a wide expanse. The key technical breakthrough besides the application of costly iron blades was the placement between the team and the plowshare of wheels, which acted as a fulcrum for applying greater pressure to the heavier plowshare and improved the machine's mobility over uneven terrain. By the tenth century it was in widespread use everywhere across northern Europe.

Northern Europe's landscape was dramatically transformed. Forests were felled, swamplands were drained, and everywhere wilderness was converted to arable cropland. The intensification of water management wrought by the heavy moldboard plow extended the footprint of mankind's intensively cultivated cropland to new climate zone and sustained one of history's great expansions of farming on rain-fed lands.

Agricultural production and productivity both soared, setting the basis for an agricultural revolution that reached its apex in the eleventh to thirteenth centuries. Notably drier and milder climate—1 to 2 degrees centigrade—in Europe from the mid-eighth to thirteenth centuries abetted the expansion. The proof of its great impact was a surge in European population, which more than doubled from AD 700 to 1200 to 60–70 million. Population density increased wherever the moldboard plow came into widespread use.

The moldboard plow also acted as a major catalyst in the structural transformation of medieval economic society. Being a powerful, but costly implement, it encouraged the cultivation of larger fields, the collective sharing of scarce draft animals, and cooperative labor among farmers. Fences separating individually owned fields came down and the collectively managed land came to be governed by peasant village councils—an early form of representative democracy—that settled disputes and rendered executive decisions about total cropland management. These councils became essential features of northern Europe's characteristic self-sufficient and self-contained village communities, or manors, which contrasted distinctively from the individualistic economic and social structures that prevailed in the scratch-plowing, drier lands south of the Loire River and the Alps.

The heavy moldboard plow became the foundation of a new three-field triennial crop rotation system that originated in northern France in the ninth century and within three centuries was common throughout northwestern Europe. Wheat or rye was planted in one of three fields surrounding the village in the autumn; the second field was planted in the spring with oats, barley, or peas; the third was left fallow in order to replenish the fertility of the soil. Farming villages often encompassed the cultivated landholdings of free peasants as well as tenant-farmers working part of the lord's manorial domain. The lord provided many general services, such as blacksmithing and the waterwheel-powered gristmill for grinding grain into flour to make the daily bread; tenant-farmers commonly were obligated to use the manorial mill for a standard one-thirteenth share of their grain or flour. The village-centric, manorial economy was integrated into the decentralized, governing feudal association of lords, vassals, knights, and peasants to constitute the signature political economic system of the era.

The agriculture bounty and population growth unlocked from Europe's waterlogged plains by the moldboard plow helped activate other of the region's latent water assets to impel further economic expansion. After 1000, Europe's long inland rivers and northern seacoasts came alive with merchant vessels, often heavily armed, transporting crops and raw material goods like timber, metals, wax, furs, wool, and eventually salted herring, among rising free commercial towns and seasonal trade fairs. On the northern seas, many of these early merchants were descendants of the longboat Norse raiders.

Major advances in the flat-bottomed cog from the eleventh century transformed the treacherous, lonely northern seas into active trade highways carrying bulk cargo with small crews of 20 or fewer. The new cog was a much larger, sturdier single-square sail vessel with a rounded bottom and an innovative central sternpost rudder that replaced the traditional long steering oar. Cargo carrying capacity of the largest cogs sextupled, reaching 300 tons by the late twelfth century. When lofty platforms for archers were added, the high-decked cog also proved to be an excellent warship. Armed convoys of cogs became the workhorse of an informal network of free German seaport cities that by the twelfth century had started to dominate sea commerce in the Baltic and North seas. Centered in Lübeck on the Baltic side of the neck of land dividing the two seas, the powerful German-centered Hansa commercial association (also known as the Hanseatic League), with its own governing membership, laws, and customs, eventually numbered almost 200 free-trading cities and towns. The Hansa domain stretched the breadth of the northern seacoast and, as inland river trade developed, up the Rhine as well. One late-joining Hansa member was Cologne, situated at the juncture of two Rhine waterways, one flowing upstream and the other downstream, and a major overland route. This intersection made Cologne Germany's largest town, albeit with a modest 20,000 inhabitants, in the fifteenth century.

Although always smaller in volume than its sea trade, northern Europe's inland river commerce created an extensive, inexpensive waterway network that galvanized economic activity in much the same manner, although to a far-lesser degree, as the Grand Canal in China. Localities built and maintained flood levees and interconnecting transport canals. In the Low Countries of modern Holland and Belgium, where extensive cropland was reclaimed by drainage, some 85 percent of commercial traffic moved by water, abetted by the use of navigation weirs and, from the late fourteenth century, canal locks. Rivermen often poled their boats downstream on busy rivers, sometimes paying exorbitant tolls for the chains stretching across them to be lowered, much like modern auto tollgates on a highway.

Buoyed by trade and agricultural bounty, the quickening economic pace expanded northern Europe's wealth to levels that eventually surpassed those of the older centers of Mediterranean Europe, which were concurrently undergoing their own commercial upswing impelled by private merchants and market economic forces. "Commerce between the tenth century and the fourteenth century became the most dynamic sector of the economy in country after country, and merchants were the main promoters of change," writes historian Robert S. Lopez. The Commercial Revolution gradually eroded the controlling power "of landowners and officials and made the market, instead of the public place or the cathedral squares, the main focus of urban life."

Urban hubs throbbing with new market activity rose to prominence starting in the eleventh and twelfth centuries. Wherever navigable waterways converged or where key river crossing or favorable harbors were established, influential urban commercial centers arose. The most vibrant cluster in northern Europe was in the Low Countries, where the navigable Rhine, Meuse, and Scheldt rivers flowed near one another. It included the ports of Ghent, the largest city with 50,000 inhabitants in the fourteenth century, Bruges, Antwerp, and later Amsterdam; other big centers included Lübeck, London, and Paris. This was mirrored in Mediterranean Europe by a cluster of large northern Italian city-states, above all Venice, Genoa, Milan, and Florence, with populations surpassing 100,000.

Due to the overriding importance of water navigation, it was no accident that the central marketplaces of Europe's Commercial Revolution developed literally on top of and alongside the bridges and quays of the leading medieval towns and city-states. Like towns, bridges underwent a major building boom from the eleventh to thirteenth centuries, often becoming each town's pulsing central marketplace. Shops and houses located around the bridges enjoyed the further medieval privilege of having at hand a common drinking water and sewage disposal source in the river below. Crowded with shops and markets were the late twelfth-century Old London Bridge across the Thames, the Grand Pont across the Seine, also featuring 13 floating water mills moored below its arches where the river flowed fastest to produce fourteenth-century Paris's daily bread flour, and the stone bridge that still crosses the Arno at Florence, the Ponte Vecchio. Many pioneering early bridges were built by monastic orders, including the famous, 20-arched, Pont d'Avignon across the notoriously flooding Rhone in southern France that was erected in the late twelfth century by the Frères Pontifes, (Brothers of the Bridge). As bridges became a practical amenity that enhanced town trade and commerce, civic authorities undertook the responsibility for building many of them. This helped revive the Roman practice of public infrastructure investment, which became a mainstay of the West's liberal, democratizing marriage of convenience between governments and private markets.

Few bridges were at the center of so much important medieval commerce as Venice's Rialto Bridge, the lone crossing over the Grand Canal at the heart of the greatest Mediterranean sea trading power of the age. The first wooden Rialto Bridge was built in 1264, replacing an old pontoon crossing. Several wooden iterations later, the late sixteenth-century stone bridge was erected, crammed then as today with two arcades of noisy shops and businesses bustling along its banks. Bakers, butchers, fishmongers, fruit and vegetable sellers, acrobats and other entertainers, and even the infirm in their beds at the hospice were conspicuous daily sights. Upon closer look, the hierarchical skeleton structure of early market capitalism itself was physically visible beneath the manifold relationships of the merchants of Venice crowded around the Rialto: the small merchants haggling over price and exchanging goods for money on the bridge within earshot of the larger wholesale suppliers who bought, sold, and signed trade and shipping contracts every morning nearby in their loggia, or meeting room—an early commodities exchange—and who then later in the day walked a few paces to the narrow bank counter stalls of the *banchieri*—the "bankers"—who settled their transactions by book entry account fund transfers and reinvested the accumulated capital profits of the marketplace in a new circuit of loans and ownership stakes in fresh speculative ventures. Many modern financial practices began in this era, including the debit and credit and double-entry format of modern balance sheet accounting and, in the fourteenth century, the bill of exchange facilitating merchants' ability to conduct business in distant locations. The era also inaugurated history's notable sovereign loan defaults—by British monarchs to their Italian bankers—with the resultant bank collapses and international financial and economic crises.

For centuries the two distinctive, rival economic realms of town commerce organized by market supply and demand on the one hand, and the immense, traditional manorial economy of barter and self-sufficient farming on the other, coexisted side by side with overlapping trade between them. Gradually, however, the more-productive market realm expanded faster and brought more and more of Europe's economic resources under its ambit, eventually relegating the manorial realm to the margins, and then the annals, of history. Enough development had occurred before the disruptive, population-decimating famines and plagues of the fourteenth century that the outline of Europe's signature model of market economies operating across a fragmented political environment was visibly emerging.

The superior competitive dynamism of the town over the manorial realm in shaping Europe's destiny was vividly illustrated by their contrasting uses of the waterwheel. On the manor, the waterwheel rarely transcended its traditional function for grinding grain into flour. Under the influence of commercial market forces centered in the towns, it was transformed into a primary agent of the Mechanical

Revolution (eleventh to thirteenth centuries) that powered the takeoff of Europe's early industries.

The waterwheel's invention in the first centuries before the Christian era ranked as one of the watershed moments in the history of civilization. In contrast to its older cousin, the ancient, animal-powered noria, or wheel of pots, that lifted water mainly to irrigate cropland, waterwheels fitted with paddles or blades turned automatically in ceaselessly flowing currents to transmit the captured water flow energy to do productive work. In effect, the waterwheel was history's first mechanical engine. It provided mankind with its first great breakthrough in harnessing an inanimate force of nature since the domestication of fire at the dawn of humanity and sailing at history's early beginnings. For some two millennia waterpower would represent the pinnacle of civilization's constructive command of nature's power.

The simple horizontal waterwheel—in which water flow turned a wheel that lay parallel with the millstone affixed above it—was used the world over, chiefly to grind bread flour. It was several times more powerful than the ancient hand mill turned at about one-half horsepower by two slaves or a donkey. Horsepower was augmented by a factor of five to six times over the hand mill by the innovation of placing the wheel vertically in the water. The vertical undershot wheel transmitted its rotating power through a camshaft and gearing mechanisms to turn multiple millstones or other devices at higher speeds of rotation, even though its reliability varied with fluctuating stream and weather conditions, such as droughts, floods, and freezes. Another major improvement—which became crucial to launching Europe's early industrial development in the late Middle Ages—was an overshot version of the vertical waterwheel. By directing a steady flow of water to fall onto the wheel blades from above, often through a millrace emanating from an artificial pond or a dammed river and regulated by a sluice gate, the vertical overshot wheel typically was three to five times more efficient than its undershot cousin and also permitted larger, more-powerful wheels to be employed. Leonardo da Vinci, who worked brilliantly on many problems of water hydraulics including canal locks, water pumps, bridges, and paddleboats as well as waterwheels, was among the earliest to argue, rightly, that the overshot wheel was the most efficient design some 250 years before engineers were able to prove why. A few exceptional overshot wheels in medieval Europe were capable of as much as 40 to 60 horsepower. Coastal regions from Venice to Brittany and Dover even experimented with ocean-tide-powered mills, although these always remained on the periphery of mainstream waterpower history.

Although ubiquitous, waterwheels until the eleventh century were commonly weak in individual horsepower and rarely used for industrial applications. The Domesday Book (1086), compiled by Britain's new Norman rulers to assess what potentially taxable assets they'd won in their conquest of 1066, recorded that south of the Severn and Trent rivers there were no less than 5,624 mills serving 3,000 settlements—or nearly two water mills per settlement. The ratio was likely similar on the even more prosperous and heavily populated Continent. Mills were widespread enough to have been taxed by Charlemagne at the start of the ninth century. Damming rivers to power waterwheels is recorded in the annals of French history from the twelfth century, with one account describing how a king hastened the surrender of a town he was besieging by breaking the dams that powered its mills. In the early fourteenth century, the Seine near Paris had 68 mills concentrated within less than one mile. Floating mills were common sights under the bridges of major cities. Virtually every suitable little stream in inhabited regions had several flour mills, often situated every quarter to half mile apart. By the debut of the eighteenth-century Industrial Revolution, Europe likely had over half a million water mills, whose enormous combined horsepower provided an indicative proxy for the advanced stage of material civilization attained in the West.

The waterwheel's greatest impact on world history was in Europe because it was there that it was most extensively applied to early industry, especially after the eleventh century. The waterwheel inspired experimentation by craftsmen with a panoply of mechanical gears, flywheels, camshafts, conveyor belts, pulleys, shifters, and pistons that seeded the foundational know-how of industrial production. The surprising technological pioneers of applying waterpower to industry were religious monasteries. Ever since St. Benedict foreswore his solitary hermit's existence and instituted the Benedictine Rule for the monastic community he established in AD 529 at Monte Cassino in southern Italy, European monks had been actively engaged in physical labor as a material and spiritual boon to their communities' purpose. Self-sufficient monastic communities played a key role in early medieval civilization by preserving ancient books and rekindling classical learning, converting pagans to Christianity and less famously, advancing and disseminating many hydraulic arts, including dike building and maintenance, swamp drainage, and bridge building, as well as the application of waterpower to myriad monastic activities.

The most ambitious monastic pioneers of waterwheel technology were the rapidly expanding Cistercians, founded in the late eleventh century, whose monasteries were consciously built near rivers to exploit its waterpower and which often housed large factories. Few individuals in history put waterpower to better use than the celebrated mystic Cistercian leader St. Bernard at his twelfth-century Clairvaux Abbey, in a valley in northeastern France. Water was drawn to the abbey from a two-mile-long millrace fed by the river Aube. Paraphrasing the description of a contemporary observer, the water first rushed to the corn mill where the wheels turned millstones to grind the grain and shake large sieves to separate the bran and the flour. In the next building, the water filled the boiler used for brewing and then drove the heavy hammers that beat the fulling cloth. After the tannery it was routed into many smaller courses, where it was employed in sawing wood, crushing olives, and providing running water for cooking, washing, bathing, and ultimately carrying away all refuse. In the twelfth and thirteenth centuries, Cistercians pioneered the breakthrough application of waterpower to iron foundries in England, France, Denmark, and Italy and were among Europe's leading iron producers for several centuries.

As waterwheel know-how migrated from the monasteries to Europe's growing commercial towns, it was employed for market-driven industrial applications. Waterpower propelled mechanized sawmills, bore wood and metals, and helped pound beer mash. In mining it was employed to crush metals, power shaft ventilation machines, and raise winches that removed buckets of mine water and excavated ores to the surface.

But it was in the seminal industries of papermaking, textiles, and iron forging that waterwheel-power technology stood out as having the most dramatic impact on Europe's economic rise. Paper mills with giant, water-powered beaters that pounded pulp migrated from Baghdad to Damascus by 1000 and to Muslim Spain by 1151. Christian Europe's first water-powered paper mill was opened in 1276 in Fabriano, Italy, where the watermark was shortly thereafter pioneered. Since papermaking required vast amounts of clean water as an input in the production process, most paper mills were located upstream of the nearby towns that might pollute it. Mass production reduced the cost of paper, stimulating the nascent commercial bookmaking industry that evolved from the monasteries in the twelfth century and the thriving centers of Islamic civilization. This paved the way for the landmark fifteenth-century invention of the printing press, which in turn helped democratize European society and reinforce the foundations of Western humanism and science through the dissemination of books and knowledge to a wider public—the original information revolution.

Clothing textiles also had a special place in European history. Textiles were one of the earliest major industries to go international, linking raw material suppliers and intermediary and finished goods producers in a web of market activity that stretched from England to northern and Mediterranean Europe. Waterwheel mechanization powered the beaters used by cloth fullers and, when the Chinese silk loom reached the West in the thirteenth century, to drive silk-spinning machines. By the fourteenth century, one silk mill in Lucca, Italy, employed an undershot waterwheel to drive 480 spindles. Eventually, the water-powered spinning of cotton and other low-priced textiles in eighteenth-century England accompanied world history's first fully mechanized factories, the earliest signature hallmark of the Industrial Revolution.

The waterwheel played a decisive role in medieval Europe's catalytic discovery of the blast furnace to smelt iron. Religious demand in the twelfth century for huge iron church bells may have provided the early impetus for the breakthrough. In the ensuing centuries, Europe's iron foundries relocated from wood-abundant forests to riversides and fast-running stream banks to tap the continent's waterpower. Waterwheels gradually supplanted the force of the smithie's arm in pounding the iron by delivering uniform strokes of giant, 1,000- to 3,500-pound trip hammers and lighter ones of 150 pounds that tapped iron into shapes with 200 strokes per minute. By the late fourteenth century, waterwheels were widely used to blast powerful drafts of air through pairs of enormous leather bellows, several feet in diameter, to heat furnaces that could run nonstop for weeks on end at up to 1,500 degrees centigrade. Iron ore heated by these stronger blasts was liquefied, enabling for the first time in Europe the casting of abundant volumes of molten iron. In short order, the water-powered blast furnace transformed iron making from a traditional, small-batch handicraft into one of Europe's earliest mass production industries. By 1500, iron production in Europe reached 60,000 tons. Soaring demand for iron nails, one of history's humblest but most useful inventions, inspired a new water-powered rolling mill in which two iron cylinders flattened iron into bars that were then mechanically cut into nails by rotary disks. At the forges, mechanical trip-hammers attached to a wooden shaft pounded large volumes of malleable fired iron into various shapes that became farm and industrial tools. Iron's marriage with the contemporaneous spread of gunpowder to fabricate firearms and cannonry, meanwhile, armed Europe's vessels and soldiers with the advanced weaponry it would use so devastatingly to subdue societies around the world.

Although comparatively backward in most other technologies and economic development, Europe by about 1150 was applying waterwheel-transmitted power to early industries on parity with the advanced civilizations of China and Islam. One of history's puzzling questions was why only in Europe this budding mechanical prowess continued to develop into the direct precursor of the eighteenth-century Industrial Revolution. Islam's failure can be explained partly by the fact that its dearth of small, year-round streams imposed upon it too great a deficiency of waterpower and internal waterway transport, and hitched its historical trajectory intractably to the plodding, overland trade network of the camel. One of China's principal hindrances was its surfeit of cheap labor, which rendered mechanical innovation less urgent, and even potentially threatening to the established social and political order as well if it reduced employment. The Grand Canal transportation network's enabling of the state to assert stronger internal command over the economy also generally blunted the innovating impetus of private market forces. Whatever the causes, the net effect was that China's vaunted technical and scientific know-how was never rigorously applied to industrial production.

Europe's natural water resources, by contrast, helped create conditions more favorable to the development of market-driven industries and pluralistic, liberal democratic states. Rain-fed, plow agriculture and myriad navigable, and energy-providing small rivers favored the rise of multiple, autonomous, decentralized regions. The natural competition among neighboring states, whetted by sea-trading merchants' freedom to choose among ports offering the most advantageous terms, strengthened the development of private property and individual political rights. In pondering the question of why parliamentary democracy and capitalism arose first in Europe, the late anthropologist Marvin Harris advanced an inverse hydraulic theory. In northern Europe, he noted,

> where there is no Nile or Indus or Yellow River and where winter snows and spring rains provide sufficient moisture for field crops and pastures, population remained more dispersed than in hydraulic regions . . .
>
> Unlike hydraulic despots, Europe's medieval kings could not furnish or withhold water from the fields. The rains fell regardless of what the king in his castle decreed, and there was nothing in the productive process to necessitate the organization of vast armies of workers . . . And so the feudal aristocracy was able to resist all attempts to establish genuinely national systems of government.

Without control over water resources, no authoritarian, centralized state could rule firmly over a great area, leaving a wider berth for independent, cooperative manorial villages and competitive market-centric towns to shape the political economic norms of society. In Roman times slavery had retarded the incentive for labor saving innovation; by the Middle Ages slavery had all but disappeared and cheap labor was scarce. The profit-seeking logic of market forces applied Europe's waterpower potential to mechanized technologies to overcome labor scarcity. Further accelerated by the competition between states, it drove European commerce and industry toward innovations that, in the end, could not be restrained by centralized command. The way was paved for developments that would eventually help drive Europe's economic rise.

First, however, the commercial and mechanical revolutions between 950 and 1350 spurred market-driven exchanges between northern and Mediterranean Europe that gradually linked the two regions into an integrated economic area. Initially, the central axis of trade between north and south was overland and concentrated on a series of seasonal fairs, which attracted merchants from all over Europe, who negotiated trade contracts based on sample goods displayed at the fair. From the late twelfth to early fourteenth centuries, the largest were six fairs held in rotation nearly year round in the Champagne region of northeastern France, astride the main roads and waterways running from the Mediterranean Sea to the North Sea, and from the Baltic to the English Channel. Yet the Champagne fairs rapidly declined in the early thirteenth century as soon as a much cheaper, faster and more reliable alternative became available—the opening of a direct Atlantic coast sea route between the Mediterranean and the north. It was this private commerce-driven, Atlantic seacoast trade route that bonded Europe's two disparate environmental zones together into a dynamic, unified marketplace that ignited Europe's rapid takeoff and the ascension of Western civilization.

The first of what would become the famous Flanders Fleet set sail from Genoa to Bruges in 1297. By 1315 regular convoys were traveling to the North Sea from Venice and Genoa. For 235 years to 1532, the Flanders Fleet sailed between Italy and the Low Countries, twin hubs of the European economy until the eighteenth century when the center moved decisively to England. In its cargo holds it transported bulky commodities of wool, raw materials, and salted herring, as well as some luxuries and spices from the Orient.

One key event in the rise of the Atlantic coast trade was the breaking of the Muslim grip over the Strait of Gibraltar. Throughout history, control of the eight-mile-wide strait had been a source of power and wealth. For centuries from antiquity, the strait known to the Romans as the Pillars of Hercules had been firmly controlled by the long-vanished city-state of mysterious origin, Tartessus. Located outside the Pillars at the mouth of Spain's Guadalquivir River, Tartessus flourished as an emporium for locally mined silver and lead and precious tin for making bronze that was imported from as far away as Brittany and Cornwall. Despite founding a trading colony close by to the east at Gades, modern Cádiz, the Phoenicians could not challenge Tartessus' Atlantic monopoly. Finally, around 500 BC, soon after Tartessus disappeared from history, the Phoenicians of rising Carthage sent an expedition under a captain named Himlico into the North Atlantic along the old Tartessian trade routes. Thereafter, for over two centuries, Carthage was master of the strait and the rich trade monopoly it conferred. With Carthage's defeat in the Punic Wars, Rome took control of the Pillars. It helped secure Rome's empire through sea power control of the mouths of the major western and northwestern European rivers and supported Emperor Augustus' sending of fleets as far as the coasts of the North Sea in his unsuccessful effort to extend Rome's frontiers from the Rhine to the Elbe. The next great civilization to profit from the long-held monopoly over the Strait of Gibraltar was Islam through its control of the land on both sides in Spain and Morocco.

The European breakthrough at the strategic strait was accomplished in 1291, when Benedetto Zaccaria of Genoa destroyed the Moroccan fleet that defended it. Zaccaria was a colorful figure, whose exploits embodied the animating spirit of Europe's early rise. Marco Polo of Genoa's archrival Venice was his contemporary; indeed, Polo was imprisoned in Genoa dictating his tales of the Silk Roads and the Orient while Zaccaria was living outsized adventures influential in European history. Over the course of Zaccaria's eclectic career—which included being a pirate in the Aegean, mercenary naval commander for several states, diplomat, crusader in Syria, ruler over a Greek island, governor of a Spanish seaport, and Europe's most powerful alum baron—his many ships put in at almost every important seaport from Flanders to the Crimea in the Black Sea. Zaccaria was a member of the upper merchant class of the Genoese republic, which had arisen as a great Mediterranean power from the late eleventh century after it and other city-states from Italy's western coast, including Pisa and Amalfi, had united to drive out piratical Muslim sea raiders from their waters.

As a youthful trader in the international wool, cloth, and color dye business, Zaccaria in 1274 had seized the opportunity to exchange naval assistance to the Byzantine Empire for the right to develop a huge, virgin, extremely high-grade deposit of alunite he had surveyed in Asia Minor. When processed, alunite provided the basis for alum. Alum was widely used in medieval times, most importantly as a color fastener in dyeing textiles and as a hardener in tanning. Because colors fastened best with the highest-grade alums, alum quality was a key determinant of the order of economic supremacy among the competing dye centers of Italy, Flanders, and England. Because alum's great bulkiness made it expensive to transport overland, comparative advantage accrued to states on the sea routes of the Mediterranean, where the era's best deposits were located. One other Asia Minor quarry had alunite deposits of superior grade to Zaccaria's; through political maneuvering, Zaccaria was able to get its exportation rights temporarily blocked—until he himself succeeded in securing an ownership interest in it. Zaccaria's huge alum refining operation featured giant processing vats that were protected on land by a fortress and at sea by cruising ships. Armed soldiers helped ensure the safe transport of the alum cargo ships once the convoys put out to sea for the textile markets. As he sought the best market price, Zaccaria was inevitably drawn northward. One of his ships got through Gibraltar and reached England as early as 1278. Ultimately, the lure of lucre drew him toward his showdown victory over the Islamic Moroccan fleet at Gibraltar in 1291 that opened the Atlantic coast to unimpeded European shipping. A naval warrior and would-be crusader to the last, Zaccaria died in 1307 or 1308, bequeathing his heirs one of medieval Europe's earliest and largest private commodity empires.

Despite the Genoese's pioneering, it was archrival Venice that ultimately profited most from the Flanders-Mediterranean sea trade. From the redoubt of its island-flecked lagoon on the upper Adriatic Sea, the Venetian republic had been one of the earliest Italian city-states to lead the revival of Mediterranean Europe from the tenth century. From its earliest roots, Venice was wedded to the sea; indeed, a great festival symbolically commemorating this marriage was consummated anew each year with the tossing of a ring into the waters. From the fifth century, when Roman citizens from the countryside fled the invading barbarians for the protection of its mucky marshes and islands, Venice's fate had hinged on its response to one of urban society's most water challenged environments. With no agriculture and sinking soils, its flat, muddy, and often waterlogged islands had to be constantly drained, built up with soil dredged from its lagoon beds, and protected from the sea tides by laboriously constructed artificial barriers. Malaria and the diseases of miasmic swamps abounded. Indeed, when Dante Alighieri's special embassy in 1321 regarding navigation rights on the Po River was unfavorably received by Venetian leaders, the great author of *The Divine Comedy* and professional diplomat was forced to return to Ravenna via malarial swamplands, from which he took fever and died.

With scant natural resources save fish and the salt of its lagoons, Venice depended from the start on commerce and sea power. By the sixth century, its flat-bottomed trading barges crawled along the rivers of northern and central Italy. In the ninth century, it ventured forth into the Muslim-dominated Mediterranean under the protective shield of the largest and wealthiest city in Christendom, Constantinople. By the tenth century, it began to emerge as a thriving sea trading power in its own right. Its ships sailed among the ports of the Mediterranean, Europe, and the Levant, exchanging Eastern luxuries like spices, silks, and ivory that arrived by sea and camel train from Islamic Alexandria for bulky Western commodities like iron, timber, naval supplies, and slaves, as well as Venetian salt and glass.

As an entirely maritime, merchant-oriented republic, Venice resuscitated the democratic, free-market traditions of ancient Athens. Yet in the favorable environment of commercially awakening medieval Europe, these Greek traditions took deeper root and flourished. They also were transplanted to other parts of Europe. Venice itself became history's longest-enduring republic—1,100 years—and one of important progenitors of modern capitalism. Its devotion was to the pursuit of profit and commerce; more than once its leaders, who actively participated in speculative ventures, defied the Latin Church, even accepting excommunication, rather than obey a papal directive that crossed its vital commercial interests.

By 1082 Venice had attained parity with Constantinople as a Mediterranean power. In that year its merchants became exempted from Byzantine tolls and received other special trade privileges when Venice agreed to provide naval help against the regional invasions of the Normans. By 1203–1204 it became master of the Mediterranean when, through astonishing cunning, calculated risk, and bravery of arms, its blind, octogenarian elected doge, Enrico Dandolo, diverted the Norman armies of the Fourth Crusade from their original target of Egypt and, against the pope's wishes, induced them to successfully besiege and sack Constantinople as repayment for Venice's furnishing of the crusaders' fleet. The Venetians succeeded where the Muslim besiegers 400 years earlier had not by capturing the Golden Horn. Its soldiers seized control of the huge windlass used to raise and lower the great iron chain across the Horn's mouth that regulated entry. Then, led by the charging Dandolo, and his banner of St. Mark, the Venetians from one side and the Normans from the others breached the walls for the first time since Emperor Constantine had founded the city nearly 900 years earlier—250 years before it would fall out of Christian hands to the Turks. Following several months of political intrigue, and a final siege and customary three-day sacking of Constantinople, Enrico Dandolo, by treaty with the Norman crusaders, took the best parts of the Byzantine Empire for Venice. Venice got three-eighths of Constantinople, including prime frontage on the Golden Horn, free-trade rights throughout the Byzantine Empire, from which its archrivals Genoa and Pisa were to be banned, and a choice string of ports stretching all the way from Venice to the Black Sea. Thus Venice was the clear winner of the Fourth Crusade, which in the end never fulfilled its purpose of assaulting Egypt or the Holy Land.

Christian Europe's new control of the Mediterranean, the rise of shipping in the northern seas, and the linking of the two regions through the Strait of Gibraltar helped stimulate a series of breakthroughs in naval architecture, navigation, and rigging that transformed European shipping from the early fourteenth century. The advent of sturdy, maneuverable, large, oarless sailing ships created vessels that, for the first time, could carry their cargo year-round in all weather. They became the direct progenitors of the world-changing transoceanic Voyages of Discovery at the end of the fifteenth century.

The adoption of the magnetic compass from China facilitated sailing in the thirteenth-century Mediterranean, which was too deep for navigation by feeling the way along the bottom as was common practice in the northern seas. From about 1280 to 1330, rigging and ship design underwent a fundamental advance. Two important ship designs emerged. Venetian shipyards began producing a large sailing ship with two, and later three, masts, and rigged with triangular, lateen sails that made it highly maneuverable against headwinds. Although the ship had oars like a traditional galley, they were used only to enter and leave port. Even larger and sturdier from about 1300 was a new model northern sea cog. Clinker-built with overlapping planks and a central sternpost rudder, the cog ultimately became the workhorse of the Atlantic coast trade. To overcome the cog's clumsy maneuverability in the Mediterranean and its problem exiting against the westerlies that prevailed at the Strait of Gibraltar due to its possessing a single square sail, the cog was enhanced with a second, or mizzen, mast rigged with a lateen sail. The Genoese, in particular, adopted this new model cog. They increased the size of its hulls so that by 1400 it could carry cargoes of alum and other bulk commodities up to 600 tons, or two to three times more than the competitive carriers of the northern Hanseatic League. The new vessels, which debuted in the Mediterranean, had much smaller crews and relied upon crossbows for defense against the ramming and boarding tactics of traditional oared galleys.

The combination of new ship design and improved navigation helped trigger a quantum leap in Mediterranean shipping volume and velocity—Italian round-trips made to the ports of Egypt, the Levant, and Asia Minor doubled from one to two per year. Instead of being forced to winter in foreign ports as had been customary for centuries, Italian fleets left for the eastern Mediterranean in February and returned in May, reloaded their holds, and departed again in early August for a Christmas return. All-weather shipping spread to the Atlantic and to the northern seas. For the first time, a coherent, price-integrated commercial network of markets served by Zaccaria's alum ships, the Flanders Fleets, and others emerged along Europe's three sea coastlines from the Baltic to the Black Sea. The enormous economic impetus helped European growth survive the multiple catastrophic setbacks of the mid-fourteenth century—colder climates, famines, peasant revolts, and finally the Black Death—which annihilated one-fourth to one-third of Europe's inhabitants. Europe's population did not return to its pre-bubonic-plague level until after 1480.

Integration by sea transport recalibrated competitive market conditions throughout the region. Baltic populations suddenly were able to preserve herring and cabbage throughout the winter with salt imported from southern Europe. Salted herring became a major export to the Mediterranean. When the Baltic herring, in one of the great ecological mysteries of history, migrated in the fifteenth century to the North Sea and within reach of Dutch fishing nets, it contributed to the concentration of commercial power in the north in the Netherlands. Another of the many natural changes that altered the course of history occurred when Bruges's harbor silted up around 1500, diverting the Atlantic coast fleets to nearby Antwerp, which happily took over as the northern center of the north-south trade. But all that paled in historical significance beside the momentous water breakthrough that lay on Europe's immediate horizon.

CHAPTER EIGHT

The Voyages of Discovery and the Launch of the Oceanic Era

The seaborne fusion of Europe's disparate northern and Mediterranean resources into a coherent, market-driven, maritime civilization comprised of many autonomous, competing states set the stage for one of history's epochal turning points—the advent of transoceanic sailing. Europe's Voyages of Discovery were highlighted by three breakthrough trips in the 1490s scarcely noticed at the time by the rest of the world—Christopher Columbus to Central America, Vasco da Gama around the coast of Africa to India, and John Cabot from England to Newfoundland in North America—that crowned a century of sea exploration by suddenly decrypting the secret code of trade winds and sea currents of the Atlantic Ocean to enable them to sail to and fro across Earth's open oceans. In so doing they converted what had always been Europe's impenetrable, storm-tossed water barrier into a dynamic navigational advantage that launched Western civilization on its historic path to global supremacy. Within only a quarter century of the Atlantic voyages, one of Ferdinand Magellan's ships completed history's first round-the-world trip. In his 1776 *The Wealth of Nations* Adam Smith heralded Columbus's discovery of America and da Gama's passage to India as no less than "the two greatest and most important events recorded in the history of mankind."

The Atlantic opening ushered in the modern era in which sea power and control of the world's sea-lanes eclipsed the importance of dominion over the land as the preeminent key to global power and wealth. The new sea-centered world system bound all the regions of the planet more closely together and created a web of international communication and maritime trade that has continually thickened and tightened in space and time into today's integrated global economy. Although the new era started around 1500, its long-term trajectories did not become clear among the world's leading civilizations for about two centuries. No civilization gained more than the West, whose maritime position on the Atlantic and superior naval power, often spurred by market economic forces, gave it access to the world's choicest sea trade superhighways. Whenever land-centered civilizations had broken beyond their frontier boundaries across narrow bodies of water or open plains or deserts, their expansions generally had been regional. Likewise, changes in control of axial sea routes in the Mediterranean and Indian Ocean mainly fomented realignments of regional power relations. But the global oceanic opening, by contrast, catapulted Europeans as an irresistible power that reordered world power balances for the next half millennium.

The fitting symbol of Europe's maritime fusion and its momentous Atlantic breakthrough was the caravel, a small Portuguese vessel that was first launched upon the seas in the first half of the fifteenth century. About 70 feet long, displacing only 50 tons, and manned by a small, 20-man crew with minimal supply needs, the caravel had been specially designed for exploration. It had an internal sternpost rudder and a sturdy, lap-jointed, rounded hull whose flatish bottom allowed it to put in at shallows and other perilous shores. The ship had three masts, often two with square sails for power and one lateen or triangular sail that facilitated nimble maneuvering through headwinds which crucially assured explorers that they had a reasonable chance of returning home as they ventured into the unknown currents and wind systems of the ocean. Both Columbus's and da Gama's fleets included caravels.

The caravel married features of both northern and Mediterranean vessels and sailing traditions. Geographically, Portugal made a natural midwife. With its fabulous natural harbor of Lisbon on continental

Europe's westernmost point of land as well as other good Atlantic harbors linked to navigable rivers, Portugal had prospered as a port of call for the coast-hugging Flanders Fleet. Without a window on the Mediterranean, and constantly menaced by large neighbors on the Iberian Peninsula that were coalescing into Spain, moreover, Portugal's survival and wealth hinged extraordinarily upon its ability to exploit the ocean. For two centuries from 1385, when its independence was secured with the help of an alliance with England, to 1580 when King Philip II successfully asserted Spanish hegemony over it, tiny Portugal had an outsized effect on world history through its pioneering of the Age of Discovery, and its ocean-crossing caravel.

The caravel's inspirational spirit was one of history's intriguing, idiosyncratic individuals, Prince Henry the Navigator. The third son of the king of Portugal and his English queen, the tall, blond-haired Prince Henry, then in his mid-twenties, in 1418 set up what amounted to the world's first scientific research institute. It was dedicated to pure discovery and sea exploration through uncharted Atlantic waters down the African coast to unknown lands. Until his death in 1460, his fortress atop the promontory at Sagres at Portugal's southern tip was home to the age's most knowledgeable assortment of sea captains, pilots, cartographers, astronomers, mathematicians, ship instrument makers, shipbuilders and other experts, all collaboratively guided by a scientific methodology unusual for the age and rare to that point in human history. Muslim astronomers and Jewish cartographers escaping religious persecution in Spain, master mariners from Genoa and Venice, German and Scandinavian merchants, visiting world travelers, and even African tribesmen, pooled their knowledge and observations. Henry's experts systematically mapped what they learned of the Atlantic and its coasts, devised methods to measure latitudes, and in general accumulated as much concrete information about the known world as possible. Every year Henry's ships were sent out on exploratory missions to bring back logbooks and charts filled with new data and observations, which were used to fill in the maps and help with planning new voyages.

In the spirit of the dawning age of the Renaissance, Henry's chief purpose was the quest for pure knowledge and discovery for its own sake. But Henry, who lived a monkish existence and is said to have died a virgin, also possessed the crusading zeal of the passing age to find the rumored lost Christian kingdoms of Prester John in East Africa. A third, initially lesser, motivation was the passion for commercial wealth stirring throughout Europe. Henry and his countrymen were tantalized by the prospect of finding the original sources of the gold, ivory and slaves that were brought by middlemen from Africa and the peppers, cloves, cinnamon, ginger and other luxuries from the Indian Ocean that might be reachable by circumnavigating Africa's cape. Indeed, public support for Henry's unorthodox enterprise at Sagres, as well as the pace of voyages, picked up dramatically after 1444, when one of Henry's explorers returned with 200 African slaves—Europe's first direct involvement in the African slave trade.

The circumnavigation and coastal exploration of Africa had been attempted, in both directions, many times over the course of history. The most famous voyage, recounted by Herodotus, was commissioned by Egyptian King Neko around 600 BC after the oracle had warned him off completing his canal linking the Red Sea to the Nile and the Mediterranean. Neko outfitted a Phoenician crew that in a spectacular voyage sailed south through the Red Sea, around Africa, up the west coast, and returned through the Pillars of Hercules three years later. The only problem was that it probably never happened. Scholars believe that such a voyage was perfectly feasible—indeed, circumnavigating Africa from east to west is technically easier than the reverse—but there is no supporting evidence and most important, it left no historical legacy if it did happen. Other east African coast explorers included the Greek Ptolemies who succeeded Alexander the Great and got as far as the Horn of Africa and figured out that the source of the Blue Nile lay in the Abyssinian highlands of modern Ethiopia. Greek sailors in Roman times reached Pemba and Zanzibar, while Roman military explorers, sent inland along the Nile to do reconnaissance for a contemplated, deep African invasion, reached the Sudd swamps and Lake Victoria. Muslim dhows went all the way south to Mozambique, but never pushed on to the unknown lands beyond the much-feared sea passage between Madagascar and the mainland.

Voyagers sailing south along the Atlantic likewise had limited success mastering Africa's western coast. The most celebrated and well-recorded was the voyage of Carthage's Hanno, who successfully established African coast colonies sometime in the fifth century BC. Precisely how far Hanno got is still a matter of dispute, but by most reckonings he reached the crocodile-infested waters of the Senegal River and beyond to Sierra Leone, and turned back before reaching the searing heat and stagnant currents of the Gulf of Guinea. Less successful was the Persian explorer Sataspes, who explained to an unsympathetic King Xerxes that he had been unable to complete his mandate to circumnavigate Africa because his ships had simply stopped while his men were attacked by hostile natives when they put in at shore: Xerxes, who had mandated Sataspes' venture in reprieve of his death sentence for violating one of the court ladies, promptly had him impaled. A Greek explorer of the late second century BC managed to get only partway down the coast of Morocco. Nor did Muslim merchants a thousand years later risk the many hazards of west coast exploration while their camel caravans already enjoyed a trade monopoly with the sub-Saharan kingdoms of west Africa. Without question, sailing Africa's Atlantic coast was dangerous, as the Vivaldi brothers of Genoa discovered in 1291 when they outfitted two galleys filled with merchant wares for trading, made it partway down the coast, and then disappeared without a trace.

So difficult was the circumnavigation feat that by late antiquity Greek writers had concluded that doing so was simply impossible—the heat was too great, the waters of the Atlantic too windless, muddy, shallow, and choked with seaweed to be sailed. Thus little was known of the western shores of Africa when Henry the Navigator, enthused by the budding European revival of Greek knowledge, Herodotus' story about the Phoenicians' circumnavigation of Africa for Pharaoh Neko, and the rediscovery of treatises like Ptolemy's, determined to undertake his explorations to open an Atlantic sea route to India. Yet the revival of classical knowledge also reconjured Greek terrors of the Atlantic, including the belief that at a certain point the waters coagulated so that ships became stuck and sailors could never return home.

Overcoming these Greek-inspired fears posed at least as daunting

a challenge to fulfillment of Henry's dream as the physical challenge itself. Among the gargantuan psychological barriers in the minds of Henry's mariners was Cape Bojador, just south of the Canary Islands on the northwest coast of Africa. Henry's explorers believed that the waters beyond this small cape, a barely noticeable bump on today's map of Africa, were so shallow and rife with treacherous currents and winds that no ship could return from it. To conquer Cape Bojador, Henry sent out 15 expeditions between 1424 and 1434, enticing captains with the promise of great rewards to push ever farther. Always they returned before passing the dreaded cape, until one finally veered boldly westward into the open ocean and then south, to find that he had successfully rounded the cape. Once the psychological barrier of Bojador was overcome, it was merely a matter of time before the rest of the African coast yielded to Prince Henry's systematic scientific methodology and explorations. By the time of Henry's death in 1460, Portuguese ships had passed Cape Verde near the mouths of the rich Senegal and Gambia rivers and reached as far as modern Sierra Leone. Fittingly, that was the year of the birth of Vasco da Gama, the man who would fulfill Henry's dream of circumnavigating Africa and in the process crack the Atlantic's final enigmas and inaugurate the great age of cross-ocean sailing.

In contrast to the navigationally simpler, reversing seasonal monsoons that governed sailing back and forth across the Indian Ocean, the Atlantic's predominant feature was three large trade wind systems that blew in the same direction all year round. The central trade wind system blew west toward the Caribbean from northwest Africa and in the summertime, to the great natural advantage of Portugal and Spain, as far north as the Iberian Peninsula where Henry's mariners easily gained access to it. Farther south, after passing through nearly windless latitudes called the Doldrums, was a second trade wind system that blew steadily from Africa toward South America. In the far north was a third belt of trades blowing west toward the New World, that in spring offered a brief, easterly moving wind that enabled an easy return sail home at the latitude of Britain. Beyond the trade wind systems, at extreme latitudes in both the Southern and Northern Hemispheres, were countervailing west-to-east wind systems; in the far south at the 40 degree latitudes they led around the African coast into the Indian Ocean with such strength that mariners called them the "Roaring Forties." Cutting across and interlinking the trade wind system were several strong sea currents, notably the warm Gulf Stream flowing from the Caribbean to northwestern Europe, and in South America, the southerly flowing Brazil Current. "Considered as a whole," writes historian Felipe Fernández-Armesto, "the wind system resembles a code of interlocking ciphers. Once part of it was cracked . . . the solution of the rest followed rapidly."

The three breakthrough voyages of the 1490s finally unraveled the interlocking Atlantic codes that Prince Henry and his successors had been trying to decipher for decades. Genoese captain Columbus's second Atlantic crossing for Spain in 1493 established viable routes outbound and home across the central Atlantic. Sailing for England, fellow Italian John Cabot's 1497 round-trip between Bristol, England and Newfoundland, Canada utilized the briefly available spring westerly later much exploited by the British colonizers of North America. Finally, and most important, Portuguese da Gama's 1497–1499 round-trip from Lisbon around the Cape of Africa to India revealed the secrets of the wind and current system of the South Atlantic, which ultimately girdled the globe. Once European sailors had broken the navigation codes of the storm-tossed Atlantic, all the world's seas suddenly became penetrable by their sturdy ships. With it came monopoly access to the wealth of the New World and an alternative, cheaper, and faster all-water route to India and the Spice Islands. The Voyages of Discovery crowned Europe's transformation into history's first, world-straddling maritime civilization.

After Henry's death, the commercial allure of Africa was sufficiently tangible for the Portuguese king to be able to lease monopoly rights on the Guinea trade in gold, ivory, slaves, and pepper to a wealthy Lisbon citizen, Fernao Gomes, in exchange for a promise of further exploration and a state share in his profits. Within five years, Gomes's profit-seeking sailors had explored a length of the African coast equal to the distance covered by Henry the Navigator in thirty years. By 1481 the economic rewards of exploration were so great, and the risk of failing so reduced, that the king granted the trading and exploration franchise to his own son, who himself soon became King John II and vigorously carried on Prince Henry's legacy. In retrospect Henry the Navigator's research institute effectively proved to be as much a precocious landmark in Europe's evolving political economic marriage between private markets and governments as it was a scientific prototype. The state, in the person of Prince Henry, effectively underwrote the front-end cost of the speculative, basic research until commercially profitable returns became foreseeable enough to attract private risk capital for further targeted development. Once actual profits materialized, entrepreneurs and governments equitably apportioned the new wealth between themselves through politically negotiated tax rates, lease fees, and other revenue-sharing arrangements. This model was very similar to the pattern of government-funded research in the West to the present day.

Europe went to the cusp of its maritime breakthrough in February 1488 when two caravels of Portuguese captain Bartholomew Diaz rounded Africa's southernmost Cape of Good Hope. If not for a rebellion among his crew after a terrible storm, Diaz would have continued on as the first European to sail into the Indian Ocean. His reluctant return to Lisbon harbor in December 1488 instead shaped a dramatic twist in the course of European history. By riveting King John II's energies on the singular ambition of a follow-up trip that would yield for Portugal the grand prize of the all-water route to India, Diaz's voyage promulgated the king's final rejection of a proposal by Christopher Columbus, who had been entreating the sovereign and his experts since 1484, for funds to sail westward across the Atlantic, where he believed India lay at a distance no more than the length of the Mediterranean Sea. King John II's experts were far more correct than Columbus in reckoning the actual, much-farther distance to India. Nevertheless, Columbus's blind faith and perseverance withstood further rejections by England and France until 1492 when King Ferdinand and Queen Isabella of Spain agreed at the last moment to outfit his voyage into the western unknown in celebration of their decisive triumph at Granada over the last Islamic stronghold on the Iberian Peninsula. His three ships departed on August 3, 1492—on the very same tide that carried away many emigrating Jews on the deadline date of their expulsion by the Spanish Inquisition—and sighted the islands of the West Indies two and half months later on October 12. Columbus returned to the New World a year later with a 17-ship fleet and 1,500 people to establish the first of many permanent Spanish settlements.

The far-reaching impacts of Spain's sweep through the New World are well recorded by history. Armed with muskets, and unknowingly with far-deadlier European diseases like smallpox and measles, Spanish conquistadores decimated the native Amerindian populations they encountered, reducing their number from about 25 million to only a few million within a century. The disease-enfeebled Aztec empire in Central America fell to them between 1519 and 1522; the South American Inca gave way between 1531 and 1535. By 1513 the Pacific was reached across land, some six years before Magellan's ships set sail from Spain on the first round-the-world sea voyage. Soon Spanish galleons were sailing the Pacific Ocean, and serving colonies that stretched from the Rio Grande on the modern Mexican-U.S. border to the River Plate dividing modern Uruguay and Argentina. While Spanish vessels discovered vital New World foodstuffs such as potatoes, corn, and squash that provided a huge boon to European population growth and health over the long run, the Spaniards' overwhelming obsession was gold and silver, which began to be exported home to the Old World in vast quantities during the 1530s. At Columbus's parting, King Ferdinand purportedly exhorted, "Get gold, humanely if possible, but at all hazards—get gold." High up in the Andes, at over 13,000 feet, the Spanish discovered a veritable silver mountain at Potosí, which filled its treasury and tempted its ambitions for many decades. Water-powered mills to crush the silver ores were introduced in the 1570s, fed by an expanding network of storage dams and feeder canals to turn the waterwheels. In 1626 one of the dams collapsed, doing so much damage that the then-declining mining operation never recovered to full capacity and striking a powerful blow against the Spanish economy.

New World bullion transformed Spain into a rich and powerful state and helped launch its Habsburg monarchs, Charles V and his son

Philip II, on their overweening quest to unify Europe as a Catholic region under their political aegis; this in turn helped stir a long period of religious and political wars and conflicts critical to the forging of modern Western society. The influx of so much bullion into the European monetary economy also fueled a great continental inflation in which prices rose three to four times by the end of the sixteenth century. The ironic, unintended net result of this inflation was a stealthy redistribution of wealth that hastened the rise of northern Europe with its bourgeois tradesmen, sea merchants, and private capitalists who could respond fastest to rising prices and unsettled the static economic and class relationships underpinning traditional, land-based aristocratic societies, including Spain itself.

To prevent Columbus's discovery of the New World from triggering a land grab war between two loyal Catholic states, the pope drew a demarcation line from the North to the South Pole and granted all new lands to the west to Spain and those to the east to Portugal. However the pope, the notoriously wanton, Spanish-born Borgia pope Alexander VI, drew the line with such a heavy bias in Spain's favor that it did not even leave Portugal sailing room to continue its African voyages. But Portugal's clearly superior naval power facilitated a swift diplomatic settlement between the sovereigns and the dividing line was relocated some 865 miles farther west through a new 1494 Treaty of Tordesillas.

The line dividing the world between Portugal and Spain cleared the way for Portugal to carry out its planned rounding of the African cape to exploit the first all-water sea trade route to India. For the task Portugal's king chose Vasco da Gama, then thirty-seven. The son of a minor official, da Gama was well qualified for a nautical and political task that was far more challenging than Columbus's. He was a skillful and disciplined sea captain, and audacious as he was ruthless and diplomatic. His voyage took him across many unfamiliar seascapes and presented complex leadership challenges onshore and offshore, including managing a crew that was out of sight of land for 4,500 miles and ninety-six days, nearly three times more than Columbus.

Da Gama's four ships departed Lisbon harbor on July 8, 1497, with stores for a three-year voyage. He was accompanied by Diaz as far as the Cape Verde Islands. Then, in order to avoid the treacherous Gulf of Guinea, he began his famous southwest detour almost as far as Brazil. This wide arc enabled him to traverse the southeast Atlantic trades and catch the strong far south westerlies that carried him back east toward the African cape. He ultimately rounded it on November 22. Vast distances of wild coastlines and unsailed seas followed. Finally, in March 1498, after an arduous voyage and a month of delay for ship repairs and rest, da Gama sailed through the treacherous channel between Mozambique and Madagascar and thereby shattered the insuperable barrier that had thwarted the advance of Muslim dhows down the African coast. Entering into the civilized sphere of the Indian Ocean, da Gama's fleet docked at the thriving, Muslim port on the island of Mozambique. The gold, jewels, spices, and silver of the Muslim merchants heartened him, as did news, which eventually proved spurious, of lost Christian kingdoms inland and up the coast. Proceeding northward, he finally dropped anchor at Malindi, one of several important ports along the Zanzibar coast near modern Kenya and Tanzania, where earlier in the century Chinese Admiral Cheng Ho had secured a giraffe for his emperor's amusement. With good fortune, da Gama secured at Malindi an expert Arab pilot—some historians believe it may have been Ahmad Ibn Madji, the most renowned Arab navigator of the era—to guide his fleet in twenty-three days across the tricky Arabian Sea. On May 20, he reached his intended destination of Calicut on India's Malabar coast. The next three months were spent in difficult diplomacy with the local Hindu ruler, to whom he explained his mission as seeking "Christians and spices." Da Gama failed to conclude a treaty with him, however, due to hostility from Calicut's established Muslim merchants and the unimpressive gifts he could offer as a foretaste of future trade benefits with Portugal.

Unfavorable winds cursed da Gama's homeward journey across the Arabian Sea. So many on board died of scurvy on the three-month voyage that he was compelled to burn one of his ships for want of a crew to sail it. Nevertheless, in summer 1499 da Gama reached Portugal in triumph. Although less than one-third of his original 170-man crew

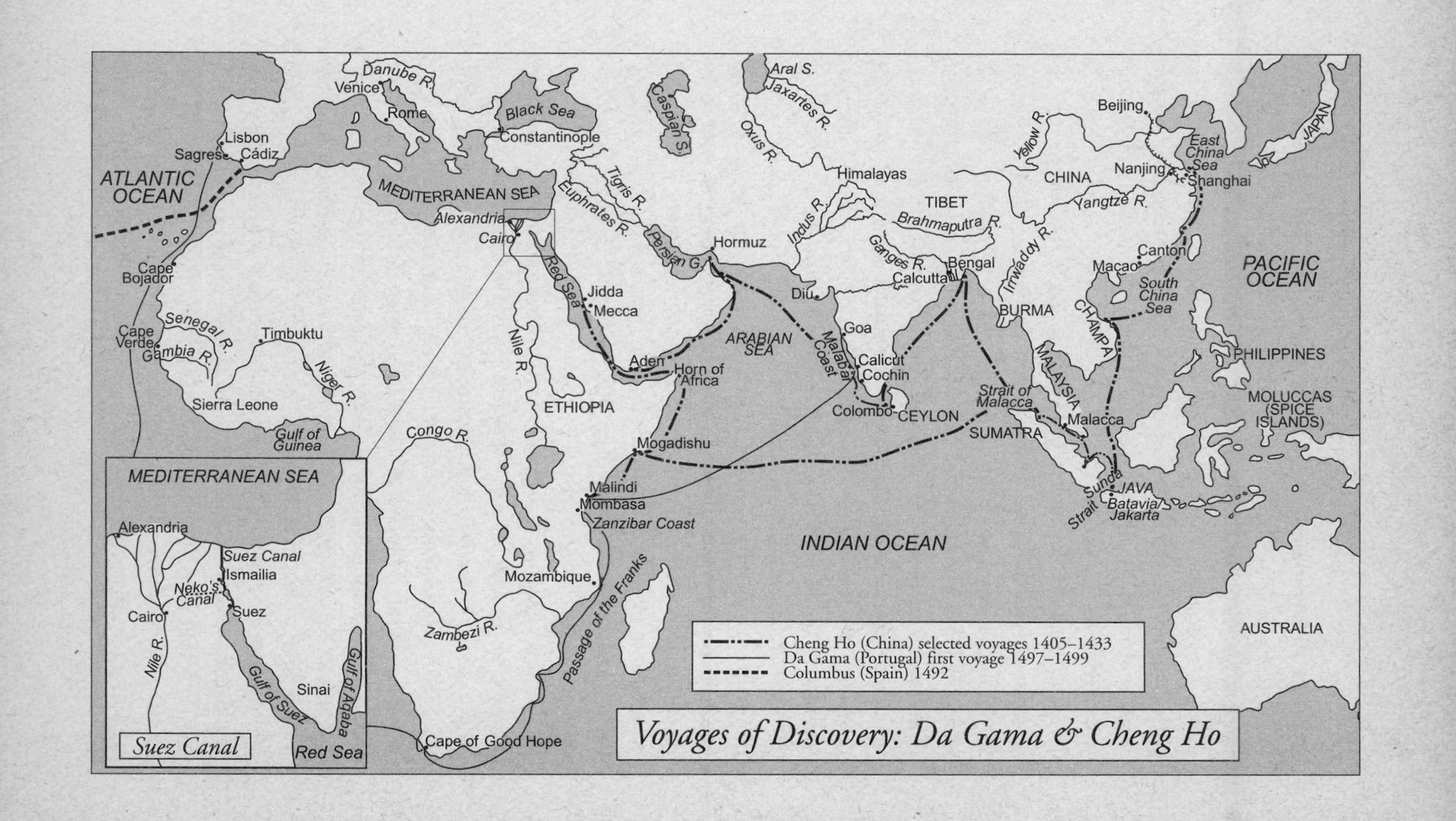

Voyages of Discovery: Da Gama & Cheng Ho
Cheng Ho (China) selected voyages 1405–1433
Da Gama (Portugal) first voyage 1497–1499
Columbus (Spain) 1492
ATLANTIC OCEAN
Lisbon
Sagres
Cádiz
Venice
Danube R.
Rome
Black Sea
Constantinople
MEDITERRANEAN SEA
Alexandria
Cairo
Caspian S.
Aral S.
Jaxartes R.
Oxus R.
Tigris R.
Euphrates R.
Persian G.
Hormuz
Himalayas
Indus R.
TIBET
Brahmaputra R.
Ganges R.
Calcutta
Bengal
Irrawaddy R.
Yellow R.
CHINA
Yangtze R.
Beijing
Nanjing
Shanghai
East China Sea
JAPAN
Canton
Macao
South China Sea
PACIFIC OCEAN
CHAMPA
BURMA
MALAYSIA
Malacca
Strait of Malacca
SUMATRA
Sunda Strait
JAVA
Batavia/ Jakarta
PHILIPPINES
MOLUCCAS (SPICE ISLANDS)
AUSTRALIA
INDIAN OCEAN
Diu
Goa
Malabar Coast
Calicut
Cochin
Colombo
CEYLON
ARABIAN SEA
Red Sea
Jidda
Mecca
Aden
Horn of Africa
Mogadishu
Malindi
Mombasa
Zanzibar Coast
Mozambique
Passage of the Franks
ETHIOPIA
Nile R.
Congo R.
Zambezi R.
Cape of Good Hope
Cape Bojador
Cape Verde
Senegal R.
Gambia R.
Timbuktu
Niger R.
Sierra Leone
Gulf of Guinea
Suez Canal
MEDITERRANEAN SEA
Alexandria
Suez Canal
Ismailia
Neko's Canal
Cairo
Suez
Nile R.
Gulf of Suez
Sinai
Gulf of Aqaba
Red Sea

returned alive, the peppers and other cargo paid for the cost of his voyage sixtyfold. Portugal's lust for the riches of the Indies was excited by the discovery that although it had little to offer in desirable traded goods it possessed one irresistible advantage that its would-be trading partners simply could not refuse—vastly superior long-range sea cannonry and a new Atlantic style of naval warfare of small crews fighting from a distance.

Long-range sea artillery stands out among a handful of military innovations that has profoundly altered the course of world history. On land, the Gunpowder Revolution altered long-standing power balances, including by breaking down the defenses of walled fortresses with large cannons, as the Ottoman Turks dramatically demonstrated in taking Constantinople in 1453. Its effects were even more far-reaching when it was applied to sea combat, which since antiquity had been based on ramming and boarding for hand-to-hand combat. An evolutionary step toward missile-launched sea warfare had occurred in the thirteenth-century Mediterranean with the intensive use of crossbows to prevent enemies from approaching and boarding. But it was in the Atlantic that sea cannonry was most precocious. The English possessed some sea artillery by the late fourteenth century, while Venetian galleys in the Mediterranean didn't carry them until the early to mid-fifteenth century.

The big difficulty was handling the cannon's tremendous recoil upon firing. Serendipitously, the sturdy caravel and its related family of Atlantic sailing vessels had bestowed one last gift upon European civilization. Its superior balance proved highly adaptable to absorbing the recoil across the deck of the heavy, long-range, mid-fifteenth-century French and Burgundian cannons. By the dawn of the history-making Voyages of Discovery, heavy long-barreled guns that could bombard with accuracy of up to 200 yards—sufficient to prevent enemies from approaching near enough to carry out traditional ramming and boarding attacks—were commonly carried aboard Portugal's seagoing vessels. "There is no doubt that the development of the long-range armed sailing ship heralded a fundamental advance in Europe's place in the world," writes historian Paul Kennedy. "With these vessels, the naval powers of the West were in a position to control the oceanic trade routes and to overawe all societies vulnerable to the working of sea power. Even the first great clashes between the Portuguese and their Muslim foes in the Indian Ocean made this clear . . . [T]he Portuguese crews were virtually invincible at sea."

The Portuguese wasted no time in pressing their naval military advantage. Armadas were dispatched almost annually to the Indian Ocean to seize freely by brute force what they had been unable to win by trade. Da Gama himself led the second armada, totaling 20 ships, which departed two and half years after his initial voyage. He expressed his cold-blooded intentions unhesitatingly upon returning to India's Malabar coast. Seizing a dhow carrying Muslim pilgrims on their way home from Mecca, he pirated the treasures on board, then burned the ship with its several hundred passengers, women and children included, locked up inside. Proceeding to Calicut, he rejected the local leader's friendship offer and instead demanded his immediate surrender as well as the banishment of every Muslim from the town. To demonstrate his seriousness, he bombarded the harbor. When two score fishermen and traders sailed out to sell him their wares, he had them immediately hung, dismembered, and sent their body parts back to the ruler with a note inviting him to make a curry of them. Upon filling his cargo holds with treasures he sailed home, but not before deploying ships to stay behind as Europe's first permanent naval force in Indian waters.

The sphere of Portuguese power continued to expand rapidly with the sailing of each new armada. To confront its growing menace to Islamic trade, the rival Egyptian Mamluks and Ottoman Turks united to send a large fleet of dhows out of the Red Sea. The decisive battle between Islam and the West for control of the Arabian Sea was fought off the Indian port of Diu near the mouth of the wide Gulf of Cambay in 1509 between heavy cannon-fitted Portuguese warships, manned by small crews, and a much larger, oared Muslim fleet. It was little contest. Portuguese broadsides decimated the enemy dhows before they could penetrate close enough with their weak artillery to execute their antiquated naval tactics of ram and board. After Diu, Portuguese hegemony over the Indian Ocean was asserted expeditiously. Goa fell in

1510. Malacca, controlling the narrow straight between Malaysia and Sumatra and access to the Spice Islands or Moluccas of Indonesia, was taken in 1511. By 1515, Hormuz, at the head of the Persian Gulf, was permanently occupied by Portugal, and Ceylon (modern Sri Lanka) was captured. The Portuguese failed only to take Aden at the mouth of the Red Sea, which was the route for supplying Alexandria, where goods were reloaded on Venetian vessels for distribution to the markets of the Mediterranean. In 1516, a Portuguese ship sailed up China's Pearl River and docked at Canton. By the mid-sixteenth century, Portugal had a chain of forts that extended from the Gulf of Guinea, around the cape and up the East African coast, across the rim of the Indian Ocean to Malacca and to the mouth of China's Pearl River at Macao. It was a stunning achievement for a nation of only 1 million people—a primacy it owed to its pioneering role in unleashing the latent power of oceanic sailing and sea power upon the world.

The effects of the sudden rise of Portuguese sea power reverberated everywhere. Power balances were upended. Trade was rerouted. The Venice–Alexandria trade monopoly with the East was shattered; within four years of da Gama's historic voyage the price of pepper in Lisbon was only 20 percent of its price in Venice. Venice's overtures to Egypt, starting as early as 1502, to reopen Pharaoh Neko's old "Suez" canal to shorten transport time and costs likewise came to naught. In 1521 Portugal felt sure enough of its position to refuse Venice's desperate offer to buy its entire stock of spice import. Venetian power never recovered. Islam's decline, too, was hastened by the loss of its monopoly over the rich Indian Ocean trade and competition from the far cheaper, faster, and safer all-water route to the Indies. Islam's overland West African trade was likewise outflanked by Portuguese ships, each of which could carry as many goods as an entire plodding, 5,000 to 6,000 camel train. Within Islam, the Mamluk Empire in Egypt and Syria, which depended mostly on the wealth of the Indian Ocean trade, soon was conquered by the Ottoman Turks. The Turks, in turn, exerted new military pressure on Europe from the east, by land and in the Mediterranean. The Turks' threatening Mediterranean advances throughout the sixteenth century forced Venice and Spain to devote great naval effort and expense to repel them. As a result, the central locus of intra-European power tilted even more decisively throughout the sixteenth century away from the Mediterranean and in favor of the insulated, northwest Atlantic sea powers.

One other noteworthy water innovation played a complementary role in maritime Europe's speedy conquest of Earth's open seas. Keeping drinking water fresh aboard ships was one of the banal, yet most frightening challenges of long-distance sea sailing. Despite countless jealously guarded formulas, there was simply no way to keep water fresh aboard ship for a long time. Explorers' first order of business upon landing at any unknown shore was finding a freshwater source. Even putting in at civilized ports didn't always guarantee freshwater in an age when drinking discolored, briny, germy, and polluted water was so much the norm that many restricted their water imbibing to alcohol-disinfected beer or wine, or to boiled hot drinks. The situation for seamen improved somewhat in the fifteenth century when Europeans developed an improved cask to keep water fresh for longer periods. Such casks enabled da Gama to make his long sea voyages to India and barely sufficed on Magellan's landmark first global circumnavigation from 1519 to 1522. As Magellan's crew wandered lost for thirty-eight harrowing days through the 334 labyrinthine miles of false bays, snowy fjords and narrow passages of the thereafter-named Strait of Magellan linking the Atlantic and Pacific Oceans, and then needed more than a hundred days to traverse the vast Pacific, which was much larger than expected, a despairing onboard diarist recorded on November 28, 1520, that the water they drank was yellow and putrid.

The wealth earned by Portugal and Spain as first movers of the Age of Discovery whet the desire of the rest of maritime Europe for a share of the prize to be had from the high seas. They did not acquiesce to the pope's assignment of the globe to Iberian, Catholic primacy. Over the ensuing three centuries of the oceanic age of sail, the intra-European struggle for supremacy was a primary force in defining the political, economic, and religious character of Western civilization, and the interlinked, colonial world-system it helped create.

One early effect of the large inflow of New World bullion to Spain was to spur its regal Habsburg rulers, Charles V and his son Philip II, to try to extend their family's mastery over many European states into a consolidated, autocratic Catholic empire by marriage or force or arms. They were harassed in this ambition by the lesser powers of England and France, whose rulers commissioned entrepreneurial privateers—state-sanctioned pirates—to plunder the gold and silver Spanish treasure ships sailing out of ports in the western Caribbean on the renowned Spanish Main. From 1566 they were joined by able seafaring privateers from the Netherlands, where a Protestant revolt for religious and political freedom against Spanish overlordship had been answered with brutal reprisals by Philip II's troops. England's Queen Elizabeth covertly and overtly supported the Dutch rebels against their common Habsburg Catholic adversary with financial aid, safe haven for Dutch privateers and on one occasion by interdicting the pay intended for Spanish troops in the Netherlands when the ships carrying it were forced by weather to dock at English ports. By 1576, the combined effect of piratical disruptions of bullion shipments, the cost of Spain's contemporaneous struggle against the Muslim Turks in the Mediterranean, and Philip's own monarchal overreach forced Spain to default on its international bank loans and to suspend payments to its troops fighting in the Netherlands. Spanish troops mutinied by sacking Antwerp, then the richest city in the Spanish-controlled Netherlands.

Private capital fled to nearby Amsterdam in the also-rebellious free northern provinces, galvanizing its rise and long reign as Europe's leading center of finance, trade and market capitalism. In 1579, the seven northern provinces united against Spain and soon formed the tiny, commerce-centric Dutch Republic. While the southern provinces of the Spanish Netherlands eventually succumbed to Spanish troops, the north successfully resisted Spain's superior army by opening dikes to defensively flood landscapes that lay below sea level and taking their battle to the northern sea-lanes, where Spain's land power advantage was neutralized by the natural exigencies of seafaring. By the 1580s the Dutch Protestant rebellion had escalated into a full-blown international struggle with its own momentum. It set up an inexorable military collision between Spain and England, whose denouement came in the celebrated summer 1588 sea battle against the Spanish Armada.

The struggle between mighty, prosperous Spain and the small, relatively poor English island-nation proved to be one of history's outstanding examples of the equalizing effect of sea power on otherwise militarily unmatched enemies. England relied exclusively on its naval prowess for its defense. Philip II, on the other hand, planned to bring to bear Spain's formidable panoply of oceangoing galleons and other ships to hold the English Channel in support of a land invasion led by 30,000 of his troops who were to be ferried across the Strait of Dover from Dunkirk near the Spanish Netherlands. The revolution in naval warfare in which battleships were employed as mobile batteries fighting from long distance with on-board cannonry was only partly complete by the time the Armada sailed. England's Royal Navy, since the time of Elizabeth's father, Henry VIII, had been in the vanguard of that revolution. Its cannons fired long-range, light 17-pound rounds through side portholes. Its naval captains commanded all on board without regard to social rank. Its sleeker and faster ships were among the most maneuverable on the seas. Spain, by contrast, lagged behind in the naval power revolution. Its large fleet carried heavy cannons that fired 50 pound shots but over a shorter range. It was far less maneuverable sailing windward, maintained an aristocratic chain of command, many swordsmen and musketeers for close-in and traditional on-board fighting.

Underlying these tactical differences, notes British historian George Macaulay Trevelyan, was a more profound "difference of social character between Spain and the new England. Private enterprise, individual initiative and a good-humoured equality of classes were on the increase in the defeudalized England of the Renaissance and Reformation, and were strongest among the commercial and maritime population." Enriched by its New World bullion, Spain remained fixedly wedded to its medieval class hierarchy, centralized political authority, army-centered military power, and a state-directed command economy anchored in traditional agriculture. The battle of the Armada between England and Spain, in short, contained within it a contest between two competing political, economic, and social tendencies for Western civilization's future.

To lead England's defense, Elizabeth turned to its resourceful privateers of the Spanish Main. Foremost among these was Francis Drake. Although second in formal title, he was first in shaping and executing England's strategic battle plans at sea. Drake in many ways personified the spirit of the rising English nation. Born of a Protestant tenant farmer in the early 1540s, his family fled a Catholic uprising and lived for a while in the hull of an old ship moored along the Thames. At 13 he was apprenticed to the captain of a small trading vessel that traveled among North Sea ports. Seeking fortune and adventure, at 23 he sailed for the West Indies. Although an early trip on which he had been second-in-command ended in financial ruin when his ship was attacked by Spanish vessels, Drake's skills came to the attention of Queen Elizabeth, who granted him his own privateering commission. Thus was launched the career of England's greatest privateer, explorer, and naval innovator.

Drake's plundering expeditions along the Spanish Main throughout the 1570s brought him such fortune and fame that in 1577 he was chosen by the queen to command the greatest raiding expedition in history—a stealthy, surprise marauding of Spanish ships and settlements in the Pacific Ocean. Drake's three-year journey in his 75-foot flagship, the *Golden Hinde,* developed into history's first round-the-world voyage completed by a single captain. Upon navigating through the treacherous Strait of Magellan—astoundingly, in only sixteen days—Drake turned northward along South America's Pacific Coast. So easy was the plundering of more than 10 tons of gold, silver, pearls, and precious stones from undefended Spanish outposts that before long Drake's booty-stuffed vessel was sailing below her watermark. Continuing northward toward Canada, Drake unsuccessfully searched for the northwest passage to the Atlantic, and for a time anchored in the San Francisco Bay area, which he claimed for England. Notwithstanding almost losing everything at one point when his ship ran aground on reefs, he returned home in 1580 by way of the Pacific, bringing with him a treaty for England to trade for spices in the Moluccas. Ignoring vigorous Spanish protests about his piratical exploits, Queen Elizabeth personally knighted him aboard the *Golden Hinde* after his return.

For several years Drake settled down as the mayor of the port of Plymouth in southwestern England—and organized an early freshwater supply system, featuring a 17-mile-long aqueduct known as Drake's Leat that lasted for three centuries. In the mid-1580s he returned to raiding Spanish interests in the New World, so successfully that it impinged Spain's international ability to borrow. When it became known by 1586 that Philip II, with the blessing of Pope Sixtus V—one of Rome's Renaissance Water Popes—finally had resolved to invade England directly, Drake, in command of some 30 vessels, led a spectacular surprise raid on Spanish shipping at ports at Cádiz, Lisbon, and Cape St. Vincent that did so much damage that it set back Philip's plans by an entire year. Yet by mid-1588, the enormous Spanish Armada of 130 ships with 8,000 sailors and some 22,000 soldiers was rebuilt, outfitted, and bound for the English Channel to spearhead the invasion of England.

The defeat of the Spanish Armada has become enshrined as one of England's hallowed national myths. In fact, the outcome more closely resembled a comical farce decided by the haphazard forces of nature than by the heroic deeds of man or, as claimed by the victors, divine intervention. Although led by its 25 formidable warships, the Spanish fleet also included many slow-sailing, converted Baltic merchantmen that had difficulty traveling upwind. As a result, the nimbler English fleet of roughly the same size had little difficulty winning through the first rule of fighting under sail: to always get and stay windward of the enemy. Yet even with this advantage, the light English cannon could not sink a single Spanish ship. For their part, the heavier Spanish cannonballs turned out to be so poorly fabricated in Spain's notoriously inferior iron foundries that they split apart on the few occasions they got close enough to strike their English targets. As a result, most of the battle was engaged with the two fleets firing futilely at one another as they floated with the tide and wind down the Channel toward the narrow Strait of Dover.

When the Spanish fleet anchored hoping to support the crossing of the army invasion force that had yet to show up, the English unleashed eight burning fire ships down the current toward the wooden Armada—one of the oldest tactics in sea warfare. To avoid catching flame, the Armada hastily cut anchor and fled in panic. Chased by an attacking fleet of English warships which itself ran badly low on ammunition, it was driven out of the Channel into the North Sea, past the troop rendezvous points and unable to sail back to it. With the wind against it, the Armada had no choice but to try to return to Spain by sailing north around Scotland and Ireland into the Atlantic. But in the rough weather of the September Atlantic, many of the ships, their crews weakened by low rations and disease, began to break up and were driven into the rocky coastlines. In the end, only half the Armada returned safely to Spain, its mission unaccomplished.

Despite its comical aspects, the battle of the Spanish Armada deserved its place among the great sea engagements that dramatically altered history. It saved England's independence, as well as saving the Dutch Republic from extinction, while ensuring the survival of northern Europe's Protestant Reformation and its seafaring, market economy and fledgling, liberal democratic states. At the same time, it presaged the decline of Spain's bid for hegemony, the shift of European power to North Atlantic states using the most advanced, mobile, long-distance artillery-based naval warfare, and the early transformation of the small English nation of 5 million, only half that of Spain at the time, into what would become the nineteenth-century colonial British Empire of 24 million. As for Francis Drake, the celebrated captain was sent by Queen Elizabeth to Panama in 1595 to seize two of its settlements for ransom. He succeeded in taking Nombre de Dios, but himself contracted dysentery and died. He was buried at sea.

With the decline of Spain and the Habsburg's bid for European dominance, the new fulcrum of European and world power in the age of oceanic sail shifted to two small maritime trading nations, the Dutch Republic and England. These states gave wider berth to private enterprise, market economy, religious and political liberty, and representative government than other European nations. During their primacy in the next two centuries, they became progenitors of modern capitalism and liberal democracy. Through their colonies and influence as global powers, their peculiar mode of political economy was exported far and wide to societies throughout the world.

The Dutch were the most precocious. A republic ruled by the merchant classes that prospered as sea trading middlemen, the United Provinces were lineal descendants of the traditions of ancient Athens and medieval Venice. Its leading seaport, Amsterdam, in the province of Holland, bore a striking resemblance to Venice. It arose in the thirteenth century from its inhospitable, muddy, often flooding lowlands—over a quarter of its territory lay below sea level—on the arduous labor and ingenuity of land reclamation involving extensive drainage, pumping, dredging, embankment building, damming, diking, and sluicing. Following the historical pattern of association between advanced water engineering and ascendant societies, Dutch land hydraulics set world-class standards during its century-long golden age and has continued to be a leader in modern times. Even more notable, the highly decentralized, democratic nature of the Dutch Republic was a direct outgrowth of the local water boards established in the thirteenth century to manage the water infrastructure sustaining the reclaimed lowlands or "polders." The success and cooperation of the local water boards—which continue to function to the present day—became an essential part of the model of governance adopted by the seven northern provinces that seceded from Spain to form the Dutch Republic in 1581.

Similar to Venice, Amsterdam's urban landscape featured its canals, which were designed in semicircular, concentric rings. At its heart, as a Dutch counterpart to Venice's Rialto Bridge, was a dam on the Amster River controlling the flows to a huge inlet of the North Sea. From the 1500s to the late 1600s, Dam Square was the leading entrepôt of the world market economy. The Republic self-interestedly championed free trade, freedom of the seas, and secure private property rights. Staple goods from the Baltic, the North Sea, the Atlantic coast, Mediterranean, and as far away as the Spice Islands were unloaded and stored in warehouses along the quays near the dam, then bought and sold by merchant representatives meeting in the nearby *bourse*. Merchant banks facilitated the trade through issuance of credit, discounting bills of exchange, and other financial instruments of modern capitalism. An early stock market developed. Gold and silver flowed in from everywhere, imparting to the Dutch Republic the significant comparative advantages of cheaper and abundant financing enjoyed by states that hosted major world financial centers. From only about 30,000 inhabitants in 1575, Amsterdam's population multiplied sevenfold to 200,000 within a century.

As a tiny state with little arable land, again much like Venice, Dutch wealth depended heavily upon its unexcelled efficiency in shipping and adding values to other states' goods. It rose by garnering control of over half of the shipping leaving or entering the Baltic Sea, and dominating the carrying trade between northern and southern Europe. Prominent in this trade was the profitable cargo shipped from Indian Ocean ports into Lisbon by Portuguese vessels.

The seminal event in the sudden rise of the Dutch Republic as a world power dates to 1592 when, in the aftermath of the Armada's defeat and the continuing rebellion in Holland, Spain's Philip II in 1592 closed Lisbon harbor to Dutch shipping. Faced with the sudden loss of access to this vital intermediary trade center, Dutch merchants resolved to undertake their own voyages directly to the emporiums of the Indian Ocean. Some 50 ships made the months-long round-trip from the Netherlands over the next ten years, laden with rich cargoes of pepper, nutmeg, cloves, mace, tea, and coffee. Private and public interests invested together in 1602 in the limited liability, joint stock Dutch East India Company, which exerted its regional monopoly trading rights like a sovereign state and for a long period was the most celebrated emblem of Western capitalism. In remarkably short order, the Dutch seized control of the trade from Indonesia's Spice Islands and the ports of Ceylon. They exploited it more effectively than Portugal ever had. Their power hinged on their domination of two strategic sea passages to the Spice Islands—the Malacca Strait between Sumatra and Malaysia, and the Strait of Sunda between Java and Sumatra on the direct sea route from the African Cape of Good Hope. In 1619 a new colonial center was established on Java at Batavia, modern Jakarta. The Dutch achievement reconfirmed the Portuguese experience of the disproportionate global influence sea power could confer on tiny states in the age of sail.

Dutch colonists fanned out elsewhere over the globe in the same period. One notable fur-trading settlement was New Amsterdam at the mouth of the Hudson River in North America, established following the failure of East India Company–hired explorer Henry Hudson to find a northwest passage to India. In the twentieth century, under its British name, New York, it became a successor of Amsterdam and London as the world center of financial capitalism. Yet perhaps the greatest legacy of the Dutch Republic's golden era was its fulfillment of the unsurpassed wealth-creating capacity of market-organized economies, and the superior military forces it could underwrite in seafaring democracies competing for primacy against traditional authoritarian kingdoms like Spain.

Where the Dutch led, the English closely followed. Unable to beat the Dutch in the Spice Islands, England successfully supplanted the Portuguese in India and rapidly expanded its colonization of North America. On three occasions between 1652 and 1674, the Anglo-Dutch rivalry for commercial dominance flamed into inconclusive warfare; the second war, in 1665, was triggered by the British seizure of New Amsterdam. By the late seventeenth century, however, they put aside their hostilities to combat a larger, mutual threat—France's bid for mastery of Europe under Louis XIV.

Between 1662 and 1683, France had dedicated itself for one of the few times in its history to building a powerful navy to complement its formidable, huge army. France's military success in subduing Europe was such that in 1689, when England came to the aid of the beleaguered Dutch, the French navy held the clear advantage against their allied strength. Financial overstretch, however, delayed implementation of the war plan at the decisive moment that its rare advantage might have prevailed. By the time Louis XIV was able to assemble 24,000 French troops and supporting ships three years later for a planned invasion of England across the Channel in 1692, the combined Anglo-Dutch navies had recovered enough strength to assert their superiority at sea. At the battle of La Hougue, in Normandy, on June 2, 1692, the French invasion fleet was utterly destroyed. With France's navy in ruins and heavy financial burdens pressing upon the his monarchal state, Louis XIV abdicated the notion of rebuilding a strong state-run navy in favor of the cheaper, traditional outsourcing of naval harassment to French privateers. Meanwhile, England embarked upon a major naval expansion. As a consequence, by the early eighteenth century Britain's navy reigned supreme among world powers; by 1730 it was as large as the next three or four navies combined.

A pivotal component in England's breakout as the world's naval superpower was its ability to obtain abundant, cheap financing for warfare from private capital markets following the Glorious Revolution of 1688. The Glorious Revolution, which brought to the throne the Dutch William III and his English wife Mary, was firmly anchored in a new, unspoken governing compact committed to Dutch-style market economics and a liberal constitutional monarchy under Parliamentary control. Financial market institutions were expanded and private investment was widely encouraged as the principal motor of the economy. The confidence of private capital markets instilled by these reforms provided England with an enormous comparative financing advantage over rival centralized monarchies like France, where debt repayment depended solely upon the whim and will of the sovereign. After so many centuries, the long association between seafaring commerce and democratic political traditions had produced an economic mechanism dynamic enough to elevate the liberal democratic model to the forefront of civilization.

England's naval dominance in the age of long-distance sail enabled it both to defend its island homeland against invasion by rival land powers as well as to win the competition among European powers for control of overseas colonial empires. Sea power conferred a vital advantage in maintaining long-distance colonial supply lines and providing safe passage to British commercial shipping. In times of tension, superior naval power enabled English ships to easily bombard or establish beachheads along enemy coastlines. By the mid-seventeenth century,

England was capable of executing its policy of blockading enemy ports in all weather conditions, bottling up rival commercial shipping, military support, or naval counterattack.

In the worldwide Seven Years' War from 1756 to 1763, known in America as the French and Indian War, superior sea power proved decisive in routing France and establishing England as the world's peerless colonial power. The 1758 seizure of the French fort at Louisburg on Nova Scotia's Cape Breton Island gave England control of the St. Lawrence River. It opened the way for British soldiers and a few colonial American allies to overcome the great bluffs to take Quebec in September 1759—the decisive event in forcing the French to abandon Canada. France's colonial ambitions in the rest of North America were demolished by a parallel British troop invasion of the Ohio River Valley. This severed the chain of forts France had been building along the key river ways of the Mississippi River basin all the way down to New Orleans in its bid to confine English settlements east of the Appalachians.

France's counterstrategy of trying to win back its lost colonies through a bold, direct invasion of England likewise was checkmated by superior British sea power in two major sea battles in the summer and autumn of 1759. The first occurred near the Strait of Gibraltar. The second and most devastating was at Quiberon Bay, in southwest Brittany, on November 20, 1759, where the French fleet that had been bottled up by a six-month British blockade at Brest tried to flee when gale conditions forced a temporary English pullback.

France's worldwide retreat after the Seven Years' War included its withdrawal from most of India, which soon thereafter became the crown economic jewel of England's colonial empire. In early 1757 Britain's superior sea power and logistics enabled the English East India Company's Robert Clive to expeditiously seize back the Bengal port of Calcutta from rebellious local Indian rulers and their French allies. The Company consolidated British colonial rule in India following a celebrated, improbable victory of Clive's 3,000 British and Indian troops over 50,000 to 60,000 French-supported Indian forces at Plassey in June 1757. Water, in the form of a monsoon deluge, played a pivotal, if unusual, role in the battle's turning point. When the heavy rains soaked and rendered useless their gunpowder, the French-backed Indian troops charged en masse toward the British position amid the mango groves on the banks of the Hughli River in the belief the British gunpowder had been similarly ruined. But the badly outnumbered British had kept their powder dry under covering. The attacking native forces were cut down and scattered amid volleys of exploding English gunpowder.

Learning the lesson of the Seven Years' War that, in the age of long-distance sail, formidable sea power was a prerequisite of great empire, France invested heavily in the following decades to bring its fleet size to parity with England's. By 1781, France was strong enough to be able to inflict an indirect defeat on England by blocking the resupply of British forces at Yorktown, Virginia, thus forcing the surrender of General Cornwallis's army to George Washington to end the American colonists' War of Independence. The balance of sea power also proved decisive a generation later in the Napoleonic Wars. With France's armies sweeping victoriously through the European continent in 1797, Napoléon Bonaparte had argued that durable French hegemony over Europe depended upon winning command of the sea and subduing England. Indeed, the showdown between France and England proved to be not only the most severe challenge to England's global leadership in the age of sail, but an historic contest for military preeminence between the greatest naval power and an invincible land army headed by the most brilliant general since Alexander the Great.

In the summer 1798, the twenty-nine-year-old Napoléon cunningly exploited England's retreat from the Mediterranean to better defend its northern ports against relentless French army pressure. Under secret orders from the French leadership about his destination, he seized Malta and conquered Egypt with a force of 31,000 troops, 400 sea transports, and 13 warships—and a boatload of Enlightenment-age scholars from many disciplines whose extraordinary mission was to study everything possible about Egypt for the sake of the pure advancement of knowledge. With these conquests, Napoléon had moved into a position to take control of the entire Mediterranean. If he could solidify his hold, he knew he would be able to dictate the fate of the Levant, the Ottoman Turkish Empire and the Red Sea route to British India. Napoléon wasted no time in personally inspecting the ruins of Neko's ancient "Suez Canal" and ordered French surveyors to study a new canal that would directly link the Mediterranean and the Red Sea. His canal plan was aborted only after surveyors calculated—mistakenly—that the Red Sea was 33 feet higher than the Mediterranean and thus would necessitate canal locks and other complex engineering.

Recognizing its grave strategic vulnerability, England turned to one of its youngest flag officers to foil France's bid for control of the Mediterranean and to restore Britain's wavering command of the seas—Horatio Nelson, forty. To England's good fortune, Nelson would prove to be as brilliant a tactician and commander at sea as Napoléon was on land. Modestly born, with personal courage in battle that was visibly displayed by his loss of an arm and an eye, and possessed of a courtesy, charisma, and panache that earned extraordinary devotion from his crew as well as the ardor of his prominent paramour, Lady Hamilton, Nelson became the embodiment of British pride in its Royal Navy. Not since Sir Frances Drake had England had so celebrated a national naval hero.

Nelson entered the Mediterranean in hot pursuit of Napoléon's forces. In a frantic search across the sea, during which in the darkness he actually sailed past his foe and thus had to backtrack to Egypt after France had made its conquest, Nelson at last caught sight of the French battle fleet in the afternoon of August 1, 1798. It was anchored in a defensive line at shallow Aboukir (Abu Qir) Bay, near Alexandria and one of the mouths of the Nile. By chance, the French ships were at the moment undermanned because the commander had sent many crewmen onshore to dig wells to restock the ships' low water supplies. Sensing he had the advantage of surprise, Nelson immediately hoisted his signal flag of preparation for attack. The Battle of the Nile, or Aboukir as it was alternately known, was probably novel in the annals of sea warfare until modern times in being fought almost entirely in the dark. Although the number of warships on each side was nearly equivalent, the more-efficient British gunners could fire twice as rapidly and more accurately than their French counterparts. Nelson took advantage of the static French position to pick off a few ships at a time while staying out of range of the others. When the dawn broke, the scale of the French disaster became visible—11 of its 13 warships had been lost.

The effects of Nelson's victory were momentous. Britain soon regained the Mediterranean sea-lanes and, with that, renewed national self-confidence and fighting spirit. By controlling the sea supply lines, British naval power rendered untenable Napoléon's position in Egypt. The great general stealthily abandoned his troops in Egypt to return to France to seize power outright. The sight of Napoléon's great army in rare retreat emboldened beleaguered European continental powers to conspire in a new anti-French coalition with England. For Napoléon the Battle of the Nile forced abandonment of his grand ambitions to sever England from India and its colonial wealth. Instead, he refocused his master strategy on gathering his naval and armed forces for a direct invasion of England itself.

This set the scene for the great sea battle of the Napoleonic Wars, Trafalgar. Although Napoléon possessed an army three times larger than England's, to invade his enemy's island redoubt he needed sufficient sea power to control the English Channel for a few days to permit a safe crossing. Over the few years following the disaster at Aboukir Bay, Napoléon had acquired the Dutch fleet by occupying the Netherlands, which also put him in control of every Continental port from the North Sea to Gibraltar. England's response had been a comprehensive blockade of French-controlled ports, a years-long effort unique in naval history. In 1805 Napoléon finally ordered his war fleets to break out of the blockades and assemble in Martinique in the Caribbean to prepare for an invasion of the Channel, where French army troops were readying. But after years of captivity in port, the fighting ability of the qualitatively inferior French fleet had deteriorated. The fleet at Brest could not break out. The fleet at Toulon, where Nelson was on watch, managed to put to sea. With Nelson hunting the French fleet back and forth across the Atlantic, the two sides finally engaged off Trafalgar Point in Spain, between Gibraltar and Cádiz, on October 21, 1805. Over a series of dinners with his captains, Nelson had devised an innovative naval tactic that would take advantage of England's more maneuverable ships and experienced crews. Instead of lining up his ships parallel with the enemy to blast away with cannon fire, he divided them into two columns, with a third in reserve, to attack the enemy line at its center. Thereby, he created two separate battles with England and enjoyed the tactical advantage in each. The English triumph at Trafalgar was definitive. Not a single English ship was sunk, and only 100 men were lost. One of those casualties, however, was Nelson himself, shot in the shoulder and chest at close range by a sniper nestled atop a French mizzen sail.

The Battle of Trafalgar ended any chance for Napoléon to invade England. British mastery at sea allowed it to impede the French army's resupplies through its strangulating port blockades, while systematically dismembering France's overseas holdings in the Caribbean, Africa, and the Moluccas. At the same time, trade along the high seas with distant countries like the United States and Russia helped England withstand the European-wide continental commercial embargo Napoléon tried to impose against it. When circumstances finally became opportune following Napoléon's debacle in Russia, English vessels were free to use the sea to transport their own strengthened troops to the Continent to help in the ultimate defeat of Napoléon, which occurred at Waterloo in modern Belgium in 1815.

The results of the Napoleonic Wars confirmed that, in the age of long-distance sail and cannonry, sea power's advantage in mobilizing the natural forces of the high seas to level the balance of power against superior, land-based armies, had soared to such a degree as to permit small, democratic seafaring states to prevail in the struggle for global dominance. The English island-nation outlasted its continental rival the Dutch Republic in part because it could concentrate its resources on its naval power without incurring the additional burden of providing for a stout defense against land invasion—which, in an interesting parallel with the fall of the ancient Phoenicians to the neighboring land empires of Mesopotamia, was the final, proximate cause of the Dutchmen's demise. Throughout the age of sail, from the Armada to Napoléon, every attempted invasion of England had failed. Even as late as the Battle of Britain in World War II, the overwhelming army, tank, long-range missile, and air force might of Hitler's Nazi Germany was unable to overcome the defensive advantages provided by British naval power and the Channel moat.

Trafalgar was the last major battle fought by wooden sailing ships. Britain would not be seriously challenged again for another century—until World War I. In the nineteenth century, British naval power became invincible because it readily applied the innovations of another entirely unprecedented development that was gathering momentum along Britain's small, rural rivers—the steam-powered Industrial Revolution.

CHAPTER NINE

Steam Power, Industry, and the Age of the British Empire

The Industrial Revolution completed the transformation of English sea power and colonial wealth in the age of wood and sail into history's first globally dominant economy and hegemonic political empire. An ongoing process that amassed spontaneous, critical takeoff in Britain's private market economy during the sixty-year reign from 1760 of King George III, the Industrial Revolution wrought such thorough changes in every aspect of human society, from daily life, social organization, and demographics to political relations, that it compared in historical significance to the irrigated Agriculture Revolution at the start of civilization some 5,000 years earlier. At the catalytic fulcrum of the Industrial Revolution were innovative applications of waterpower. This featured not merely new uses for traditional waterwheels, but above all the breakthrough use of water in a previously unexploited form—steam.

Geographically, Britain had been richly endowed with both sea and inland water resources that could be propitiously exploited by the technology cluster of the age. Thus while its naval and merchant fleets were taking advantage of the island-nation's many excellent harbors, long indented coastlines, defensive sea moat, and the favorable direction of the currents and winds in the English Channel, early industrial entrepreneurs were starting to exploit the rural interior's many fast-running, perennial, small rivers and streams that were both easily navigable and capable of generating substantial power by waterwheels. Indeed, Britain in its nineteenth-century glory may have commanded the globe by its rule of the ocean waves, but the economic might upon which its empire rested flourished principally upon its inland waterways.

England's Industrial Revolution was born in two, overlapping phases in the small industries that developed alongside the small rivers of Britain's rural Midlands region. Centered in Lancashire, the first phase was driven by the emergence of the waterwheel, and then steam-powered, cotton textile factories in which traditional home handicraft work was reorganized into a standardized, mechanized manufacturing system of specialized functions performed at a central location. The second phase, which gathered momentum later and depended entirely on the energy outputs achievable only by steam, was centered in Shropshire on the production of cast iron from which the heavy industries of the late nineteenth century derived. By combining the power of steam and iron with its superior navy, British sea power became invincible and extended its dominance from coastlines up into the waterway interiors of foreign states. Britain's accelerating industrial productivity, economic wealth, and widening distribution of income to the middle classes became a self-propagating, dynamic phenomenon that completed the free market's historical migration from the seafaring trade periphery to the very center of world economic society.

As so often in history, necessity was the mother of the great innovations that sparked the Industrial Revolution. For nearly two centuries from the days of Shakespeare, Drake, and Queen Elizabeth to the eve of the American colonies' War of Independence, even while its navy was vanquishing its foes on the high seas, England suffered at home from an acute fuel famine caused by the early depletion of British forests. Thus while France and its continental rivals enjoyed ample wood resources, England struggled with shortages that steadily drove up the cost of wood charcoal and timber needed to heat English homes, fire its cannon-producing iron foundries, and construct ships for its navy. The fuel famine was exacerbated by the fact that Europe was still in the throes of the Little Ice Age (mid-fifteenth to mid-nineteenth centuries) when English temperatures were 1–2° centigrade colder than in the early twentieth century, cropland and woodlands were receding, and the Thames often froze over.

As a substitute for costly firewood, Britain intensively mined the plentiful seams of coal that lay near the surface of many parts of the Midlands and Northern England. Although coal could provide heat, only wood charcoal burned hot enough to reduce iron ore in blast furnaces to make iron. Thus when Abraham Darby, an ironmaster at Coalbrookedale on the Severn River, in 1709 independently reinvented a process long ago discovered in China for converting coal into coke that could be used to fire blast furnaces, it raised hopes that coal would also deliver England from the fuel bottleneck causing the nation's chronic iron shortage.

Yet two mundane obstacles continued to impede England's relief from the fuel famine. First was simply the difficulty of transporting huge volumes of coal from the mining districts. Packhorses and carts moved slowly, unreliably and expensively on the poor, muddy roads. Some coal near coastal cities moved by sea for use in the coal burning hearths of London and other seaports. But to meet the great demand from the nation's growing interior industrial regions, another transportation solution was needed. The second obstacle was that as miners dug deeper shafts to extract more coal, their excavations hit the water tables. To remove water they dug drains into the hillside and used lift pumps powered by horses or, if a suitable source of running water was available, by waterwheel. But the deeper they mined below the water table, the greater volumes of floodwater made it ever harder for them to excavate enough coal reserves to fulfill even the country's most basic, rising demand.

Thus Britain's fuel famine continued unabated. As late as 1760, the high cost and shortage of coal and timber was forcing the country to import half its iron from foundries in virgin forests of Sweden and Russia. One-third of its new shipbuilding was being outsourced to shipyards on the timber-rich eastern seaboard of its American colonies on the eve of the American Revolution. Unless relief came soon, Britain's incipient Industrial Revolution faced premature stultification.

The endurance of its still modest empire would be in doubt. The fuel famine was leaving "many domestic hearths cold," comments English historian Trevelyan and "if the old economic system had continued unchanged after 1760, it is doubtful whether the existing seven millions could have continued much longer to inhabit the island in the same degree of comfort as before."

England's Industrial Revolution was rescued by two catalytic water engineering breakthroughs. The first was an unforeseen inland transport canal building boom, spontaneously generated and financed entirely by England's burgeoning private sector, which swiftly created a national waterway network unique in the world outside China that greatly stimulated economic expansion. The pioneer of Britain's canal age was a youthful aristocrat, Francis Egerton, the Duke of Bridgewater, who financed and built the short, but influential Bridgewater Canal between 1759 and 1761. Bridgewater's inherited estates had included a large coal mine, from which he earned a contentedly handsome income. However, when Bridgewater lost his much-beloved wife to another man, he rechanneled his ardor to the personal dream of building a transport canal from his colliery to the growing industrial mills of Manchester. He calculated that such a canal would enable him to slash the price of his coal in half and thereby garner him a much-larger share of the local coal market.

Although Manchester was only 10 miles distant, the landscape posed complicated hydraulic challenges for the construction of a canal. The terrain was hilly. There were few local streams to provide enough water to float barges. But he had little doubt as to the canal's technical feasibility because as a young man on his continental Grand Tour, Bridgewater had visited France's Languedoc region and witnessed firsthand the most wondrous canal in European history, the Canal du Midi. Built between 1666 and 1681 with the endorsement of King Louis XIV, who was then trying to expand the French navy, the 150-mile-long Canal du Midi created a secure, domestic inland waterway shortcut that united France's Atlantic and Mediterranean coasts without any need to sail through Gibraltar and around Spain. The interior region it crossed soon throbbed with new economic activity. The canal was an engineering marvel, later hailed by Voltaire as a glorious achievement, that ascended and descended a summit elevation of 620 feet above sea level; accomplishing this feat required some 328 structures, including 103 locks, dams to furnish water for the channel, the bridges, and a 500-yard canal tunnel, Europe's first. Bridgewater thus knew that his more modest canal, although difficult, could be built. Moreover, he was ready to risk his entire fortune to accomplish it.

One main challenge of building any canal was to maintain slow water current so that barges could be towed along easily in either direction. One engineering approach was to follow nature's topographical contours, as China's founding Ch'in dynasty had done 2,000 years earlier with the Magic Canal. But that required an accommodatingly gentle landscape and a much longer canal. The ancient Assyrians and Romans had devised aqueducts to maintain steady, gravity-flow gradients across hilly landscapes. A more modern solution, widely used in hilly regions, was to artificially divide the canal into a series of stepped segments with an adjustment mechanism to raise or lower canal barges between the differing water levels of each segment. Through early medieval times, societies had often erected single, flat gates—flash locks—between canal segments. As a downstream canal barge approached, the lock gate was opened and the boat flashed through the resulting rapids. Travel upstream was much harder, requiring manual hauling of barges with ropes and windlasses by labor gangs of men and animals; often the barge was dragged up a slipway from the waterway below the level of the lock and returned to water at the higher level. China's monumental Grand Canal had originally relied wholly on simple flash locks and massive amounts of brute manpower to lift boats, dredge channels, and maintain embankments until the 984 innovation of the double-gated pound lock which exploited water's buoyancy properties to lift or lower barges as water filled or was released within the two gates. In Europe, navigation canals with locks had been built in the Low Countries in the fourteenth century. Leonardo da Vinci took a keen interest in Milan's extensive network of canals, including the 18-lock Berguardo Canal of 1485 that ascended 80 feet, and designed a version of the modern chamber lock with efficient, upstream-facing gates that closed in a V shape so that the water pressure from the downstream current helped seal the lock against wasteful leakage. Leonardo's design facilitated the building of a stairway of successive chamber locks capable of climbing and descending long, steep hillsides. By the seventeenth century, canal engineering leadership had passed to France, which was still confident enough of its hydraulic skills in the nineteenth century to take on the building of the Canal du Midi's historical heir, the Suez Canal.

Thus when the Duke of Bridgewater embarked on his canal, native English know-how still lagged far behind. As a chief engineer Bridgewater selected James Brindley, a self-taught expert who learned his skills by wit and experience garnered at various jobs since youth, including as a machinery maintenance man at regional water mills. Brindley and Bridgewater opted for a no-lock design with an aqueduct bridge to maintain the steady gradient. To supply the water needed to fill the canal, they also tapped the troublesome, high water tables that often flooded the coal mine and had to be drained anyway. Through his ultimate success on the Bridgewater Canal, Brindley would come to epitomize the new, self-made man of the era and eventually be renowned as Britain's leading canal engineer. Yet while the 400 laborers were toiling on the pioneering canal, faith in its success at times sank so low that it seemed as if Bridgewater would run through his personal fortune before the technical challenges were overcome. Many whispered that he had gone mad. At the lowest moment, banks in Manchester and Liverpool refused to honor the Duke's £500 check. In the end, however, the success of the Bridgewater Canal became one of the inspiring marvels of the age. People traveled long distances to see it, especially its three-arched, elevated segment that transported 12-ton coal barges 200 yards above the River Irwell. Out of public view was another of the canal's marvels—many miles of underground channels dug deep into the mine itself to expedite the loading of freshly excavated coal at source onto barges. In the end Bridgewater's coal was delivered to Manchester even more inexpensively than the Duke had originally calculated. Soaring coal sales made the Duke of Bridgewater one of England's richest men.

The Bridgewater Canal ignited a frenzy of English canal building. Within a mere few decades, the private-canal boom created an extensive, economical inland waterway network to move coal by barge among all the mining and industrial districts of the Midlands, North, Thames Valley, and the nation's seaports. Bridgewater and Brindley themselves went on to finance and construct a canal from Manchester to Liverpool and one from the Trent River to the Mersey, which cut a waterway across Britain's middle from the North Sea to the Irish Sea. In all, the canal frenzy added 3,000 miles of navigable inland waterways to Britain's existing 1,000-mile network.

That the canal boom was driven by private barge and toll operating companies that raised capital from London's growing financial markets helped excite the capitalist "animal spirits" across Britain's burgeoning laissez-faire economy, whose still unfamiliar, self-regulating "invisible hand" and wealth-creating mechanisms were just beginning to be expounded, most famously by Adam Smith in *The Wealth of Nations.* The canal boom whet the risk appetite of London's financiers to underwrite further industrial investments, which in turn stimulated greater capital accumulation and a new cycle of low-cost lending and investment. A half century later Britain's canal boom would be replicated—with remarkably similar, stimulating effects—across the Atlantic in the eastern United States with the completion in 1825 of the 363-mile-long Erie Canal.

The first rapid improvement in British transport since Roman times provided by the new canal network was a necessary, but not a sufficient condition for overcoming the country's fuel famine and launching its Industrial Revolution. Britain first needed an antidote for the mine flooding that constrained the volume of coal that could be excavated. Removing floodwater was a problem that had long vexed mine operators in the tin regions of Cornwall as well as the coal fields of the Midlands and the North. Among the many methods tried, the most effective was the application of waterwheel-powered pumps. Yet waterwheels required a dependable nearby source of swift running water, something lacking at most British mines. Thus another solution was needed. In the event, the invention of the steam engine, spurred by profit-seeking market forces, would deliver the seminal breakthrough that unlocked Britain's natural resource wealth and propel the creation of a radically new society based on mass industrial power and production.

Although water was the only common substance on Earth to exist in nature as a liquid, a gas, and a solid over a wide temperature range, civilizations heretofore had utilized it primarily in its liquid state. The expansive force of steam—water in its heated, gaseous form—had been understood in antiquity going back to Hero of Alexandria nearly two millennia earlier. Leonardo da Vinci had drawn sketches of a theoretical steam-powered bellows and cannon. Yet no one had actually tried to apply the scientific knowledge about steam to a practical technology. Only in the late seventeenth century, following new scientific discoveries about atmospheric pressure, did Denis Papin, a French-born physicist who had worked in London with physicist Robert Boyle, invent a practical steam pressure cooker and write conceptually about some of the basic designs upon which the first steam engines would be based. In 1698, a British military engineer, Thomas Savery, built the first functioning steam pumps to remove water from the Cornish tin mines, although they were highly unstable and tended to explode.

Credit for building history's first successful atmospheric steam engine belonged to Thomas Newcomen, a Dartmouth blacksmith who worked for a time forging Savery's pump. Installed at a coal mine in 1712, the Newcomen engine lifted about 10 gallons of water 153 feet at each stroke. The Newcomen steam engine, however, had many drawbacks that limited its usefulness. It was huge—requiring its own two-story-tall building to house it—and it generated little more horsepower than a good waterwheel. Moreover, it burned prodigious amounts of coal to heat the water into steam, and thus could be profitably used only at a coal mine. It was also thermally very inefficient. With the coal market still impeded by the pre-canal-boom transport bottleneck, Newcomen's engines diffused too slowly to overcome the coal shortage and Britain's fuel famine. By 1734, there were less than a hundred pumping water from English coal mines.

The historic breakthrough that would power the takeoff of the Industrial Revolution awaited the invention of a steam engine superior to Newcomen's. That steam engine was first patented in 1769 by a thirty-three-year-old Scotsman, James Watt, and began to be installed commercially in 1776. One of the first ways it was put to use was pumping water from a coal mine, where in its first hour it emptied a pit containing 57 feet of water. About the same time, another was put to work powering the bellows that forced air into the iron furnaces of prominent ironmaster John Wilkinson, foreshadowing the momentous impact it was soon to have on iron casting. The surpassing historical import of his steam engine made Watt one of the most celebrated luminaries of the remarkable eighteenth-century Scottish Enlightenment that included philosopher David Hume, political economist Adam Smith, geologist James Hutton, chemist James Black, and poet Robert Burns. Watt also became the object of an adulatory biography by no less than late nineteenth-century industrial-age U.S. steel baron (and Scottish emigrant) Andrew Carnegie, and was immortalized for all posterity by the naming of a common measure of electrical energy after him—the watt.

A mathematical instrument maker and land surveyor with a scientific mind and conversant with many scientists at the University of Glasgow, Watt was the son of a shipbuilder, architect, and maker of nautical devices. By his midteens Watt was making models of instruments, and soon went off to London as an apprentice instrument-maker. He returned to Glasgow and in 1757 opened his own instrument-making shop at the university. In 1763, he was asked to repair the university's lab-scale model of the Newcomen engine. Struck by the engine's wasteful loss of four-fifths of its heat, Watt began to investigate how to improve its efficiency. The heart of the Newcomen steam engine was a cylinder, which when filled with steam pushed out an attached piston. Cooling the cylinder condensed the steam into water, creating a vacuum. The piston was pushed back into the cylinder by atmospheric pressure. Anything attached to the piston, such as a pump, could thus move up and down to perform useful work. Through trial and error experimentation, Watt discovered that the main source of the Newcomen engine's inefficient waste of steam was the need to directly cool the heated cylinder between each stroke. His inspirational breakthrough came in 1765: using a separate condenser for the steam allowed the cylinder to remain continuously hot, more than doubling the efficiency of the engine. He soon had a design that quadrupled the Newcomen engine's output. Watt's design was also far more compact, and therefore maneuverable, than Newcomen's behemoth.

Watt's revolutionary modern steam engine accelerated mining of British coal and tin by removing floodwater, while the discharged mine water was put to productive use as a supplemental source of supply for the new transport network of canals. Indeed, Watt's decade-long delay from design to installing his first commercial steam engine was due, in part, to his being frequently called away to work as a land surveyor during the canal boom that had just begun. He had commercial and technical problems too. His first business partner, John Roebuck, who had commissioned him to build a steam engine to help pump water from his coal mine but never got much benefit from it because the iron workmanship available in Scotland at the time proved inadequate to the precision, quality machining required by its large cylinder and tightly fitting pistons, went bankrupt. Eventually, Roebuck sold his partnership interest in 1775 to Matthew Boulton, a well-to-do buckle and button maker from Birmingham to where Watt had relocated in search of more skillful iron making. Watt soon found the necessary quality in the shop of ironmaster John Wilkinson, who produced Watt's precision cylinder with the new boring machine tool he used to make cannons for the British navy. Armed with a twenty-five-year patent extension granted by an act of Parliament, Boulton and Watt began business in 1775.

The partnership of Boulton and Watt was one of the most remarkable in business history. Boulton's business acumen complemented Watt's technical ingenuity. Boulton's integrity allowed both to prosper, without the rapacious exploitation of the inventor by the entrepreneur so common in early business history. Indeed, Boulton went to great pains to shield his easily distressed and risk averse partner from the many financial stresses of the business, especially in its first dozen years. Both became active members of Birmingham's famous arts and sciences Lunar Society, and in 1785 were elected fellows in the Royal Society.

Boulton brilliantly promoted their steam engines with the flair of a natural-born pitchman: "I sell here, Sir, what all the world desires to have: Power." In a foreshadowing of the organizational science-industrial systemization of the innovation process, the scientist Watt devised new steam engine designs to meet market opportunities identified by Boulton. On June 21, 1781, Boulton exhorted Watt to devise what would become the second great development in the Watt engine—rotary motion. Urgently, he wrote to Watt: "The people in London, Manchester and Birmingham, are Steam Mill Mad . . . I dont mean to hurry you into any determination, but I think in the course of a Month or two we should determine to take out a patent for certain methods of producing Rotative Motions." The original engine's up and down motion was fine for pumping water out of mines or from rivers to deliver larger volumes of freshwater to growing cities. But Boulton saw a much larger market arising among the new waterwheel-driven cotton and other manufactories springing up around crowded waterwheel sites on England's small rivers. The convergence of Watt's 1782 rotary steam engine with the independently arising factory system of mechanized production hurtled England's Industrial Revolution into accelerated gear.

The factory system was the direct descendant of the water-powered, medieval Mechanical Revolution. History's first modern factory was the water frame cotton spinning machine mill opened in 1771 at Nottingham by Richard Arkwright. Its nine horsepower waterwheel drove 1,000 spindles and turned out cotton thread of superior quality, and in far greater volume and productivity than any traditional, home-based handicraft producer. It was an immediate success. Arkwright was not an inventor, but a guileful entrepreneur with a genius for organizing production and sales and in raising capital. As an enterprising Lancashire wigmaker and traveling barber in the textile districts where small batch work was still done from cottages, he had spotted his main chance with the water frame—and was not about to miss it. The textile industry had been undergoing rapid growth from the combined stimulus of design improvements in hand-powered spinning machines and the availability of cheap raw cotton from India. Arkwright's water frame was inspired by the model of an early eighteenth-century Darby silk-stocking factory powered by a 13-foot-diameter waterwheel and spinning machines whose design had been pirated from Italy. The water frame was the brainchild of an obscure inventor whom Arkwright promptly discarded and severed from any share of his ample rewards, which included becoming one of England's wealthiest men and the honor of knighthood.

Once established in business, Arkwright moved adroitly to dominate the booming early English cotton industry. Within a decade, his water-powered factories were running day and night with several thousand mechanically driven spindles and over 300 workers, most of whom were docile, low wage women and children performing specialized functions overseen by a few male supervisors. Despite aggressive sanctions against British technology exports, the early Arkwright water frame mill design was soon smuggled into the United States in the lucid memory of a fortune-seeking English textile employee where it became the technology backbone for launching New England's vibrant waterwheel-powered textile industry. The steam power threshold in textiles manufacturing was crossed in 1785 when Arkwright installed a Boulton and Watt engine to open the world's first steam-powered cotton mill. In 1790 Samuel Crompton's even more powerful, larger, hybrid-function spinning mule was adapted for steam power, and became the new standard for cotton spinning factories.

The transition from flowing waterwheel to steam-driven factories wrought a fundamental change in the way society was organized. Early water-flow-powered factories had been located at remote, rural sites where waterpower could most easily be exploited year-round—indeed, English streams had a comparative advantage over would-be competitors in southern Europe where similar-sized streams tended to run low or dry in summers. Labor came to these rural factories, often in the form of live-in child workers from foundling homes and workhouses. With the application of steam, everything changed. Factories moved out of the rural stream valleys and relocated in towns and cities, closer to their markets and where the key inputs of wage labor and coal were abundant and cheap. Steam, in short, brought industrial urbanization. Huge textile factories sprang up. Manchester, which had only two factories in 1782, had 52 two decades later. English cotton output surged, while its production costs and selling prices plunged.

Cotton exporters soon gained a stranglehold on markets throughout the world. Thanks to factory production, by 1789 British factories using Indian cotton were able to produce goods less expensively than Indian hand weavers themselves. In this manner, the rise of Britain's steam-powered factory system became interlinked with the political economy and militant expansion of British colonialism. The voracious intensity of the manufacturing expansion and its impact on local and distant societies can be apprehended in the fact that in a mere thirteen years between 1789 and 1802, British raw cotton imports for spinning accelerated twelvefold, from 5 million pounds to 60 million pounds, versus a fivefold growth over the preceding ninety years. The need to secure overseas raw material supplies and end-markets for the tide of British-made goods inexorably became a central focus of official British government policy and motivation for many nineteenth-century British naval missions.

The steam engine's catalytic effect on the expansion of the factory system of production was spectacular. Almost overnight, the production of goods was transformed from a centuries-old handicraft industry performed largely by individuals at home into a collaborative, standardized and mechanized system performed at a common factory location by large employee teams on precise time schedules. From about 1780, industrial growth in Britain startlingly quadrupled from an average 1 percent to 4 percent per year, and remained at that elevated level for about a century.

Watt worked continuously to improve the steam engine, experimenting with steam pressures, valves, and cylinder designs—and trying to stay one step ahead of patent infringers, including John Wilkinson, who pirated around the edges of Boulton and Watt's design. In 1788, at Boulton's suggestion, Watt added a governor to automatically regulate engine speed, and in 1790, a pressure gauge. By the end of the eighteenth century, Watt's steam engine was far more powerful and fuel efficient, as well as smaller and more portable, than the one they had begun selling less than a quarter century earlier. The average engine generated about 25 horsepower, although many were capable of up to 100 horsepower. Watt, sixty-four, retired as contented, healthy, wealthy, and celebrated in his own time as any man could reasonably hope to be when Boulton and Watt's original twenty-five-year partnership disbanded in 1800. He died in 1819, at age eighty-three.

In all nearly 500 Boulton and Watt steam engines were sold by 1800. The uses to which they were put provided a time capsule snapshot of the era's most dynamic activities. A large number were used to pump water out of coal and tin mines. Others were used to drive the bellows in blast furnaces that were producing Britain's fast rising output of high-quality cast iron. Most frequently, by the end of the century, they directly powered factories for cotton, wool, beer, flour, and china. In 1786, all London came to marvel at the spectacle of two steam engines driving fifty pairs of millstones at the world's largest flour mill. Many of the original Boulton and Watt engines were used to raise ever larger volumes of river water for delivery to expanding urban water supply systems.

The delivery of freshwater for drinking, sanitation, and other domestic uses had been an increasingly critical challenge as cities grew in population size and density. Waterwheel pumps installed upon urban rivers in the seventeenth century represented the first advance in domestic water supply provision in Europe since the aqueducts of ancient Rome, even though polluted water and insufficient pumping power remained constant problems. The Seine at Paris had a single undershot-waterwheel-powered pump below the recently built Pont Neuf as early as 1608, and added another at the Pont Notre Dame in 1670. The largest and most famous waterworks of the seventeenth century was installed in 1684 on the Seine to serve Louis XIV's royal fountains and gardens at Versailles. It featured 14 undershot wheels, each nearly 40 feet in diameter, turned by water derived from a dam in the Seine and powering 259 pumps that raised 800,000 gallons of water per day more than 500 feet in three stages. The Thames had a waterwheel pump under the London Bridge as early as 1582. But it was destroyed in the Great Fire of 1666—disastrous fires being another constant peril of urban life until the application of steam pumps to firefighting.

Steam power first had been applied in 1726 on both the Thames and the Seine with the placement of early Newcomen engines. Much larger Newcomen engines were added in London after 1752. Yet the engine's inefficiency goaded John Smeaton, the father of modern civil engineering, to conduct methodical scientific investigations into ways to enhance it—in much the same spirit of the age that impelled Watt soon thereafter to explore ways to improve the Newcomen engine. One of the earliest Boulton and Watt steam engines was installed at London in 1778 and pumped water through the city's network of wooden pipes for distribution to households three times per week. Watt steam engines tripled the average daily water supply of water-starved Paris from about one to three gallons per person after the Périer brothers, Jacques and Auguste, installed powerful steam pumps at two locations on the river in 1782 that lifted the river water 110 feet. Reflecting a repetitive pattern of history, first served was the wealthy Saint-Honoré district, while Paris's 20,000 omnipresent water carriers, who toted two buckets per delivery 30 times per day were left to anxiously contemplate the inevitable demise of their long-lived profession and livelihood. In America, Philadelphia earned admiration from visitors for its Fairmount Waterworks on the Schuylkill River, opened in 1815 in response to the public clamor against the city's industrially polluted, fetid, and insufficient water supply. Fairmount soon became the most profitable enterprise in Philadelphia. Its steam-powered pumps were designed by native son engineer Oliver Evans, based on a high-pressure system that Watt himself had eschewed as too dangerous. Water was pumped up to a hilltop reservoir and distributed by gravity flow throughout the city in log and cast-iron pipes. Evans's steam engines remained in use only until 1822, however. Due to explosions that shut down the water system, they were replaced by a battery of lower-tech, but more reliable, waterwheels, and after 1860 by water turbines.

Just as revolutionary as the application of steam power to factories was its use in powering the bellows that heated Britain's blast furnaces. Steam power facilitated the mass production of high-quality, inexpensive cast iron—which quickly became the great building material of the industrial age. Until then, limited forged iron supplies had been reserved mainly for making British naval cannons and other vital equipment. Steam power and iron had a dynamic synergy that galvanized a virtuous circle of self-reinforcing economic expansion that made them the core technology cluster of the second, mass production phase of the Industrial Revolution. Steam power helped cast more iron; more iron produced durable devices and applications to which steam power could be applied. With its blast furnaces working at full capacity, England's iron production soared more than twentyfold to nearly 1.4 million tons in the half century from 1788 to 1839.

The synergy between steam and iron was displayed in Boulton and Watt's interrelationship with iron master Wilkinson, who both fabricated key precision parts for Watt engines and used one engine to drive his own influential iron bellows. Wilkinson also employed a 20 pound, steam-powered hammer to pound his cast iron at 150 strokes per minute. Wilkinson's many innovative iron applications included the first iron-hulled river barge in 1787, which carried coal and iron along the Severn River. He built the first iron bridge across a river—the Severn at Coalbrookdale—and a steam-powered threshing machine. His major client was the British military, which depended upon his large furnaces for building the cannon and artillery used by Horatio Nelson and others to defeat Napoléon. Wilkinson kept experimenting with iron right to the very end of his life—he even designed the iron coffin in which he was buried.

Paradoxically, another of the important early uses of the steam engine, including by Boulton himself at his small metal goods factory at Birmingham, was to lift water to accelerate the turning rate of conventional waterwheels. The power output of waterwheels was vastly enhanced by the supplemental flows lifted by steam-power and the design of large, all-iron wheels. By the early nineteenth century, the most powerful waterwheels generated a stunning 250 horsepower—and remained more cost effective than coal burning steam engines.

In the 1830s, the power generated by falling water was significantly augmented by the French invention of the hydraulic turbine. In latter nineteenth-century America, for example, the Mastodon Mill on New York's Mohawk River generated 1,200 horsepower by taking water into its giant turbines through 102-inch-diameter pipes to drive 10 miles of belts, 70,000 spindles, and 1,500 looms. which produced 60,000 yards of cotton per day. Thus the use of waterpower continued to grow alongside steam. Only after the mid-nineteenth century did steam visibly supersede waterpower.

The harnessing of steam energy shattered the waterwheel-power barrier that for 2,000 years had been the ceiling of mankind's command over Earth's inanimate energy resources. Steam utterly transformed the speed, scale, mobility, and intensity of man's material existence. The fundamental nature of human society was reshaped, and propelled history in entirely new, previously inconceivable directions. The overwhelming benefits accrued first to the West, whose economic trajectory took off with seemingly magical force.

Within a few decades, steam power was propelling iron locomotives, riverboats, oceangoing gunboats, large dredgers, and earthmoving equipment. The face of Earth was literally resculpted by immense hydraulic civil engineering undertakings. Mass production factories swallowed handicraft trade. Small cities first born in the ancient, irrigated agricultural civilizations became giant metropolises. The most amazing transformation of all was that for the first time in human history the prodigious wealth created by the intensified uses of water and other productive resources outstripped the record-shattering eruption in human population—causing individual living standards, as well as individual health and longevity, to perceptibly rise from one generation to the next.

Except for relatively short-lived, localized spurts, nothing like it had ever before happened in human history. All previous economic gains had been so slowly accrued that they were visible in retrospect only as a gentle increase in the supportable level of a society's population.

Changelessness had been the enduring condition of daily life, from birth to death, century after century. As recently as the three centuries from 1500 to 1820, for instance, average world economic production per person rose a mere 1.7 percent per *century.* Over the next eighty years of early industrialization, by comparison, it nearly doubled, then quadrupled again in the late twentieth century. This unprecedented leap forward in individual living standards from 1820 to 2000 occurred even as overall world population was soaring from 1 to 6 billion. With the sudden explosion of economic wealth, a revolutionary new social concept infiltrated human politics, economics, and society—an expectation of progress.

That the stupendous break from all previous growth trend lines itself was accompanied by a stunning enlargement in the accessible supply of freshwater was no historical anomaly: in every age from the advent of irrigated agriculture, rising civilizations seem to have experienced contemporaneous, quantum leaps in the availability or exploitability of their water resources. The Industrial Revolution intensified this pattern. From 1700 to 2000, freshwater use grew more than twice as fast as human population. In the twentieth century alone, world water use would multiply ninefold—comparable in impact on society to the thirteenfold increase in energy use. Indeed, the unprecedented prosperity and population growth of the industrial age was driven as much by voluminous use of freshwater as it was by cheap fossil fuel energy. The augmented supply, in turn, stimulated still greater demand for new and existing uses of water.

As in all great water breakthroughs, steam transformed water's extraordinary, latent catalytic energy potential into productive use. But steam power's impact was exceptionally seismic because it leveraged further innovations throughout all man's primary realms of water use—economic production in industry, agriculture and mining; domestic uses for drinking, cooking, and cleanliness; transportation and strategic advantage in commerce, communication and naval power; and not least in energy generation itself, where it set in motion a cascade of advances that exponentially multiplied mankind's ability to tap nature's energy for his purposes.

Just as river irrigation came to be the defining fulcrum of ancient hydraulic states, steam power stamped an indelible imprint on the essential character of modern industrial society. The mobility of steam power freed man for the first time in history to deploy significant power anywhere and anytime. Paradoxically, this both democratized society and deepened fundamental pillars of hierarchical control. On the one hand, small-scale steam power promoted decentralization, diversity of activity, and pluralism of interests. On the other, within established sectors, it enabled steam power Haves to better exploit economies of scale and amass oligopolistic concentrations of economic power and wealth. In warfare, the cost-benefits of steam power were less ambiguous. It fostered greater state control over organized violence and the rise to preeminence of more solidly entrenched nation-states.

Thanks to steam people moved much faster and farther than ever before imagined. The farthest distance a man could cover in a single day from antiquity to the mid-nineteenth century by sail, oar, or horseback had been 100 miles per day; suddenly, steam power enabled him to traverse 400 miles per day by ship or rail. The pace of communication, trade, and large-scale human movement between places accelerated. Thus began the historic defeat of distance that marked the transportation and communication revolutions and evolved into the oceanic, intermodal sea-to-rail, containerized shipping and telecommunications web of the twenty-first century, cornerstones of the integrated information age society.

Richard Trevithick built the first steam locomotive, or "Iron Horse," in 1802 in Shropshire. When long iron bridges were engineered to carry trains over rivers and other landscape barriers, steam engine railroads superseded canal and barge transport systems in transporting coal, other freight, and people across continents. The U.S. transcontinental steam railroad drove its final, golden spike at Promontory Point, Utah, on May 10, 1869. The fabled Orient Express made its debut run from London to Paris to Istanbul in 1888.

In water transport, wood and sail was superseded by a more tightly interlinked oceanic era of iron and steam. American Robert Fulton ordered a Boulton and Watt steam engine to power the maiden voyage of his 100-ton steamboat, the paddle-wheel driven *Clermont,* up the Hudson River in 1807, which opened the era of commercially successful river steamboats. Fulton's was not the first river steamboat, nor even the first in America. In 1778, the eccentric, ill-starred American inventor John Fitch sailed a ship named after himself on the Delaware, but failed to establish a successful business model. Soon steamboats were servicing America's Great Lakes and Mississippi; Europe's wide rivers like the Rhine, Danube, Rhone, and Seine; and appeared in the Mediterranean, the English Channel, and the Baltic Sea. In 1819 the *Savannah,* powered by a 90-horsepower engine that turned a collapsible paddle wheel, became the first steam-power-assisted ship to cross the Atlantic, covering the distance in twenty-seven and a half days, though using her engine only eighty-five hours. Regular transatlantic service started in 1838. A journey that routinely took sailing ships two months required only nine days by fast steamer in 1857. By 1866, after ten years of effort, Cyrus W. Field successfully laid a communications cable under the Atlantic; by 1900 there were 15 cables on the Atlantic floor, facilitating intercontinental exchanges. Without such developments it was inconceivable that 55 to 70 million Europeans could have emigrated to the Americas, Australia, and elsewhere from 1830 to 1920, relieving the chronic labor shortages that threatened to choke off America's westward frontier expansion and Europe of its excess industrial unemployed who threatened domestic uprisings such as those of 1848.

The great age of the ocean steamer arrived after 1870 with the development of the screw propeller in the 1840s, compound engines in the 1850s, steel hulls in the 1860s—and the opening of the Suez Canal in 1869. Steamships between China and Europe, for example, shipped three times as much cargo in half the time taken by a sailing ship. A worldwide steamship network evolved that regularly carried grain to Europe from America's Great Plains, Argentina and Australia, while wheat, indigo, rice and rubber moved into Europe through Suez from India and Southeast Asia.

As with previous water transportation innovations, the cheaper cost of steam power helped realign geopolitical world balances. Steam made every society on Earth a potential raw material supplier, as well as a potential market, for Europe's fast-growing industries. A relationship of subservient interdependence evolved between colonial satellites and their European masters. Outside Europe, diverse, self-sufficient subsistence economies driven by land-owning peasant farmers gave way to large, specialized single crop plantations manned by sharecropper labor producing chiefly for export to Europe and economies dependent upon imports of previously self-made goods. In the new world economic order that came into being, an ever more dominant and richer Western industrialized manufacturing center was supplied by a colonial periphery with unskilled labor and few developmental paths to garner a growing, relative share of the world's increasing wealth.

The revolutionary effects of the 1869 opening of the Suez Canal, with its main channel reserved for steamships and only a small freshwater side channel for sailing vessels, intensified the interlinking of this colonial world order. British steamboats using the canal could travel to India in only three weeks compared to the three months it took a generation earlier to sail around Africa's Cape. As a result, within a year of the canal's opening, Indian wheat was being exported in large volumes to England. British manipulation of land tax policies in India helped maintain wheat exports even during the acute Indian famine of 1876–1877. By the 1880s, some 10 percent of the world's grain exports were coming from India. Thanks to Suez and steam railroads, England became the first power in history to unite the entire Indian subcontinent. To consolidate its grip on its export breadbasket, England did as ruling powers have done throughout history by expanding irrigation investments. Old Muslim hydraulic works, such as the Cauvery Dam in South India and the Jumna canals near Delhi were restored, followed by the enlargement of the footprint of irrigable cropland along the Indus River. To defend the Suez Canal, British engineers who had been trained in India transferred their expertise to the Nile after the establishment of unofficial British rule in Egypt in 1882.

The opening of the Suez Canal loosed the full force of Europe's superior steam and iron navies upon the rest of the world. The ensuing clash of civilizations rendered transparent the West's rise to dominance that had begun with the Voyages of Discovery. Within Europe, England rose to unchallenged primacy through the wedding of its steam-powered industry and naval leadership. Its global economic, colonial, and naval Pax Britannica lasted nearly a century. As early as 1824–1825, British steam gunboats sailed up the Irrawaddy River to subdue Burma. Design improvements over the following two decades transformed such steamboats into lethal, iron-hulled river armadas that could penetrate deep into the heart of enemy country, where previously sailing ships had been confined to bombarding from the shorelines. Just as China had been rudely awakened to its relative decrepitude after four centuries of somnambulant isolation by the appearance of invincible British gunboats sailing up its rivers to impose free trade in Indian-cultivated and British-transported opium in the first Opium War of 1839–1842, American gunboats forced open long-internationally sealed Japan to free trade on Western terms when Admiral Matthew Perry's "black ships" steamed into Tokyo Bay in July 1853. Japan's response to this national trauma was the catch-up industrialization of the Meiji Restoration. The dramatic superiority enabled by Western industrialism posed traumatic, long-term challenges for subordinated Islamic societies, whose leaders were left to ponder whether to respond by trying to imitate Western ways at the one extreme, or to seek renewal by turning inward to religious neofundamentalism at the other.

Maintaining naval superiority was the central focus of British policy throughout the Pax Britannica era. By applying steam and iron innovations to design, warships in the early nineteenth century quickly evolved into speedier, heavily armed carriers for ever-more powerful and accurate long-distance artillery. From a matter of yards at the time of the Armada, English warships' gun ranges reached three miles by 1900; by World War I, the distance had trebled to nine miles. With the advent of aircraft carriers by World War II, missile ranges were extended to hundreds of miles and guided by mobile bombers. An earlier innovative vessel, the submarine, first used through oar propulsion by the Dutch as early as 1620 in the Thames, became a lethal weapon by World War I from the combination of the invention of the torpedo (1866) by

British engineer Robert Whitehead and the design integration of electricity and iron. Torpedo ranges multiplied tenfold to over one mile in the nearly forty years from 1866 to 1905 and nearly tenfold again to almost 11 miles less than a decade later. By the late twentieth century, guided intercontinental ballistic missiles launched from submarines could span several thousand miles—literally crossing oceans—and were capable of delivering civilization-incinerating nuclear warheads.

Through the late nineteenth century, British naval strategy was focused on maintaining superiority over the combined power of the Franco-Russian alliance, while still being able control the key points in the Mediterranean. That changed when newly militarized and industrialized Germany entered the naval armaments race. Britain's response—its last hurrah as a naval superpower—was the 1906 *Dreadnought*. Fitted with oil fuel and huge turbine engines and fortified with alloyed steel, the *Dreadnought*-class battleship set a new world standard with its combination of speed—10 percent faster than any rival—and long-range, accurate, heavy firepower. Although the *Dreadnought*'s advantages were short-lived, they enabled Britain to control the critical Atlantic sea supply and communication lines that helped it win World War I. At the start of the war in August 1914, for example, British ships hoisted up and cut Germany's five transatlantic cables, compelling the Germans to revert to wireless telegraph communication that was much easier for the British to intercept. One of those intercepted communications, the famous 1917 Zimmermann Telegram offering an alliance with Mexico, proved pivotal in bringing America into the Great War on Britain's side. By World War II Germany's *Bismarck* set the new leading technical standard in battleship power, armed with radar gunnery control; only an all-out, desperate British hunt in 1941 succeeded in sinking it at painfully high expense before it could tilt the balance of power in the open seas. Yet in World War II, even battleships like the *Bismarck* were being eclipsed by an altogether new class of naval weapon, the aircraft carrier, and a new superpower on the high seas, the United States.

The shift of naval superiority to America, the propitiously situated, continental-island nation whose navy sat astride Earth's two largest oceans, the Atlantic and the Pacific, represented the completion of the slow, fitful migration of history's central naval axis from the Mediterranean and the Indian Ocean in ancient times to the Atlantic in Europe's heyday, and westward again toward the Atlantic-Pacific bridge in the twentieth century, and finally to a truly world-spanning network in the twenty-first century's integrated, global era.

History's great hydraulic projects often heralded turning points of world power. So it was with each of the interoceanic canals built at Suez in 1869 and Panama in 1914. Both strategic waterways were civil engineering tours de force of their day, possible only with steam age machinery. Both had a world-changing impact on global commerce and balances of power. Suez proclaimed the apogee of the Pax Britannica. Panama signaled the transfer of leadership to America. In 1870, Britain accounted for some one-fourth of world commerce and 30 percent of total industrial production. Its wealth was reflected in its population, which had tripled in the century to catch up to its historically larger rivals France and Spain. By 1914, however, the United States and Germany both had caught up economically.

From the moment of its extravagant opening on November 17, 1869, the 101-mile-long Suez Canal directly linking the Mediterranean to the Red Sea, and thence onward to the Indian Ocean, became the strategic aorta of the British colonial empire. Ironically, Britain had originally opposed the private, French-built canal project. After all, a mere three generations had passed since Nelson defeated Napoléon's Mediterranean fleet at the Nile and with it, France's bid to undermine Britain's grip on its vital route to India. The British were still suspicious of French intentions, and they felt well-served by the status quo—travel time to India from London had already been reduced to less than a month by steamship and the steam railway link between Alexandria and Suez.

Napoléon's engineers who inspected the ruins of Neko's ancient "Suez" canal links via the Nile had abandoned their plans for a direct canal link between the Red and Mediterranean seas through a technologically simple, open cut channel when they erroneously measured a significant altitude difference between the two seas. In 1832, these old Napoleonic plans came into the hands of an experienced French diplomat in the region, Viscount Ferdinand de Lesseps. He became seized by the vision of building the Suez Canal. More accurate surveys soon revealed that the sea levels were in fact similar and that an open cut channel without locks had been feasible all along.

De Lesseps plan originally got no traction with Egypt's powerful, ambitious, modernizing, and militaristic-minded ruler, Muhammad Ali, who served as the Ottoman Turks' viceroy but in reality was all but autonomous of Constantinople. A native Macedonian and small tobacco dealer who liked to boast he was born the same year as Napoléon, to whom he seems to have fancied himself a would-be Muslim counterpart, Muhammad Ali originally came to Egypt as part of the Ottoman force resisting the French general and himself was saved from drowning by British troops after being driven in retreat into the sea. Within a few years he consolidated political power; his signature act was the summoning together and ruthless mass murder of his Mamluk opponents. He schemed and adventured militarily, initially to ingratiate himself to his Ottoman suzerains, and ultimately toward his never-achieved goal of establishing his own sovereign dynasty, and regional empire, in Egypt. Muhammad Ali vehemently opposed de Lesseps' canal because he foresaw, rightly, that it would entangle Egypt, and his dream of independence, in European great power affairs.

De Lesseps finally got his chance in the mid-1850s when two of Muhammad Ali's successors, Said and Ismail, inverted their forefather's political calculus and endorsed the canal as a means to physically and legally separate Egypt from Ottoman overlordship and to relaunch Egypt's imperial glory. De Lesseps set up a private company to build the canal and operate it for 99 years. English investors were offered shares, but without support from their government, refused to participate, leaving mainly 25,000 French investors with a majority and Egypt itself with 44 percent.

Any British government calculation that it could kill the project politically by its opposition proved erroneous when matched against the extraordinary organizing talents, energies, and determination of de Lesseps. Building the Suez Canal took ten years, nearly twice as long as de Lesseps had projected. The high cost nearly bankrupted the Egyptian government. As in ancient Egypt, coerced peasant labor was employed. Work was delayed by cholera outbreaks that killed over half the workers in the first few years, by labor unrest, and by the sheer inadequacy of the traditional hand tools of pick, shovel, and dirt-removal baskets to do the dry excavation. Only by calling in huge steam-powered dredges and shovels operated by imported, skilled European workers was the job finally brought to completion.

Its opening in November 1869 was one of the great occasions of the nineteenth century. Determined to show that Egypt belonged within modern European civilization, the Viceroy spared no expense from his depleted Treasury. Some 6,000 guests were invited, all expenses paid. The emperor of Austria and other royalty, artists such as Emile Zola and Henrik Ibsen, and other luminaries of the age were among the headline attendees. Thousands lined both sides of the canal to cheer the processional yachts. De Lesseps himself was hailed by all as the great "Engineer"—even though he had no technical background whatsoever and his great achievement was as an enterprising impresario. An opera house in Cairo was constructed and Giuseppe Verdi commissioned to write an opera for the opening—thus *Aïda*, although not performed until two years later, was conceived; its story of star-crossed love between an Egyptian officer and an Ethiopian princess to whom he betrays Egypt's plan to invade Ethiopia tugged at the Egyptian nation's historical nightmare about losing control of the waters of the Nile, some 85 percent of which originated in Ethiopia.

Britain recognized its strategic interest in the canal from the moment it opened. Thus in 1875, when financial burdens induced the Viceroy—who was then engaged in a costly and disastrous war of imperial aggression in Africa in which he would be humiliatingly trounced by Ethiopia—to offer to sell Egypt's 44 percent stake in the canal company to England for the large sum of £4 million, Prime Minister Benjamin Disraeli acted swiftly to secure funding from the Rothschild banking family to buy it. Egypt's financial woes continued, however.

Political crises ended in a military coup by anti-European nationalists that seemed to forebode Egyptian foreign loan defaults, and threats to the physical welfare of 37,000 resident Europeans as well as to the control and operation of the canal itself. British prime minister William Gladstone, formerly an ardent critic of Disraeli's canal shares purchase, ended up doing a total policy about-face. Acting diplomatically with France, and unilaterally by force, in the summer of 1882 he moved under the thin pretext of restoring legitimate order to suppress the nationalists. Alexandria was bombarded by British forces, while a surprise cavalry charge trounced the nationalists' much larger army in just 35 minutes. The canal was secured. The occupying British forces, however, never left. Despite regular reassurances year after year that its occupation was merely temporary, Britain remained to unofficially rule the country through half the twentieth century.

Britain quickly understood what all previous rulers of Egypt had learned: that to govern the country, one had to control the Nile waters. Accordingly, the British promptly focused on imposing their might over the length of the White Nile from its source near Lake Victoria to the Mediterranean. Sudan, Kenya, and Uganda were all subdued. British engineers were brought in from irrigation projects in India's Punjab to help design waterworks throughout the Nile basin to maximize river water flow volumes and agriculture in Egypt. Reforms begun by Muhammad Ali in the first half of the 1800s came to fruition by the end of the century in a modernized network of dikes, sluices, and canals to provide Egypt with its first, fully operational system of perennial irrigation that yielded two and sometimes three harvests per year. It was the first significant change to the one crop basin agriculture that had existed since the dawn of Egyptian civilization almost 5,000 years earlier. Egypt's population surged from four to 10 million, twice as many as its three-millennia ceiling. British engineers had less success in the backbreaking efforts made to augment the White Nile's flow to Egypt by cutting through or diverting the meandering river through Sudan's huge Sudd swamps, where so much water was lost to evaporation. All British hydraulic engineering efforts were dwarfed, however, by their momentous achievement in 1902 of the first Aswan Dam, then one of the largest and most sophisticated dams in the world. The low dam, as it is now called, was unique in water history in allowing the passage downstream of the river's fertile silt during the early floods through low-level sluices. As a result of the dam, irrigated acreage and agricultural production in the Nile Valley and delta soared, feeding political stability and the continuous swell of Egypt's population.

Britain's seizure of the Suez Canal and the Nile basin also triggered a new phase of colonialism known as the Scramble for Africa. Across the continent, European powers engaged in a free-for-all military land grab. In 1898, England's military campaigns in the Nile basin nearly led to comical war with France, known to history as the Fashoda Incident. Its genesis was an 1893 proposal made by a French hydrologist and former schoolmate of the president of France to erect a French dam on the White Nile at Fashoda in Sudan. On paper, the dam promised to deliver, in a single master stroke, control of the Nile and Egypt's fate into French hands, while checkmating British expansion into East Africa as France completed its own Atlantic to Indian Ocean colonial run. The French establishment was smitten by the diabolical brilliance, and romantic flourish, of what would have been exposed by any realistic assessment as an utterly fanciful quest. For starters, there was hardly a stone within a hundred miles of Fashoda with which to build a dam. For another, also unknown to French hydrologists, the While Nile provides less than one-fifth of the Egyptian Nile's water and almost none of its precious silt; thus impeding its flow, even if achieved, could not have had the intended dramatic effects. Nevertheless, in June 1896 an expedition was launched from Marseille and a tiny, intrepid band of French officers and Senegalese foot soldiers began their arduous 2,000-mile, two-year trek across Africa, up the Congo, and through the thick Sudd swamps to seize Fashoda. They carried with them 1,300 liters of claret, 50 bottles of Pernod, and a mechanical piano.

The comical became a sublime farce when the British became so alarmed at the Frenchmen's scheme that they deemed it an importance to conquer Sudan to secure the river. Thus they dispatched an army under General Horatio Herbert Kitchener. In September 1898, two weeks after destroying the Islamist Mahdi state near Khartoum,

Kitchener arrived at Fashoda to face off against the Frenchmen. On the French side were a dozen officers and 125 Senegalese colonial soldiers; arrayed against them were no less than 25,000 British troops, artillery, and a fleet of steam gunboats. Kitchener advised the French to leave. No shots were fired. The two opponents fraternized, and even shared some of the French wine. Diplomatically, however, Fashoda ignited an explosive, several-month international incident that nearly triggered a wider war between the nations—a colonial-era Cuban missile crisis—due to the national humiliation France felt over the lopsided confrontation. In the end, the French discretely withdrew—while the British army band played the French national anthem—and with equal discretion the British removed the name *Fashoda* from African maps. Both sides put renewed cooperative emphasis on their mutual larger national interests, including the welfare of the Suez Canal. As if to symbolize their newfound comity, in 1899 a monumental, more-than-30-foot-tall statue of de Lesseps, with his right arm outstretched in welcome, was erected on a huge pedestal at the entrance of the canal at Port Said—making a towering impression, similar to that of the Statue of Liberty in New York Harbor.

The interests of both nations were served in the world wars by the denial of Suez Canal access to German shipping, despite the international convention that designated it an international waterway open to all. But in 1956 the Suez Canal became the instrument that extinguished both nations' final imperial pretensions and inaugurated the era of Cold War politics in the Middle East. In one of the greatest blunders of American postwar foreign policy, Eisenhower administration secretary of state John Foster Dulles inadvertently opened the door to Soviet Union influence in the region and fanned the flames of anti-Western pan-Arabism.

The "Suez Affair" began in 1952 with the rise to power in Egypt by military coup d'état of the charismatic Colonel Gamal Abdel Nasser. With tacit approval from the postwar superpower, the United States, Nasser negotiated the British withdrawal of troops from the Canal Zone, which was completed in the summer of 1956. Nasser's supreme ambition was to build a giant dam on the Nile River at Aswan that would vastly increase irrigation and electrification in impoverished Egypt. It was a project of such monumental economic and symbolic political importance, heralding a renewal of Egyptian control over the Nile like that exerted by the Pharaohs of its bygone ancient civilization, that Nasser himself likened it to a modern pyramid. Concurrent to negotiating the exit of the British from Suez, Nasser thus also sought financing from the West for the enormously expensive Aswan high dam. Dulles, like British and French leaders, deeply distrusted Nasser. He disliked him personally as well. Above all, Dulles could not abide Nasser's effort to steer a neutral policy in the Cold War, and resented the Egyptian's efforts to negotiate between the West and the Soviet Union, which was eager to establish itself as a strategic power in the Middle East. In the fall of 1955, Dulles had been both shocked and upset when Nasser, after being stonewalled in a request for American arms, made a massive military purchase from the Soviet bloc, including 200 warplanes and 275 tanks—an alarming action that promptly accelerated war planning in Egypt's neighbor and enemy, Israel.

In the winter of 1955–1956, Dulles elected to keep Nasser grounded in Western influence by agreeing to a substantial loan and grant package from the World Bank, the United States, and England for the high dam at Aswan. When Nasser bridled that the stringent terms, which included the monitoring of the Egyptian economy by the World Bank, were insulting and patronizing, however, Dulles didn't budge. He knew the Soviets had offered to build the dam at Aswan. But he didn't believe they had the technical capability to accomplish it. He thought they, or Nasser, were bluffing. So he waited Nasser out. After several months Nasser indeed capitulated to the terms and dispatched the Egyptian ambassador on July 19, 1956, to Dulles's office atop the State Department to conclude the deal. But Dulles had been growing increasingly sour toward Nasser in the meantime. In what, at best, was a confused policy, and at worst a blundering miscalculation, Dulles wanted to give Nasser, and the Soviets, a further comeuppance by strongly asserting American dominance in the region. Despite urgings from top British officials to equivocate and "play it long," Dulles began to inform the

Egyptian ambassador why the United States was not able to support the Aswan deal at that moment. The ambassador grew agitated. He pleaded with Dulles not to withdraw the offer, informing him, with a tap on his pocket, that Egypt had an alternative financing offer ready for signing from the Soviets. It was plain he preferred the Western deal. Yet Dulles was irritated at what he viewed as blackmail for better terms. He sniffed, "Well, as you have the money already, you don't need any from us. My offer is withdrawn!"

An infuriated Nasser not only signed with the Soviets to build the Aswan Dam. One week later, on July 26, he did something wholly unanticipated by Dulles—he unilaterally nationalized the Suez Canal. Tolls, he predicted, would pay for the dam within five years. He secretly gave the signal to begin seizure of the canal in a fiery speech before a large throng in Alexandria by using a prearranged code word—"de Lesseps."

It was a move that changed the history of the twentieth century. While President Dwight Eisenhower and Dulles equivocated in their response over the fear of igniting a wider Middle Eastern war, to England and France it was an intolerable threat to let Nasser "have his thumb on our windpipe," as British prime minister Anthony Eden put it. Beyond anger at Nasser's impudence, England and France feared the economic costs if Egypt used the canal to hold hostage the lifeline of oil shipments transported in tankers from the Middle East to Europe, as well as the imitative repercussions in their other restive colonies. They began colluding to seize back the canal and depose Nasser. To execute their plan, they enlisted Egypt's archenemy, Israel. Israel, which had fought against Egypt and other Arab states in its war for independence (1948–1949) and in 1956 was still denied use of the canal and was furthermore without shipping access to the Red Sea and Indian Ocean due to an Egyptian blockade of the Gulf of Aqaba at the Strait of Tiran, readily cooperated. On October 29 Israeli paratroopers, led by future prime minister Ariel Sharon, dropped into the Sinai peninsula 25 miles from the canal and began advancing upon it, while simultaneously seizing control of the Strait of Tiran at the eastern tip of the Sinai. Following their script, England and France feigned the role of neutral peacemakers interested in safeguarding the integrity of shipping through the canal. They demanded an immediate cease-fire and withdrawal by both sides to 10 miles from the canal. The Israeli troops froze. Egypt, which was the actual target since only its troops were within that proximity, did not. Reminiscent of 1882, England and France bombarded Egyptian air bases and landed "peacekeeper" troops that occupied the northern part of the canal when Nasser refused to withdraw.

But the world was entirely different than in 1882. Cold War politics and postwar independence movements had eclipsed the colonial imperial axis at the center of world power relationships. The Soviet Union threatened to intervene on Egypt's behalf. Egyptian forces managed to block oil shipments through the canal. Investors around the world began dumping the British pound sterling, driving England into a financial crisis. Dulles and Eisenhower, the ultimate global power broker, felt personally betrayed by the secret Anglo-French collusion. Upon hearing the news of the allied attack, Eisenhower called Eden and said, "Anthony, have you gone out of your mind? You've deceived me."

Fearing a new Cold War blowup—the abortive Hungarian revolution had also just occurred—Eisenhower decided that England and France had to withdraw. He obtained England's capitulation by threatening to block an emergency International Monetary Fund jumbo financial loan package to save the wobbling sterling. British troops were already halfway down the canal when they were recalled on November 7. France was enraged, but could not stand alone. Concluding from the experience that England would always stand with America ahead of France and continental Europe in a crisis, France became the prime mover of the launch, within months of Suez, of the six-country Continental European Common Market—without England. French leaders managed to deny England membership in the forerunner of the European Union until 1973. A U.S.-led resolution, backed by the Soviet Union, condemning Israel and humiliating England and France, led to the first-ever deployment of a UN peacekeeping force, 6,000 blue helmet soldiers were sent to Suez and Sinai.

The withdrawal of England, France, and Israel put a triumphant Nasser back in charge as the canal reopened, adulated by a suddenly newborn pan-Arabist movement, and gave the Soviets their first important foothold in the Middle East. Before year's end in 1956, de Lesseps's statue at the entrance to the Suez Canal was torn down with the incitement of angry Egyptian mobs. The Suez debacle hastened the setting of the sun on the last important vestiges of the global colonial empires of England and France and brought to a close the world political era that had reached its apogee with steam-powered industrialization.

The heyday of the coal-burning steam engine lasted until the end of the nineteenth century. By then it began to be superseded by new forms of energy developed to meet industrial society's insatiable demand for greater and more manageable power. The technical limit of the classical steam engine peaked out at about 5,000 horsepower. This was inadequate for generating electricity in a fast-spinning dynamo—a versatile, important new form of energy that replaced much steam-powered machinery and created an entire new cluster of technological capability. Oddly enough the electrical age was initially powered by the reengineering of old-fashioned energy derived from falling water. Water turbines, in which water falling from high elevation flowed through fixed channels to turn finlike blades, were more efficient descendants of the waterwheel. Through most of the nineteenth century, turbines simply replaced waterwheels where great horsepower was required. Eventually their power efficiency surpassed that of steam engines. When the electrical age began following Thomas Edison's 1879 invention of the lightbulb, water turbines were the most effective way of generating electricity. The world's first big hydroelectric power station, featuring water turbines, was built in 1886 in America at Niagara Falls. Within a decade it was running ten 5,000 horsepower pressure water turbines to produce electricity. By 1936 the Hoover Dam had water turbines generating 134,000 horsepower, or 100,000 kilowatts.

Steam turbines also rapidly gained efficiency and soon became a major source of electrical generation in conjunction with the burning of fossil fuels. Steam turbines that generated a mere 1,600 horsepower in 1900 were producing three times as much a decade later. A series of steam turbines generating 68,000 horsepower powered the transatlantic liner *Lusitania* in 1906, heralding its wide application in high-speed ships. Gradually, steam turbines in thermal power plants burning coal, natural gas, or oil became viable alternatives to hydroelectric plants.

Hydroelectric power created its own political, economic, and social revolution. Electricity was highly transportable, and it could power cities and factories far from the water site where it was located. Nations poor in coal for steam power but rich in mountain water flows—or "white coal"—suddenly gained access to an energy resource that allowed them to enter the industrial age. This was dramatically illustrated in mountainous Italy, which not only became an industrial power, but also a viable nation-state, thanks to hydroelectricity. Abundant hydroelectricity liberated Italian steam-powered industry from the crippling burden of having to pay up to eight times more for scarce coal than its English counterparts. The newly unified Italian state built its first small hydroelectric plant in 1885. By 1905 it used more hydroelectricity than any other country in Europe. Nearly all of Italy's electricity came from hydropower as late as 1937. Dams were erected throughout the glacial Italian Alps, and its northern alpine lakes were used as reservoirs for generating hydroelectricity. Milan became the world's second city with electric street lighting. Hydroelectricity thus added a modern chapter to Italy's legacy of water history dating back to the large-scale drainage projects and aqueducts of ancient Rome. The hydroelectric boon, as well as earlier land drainage and irrigation projects in Lombardy, came at a timely moment to legitimatize a regionally fractious nation-state that formed in stages around 1870, against the many who doubted that it would hold together.

The diffusion of hydroelectricity helped spread the Industrial Revolution throughout Europe, America, and belatedly to other parts of the world. By 1920 environmentally clean and renewable hydroelectric plants generated some two-fifths of electricity in America; by 2000 hydropower still generated nearly one-fifth of world electricity. As the good hydroelectric water sites grew scarce and steam turbines improved, more electricity was generated in large thermal plants using fossil or nuclear fuels. These power plants used water not just for the steam turbines but also as coolant. As a coolant, massive volumes of river water were sucked in for brief periods to absorb huge amounts of heat, and then, after recooling, were discharged back into the river. As most auto motorists knew, water was similarly used as a coolant in the important innovation of the oil burning internal combustion engine. The exponential growth of power man extracted from water reached astonishing amounts by the late twentieth century. State-of-the-art water turbines installed at large dams were capable of more than 1 million horsepower (750,000 kilowatts); steam turbines were even more powerful, up to 1.7 million horsepower (1.3 million kilowatts).

The progression of water use in energy production—from simple, directly channeled water current to turn waterwheels, to steam engines heated by coal, to spinning turbines that generated electricity from cascading torrents or steam pressure, to coolant for large nuclear and fossil fuel plants—itself highlighted a prominent feature of water's role in human history: with each new technology cycle, its uses were always evolving and expanding. Starting from the steam engine's role in extracting coal to overcome the fuel famine, and coal's provision of the combustive material to generate steam, water, and energy in many of its forms have become symbiotically interlinked partners in the key processes that powered much of the twentieth century's extraordinary industrial expansion. No nation exploited the water-energy resource nexus so beneficially and on such a large scale as Britain in the nineteenth century. In the twentieth century that mantle passed to America.

Before England or America could take full advantage of the beneficial opportunities of their new industrial societies, however, they had to find an effective response to the first of several major environmental challenges created as a by-product of industrialism—urban water pollution.

PART III

Water and the Making of the Modern Industrial Society

CHAPTER TEN

The Sanitary Revolution

The summer of 1858 was one of the hottest and driest ever in London history. In the first fortnight in June, sweltering heat caused a putrid stench to rise from the pools of stagnant sewage that choked the river Thames—headlines in the British press dubbed it the Great Stink. Inside the Houses of Parliament overlooking the river, behind the heavy lime chloride-soaked window draperies, members who had dithered ineffectually for decades over the worsening unsanitary water and sewerage conditions suddenly discovered that they had no escape from the assault of the stench, which demanded their immediate response. Adding urgency to their concerns was trepidation that the miasmas emitted from the Thames were putting their own personal health in mortal danger, since the prevailing medical theory of the day held that diseases were communicated by such foul air. The Great Stink, in short, succeeded in riveting politicians' attention in a way that all the many years of appalling manifestations of London's mid-nineteenth-century sanitary crisis had not.

In the previous decade alone, two cholera epidemics had killed over 25,000 Londoners. Throughout the city, sewage and human waste regularly seeped from cesspools into wells and was flushed into the Thames from which it was promptly pumped back up in Londoners' drinking water. People were consuming their own sewage. Yet even contaminated water supply was insufficient to slake the water famine among the city's fast-growing population. Street taps that each supported 20 to 30 crowded houses often were opened to sell water only one hour per day three days per week. Not surprisingly, Londoners' daily existence was afflicted with chronic illness, shortened lives, and infant mortality that claimed some 15 of every 100 children within their first year. Even the vociferous gathering cry to reform London's water and sanitary system, echoed by such celebrated figures as author Charles Dickens and scientist Michael Faraday, had been insufficient to rally MPs to empower an effective central public municipal authority to meet the long-brewing crisis.

An environmental by-product of the urbanization that accompanied early industrialization, the Great Stink was more than a mere nuisance or embarrassing advertisement for the social virtues of the British Empire's vaunted liberal market democracy. It threatened the very sustainability of a sufficiently healthy surplus force of wage labor to cheaply man the new factories. Given Parliament's unresponsive record, when the weather suddenly cooled on June 17 to provide a break from the Great Stink, the *Times* of London bemoaned: "What a pity it is that the thermometer fell ten degrees yesterday. Parliament was all but compelled to legislate upon the great London nuisance by the force of sheer stench. The intense heat had driven our legislators from the portions of their buildings that overlook the river. A few members, indeed, bent upon investigating the matter to its very depth, ventured into the library, but they were instantaneously driven to retreat, each man with a handkerchief to his nose. We are heartily glad of it."

Fortunately the Great Stink did not abate so quickly in the hot summer of 1858 as to dissipate from the politicians' agenda. On July 15 House of Commons leader Benjamin Disraeli—the same Disraeli who two decades later as prime minister would boldly commit England to buying shares in the Suez Canal—took the floor and introduced the overdue mandate and funding legislation to purify the Thames's water and construct a proper sanitary sewerage system befitting the world's leading city. So it happened that after years and years of fruitless debate, the reform legislation passed in just eighteen days. It proved to be the watershed turning point of the mid-nineteenth-century Sanitary Awakening. The awakening triggered a public health and environmental revolution that in the twentieth century resulted in the virtual abolition of the age-old scourge of infant mortality, the breakthrough scientific germ theory of disease, a quantum jump in human longevity, an unprecedented explosion in urban and total global population, and an enlarged, proactive state role in the governing compact between democratic governments and free markets.

From the outset, industrial steam power and large manufactory production had promoted rapid urban concentration. Within a century, populations in factory towns like Manchester, Birmingham, Leeds, and Glasgow surged five- to tenfold. Great political metropolises like London and Paris also swelled. London made history by the turn of the nineteenth century by surpassing ancient Rome's threshold of 1 million inhabitants, then tripled in the next sixty years to nearly 3 million. The swollen densities simply overwhelmed the urban sanitary and water supply infrastructures built for an earlier age. For nearly all of human history, cities had been unwholesome, disease-infested death traps reliant upon migration from the countryside to replenish the decreasing natural reproduction of their indigenous population. By the mid-nineteenth century, the historic urban challenge demanded an innovative response lest the Industrial Revolution and the liberal democratic institutions allied with it choked on the waste of their own astonishing growth.

Throughout history, water's life-giving indispensability had always been double-edged. On the one side, drinking two to three quarts of clean freshwater daily sustained each person's existence; several gallons enabled healthy cooking; about 10 to 20 gallons were needed for minimal hygiene. Yet simultaneously, drinking contaminated water and exposure to stagnant water bearing an infiltrating army of diseases also was the main source of human illness, abbreviated life spans, and physical miseries. The greatest waterborne killers of all ages were dysentery (a.k.a the bloody runs) and common diarrhea. Mankind's transition from hunting and gathering to irrigated agricultural civilization had notably worsened average individual health and longevity by increasing man's exposure to sitting pools in irrigation canals bearing malaria, yellow fever and dengue-transmitting mosquitoes, schistosomiasis, and guinea worm. Cities and intensifying industrialization elevated to prominence the deadly waterborne diseases that spread through unsanitary conditions, above all pandemics of cholera and typhoid fever.

Without understanding the science of waterborne disease, societies throughout time were cognizant of the linkage between water and illness. Almost everywhere consumption of freshwater was subject to precautionary social customs. Few drank it cold and untreated by choice, unless it came from a prescribed source. The fifth century BC Greek "father of medicine," Hippocrates, who made civilization's first systematic effort to associate disease and the environment, recommended boiling water to eliminate particles that darkened its clarity and polluted its taste. From ancient times the Chinese consumed hot tea and boiling water, which was widely sold on city streets by vendors. Chinese wise men believed that water possessed special qualities depending upon its origin: early spring rainwater was considered beneficial, water from storms was dangerous, water from melted winter frost or hailstones or obtained from cave stalactites was medicinal. In an admonition that accorded with both Hippocrates and modern science, any suspect water was to be boiled. Snow water in many parts of the world was a prized luxury shipped to royalty and coveted by those who could afford it for its curative powers. The ancient Romans likewise esteemed the drinking qualities of water from preferred springs, such as the source of the Aqua Marcia aqueduct that gained its natural purity and coldness by filtering through the porous limestone hills near suburban Tivoli. In medieval and modern times, Frenchmen appreciated pristine fossil water preserved for eons under high pressure in underground aquifers that jetted to the surface of its own accord when struck by a drill; such artesian wells took their name from their initial discovery in 1126 at Artois.

Tea, coffee, and chocolate, perhaps because they were consumed hot, were considered medicinal when they were first introduced to Europe from China, the Islamic Near East, and Mexico in the aftermath of the Voyages of Discovery. Around the same period, another disinfecting drink, alcohol distilled from grains, became popular. The Greeks and Romans had had primitive versions of distilled spirits in antiquity and stills were operating in the ninth century and thereafter in Europe.

Modern distilled alcohol was recommended by doctors and apothecaries for its medicinal qualities; its popularization two centuries later was attended by a visible rise in public drunkenness. To help purify suspect water when better alternatives were unavailable a few drops of vinegar were sometimes added as a makeshift home remedy. Wine drinking, of course, had been popular in semiarid Mediterranean habitats from Greek and Roman times. Present-day Italians often mixed their wine and water. Hot sake, or rice wine, was long imbibed in Japan. The most ancient of all common ways of safely consuming one's daily dose of purified water was beer. Beer drinking redounded to the health of ancient Babylonians, Egyptians, and Shang dynasty Chinese, and much later to northern Europeans. From the mid-nineteenth century, the urban rich often took the additional, expensive precaution of buying water that had been filtered of its crudest and largest foreign particulates.

Until the Sanitary Revolution, ancient Rome had represented history's zenith in urban freshwater supply, hygiene, and sanitation. Public sanitary amenities had been known before Rome, but never on such a large scale and usually limited to enjoyment by the upper classes. The fifteen centuries after Rome's fall, by contrast, were the regressive, sanitary dark ages. The conditions of public waterworks likewise heralded the fortunes of Rome's successor cities. Byzantine Constantinople began its gradual descent from the sixth century AD in the same period that its attention shifted from building new water-storage dams, aqueducts, and giant cisterns to fortifying its existing supplies against the many sieges that followed. The city's magnificent sixteenth-century Turkish Islamic revival was underpinned by an expansive burst of aqueduct-building and hydraulic renovations that followed the Ottoman conquest of 1453. Likewise, Roman civilization's republican Italian descendant, Venice, was never able to grow to more than a fraction of Rome's size in part due to its chronic scarcity of freshwater. Although it built elaborate rain catchments half-filled with fine sand to filter and decant captured precipitation, which dripped through into wells in the public square, Venice faced constant shortages during dry spells and when storms caused briny water to seep from the lagoon into the wells. As a result, a fleet of water boats ferried freshwater daily from the mainland.

Elsewhere in Europe, existing Roman infrastructure and hydraulic technology fell into desuetude. Christian Europe in the Dark and Middle Ages relied upon the most primitive water supply techniques, drawing water from local wells, springs, and rivers. Chamber pots were emptied straight out the window or into underground cesspools, whose foulness seeped out into city streets and water supplies. Conditions were worst in northern Europe, where the wet climate and lack of good drainage or sewage disposal systems all but guaranteed pollution seepage into drinking sources. Antipollution regulations existed, but were often ineffective. An eyewitness in late eighteenth-century Paris reported that the Seine, the city's main drinking source, was particularly noxious the three times each week when dyers dumped their dye into one of its tributaries.

With the rapid urban growth of early industrialization conditions worsened. Heaps of rotting refuse, mixed with accumulations of human and animal excrement and urine, produced ungodly odors that overwhelmed olfactory sensibilities. "Whole quarters were sometimes without water even from local wells," writes Lewis Mumford, historian of the city. "On occasion, the poor would go from house to house in the middle-class section, begging for water as they might beg for bread during a famine." As freshwater became too precious for anything but drinking and cooking, personal hygiene deteriorated. Public bathhouses, which in the Roman tradition had remained popular into the fifteenth century, gradually degenerated into houses of prostitution and were closed by industrial times.

It was the new industrial towns of enlightenment Scotland and northern Britain that responded with the greatest alacrity to the sanitary and freshwater-supply challenges of the early nineteenth century. Scotland resurrected Rome's public water supply ideals by impounding water behind dams and instituting the first modern water filtration systems. James Watt took a particular interest in Glasgow's waterworks, which pumped water in a cast-iron pipe under the River Clyde with the assistance of six steam engines. Edinburgh tapped new springs and built a new dam and aqueduct so that by the mid-nineteenth century its six reservoirs provided each resident up to 30 gallons of wholesome springwater daily. Northern British industrial towns followed. By 1850 they had erected a dozen water supply dams to combat their shortages. Lagging far behind, however, was the world's urban leviathan, London, all but paralyzed before a mounting sanitary crisis.

It was in the capital city at the very heart of the globe-straddling British Empire that urban living conditions grew so abominable that they provoked the Sanitary Awakening and public health revolution that eventually spread worldwide. Originally settled by the Romans, London had inherited a network of pipes that connected its public fountains and baths to one of the Thames's tributaries. By medieval times water was being drawn from numerous wells, the Thames, and its tributaries, such as the Fleet and Walbrook, which today still course out of sight beneath London's streets. Water was distributed to individual homes in buckets by an industry of private water carriers who by 1496 were indispensable enough to claim their own guild. Pipes of clay, lead, and hollowed-out elm trunks conveyed some water throughout London. Publicly drawn water was dispensed free of charge to householders, but business users such as brewers, cooks, and fishmongers were charged a plumbing maintenance fee. London's first and only major long-distance water conveyance project, launched as a private venture, was initiated in 1613 to meet Elizabethan-era population growth. By 1723 there was enough water coming from the countryside for one of the half dozen private water companies to proudly fulfill its business pledge of providing water three times per week for three shillings a quarter. Because the Thames lay well below the elevations where it needed to be delivered in London, the proportion of water consumed from the Thames itself greatly increased with improved pumping technologies and soaring population. The first waterwheel pumps had been installed under London Bridge starting in 1582; steam pumps were used from 1726, one of the earliest applications of the newly minted Newcomen steam engine.

Yet neither waterwheel nor steam engine could overcome the perennial shortages of water quantity and deteriorating quality. The trebling of London's population in the first sixty years of the nineteenth century outstripped the growth of available water supply. Simultaneously, the commensurate increase in dumped sewage turned the Thames into an increasingly toxic soup. The net effect was that pumping stations on the river were doing little more than recycling increasingly poisonously polluted water to an ever more desperately thirsty London populace. As early as 1827 an irate pamphleteer created a political stir by taking umbrage with the close proximity of water intake pumps to sewer outfalls. He described the state of the Thames as "charged with the contents of more than 130 public common sewers, the drainings from dung hill and laystalls, the refuse of hospitals, slaughter-houses, colour, lead, gas and soap works, drug mills and manufactories, and with all sorts of decomposed animal and vegetable substitutes." In 1828, the Chelsea Waterworks Company introduced a pioneer filtration system to try to eliminate the largest foreign particulates; private water companies also later moved intake valves farther upstream away from the thickest pollution. Nevertheless, the rapidly dying fisherman's trade testified that they were fighting a losing battle. The last salmon caught from the Thames was in 1833.

The strong tidal nature of the Thames magnified the unsanitary brew that created the Great Stink in 1858. The level of the Thames ebbed and swelled dramatically with the shifts between low and high tides. At highest tide, river water backed far up into the sewage drains under London's streets, which were as much as 30 feet lower than the high tide mark. The sewage drained out as the tide receded, but before escaping was pushed partway back up again by the next rising tide. Thames sewage thus oscillated back and forth around London, turning pestilent with exposure, before gradually drifting out with the downriver current at lowest tides toward the sea.

Other deteriorating environmental conditions also converged in the Great Stink. For centuries Londoners disposed of their personal waste in cellar cesspools that frequently spilled over. On October 20, 1660, famous diarist of London life Samuel Pepys notated: "Going down to my cellar . . . I put my feet into a great heap of turds, by which I find that Mr. Turner's house of office is full and comes into my cellar." By

1810 London had an estimated 200,000 cesspools, one for every five residents. Some cesspools were emptied for a fee by nightsoil men, who sold the waste as fertilizer to countryside farmers. But the high cost of nightsoil removal—one-third a workman's weekly wage—discouraged free-market forces from expanding this constructively sanitary practice. Schemes to improve London sanitation by commercializing nightsoil collapsed utterly in 1847 when guano, solidified South American bird droppings, became available to English farmers as a cheaper and more pleasantly applied fertilizer. Thus the volume of cesspool discharge, and the commensurate stench of London and the Thames, continued to grow.

Ironically, the crisis also worsened from the rise of one of sanitary history's milestone achievements, the modern flush toilet, in the first half of the nineteenth century. The modern toilet originated with English poet and inventor John Harington, who in 1596 created it as a "necessary" for his godmother, Queen Elizabeth, whose high regard for cleanliness in an unsanitary age was reflected in her purported declaration that she bathed once a month "whether I need it or not." Harington's toilet had two of the three basic elements of the modern flush toilet—a valve at the bottom of the water tank and a system to wash down the waste. Yet Harington built only two toilets in his lifetime—one for his own home and one for Queen Elizabeth's palace at Richmond. Two centuries passed without notable development until 1775 when Alexander Cummings, a watchmaker by trade, invented an improved version of Harington's toilet. Successful commercialization began three years later when another self-taught inventor, Joseph Bramah, began selling toilets with an improved valve design; by 1797 he had sold over 6,000.

The third element of the modern toilet, a reliable flushing mechanism, is commonly associated with one of history's subculture folk heroes, Thomas Crapper. Contrary to popular lore, Crapper did not invent the toilet and was never knighted. What Crapper did do was to obtain a patent for an effective flushing mechanism that fulfilled his toilet's advertised promise of "a certain flush with every pull." From 1861 to 1904, Crapper's successful London plumbing business sold flush toilets with his name branded on them. His name captured the fancy of American soldiers returning from World War I, who immortalized Crapper in folklore by using it as a slang expression for the toilet, and possibly, in abbreviated form, as a verb to describe its purpose.

Toilet use in London became noticeable after 1810 and accelerated rapidly after 1830. Toilet flushing caused London's water usage to surge—as much as doubling between 1850 and 1856 alone. The increased flow washed the waste from cesspools and sewers into the Thames, whose odor grew more putrid. At high tides, the waste backed up through the antiquated sewer lines into house basements.

The use of toilets directly connected to the sewer system, mandated by the government in 1848, had been championed by a growing sanitary reform movement that had arisen following the outbreak of London's first cholera epidemic in 1831–1832. A leader of that movement was Edwin Chadwick, a lawyer and lifelong gadfly for social reform, whose influential *Report on the Sanitary Condition of the Labouring Population of Great Britain* (1842) had highlighted the link between unsanitary conditions and the disease-ridden and debased social conditions of the urban poor. To alleviate squalor, Chadwick advocated a completely new system of water and sewer pipes that would both provide abundant, clean freshwater and remove sewage far from human habitation. Aware that another disastrous cholera pandemic was headed toward England, Parliament in 1848 created a central board of health, with Chadwick at its head, to rebuild the nation's sanitary infrastructure.

At the time, no one knew what caused cholera. The establishment view was that the disease was probably transmitted through foul smells; hence Chadwick's rationale for flushing the malodorous waste away from underneath residential streets to the river. Florence Nightingale, who gained fame nursing those afflicted during the terrible cholera epidemics of the era, was a firm believer to the end of her life in the prevailing miasmatic theory of disease.

In hindsight Chadwick's sanitary policy prescription was farsighted. But its sequencing of flushing the sewers into the Thames as the first step before providing clean-drinking-water pipes proved to be tragically misguided in the devastating cholera epidemic of 1848–1849 because it misconstrued the nature of cholera. Chadwick's sequencing was challenged, unpersuasively to decision makers, by an inspired young London anesthesiologist named Dr. John Snow, who advanced the prescient theory that cholera was transmitted through contaminated water. Flushing the sewers into the Thames increased the mix of feces and drinking water, and thereby, he maintained, spread the epidemic, rather than helped contain it.

Cholera was the first rapidly spreading global disease and the most feared of the nineteenth century. A victim contracting the bacteria in the morning could be dead of its horrifying symptoms from acute dehydration by nightfall. Sudden stomach cramps, intense diarrhea, vomiting, and fever declared the disease. The face grew haggard and sunken and the skin became black and blue from rupturing capillaries. Death came from collapse of the blood circulatory system. Often one-fifth to half of those contracting the disease died.

Cholera emerged in 1817 from the delta of the Ganges River near Calcutta. It spread rapidly around the world in half a dozen pandemics, some of which leapfrogged continents as fast as a steamship could travel. It voyaged in contaminated drinking water casks aboard ship and in the fecal secretions of its victims. It spread readily between the leaky sewers and wells and in the foul drinking, cooking, and bathing water pumped up from polluted rivers like the Thames. Soldiers carried it into battle and spread it as they marched. Cholera usually showed up first in port cities, spreading rapidly along rivers, canals, and commercial routes.

The first pandemic spread in Asia, but did not reach Europe. The second emerged in 1826 from Bengal and became truly global. It struck Moscow in 1830, killed 100,000 in Hungary in 1831, hit the Baltic by 1831, and jumped by ship to England. Cholera doomed thousands in London and Paris in 1831–1832. Quarantines failed to do anything but add to the material deprivation of the cramped urban poor, who were the most afflicted due to the abysmal state of their hygiene. Riots broke out in Paris. Doctors were stoned by half-crazed mobs. In London they were accused of murdering victims in order to dissect their corpses. By 1832 the pandemic reached Ireland, then crossed the Atlantic with emigrants to terrorize Montreal and Quebec. It migrated south to the United States, striking Detroit and towns along the Erie Canal. New York became a graveyard of tolling church bells and citizens fleeing for the pastures of northern Manhattan. By 1833 the cholera reached Mexico. Pilgrims on the hajj to Mecca, first struck in 1831, carried the disease back to Islam's far-flung global homelands. Some 13 percent of Cairo's inhabitants were decimated.

The cholera pandemics that ravaged London in 1848–1849 and again in 1853–1854 added fervor to the debate over the disease's cause. In a celebrated case of medical sleuthing, John Snow, determined to find hard evidence to back up his waterborne theory of cholera, tracked a disproportionate number of cholera cases in the latter outbreak to a single free public well pump at Broad Street, not far from his own Soho medical office, that was widely used by the neighborhood's overcrowded, poor residents. Subsequent research revealed the proximity of the well to a potentially contaminating sewer. Snow persuaded a local governing body to remove the pump handle to prevent further contagion. But he could not persuade the special government committee investigating the cholera epidemic, who saw potential miasmatic causes as well. Snow continued to press his pioneering work to the end of his short lifetime. He died prematurely in 1858, the year of the Great Stink, at age forty-five.

Parliament's political will for sanitary reform had oscillated with the outbreaks of cholera. Yet even the tens of thousands of cholera deaths within five years during the midcentury epidemics did not provide enough impetus to overcome the entrenched nexus of vested local interests and liberal market economic ideology of those opposed to any centralization and enlarged public role for London's fragmented municipal government. The increasing foulness of the Thames and fear of the next pandemic, however, was a constant reminder that reform's opponents could offer no viable remedy of their own.

The mid-nineteenth-century sanitary crisis was an early manifestation of an inherent dilemma in the industrial market economy: it had no automatic, internal mechanism to restore a healthy equilibrium to natural ecosystems polluted by the unwanted by-products of growth, even though such environmental sustainability was a necessary condition of its continued productive expansion. In ancient Rome the sanitary welfare and public order had been provided by state bread doles and the public construction of aqueducts. In England the liberal democratic competition among pluralistic interests, under the duress of urgent crisis, ultimately produced an accountable municipal body with sufficient authority to provide for the common public good. The final triggering event for this reform was the Great Stink, which Parliamentarians, led by Disraeli, personally could no longer ignore.

Once empowered, London's Metropolitan Board of Works expeditiously built a world-class model urban sanitary and water supply system. Under the direction of its longtime chief engineer, Joseph Bazalgette, a sophisticated network of intercepting sewers was built under London, part of which ran parallel along each side of the Thames, to reroute the waste far downriver from central London. At certain low-lying areas the sewage had to be lifted to join the gravity flow of the rest of the system. To house part of the network, as well as provide for the underground railway, gas lines, and other familiar modernizing features of late Victorian London, three river embankments were constructed between 1869 and 1874. Another innovation was to build the sewers and tunnels with little-tested Portland cement, which proved both admirably resistant when submerged in water and able to withstand three times more pressure than traditional Roman cement.

Validation for the new sewerage system came swiftly. In the cholera pandemic of 1866, the only afflicted communities in London were those not yet fully connected to the new network. London was never again afflicted with cholera. The 1866 experience tilted the tide of official opinion in favor of Snow's hypothesis that cholera was indeed communicated through contaminated water. Final doubters were quelled by the dramatic 1892 experience in the German city of Hamburg, where one side of the street, which drew its water unfiltered from the Elbe, was devastated by the cholera outbreak while residents on the other side of the street, who drank filtered water, were entirely spared. By then, German scientist Robert Koch had already announced his 1883 discovery of the waterborne cholera bacillus during an outbreak in Egypt.

Koch's isolation of the cholera bacillus, buttressed by the contemporary research of Louis Pasteur and other pioneer bacteriologists, was a cornerstone of the landmark germ theory of disease and the stupendous public health breakthroughs of the twentieth century. Koch won the Nobel Prize in 1905. By 1893 a cholera vaccine had been developed and inoculations quickly became commonplace. The cholera breakthrough was rapidly replicated with cures for other major bacterial diseases. Typhoid fever—another waterborne filth disease whose epidemics afflicted urbanizing cities throughout the nineteenth century, and in 1861 claimed the life of Queen Victoria's husband, Prince Consort Albert, and later nearly killed her son and future king Edward—was brought under control with an effective vaccine (1897) and the same sanitary reforms that eradicated cholera. Following the stunning success of U.S. doctors in eradicating endemic, mosquito-borne yellow fever during the construction of the Panama Canal, a worldwide assault on the disease was launched in 1915 by the newly created Rockefeller Foundation; by 1937 a new, inexpensive vaccine all but eliminated the dreaded disease as a world health problem. Global malaria control became a target in the 1920s. Initial success came with drainage, and after World War II the widespread use of pesticides, such as DDT. All in all, the virtual elimination of many communicable diseases through the combination of improved sanitary and environmental conditions, antibiotics, and vaccinations caused average human longevity to leap stunningly by twenty years between 1920 and 1990 and doubling from the pre–Sanitary Awakening age. Infant mortality plunged, falling to half of 1 percent in the United Kingdom and most of the industrialized world by the early twenty-first century—a twentyfold improvement from the mid-nineteenth century.

The Sanitary Awakening and acceptance of the germ theory of disease also spurred England to take important further actions to ensure that London's water supply was both ample and clean. The guiding principles were that water should be drawn from the cleanest available source, purified, and protected against contamination during distribution. Although the Thames remained London's main supply of drinking water, it was supplemented by underground and upland river sources. Filtration plants were built to eliminate impurities through various methods, including traditional, slow sand filtering and, after the 1890s, rapid filtration of water pretreated with coagulants. Another key turning point was achieved with chlorination of water supplies from the early twentieth century. To purify water of germs, other chemical and heat disinfectants were applied, including copper, silver, ultraviolet light, and powerful ozonization processes. Sewage was jettisoned far from population centers into bodies of water under the catchy, good housekeeping guideline of societies everywhere that "the solution to pollution is dilution." From the late nineteenth century, London ceased discharging its sewage into the Thames and instead carried it on barges to be dumped in the ocean.

By 1900, England had turned the corner on improving public sanitation and health. Very gradually, the Thames recovered. Even the fussy salmon reappeared in the river in 1974 after a 140-year hiatus. By 2007, London had some 14,000 miles of sewers and was preparing its first major upgrade, featuring a 20-mile-long sewage storage tunnel under the river, since the original Victorian-age network because the old sewer system could no longer handle a population that had grown to 8 million.

England's sanitary revolution triggered a virtuous cycle of competition among industrialized democracies to improve water supplies and public health. By 1920, residents of almost all the world's rich industrial cities in Europe and North America enjoyed abundant, clean freshwater for drinking, cooking, and washing. For the first time in 5,000 years, cities became generally self-sustaining habitats for human populations. Typhoid and yellow fever outbreaks, and some cases of great, deadly fires, induced several eastern American cities to act contemporaneously with Scotland and northern Britain to provide water for sanitation, drinking, and firefighting. By 1860, 12 of the largest 16 American cities had municipality-run water supply systems. At the turn of the twentieth century, Chicago achieved America's most ambitious civil engineering project until the Panama Canal—the reversal of the flow of the Chicago River. By reversing its flow, the river no longer evacuated sewage into the city's Lake Michigan drinking supply, but instead carried it downstream to be diluted in the Illinois and Mississippi rivers. Death from waterborne disease fell sharply in America and became negligible by 1940.

Contemporaneously, sewage treatment plants became commonplace. In one of the unsung achievements of modern society the effluent of fully treated wastewater was often wholesome enough to be safely consumable as drinking water, although almost nowhere in the world did cities dare to actually do so. After the three steps of state-of-the-art sewage treatment—filtering our solids, breaking down the remaining organic matter with microorganisms, and applying chemical disinfectants to kill remaining bacteria—the quality of the discharged water was often superior to the bodies of water into which it was discharged. Rather than being dumped in the sea, London's sludge today is incinerated through a bed of sand at 850°C—with the recovered heat used to power electricity-producing steam turbines that drive the treatment plant, and the excess energy sold to Britain's electric grid. The final, released wastewater is measurably cleaner than the water in the Thames.

By enormously increasing the supply of clean freshwater resources, the sanitary revolution played a pivotal role in sustaining the urban ecosystems at the heart of industrial civilization. Without it, the momentous, rapid shift of humanity from the farming countryside to the industrial cities would have been impossible. In 1800, only 2.5 percent of the world's population, or about 25 million people, lived in cities. In 2000, nearly half the world's 6 billion people did so. Urban concentrations became immense: 29 megacities held over 7 million compared to only six cities in the world with 500,000 two centuries earlier.

Western liberal democracies' success in delivering ample freshwater and sanitary services to its citizens provided one of its important comparative economic and politically legitimizing advantages over its Cold War rivals. The communist world's authoritarian, command economy states, in contrast, lagged notoriously far behind in providing sanitary and other kinds of environmental health—both a leading indicator and a causal force of their relative decline. Shortly before the collapse of the Soviet Union in the late 1980s, for example, the Moscow River received untreated nearly all the sewage of the capital city, rendering it virtually an open sewer reminiscent of the Thames during the Great Stink. In the same period, some 90 percent of Chinese cities had no wastewater treatment at all. Worse still were conditions in the third world, which could best be compared to those of Europe in the mid-nineteenth century, with 90 percent of all sewage and 70 percent of industrial wastes dumped into streams and lakes without any treatment at all at the dawn of the twenty-first century.

That the influential pacesetter in leading the response to the sanitary environmental challenges of early industrialism had shifted from Britain to the United States was not coincidental. It reflected the continuing historical shift westward in power across the oceans to the United States, which became the world's most prolific, productive, and innovative manipulator of water by the early twentieth century.

CHAPTER ELEVEN

Water Frontiers and the Emergence of the United States

The global ascendance of the United States closely paralleled its mastery of its three disparate hydrological environments: its rainy, temperate, river-rich eastern half, dominated by the continent's arterial Mississippi River; its predominantly arid, drought-prone, Far West extending to the Pacific Ocean from the 100th meridian of the high Great Plains; and its frontage on the sea-lanes between the world's two largest oceans. By fusing these diverse water frontiers into a coherent national political and economic realm, America leveraged its favorable geographical location and the abundant natural resources of its vast island-continent to become civilization's world superpower in the twentieth century.

Like other great states' rise to power, America gained command of its native resources for the main conventional uses of water as well as made innovative responses to special challenges that mobilized water's inherent transformational powers to produce spectacular breakthroughs that defined the age. The first phase, which was fully realized by the end of the nineteenth century, featured the westward expansion of its frontier from the coastal states east of the Appalachian Mountains throughout the rich farmland of the Mississippi River Valley as far as the beginning of the dry belt in the Great Plains of Kansas and Nebraska. Development was activated mainly by the application of "Yankee ingenuity" to existing European economic technologies. This enabled America to take advantage of the region's abundance of lakes, rivers and fast-flowing streams, rich farmland, wooded forests, and long indented coastlines, and to compensate for the young nation's shortages of labor, capital, and technical expertise. Waterwheels and early water turbines powered the rise of homegrown factories and later provided the key to exploiting America's huge potential hydroelectricity. The enormous farm and raw materials wealth of the Mississippi Valley heartland was unlocked by the advent of river steamboats and canals, which created an inexpensive, long-distance inland water transport network linking the markets of New York, Pittsburgh, Chicago, and New Orleans at the mouth of the Mississippi. Steam locomotives thickened and extended the transport web across the continent by 1869, adding momentum to America's industrial ascent. By the late nineteenth century, when the age of iron and steam was superseded by the mass production technologies of steel, electricity, petroleum, and the internal combustion engine, American industry was the most productive in the world.

While America rose to power on its eastern resources, it truly distinguished its destiny as world superpower by overcoming and harnessing the latent potential of water obstacles within its two other hydrological frontiers. Its global primacy was first declared through its completion in 1914 of the grandest water engineering challenge of the age—the Panama Canal. At a stroke, the Canal established America as the commercial fulcrum of maritime world trade, launched the sea power of its increasingly formidable "big stick" navy across two oceans and quickened the linkages between its underdeveloped Far West and its productive eastern economy.

Even greater impetus was generated by the water innovations that transformed its inhospitably arid, virgin, western frontier lands into a cornucopia of irrigated agriculture, mining, and hydroelectric-powered industry. The original Boulder (later renamed Hoover) Dam on the Colorado River provided the technology prototype for the giant, multipurpose dams erected worldwide in the twentieth century that facilitated the extraordinary prosperity of the agricultural Green Revolution and global industrialization. Midwestern farmers created a breadbasket

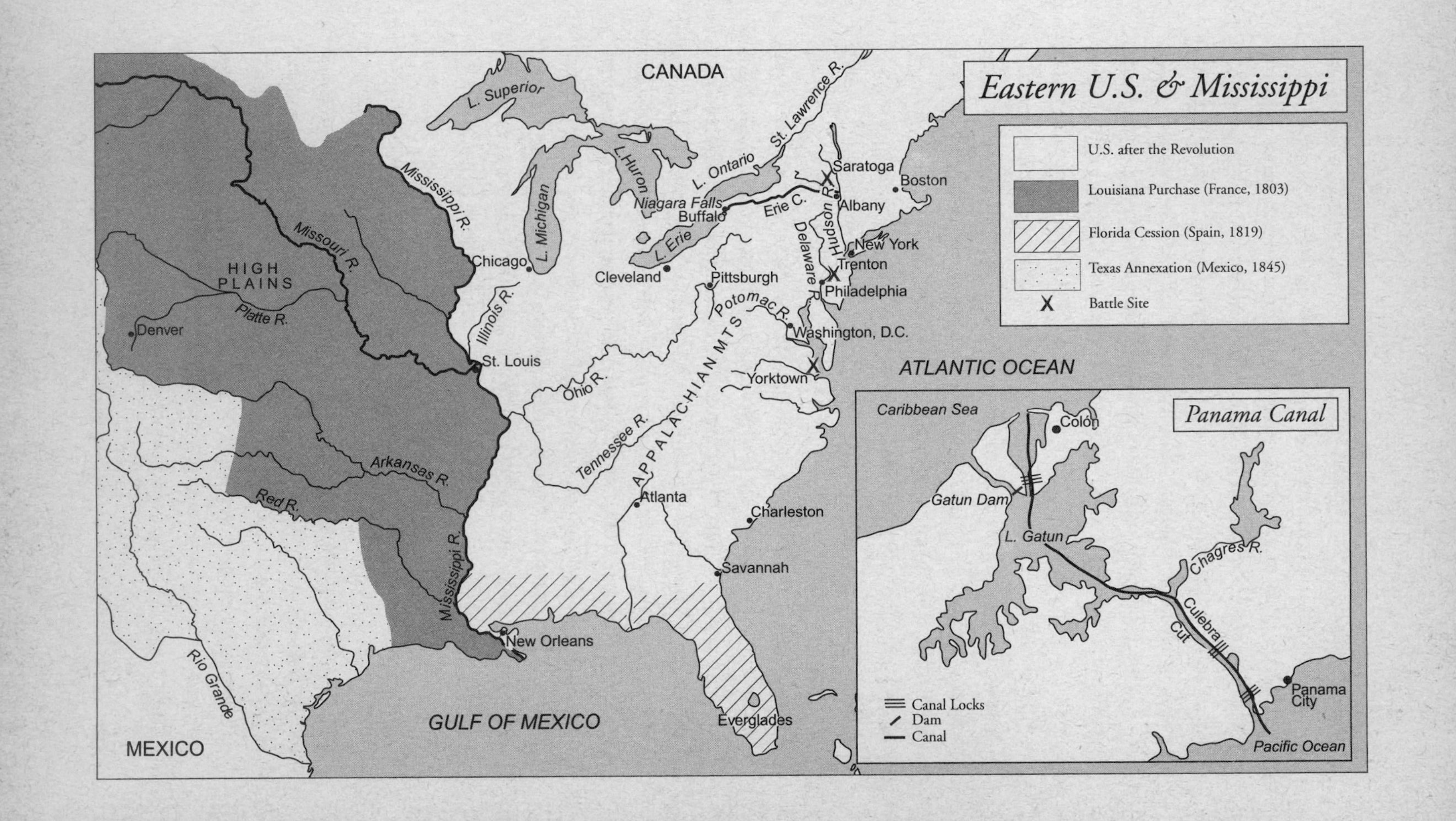
Eastern U.S. & Mississippi
U.S. after the Revolution
Louisiana Purchase (France, 1803)
Florida Cession (Spain, 1819)
Texas Annexation (Mexico, 1845)
Battle Site
CANADA
L. Superior
L. Michigan
L. Huron
L. Ontario
L. Erie
St. Lawrence R.
Niagara Falls
Buffalo
Erie C.
Saratoga
Albany
Boston
Hudson R.
Delaware R.
New York
Trenton
Philadelphia
Pittsburgh
Cleveland
Chicago
Mississippi R.
Missouri R.
HIGH PLAINS
Platte R.
Denver
Illinois R.
St. Louis
Potomac R.
Washington, D.C.
Yorktown
Ohio R.
APPALACHIAN MTS
Tennessee R.
Arkansas R.
Red R.
Atlanta
Charleston
Savannah
Everglades
New Orleans
Rio Grande
MEXICO
GULF OF MEXICO
ATLANTIC OCEAN
Panama Canal
Caribbean Sea
Colón
Gatun Dam
L. Gatun
Chagres R.
Culebra Cut
Panama City
Pacific Ocean
Canal Locks
Dam
Canal

from a dust bowl when they became capable of tapping the enormous water wealth hidden away in the huge fossil aquifer—an underground lake the size of Lake Huron—submerged deep beneath the high central plains through the advent of more powerful pumping and irrigation technologies. By the 1940s, in short, America was exploiting its ample natural water resources in a more intensified and enlarged manner than any society on Earth—a reliable leading indicator and catalyst, in every age of history, of robust prosperity and civilization.

Water was a key strategic determinant of America's victory in its War of Independence from England in the late eighteenth century. By fortuitous timing the American Revolution occurred at the end of the age of sail and prior to the beginning of naval steam power. This effectively enlisted the sea itself as a natural ally of the secessionists and minimized England's greatest military advantage—sea power. The arduous, six- to seven-week transatlantic sail protracted British supply lines and complicated execution of its command and control. Every British soldier, every cannon and musket, and every food ration had to be shipped 3,000 miles across the sea. Had the British already possessed steam gunboats, they could have easily maneuvered up America's inland rivers to impose their will militarily upon the inland population as they did from the 1820s onward in Burma, India, and China. Instead they were confined to their more cumbersome sail age tactics, such as blockading harbors, raiding and seizing seaport cities, patrolling traffic on the high seas, and convoying troops and supplies among coastal ports. Suppressing the rebels required large troop deployments over expansive distances and rugged interior terrain, exploited by the American Continental Army's hit-and-run tactics, which would have been a challenge even for the brilliance of Napoléon. The British army of the day was not up to the task, militarily or logistically. England therefore pinned its main hope of victory on rallying the active supply and intelligence support of colonial Loyalists.

Three of the Revolutionary War's decisive battles, in fact, turned on the control of strategic waterways—Washington's surprise attack across the Delaware on the British garrison at Trenton at Christmas 1776, Burgoyne's surrender at Saratoga on October 17, 1777, following Britain's failure to secure the Hudson River, and Cornwallis's final surrender at Yorktown on the Chesapeake Bay four years later when combined French and American forces cut off the British army from naval resupply or escape route. Even in precipitating the Revolutionary War, water had figured symbolically in awakening the public imagination on both sides of the Atlantic. On December 16, 1773, colonial radicals, thinly disguised as Mohawk Indians, dumped 342 chests of tea belonging to the East India Company into Boston harbor to incite against the company's tea monopoly and British taxation. With the war enjoined in earnest in the summer of 1776, the central theater focused on strategic New York, America's second-largest city after Philadelphia with 22,000 inhabitants. In addition to having an excellent harbor for supplying and deploying troops, New York was a vital choke point from which an army could strike east into New England, north up the Hudson River Valley, or west into New Jersey. General George Washington put his army's best efforts into holding New York and was nearly destroyed in failing. The British made the city its central base of operations throughout the war.

Defeated at New York, Washington's Continental Army was forced to retreat through New Jersey. In early December 1776, Washington saved his beleaguered troops from destruction by escaping across the Delaware River into Pennsylvania. Before crossing, he gathered all the boats he could find on the Jersey side. Since there were no nearby bridges across the river north of Philadelphia, British forces had no ready means to give chase without boats. Thus the defensive barrier of the Delaware River, and the onset of winter, deprived the British of early victory. Nevertheless, Washington's defeats had demoralized his troops, whose enlistment tours were set to imminently expire, while sympathetic colonists began to succumb in large numbers to British offers of clemency. This desperate reality drove Washington to make an inspired gamble. On the frigid night of December 25, 1776, he ordered 2,400 weary, underclothed soldiers, horses, and 18 cannon to be ferried back across the icy Delaware into New Jersey. They started at 7:00 p.m., and the ferrymen worked all through the hours of darkness. By sunrise all were across, marching through the sleet and rain toward Trenton. In one of the most celebrated victories in American history, their surprise attack caused the surrender of 900 unprepared German mercenary troops employed by the British, along with their six cannon and 1,200 small arms. No Americans were killed. Only four were wounded, while two froze to death marching to the battle. The victory's effects were electrifying. Reenlistments promptly increased and new troops surged to Washington's side. Sentiment among wavering colonists was buoyed. The Revolution survived to fight another season.

When the war resumed after the winter, the British ministry in London launched a new campaign to subdue the rebels. Two British armies, one moving south from Canada and the second moving north from New York, were to execute a pincer movement to take control of the strategic Hudson waterway. By controlling the Hudson, the British could sever the radicals of New England from the rest of the colonies. But slow and poor communication across the Atlantic from London hindered British execution. While 8,000 troops under General John Burgoyne launched the campaign from Canada, General Sir William Howe's New York force became diverted by its seizure of Philadelphia and did not remobilize in time to close the southern pincer. Fighting alone, Burgoyne's army became seriously encumbered by patriot resistance and its own heavy supply logistics. As it marched across the wild Hudson Valley terrain, rendered even less passable by the rebels' systematic destruction of bridges, tree felling and stream diversions, Burgoyne's army was slowed at times to covering but one mile per day. It had to rebuild over 40 bridges along the way. Nor did it help that Burgoyne, known as "Gentleman Johnny" for his playboy ways, traveled with a personal entourage some three miles long, including 30 personal baggage carts, his mistress, and numerous bottles of claret and port. Patriot troops led by General Benedict Arnold and Vermont's Green Mountain Boys inflicted defeats that degraded Burgoyne's forces, whose Mohawk Indian allies began to slip away. Finding few Loyalists en route and without Howe's army in rendezvous, Burgoyne's supplies and manpower ultimately gave out. Trounced in two bloody battles near Saratoga, Burgoyne surrendered his 6,000 man army on October 17, 1777.

The Americans' triumph reverberated across the Atlantic. Persuaded that the rebels had a chance, France's Louis XVI finally acceded to ambassador Benjamin Franklin's entreaties for a Franco-American alliance. In a late bid to head off just such an alliance, the British government prepared to offer the colonists everything they had asked for prior to the Declaration of Independence, including freedom from loathed Parliamentary taxation. But the conflict had escalated too far. On February 6, 1778, the French and American governments signed treaties dedicated to obtaining British recognition of America's independence. Within two years other European nations joined what amounted to world war against England. In England itself the war was growing costly and politically divisive. But the autocratic King George III overruled all Parliamentary dissent—he insisted that the rebels be put down firmly and, if necessary, ruthlessly.

The whole New World was now in play, including the fabulous wealth of the Caribbean islands, which were more preoccupying concerns to England and France than the fate of the relatively poor American colonies. Over the next couple of years, a military stalemate prevailed. Washington lacked the naval power to evict the British from New York, an effort the French were reluctant to undertake with their forces. The British, meanwhile, facing a hostile populace avoided undertaking unsupported invasions deep into the interior. The situation changed in the summer of 1780 when the British under General Lord Charles Cornwallis took Charleston, South Carolina, by amphibious assault. Cornwallis's plan was to take the Carolinas and Virginia, where Loyalist sympathies were stronger, and to try to overwhelm the rebel colonies from the South. The Continentals, with substantial difficulty, managed to impede Cornwallis's campaign. In the spring of 1781, Cornwallis began building a base at Yorktown, Virginia, a Chesapeake peninsula flanked by the York and James rivers, where he planned to receive supplies and troop reinforcements by sea. The Americans and French, however, seized it as a chance to entrap him.

The denouement of the Revolutionary War began with France's decision to wholeheartedly invest its navy at Yorktown. Having learned its lesson about the importance of sea power from its disastrous defeat in the Seven Years' War, France had been straining its Treasury for a decade to build up its navy. By 1781, it had attained near parity with England's. A French fleet under Admiral François de Grasse sailed from the Caribbean to blockade the Chesapeake Bay from British resupply. After an elaborate feint at invading New York that successfully misled the British as to the rebels' genuine target, a large force of American and French troops were dispatched to confront Cornwallis. The crucial sea battle was fought in the afternoon of September 5, 1781. De Grasse's 24 French warships outgunned 19 British ships, which sailed away to New York for repairs. With the Chesapeake blockaded, George Washington and the French generals, with a combined force twice that of Cornwallis's nearly 8,000 troops, amassed for the siege of Yorktown. It began on September 28. Artillery bombarded the British position. Several raiding parties struck, one led by Washington's favorite, Lieutenant Colonel Alexander Hamilton, later U.S. treasury secretary and influential Federalist. Although casualties were light, Cornwallis realized his plight was hopeless when his two small wooden forts were overrun and his army's escape attempt across the York River failed in a squall. He was cut off from both naval resupply and retreat route. On October 17, four years to the day after Burgoyne's surrender at Saratoga, he put up the white flag to discuss terms. Two days later he surrendered his entire army. The war was over.

The Revolutionary War was a decisive victory for the spread of liberal democracy. An influential republic was born in North America. In England the war subdued King George III's bid to reassert personal monarchal authority, with the result that full Parliamentary government was permanently established in Britain on the eve of Britain's influential age of empire. On the European continent, too, Louis XVI's monarchy soon was toppled in 1789 by a French revolution inspired partly by American success and the repercussions of the treasury's depletion in having spent so much to assist them.

• • •

Although the war had been fought along the eastern seaboard, a principal battlefield of contention was the land west of the Appalachian Mountains leading to the expansive Ohio and Mississippi River valleys. For the founding fathers, the western frontier was the gateway to America's destiny. British policy had long aimed at fencing in the colonists east of the Appalachians where they would stay more closely linked to England's maritime empire. But native colonial leaders saw the beckoning Mississippi Valley farmland and intersecting navigable rivers beyond the mountains as the vital keys to a prosperous western empire of America's own. George Washington and Benjamin Franklin were among many land speculators in the western farming frontiers who believed it would be the main source of America's wealth for years to come. Thomas Jefferson envisioned the expanding frontier of self-reliant, individualistic, yeoman farmers as the quintessence of America's democratic character. The free land of the frontier was a safety valve that would attract settlers westward when urban conditions grew too squalid, thus protecting America's uniqueness and inhibiting the transplantation of Europe's class-based and industrial inequities to American soil. Jefferson and others also imagined an American continental destiny that fulfilled the old dream of a Pacific trade route to the Orient. Although the early U.S. political landscape was split by the sharp divide between industry-promoting Hamiltonian Federalists and agrarian Jeffersonian Republicans, both sides were united in the importance of securing western expansion.

At the heart of fulfilling America's westward destiny was control of the continent's mighty river, the Mississippi. The Mississippi was America's Nile, its twin rivers, its Indus and Ganges, its Yangtze and Yellow rivers. By length, the Mississippi-Missouri system was the world's fourth longest at 3,740 miles, after the Nile, the Amazon, and the Yangtze. Only seven rivers carried a greater volume of water. Most important was its fertile valley that stretched some 1,250 miles across America's heartland between the Rocky and Appalachian mountain chains, north to Canada and south to the Gulf of Mexico. The Mississippi valley was double the size of the valleys of the Nile or the Ganges, and 20 percent bigger than north China's Yellow River valley. It encompassed over two-fifths of the continental United States, draining hundreds of rivers and streams, including the Ohio, the Missouri, the Tennessee, the Platte, the Illinois, the Arkansas, and the Red. In the world only the Amazon and the Congo had a larger drainage basin. It alone could sustain American ambitions for a national empire.

The Mississippi was both an arterial and a flooding, irrigable river. Its navigability offered the potential of a natural inland waterway transport network that could almost instantly integrate the scantly populated nation over unimaginably expansive distances that were virtually impassible overland. Especially from its western tributaries, the "Big Muddy" carried more silt than all but six world rivers that it spread as a thick fertile residue over midwestern farmlands during its many floods. At the confluence with the Ohio River, the Mississippi doubled in size and often stretched a mile and a half from shore to shore as it meandered through the floodplain in its lower reaches to its exit in the Gulf of Mexico near New Orleans.

The Mississippi was also a highly complex, turbulent river, driven by the peculiar dynamics of its size, diversity of its four component parts, deep channels, snaking pathways, tidal forces from the Gulf, multiple currents and velocities, and its penchant for massive flooding and unpredictable changes of course. An unusual feature of the lower Mississippi was that for its final 450 miles its riverbed lay below sea level to a depth of 170 feet at the port of New Orleans. One effect was that the higher water tumbled over itself at a much faster speed than below, sometimes generating such an enormous force that it sheared away and broke through entire riverbanks and the levees that man erected to try to contain it. At its mouth, the sediment loads spewed out by the river created gigantic sand bars that sometimes blocked shipping access to the Gulf of Mexico for weeks and months at a time.

After 1850, the American government toiled relentlessly, much like the river irrigation civilizations of antiquity, building and rebuilding thousands of miles of levees and spillways in an effort to control the lower-Mississippi flooding and enhance its navigability. But every so often a surging flood overwhelmed man's best artifices. The giant flood of 1927, driven by months of heavy rains, turned the Mississippi into an immense continental drain that obliterated the entire network of flood control levees created by the U.S. Army Corps of Engineers. Cities and farmland were submerged across the expanse of the Mississippi Valley in one of the most devastating natural disasters in U.S. history. Half a million people were displaced. The force of the flow was so strong that it temporarily forced the mighty Ohio to flow upstream. At its mouth, New Orleans was saved from destruction only by dynamiting levees upriver to divert the floodwaters in other directions. The flood exposed the Army Corps' disastrous error in trying to control the powerful river only with levees and without extensive reservoirs and cutoffs. The water engineers rebuilt, this time creating large spillways to accommodate the river's extreme swells and reduce the pressure on the levees. The river was also straightened, heavily dammed and channelized so that nearly half its flow was contained within man-made barriers; at the same time, however, over 17 million acres of surrounding wetlands—now understood to be a critical, natural, buffering sponge against floods—were filled in for development. Once again, another great Mississippi flood in 1993 defied man's ablest efforts to manage nature. Although only about one-third the force of the 1927 disaster, the 1993 flood nevertheless overflowed the levees, and turned 1.2 million acres of its natural midwestern floodplain into a giant lake that did not recede for months.

America first gained access to the Mississippi in the 1783 Peace of Paris that ended the War of Independence. Throughout the long, complicated multinational negotiations, America's main representatives, John Jay, Benjamin Franklin, and John Adams, held out obdurately, and when necessary inveigled, to obtain U.S. domain over the lands west of the Appalachian Mountains as far as the Mississippi River. In the end, the parties agreed that the independent United States was to be bounded in the west by the Mississippi and in the north by British Canada. The southern boundary was Spanish Florida, whose elongated panhandle extended across the entire Gulf of Mexico, including the strategic mouth of the Mississippi at New Orleans. The Spanish holdings formed a land bridge from the western side of the Mississippi to the Louisiana territory that France had reluctantly ceded to Spain in the 1763 settlement of the Seven Years' War.

The Peace of Paris left unresolved several territorial and commercial disputes that within a decade flared into renewed contentiousness with Britain. To avert war with England, and to safeguard the basis for America's western expansion, Washington's government in 1794 signed a controversial treaty that resolved many outstanding territorial and commercial disputes and formalized U.S. commerce with the British West Indies. The treaty paid the unanticipated dividend of encouraging Spain, which feared ulterior American and British designs on its Louisiana possession, to seek appeasement by agreeing to withdraw from many contested positions east of the Mississippi and granting America free navigation rights on the lower Mississippi and transit at New Orleans, the coveted gateway port to the Gulf and the Caribbean. Postrevolutionary France, which had resumed its superpower rivalry with England with the outbreak of a prolonged, new period of warfare during which Napoléon Bonaparte came to power, however, viewed the 1794 treaty as a blatant tilt of America's professed neutrality in favor of England. Franco-American relations deteriorated. By 1798, the two former allies were fighting what was in effect an undeclared naval war in the Caribbean.

This was the belligerent backdrop to a series of dramatic events that unexpectedly led to one of the greatest land deals in history—the 1803 Louisiana Purchase. With it, America secured control of the whole Mississippi Valley. Yet in an age when colonial territories continuously changed hands with each shift in the balance of power among the great European nations, America's legal grip on the mostly unoccupied Mississippi Valley remained vulnerable to reversal. Indeed, the grandiose plans of Napoléon, then rising to supreme power in France, included the establishment of French dominion over America's west, Florida, and Canada. In expectation of a French invasion of America, America began actively fortifying its navy and army at the end of the eighteenth century. Invasion fears temporarily abated after Nelson's victory in 1798 at the Battle of the Nile trapped Napoléon's army in Egypt. But they soon rekindled following a failed U.S.-French peace negotiation and the revelation in 1801 of the secret retrocession to France by Spain of its rights to Louisiana. War fears heated in 1802 when Napoléon deployed tens of thousands of French troops to nearby Haiti to suppress the slave rebellion in France's prized sugar- and coffee-producing colony. They became febrile later that year when Spain suddenly rescinded transit rights to U.S. traffic at New Orleans, effectively closing America's Mississippi River access to the Caribbean.

Although a famous Francophile, President Thomas Jefferson became so alarmed at Napoléon's designs on America that in April 1802 he wrote to America's ambassador in France, Robert Livingston, exhorting him to negotiate a solution because, "The day that France takes possession of New Orleans . . . we must marry ourselves to the British fleet and nation." With war tensions rising, Jefferson a few months later dispatched specific additional negotiating instructions to his ambassadors in Paris. The United States, he instructed, was to offer up to $10 million to buy New Orleans and the Floridas outright. The price for New Orleans alone was $7.5 million. If France still refused to sell, they were to try to negotiate a perpetual right of transfer. All failing, Jefferson said, the American emissaries were to begin secret communications with England for the closer entangling alliance he fervently hoped to avoid.

Through no action on America's part, circumstances at this juncture unexpectedly broke to the young nation's advantage. The large French campaign to suppress the Haitian slave rebellion went badly awry. Rebel resistance was part of the reason, but far more debilitating to France's 33,000 troops was the yellow fever epidemic inflicted upon them by the waterborne mosquito indigenous to the Caribbean tropics. Thousands died or became too enfeebled to fight. It was not the first nor the last time that waterborne diseases altered the course of history. Napoléon was forced to abandon Haiti, and with it his dream of rebuilding France's New World empire, lest it compromise his grander strategy of invading England. According to Napoléon's new political calculus, Louisiana would be better placed in American hands than to be left vulnerable to British seizure.

Thus on the very day Napoléon broke diplomatic relations with England, April 11, 1803, Napoléon's minister, Talleyrand, stunned U.S. ambassador Robert Livingston at their day's meeting by suddenly inquiring, "What would you give for the *whole* of Louisiana?" Livingston, regaining his composure, offered $4 million. "Too low!" Talleyrand said. "Reflect and see me tomorrow." By April 30 the deal was done: for about $15 million, America got all of the Louisiana territory, including New Orleans, a claim on part of the western panhandle of Florida, and some of Texas. The entire fertile Mississippi River Valley—the key to the midwestern empire—was now officially within America's grasp. Transfer to the United States was expeditiously completed in December 1803, the same month France withdrew its last troops from Haiti, which soon became the world's first independent nation created by former slaves.

Yet Jefferson did not depend on legal negotiation alone to win America's claim on the western frontier. Several months *before* the Louisiana Purchase was agreed, he took steps to launch the famous, 50 man Lewis and Clark expedition of 1804–1806 to follow the Missouri River in search of a northwest water passage to the Pacific. Establishing viable water transportation routes, Jefferson reckoned pragmatically, would stimulate settlement and trade, and win possession of the unsettled territory by the de facto force of occupation. Soon thereafter he sent other, less-celebrated expeditions to explore the courses of the Red and Ouachita rivers and the source of the Mississippi; the latter went astray and instead tracked the Arkansas River to its headwaters in the Rockies.

Although by the early nineteenth century the western Mississippi Valley frontier beckoned with promise, it remained an unsettled wilderness. The majority of America's population of almost 4 million still lived along the eastern seacoast. There was no ready transport route through the Appalachian Mountains to join the two regions by commerce, emigration, and common political destiny. Almost immediately after the Peace of Paris was signed, George Washington returned with urgency to the project that had absorbed him before the Revolution—to convert the rocky Potomac River into a navigable waterway that would become the primary gateway through the mountains to the west. As a statesman, Washington was preoccupied by the necessity of opening inland navigation to bind the western settlers to the United States instead of to the British or Spanish to the north and south. As the largest landowner along the Potomac and owner of over 33,000 acres of prime bottomland in the Ohio River Valley, Washington stood to profit handsomely from such a waterway. In 1785, after personally surveying the Potomac route and looking for a way to connect it to the Ohio, he became president of the Patowmack Canal Company. With the backing of influential fellow Virginians who stood to benefit most, he raised capital from private investors to build the project. In the end, he failed. The stony, waterfall-strewn, and sometimes too shallow Potomac proved too technically challenging. In 1788 Washington retired from active management to take up his new job as first president of the United States.

Settlement of the western farming frontier thus proceeded modestly. Yet the new nation had fertile, rain-fed cropland in the east as well as other water resources to launch its growth. From the outset, America's long coastline and its British heritage had bred a vigorous maritime culture. Whaling and fishing were major activities in the northeast. Dried fish and whale oil, timber from America's virgin woodlands, and agricultural surpluses from the middle colonies were traded up and down the Atlantic seaboard through the many good natural harbors from Boston to Baltimore to Charleston, into the Caribbean, and across the ocean to southern Europe. Often they were transported in American-built ships. New England's abundance of tall trees, notably the 120-foot white pine that was ideal for cutting sturdy, single-piece masts, had made it a major shipbuilding center for over a century. Indeed, Britain's early deforestation had spurred English orders from colonial shipbuilders; on the Revolution's eve, some one-third of Britain's fleet came from U.S. shipyards.

With a few notable exceptions such as Alexander Hamilton, most of the founding fathers quite reasonably believed that there was little prospect for rapid industrial development in a country so rich in farmland

and raw materials and so short of capital, labor, and technical expertise. One comparative advantage America could exploit, however, was the cheap waterpower from its large number of swift-flowing rivers and streams. Indeed, settlement patterns since colonization had tracked good waterpower sites as well as navigable water routes. Many inexpensive wooden and cast-iron waterwheels were built from local supplies to power town gristmills and sawmills and the bellows and trip-hammers of the surprisingly widespread industry of small iron forges and foundries. Due to the availability of cheap waterpower and wood charcoal fuel, in fact, colonial ironmongers were actually producing more total pig and bar iron than England, and one-seventh of total world output, at the time of the Revolutionary War. Nevertheless, there was scant reason to think such rudimentary beginnings could provide the springboard of a homegrown American industrial revolution.

One unlikely motor of the U.S. industrial revolution was in textile manufacturing, notwithstanding Britain's growing global dominance in inexpensive, high-quality textiles produced in its state-of-the-art, steam-powered factories. To protect its textile technology monopoly, moreover, Britain vigorously enforced sanctions against machinery exports and the emigration of skilled textile workers. America's industry got its start because one ambitious young Englishman defied British sanctions to seek his fortune across the Atlantic. Samuel Slater had worked for years as a teenage apprentice and risen to the job of overseer in the textile mill of one of the partners of cotton manufacturing tycoon Richard Arkwright. A skilled technician with a gifted memory, Slater managed to memorize the design of the entire Arkwright factory. He disguised himself as a simple farm boy, and in 1789 sailed for America. Straightaway he entered business as a partner of a wealthy Rhode Island merchant, Moses Brown, who had been vainly trying to build an efficient cotton factory. Within a year, Slater reproduced an Arkwright-type mill in Pawtucket on the Blackstone River. But when opening day came, the machinery failed to function properly. It turned out that Slater had failed to remember the correct angle of the carder teeth. Following a tedious adjustment, America's first automated cotton mill was operational. It had three carding machines and a spinner with

72 spindles—tiny in comparison to the 1,000 spindles at Arkwright's original 1771 mill—and was run by a workforce of nine children, aged seven to twelve. By 1801, the profitable mill was powered by the falls of the Blackstone River with over 100 employees.

Many of the millwrights and workers who trained under Slater spawned a new generation of water-powered cotton mills. Most, however, failed because they couldn't compete with imports from English factories. America's infant textile industry was saved from premature demise by a radical change in business conditions caused by President Jefferson's 1807 imposition of a foreign trade embargo. The embargo was intended to discourage the ongoing seizure of neutral American vessels on the high seas by both France and England during an escalation of the Napoleonic Wars. Like many embargos, it had unintended consequences. Exports and imports, including British textiles, froze. Any domestic American manufacturer able to make goods that substituted for the absent imports, suddenly could earn handsome profits. In 1809 alone, some 87 new cotton mills were built, nearly sextupling the 15 in existence. It was one of the ironies of U.S. history that it was the embargo policy of the agrarian champion Jefferson that galvanized the American industrialization he feared and that had been so ardently championed by his arch political opponent, Hamilton.

Yet early industry couldn't have so responsively taken root had not England's free enterprise culture already been transplanted in American soil. Indeed, America's water geography further invigorated that culture. Its sea coastal economy linked it to Europe's maritime free-market trading traditions. Its temperate, rain-fed, and small-river-rich landscapes promoted self-sufficient, independent communities with the wherewithal to safeguard its private property rights from any excessive commanding impulses of the central government, which in any case, as a practical matter, wished to encourage market entrepreneurialism in the economic realm. The rustic necessity of clever tinkering to make things work within the available physical resources and limited human labor at hand furthermore bred a distinctive practical innovativeness tailored to American conditions. This "Yankee ingenuity," whet by the incentive of large pecuniary reward, yielded many original industrial inventions that spurred private enterprise. In 1787 Oliver Evans—who later invented the high-pressure steam engine that pumped Philadelphia's waterworks and a gigantic steam-powered amphibious dredger—built a fully automatic, water-powered flour mill that processed wheat into flour with virtually no need for human labor; by 1837, 1,200 automated factories were at work on the frontier west of the Allegheny Mountains. What Evans did for flour, Eli Whitney did for cotton with his 1793 invention of the cotton gin that cleaned cotton with fifty times greater efficiency than by human hand and could be powered by water or animate force; overnight cotton became a thriving cash crop of the American South, reviving the waning institution of slavery in order to meet the surging demand for raw cotton. No innovation of Yankee ingenuity had so far-reaching, long-term impact as Whitney's 1801 machine tool that produced standardized and therefore interchangeable parts—the core technology of the mass production methods that became the signature of American industry.

In textiles, the next important entrepreneurial breakthrough was made by Francis Cabot Lowell. While on a two-year family sojourn in England, Cabot, a well-to-do merchant from a prominent New England family, took special interest in his visits to the cotton mills of Birmingham and Manchester. At each stop he endeavored, like Slater before him, to remember as many details as he could about the mills' layout and design of its machines. Upon his return home, Lowell raised capital from an association of wealthy Boston families and hired a master mechanic, Paul Moody, to help him build America's first power loom to weave cloth. His 1813 spinning and weaving mill on the Charles River near Boston became America's first integrated cotton factory, turning raw cotton into finished cloth. It was so successful that within a decade it became the model for the first planned industrial town. Located northwest of Boston at the junction of the Merrimack and Concord rivers near a 30-foot waterfall with enough waterpower to drive its factory system on a massive scale, the new town was named Lowell by the founding entrepreneur group in memory of Lowell, who had died in 1817 at age forty-two. At its peak in the late 1840s, Lowell's complex of 10 major factories employed 10,000 workers. Its waterpower system included six miles of canals, dams, and reservoirs and the extraction of more than 10,000 horsepower from the falls. The company became the largest cotton producer in a national industry that by 1840 had grown to 1,200 factories with 2.25 million spindles. The village had become an industrial city of over 20,000 inhabitants. As late as 1870, cotton goods were still America's second largest industry, surpassed only by indispensable flour mills that produced Americans' daily bread.

What made Lowell's factory system the object of wide admiration of Europeans for whom it was a fixed stop on their American tour, however, was not its productive output but its unique approach to labor management relations. The high-minded Lowell, influenced by utopianist Robert Owen and the ideals of nineteenth-century New England, conceived his factory system to demonstrate that profitable industry need not be accompanied by the satanic conditions of squalor, filth, poverty, illiteracy, and moral depravity that characterized the crowded British mill towns he'd seen. To attract enough farm girls in a chronically labor short, rural environment to his factory, he offered good living conditions and high enough wages to allow them to be able to save for a small dowry after two or three years. At the company town of Lowell, the neatly dressed girls lived in chaperoned boarding houses, with no more than two to a bed, around a square landscaped with trees and shrubs. Although their lives were regimented and they worked twelve hours per day, six days per week, they were edified by a literary weekly, company-organized lectures, and religious instruction. In his 1842 American tour, famous British factory system critic Charles Dickens extolled the Lowell system's virtues.

Yet Lowell's novel industrial labor relations proved to be less enduring when challenged by the fierce realities of free-market competition. Living and work conditions did not keep pace with business expansion and the company's quest for profits. Early strikes in 1834 and 1836 were crushed. From the 1840s, the Lowell girls were replaced by vast numbers of unskilled, illiterate, low wage, and compliant European immigrants who began swarming across the Atlantic with the advent of ocean steamers in search of better lives. In the twenty years leading up to 1840, the number of immigrants rose ninefold to 90,000 per year. In 1850, 300,000 came over; in 1854, nearly 500,000. By the mid-nineteenth century, foreign immigration had ended America's chronic labor shortage. Domestic businesses had accumulated ample capital to finance large-scale investment. Technical expertise, too, was available. Other bottlenecks to growth, including transportation, had been alleviated. America was beginning the industrial takeoff that fully flourished after the Civil War.

In contrast to the steam-driven industries of England, the U.S. industrial revolution was distinguished by its heavy emphasis on the inventive exploitation of America's waterpower. Experimentation with waterwheels and power designs steadily improved horsepower output and ultimately surpassed the limits of the steam engine. Lowell's textile industry was a creative fulcrum of the development of the seminal water turbine, a derivative of the waterwheel, which harnessed greater energy from falling water by channeling it through enclosed passages to spin finlike rotary blades. By the mid-nineteenth century, water turbines were used to drive sawmills and the elaborate gears, camshafts, pulleys, and belting of large textile mills. As early as the 1840s, one of Lowell's textile companies on the Merrimack River began using turbines capable of 190 horsepower. A watershed innovation was made by James B. Francis, chief engineer of Lowell's waterpower works. Through the combination of methodical scientific analysis, theory, and testing, and the expert craftsmanship of Lowell's famed machine shop, Francis produced a highly efficient, new class of turbine design in 1848. The water turbine's heyday arrived in the late nineteenth century when an evolved iteration of the Francis turbine with its rotating shaft attached to a dynamo proved to be the most effective motor for the mass generation of electricity.

Ever since British scientist Michael Faraday's 1831 discovery that electricity could be produced by rotating magnets inside of copper coils, inventors had been seeking ways to tap the awesome potential of the new energy source. Morse's electric telegraph in the 1840s revolutionized communications and linked continents. The industrial electricity age arrived with the advent of the modern electricity-generating dynamo and early applications like Thomas Edison's lightbulb and

Werner Siemens's electric streetcar in the last quarter of the nineteenth century. But to achieve takeoff required a means to generate massive amounts of electrical power. It was in the ensuing quest to extract hydroelectricity from large waterfalls, notably the pioneering effort at Niagara Falls in the 1880s and 1890s, that the water turbine found its great historic application. Just after the turn of the century, the Niagara Falls Power Company was generating hydroelectricity from 5,500 horsepower Francis turbines that spun under 135 feet of water; two years later Francis turbines capable of generating 10,000 horsepower were being built.

Electricity was the only form of energy that was easily stored and transmitted over long distances. Wherever it was applied on a large scale, it transformed almost every aspect of human life. Cities were illuminated; homes eventually got washing machines, telephones, and radios. Refrigeration allowed food to be stored longer and transported over long distances. Transportation accelerated. Precision increased. Compact electric engines were fitted on all sorts of work products to boost productivity. Entire new industrial sectors came into being. Aluminum, for example, could be cost-effectively extracted from its ores and refined only with massive amounts of electricity. Countries rich in hydroelectric power, such as the United States, Canada, and Norway, became world leading aluminum producers. Cheap aluminum, in turn, spurred advances in manufacturing airplanes, ships, and cars. Along with steel, petroleum, and the internal combustion engine, electricity became one of the dynamic foundations of the mass production industrial revolution that superseded the age of steam and iron.

Hydroelectricity helped elevate America to world leadership in the new industrial age. Between 1907 and 1929, U.S. nonfarm homes with electricity multiplied tenfold to 85 percent. By 1930, Americans were consuming more electricity than everyone else in the world combined, as well as providing half the world's industrial output. Hydroelectricity's legacy in America's history loomed even larger than that because it was one of the core technologies that ultimately unlocked the wealth stored in the nation's arid, western hydrological frontier. Nature offered few sites like Niagara between Lakes Erie and Ontario with the high falls and steady, year-round flow suitable for mass production of hydroelectricity. Around the turn of the century, however, industrial technologies converged to provide an artificial, man-made alternative—giant concrete dams. The pioneering, giant, multipurpose Hoover Dam on the Colorado River in the southwestern United States, completed in 1936, provided flood control, irrigation, and vast amounts of hydroelectricity through each of its 100,000 horsepower Francis turbines. It and subsequent similar giant dams modeled upon it became the critical infrastructure of America's development of its Far West. Hoover was the largest hydroelectric facility in the world; at the end of the twentieth century, its 17 upgraded Francis turbines were capable of producing some 2.7 million horsepower of clean, endlessly renewable energy.

Other turbine designs and applications also helped to foster America's rise. The screw propeller turbine, a derivative of the Francis turbine, powered fast naval vessels. As good hydroelectric sites grew scarce, steam turbines powered by fossil and nuclear energy plants instead of falling water evolved to produce mass electricity alongside rivers, whose water was used as coolants. By the end of the nineteenth century, the temperate, rainy, river-rich half of the United States from the East Coast through the heart of the Mississippi Valley, was on the verge of overtaking Britain as the world's greatest economic power. Yet this historic transition in the global balance of power could not have occurred without a prior revolution in transportation that forged America's many navigable rivers into an interconnected, inexpensive, inland waterway network. The water transport network was galvanized by two developments in water history in which America not merely followed, but took the lead from industrial Europe in applying new technologies—steam-powered, wooden riverboats, and modern, long-distance canals.

George Washington's unfulfilled ambition of a Potomac passage through the Alleghenies had faltered partly due to the lack of a commercially viable steamboat capable of traveling upstream. Washington's hopes had been pinned on a steam-pump-powered boat built by Virginian James Rumsey in 1784, a year after the French had pioneered early steamboats on the Saone River. In 1787, the eccentric, ill-starred American silversmith and inventor John Fitch launched a stern paddle-wheel-driven steamboat on the Delaware—even giving rides to members of the Constitutional Convention then meeting in Philadelphia—but its commercial viability was undermined by unreliable boilers and an inability to stay on schedule. In 1802, the first successful, practical steamer gave very brief service as a canal tug that crossed Scotland. The era of commercial river steamboats began full throttle in August 1807 on the Hudson River when American Robert Fulton's 149-foot, twin-side paddle-wheeled *North River Steamboat,* later popularly called the *Clermont,* completed the 150-mile journey from New York to Albany in only 32 hours at a time when competing sail-powered sloops required four days.

A failed painter, self-made engineer, astute calculator of business costs, and ambitious schemer, Fulton was living in Europe in pursuit of his painting career when he encountered the burgeoning ferment over steamboats. Soon, he gave up painting, wrote a treatise on the efficiency of inland transport using small canals and invented a submarine that he tried to sell to Napoléon for his war against England. His main chance came in Paris in 1801, at the age of thirty-six, when he met Robert Livingston, the U.S. minister to France who soon was to negotiate the Louisiana Purchase, and who had presciently obtained a twenty-year monopoly for steamboat navigation in New York before taking his post. The two men formed a partnership. Managing to obtain a 24 horsepower steam engine from Boulton and Watt, Fulton returned to America and built his historic steamboat. It was financially successful from its first run. Fulton and his associates soon launched a steamer on Pittsburgh's Monongahela River that made the nearly 2,000 mile downstream trip on the Ohio and Mississippi to New Orleans in only two weeks. By 1815 the first riverboat was able to steam its way against the Mississippi current to complete the return trip in about four weeks. Previously crops and other goods had been floated downriver in one-way wooden flatboats that had to be dismantled and sold for lumber at journey's end. The era of Mississippi shallow-draft riverboats and two-way western river cargo transport had begun. The U.S. Army Corps of Engineers started its career assisting navigation by removing snags and sandbars from the major rivers. Within five years some 60 steamers plied western rivers; by 1840, 536. Freight costs plummeted and commerce boomed throughout the Mississippi River Valley. By the end of the decade, western river steamboats were carrying freight comparable to that of the entire British Empire.

The boom in western river steam commerce could not have occurred without the conquest in the meantime of the long-insuperable obstacle—the absence of a trans-Appalachian water route. Without such a passage, the western rivers remained unconnected to the vibrant industries, farming and markets of the East, where the vast majority of Americans still lived and traded along the seacoast, effectively hemmed in by the Appalachians. Yet at the start of the river steamboat age, the grand engineering breakthrough that unified America, unlocked the wealth of the Mississippi Valley, and recast its historical orientation from north-south toward western expansion was on the verge of being launched—the Erie Canal, built from 1817 to 1825.

While George Washington was dreaming of a Potomac waterway to the west, he was keenly aware that a rival group was striving to transform New York's Mohawk River into the main western gateway. The Mohawk had the advantage of being the only American river that cut through the mountains, which it did via a 500-foot-deep gorge. Although it had many waterfalls, rapids, and rocky shallows like the Potomac, it had an invitingly gentle slope for most of its route. If extended to Lake Erie south of Niagara Falls, it offered a potentially fabulous trade water route from New York City and the Atlantic to the Great Lakes and, with some adaptation, down the Mississippi to New Orleans. As with the Potomac, the initial effort to convert the Mohawk into a navigable waterway failed. But the New Yorkers hit upon another scheme—a 363-mile canal alongside the river. It was an audacious, grandiose plan. One early advocate was Robert Fulton. In 1807, he meticulously (and with remarkable accuracy) calculated the costs and rewards of building it for the federal government's report on how to improve U.S. inland transportation. Fulton enthusiastically concluded that "when the United States shall be bound together by canals, by cheap and easy access to market in all directions, by a sense of mutual interests arising from mutual intercourse and mingled commerce, it will be no more possible to split them into independent and separate governments."

Nevertheless, on the eve of its construction, little more than 100 miles of canals had been built in the entire United States. While proponents pointed with encouragement to the successful model of Europe's epoch-making canals, the truth was that the 1761 coal-carrying Bridgewater Canal that spurred England's canal craze had been only 10 miles long, while France's more complex, 150 mile, 1681 Canal du Midi joining the Atlantic and Mediterranean, had been built across a civilized, heavily populated landscape. The much longer Erie was to traverse vaster expanses of untamed, barely inhabited wilderness. Large government financial support was needed. Thus when the project was presented in January 1809 to President Jefferson, the usually visionary leader passionate about promoting improved internal navigation, dismissed it as a wonderful but unrealistic idea that was a century ahead of its time. "It is little short of madness to think of it at this day," he told his disappointed visitors.

The project might have died at that point had not an outstanding New York politician, De Witt Clinton, stepped forth to become its champion. A longtime mayor of New York, U.S. senator, scion of a leading political family, and later to be New York governor, Clinton had been persuaded to steam up the Hudson in early July 1810 on Fulton's *Clermont* to embark on what would be a revelatory, nearly two-month round-trip expedition between Albany and Buffalo to study the feasibility of the canal. He came back inspired to make the Erie Canal the capstone of his political career. Over the next seven years he determinedly overcame all political obstacles, technical doubters, and the disruptions of the War of 1812 to win the backing of the New York legislature for the state-financed canal. Finally on July 4, 1817, three days after Clinton himself had been elected governor of New York, ground was broken on what his many critics called "Clinton's ditch."

Building the four-foot-deep, 40-foot-wide canal and mule towpath was an immense technical and financial challenge. The work was done in three stages, entirely by hand labor, horse, oxen, and blasting powder. European canals provided engineering experience to cross rivers and build locks. But clearing and excavating through hundreds of miles of thick, forested wilderness was an utterly novel challenge. Ingenious solutions were found by work teams on the job through trial and error to expeditiously fell trees, remove stumps, and slice through the tangle of tree roots by the use of plows. When common quicklime proved unstable for lining and sealing the culverts, locks and aqueducts, the engineers found a source of inexpensive New York state limestone that acted like waterproof Roman cement when it hardened. By the fall of 1819, the midsection of the canal, which ran through the state's lucrative salt-producing region, was finished and filled with water. The first tolls were collected along the finished stretch in July 1820.

The moment of financial truth for the canal came with the Panic of 1819, and the related bank lending contraction and national economic depression, which had initially been triggered by the depleted Treasury's urgent 1818 calling in of $3 million in gold in order to meet a large debt payment due to France for the Louisiana Purchase. From the start, many of Erie's doubters contended that Clinton's $6 million budget was simply beyond the wherewithal of the state and nation's limited capital. But the steadiness of state financing enabled work on the canal to continue without interruption, insulating New York from the worst of the national depression. The national economic collapse, moreover, caused capital market borrowing rates to plunge. With the canal increasingly looking like a viable project and few other attractive issues being offered, demand for New York's new issue of canal bonds surged, lowering the cost of the project. Previously cautious large investors jumped in. So did British speculators, as the frenzy for Erie paper spread overseas; by 1829, foreigners held over half the canal's $7.9 million in obligations.

By 1821, 9,000 men were working to build out the midsection in the canal in both directions. At each extreme it faced its toughest geographical challenges. To negotiate the steep gorge that sliced through the Appalachians, an aqueduct was built some 30 feet above the rushing white water and falls. Farther east, the steep descent toward the Hudson River was managed by the construction of long aqueducts, one supported by 26 piers, which crisscrossed the river. The greatest challenge of all lay at the western end, where the canal had to surmount the precipitous six story high cliffs leading to the escarpment from which Niagara Falls poured 17 miles away. At the end of two and a half years, workers had completed five flights of giant, 12-foot locks and sculpted a canal and towpath channel through seven miles of solid rock face. When it was finally finished in October 1825, the Erie Canal was the nation's marvel. Across its 363 miles it had 83 locks and 18 aqueducts that carried 50 ton, mule- or horse-pulled freight boats up and down over 675 feet.

The two-week-long dedication was led by Governor De Witt Clinton, who paraded through Buffalo before setting out on a triumphal week-long journey on the canal to Albany—a trip that had taken him thirty-two days by land fifteen years earlier. At Albany he and his entourage boarded the Hudson steamer for New York. At the mouth of New York harbor, Clinton performed a symbolic wedding of the waters by pouring water from Lake Erie into the Atlantic. Other dignitaries stepped forward to pour water bottled from 13 of the world's great rivers—Ganges, Indus, Nile, Gambia, Thames, Seine, Rhine, Danube, Mississippi, Columbia, Orinoco, Rio de la Plata, and Amazon.

The Erie Canal was an immediate, heroic success. At four cents a mile, it slashed freight transportation costs by 90 percent overnight. Toll collections surged as some 7,000 boats plied the canal the very first year; within just twelve years, the entire canal debt had been paid back. Relieved of the stultifying burden of transportation costs, midwestern wheat, corn, and oats quickly won new markets throughout the East Coast and across the ocean in Europe. Farm production soared, prices plunged, and settled cropland throughout the Mississippi Valley expanded. Georgia's governor was shocked to find that New York State wheat sold in Savannah for less than wheat from central Georgia. Philadelphians awoke to discover that the cheapest route to Pittsburgh was up the Hudson, across the Erie Canal and south from Lake Erie by canal or wagon. As world markets opened wide to America's interior beyond the Appalachians, east-west commerce in other goods also took off at an astonishing pace. Between 1836 and 1860, total tonnage on the Erie Canal multiplied thirty-one-fold.

The Erie Canal's catalytic impact was even more dramatic than its opulent profitability. Over 100 new canal projects were launched across America within a year of its opening. By the late 1840s, a full-fledged canal boom had produced more than 3,000 miles of canals—three-quarters, following Erie's lead, financed with public funds—to connect America's navigable rivers into an integrated, inexpensive highway extending from New York to the Gulf of Mexico. Canals joined the Great Lakes to the Mississippi via the Illinois River in the west and via the Ohio River in the east. In the 1830s a canal spur also joined Erie to the head of the Chesapeake Bay, while early steam railroads linked canals across terrain that canals didn't traverse. Chicago, Cleveland, Buffalo, Cincinnati, and Pittsburgh became bustling inland port cities. The marriage of steamboats and canal-linked rivers quickened farming and burgeoning industrial activity throughout the Midwest.

While the Erie-ignited canal boom had many parallels with England's forty-year canal craze following 1761 that transformed the Midlands and northern England into the birthplace of the Industrial Revolution, its dynamic impact was more fittingly compared to that of the Grand Canal upon medieval China. Like the Grand Canal, America's Erie-inspired waterway network united a continental-sized nation-state challenged by splintering regional divisions, geographical impediments, slow travel and communication, and divergent economic and social organization. Inland navigation routes supplanted each country's reliance on coastal sea commerce, extended the reach of the central government, tightened webs of mutual economic interest, and spread common political and cultural discourse over a wide landscape. As the Grand Canal invigorated China with a stronger south-north orientation, Erie and its offspring reoriented America from its traditional north-south axis toward the more unifying east-west continental destiny envisioned by Jefferson, Washington, and Franklin. With a traversable route through the Appalachians, settlers began to pour into the western lands. By 1840 two out of five Americans lived west of the mountain barrier. National economic growth accelerated notably after Erie's opening. In the first quarter of the nineteenth century, the economy expanded on average about 2.8 percent per year; from 1825 to 1850, 4.8 percent annually—the fastest growth spurt in all of U.S. history. By the 1840s the inland transportation infrastructure, influx of immigrant labor, and large accumulations of domestic capital had structurally transformed America into a nation that was ready to embark upon its great economic takeoff.

The Erie Canal boom also created the platform for the frenzy of steam railroad building that succeeded it. The speedy, all-season railroads effectively extended the waterway network, and finally surpassed it as America's leading freight hauler after midcentury. Like the medieval Islamic camel caravans, the steam and iron railroads ultimately provided the means for crossing the expansive desert and mountain frontiers of America's Far West and fusing its resources with the mass production industrial economy of its temperate, eastern half. Steamships carrying iron ore from the western Great Lakes met up with railroad and canal barge shipments of coal to mass-produce various shapes of carbonless iron, or steel, at huge steel mills from the 1870s. As iron improved upon weaker wood, incredibly strong yet resilient steel improved upon iron that tended to bend and snap. Steel—which used water prodigiously for cooling—became the basic building material of the most dynamic industries and cities of the new era. Steel cable wires were used to support the landmark 1883 Brooklyn Bridge, then the world's longest suspension bridge. In combination with energy harnessed from petroleum, which began to be drilled in significant volumes from the 1860s, and the electric dynamo, steel provided one of the core technologies of America's rise to world economic leadership in the second great phase of the Industrial Revolution.

The transportation revolution ignited by the Erie Canal also shaped the urban hierarchy of the nation. As the main commercial gateway between the world and the fertile, huge U.S. interior, New York became America's leading city. Its port, which had ranked fourth in tonnage in the colonial era behind Philadelphia, Boston, and Charleston, swelled with traffic equal to all other U.S. ports combined by the mid-nineteenth century. Likewise, its small capital market of brokers—24 of whom had formalized a trading agreement among themselves in 1792 under a buttonwood tree on the street where the city's Dutch founders had erected a protective wall in 1653 and had continued to meet in various local taverns until 1817—formed the forerunner of the New York Stock Exchange, and in 1825 moved into a fixed location. The post-Erie fever in canal and railroad securities made it America's premier center of finance. Between 1815 and 1840, the city's population tripled to 300,000, then more than doubled to about 700,000 by 1850.

To the west, the canal and steam railroad booms similarly vaulted Chicago on Lake Michigan to regional metropolitan leadership. From a population of only 350 in 1833, Chicago grew rapidly on grain and raw materials freight connecting to the Erie Canal, then even faster after the 1848 opening of the 96-mile Illinois and Michigan Canal linking Chicago with the Mississippi. By 1850 it had 30,000 inhabitants. The city's destiny as the key midwestern transit point for the crops, livestock and raw materials of America's interior was solidified from the 1850s when Chicago became the central hub for the steam railroads. Travel time to New York fell from three weeks by boat to three days by train. Slaughterhouses, meat-packing enterprises, and transit by refrigerated railway cars soon followed. By 1890 Chicago was the nation's second largest city.

Rapid urbanization evoked a sanitary awakening in America, as it had in Europe. New York led the way. Manhattan island, surrounded by undrinkable, brackish rivers and possessing only one wholesome freshwater source on its low-lying, settled southern tip, entered the era as one of the most parched cities in the country. Reflecting its rise in status, by the 1880s it had built the world's most abundant urban water supply system providing each of its over 1 million inhabitants with a lavish 100 gallons of water per day—a modern urban incarnation of ancient Rome in its glory days.

From the colonial era through the early decades of the American republic, New York City's public water had been so notoriously bad that residents seldom drank it untreated. Instead they consumed it for breakfast as warm beer, in disinfecting alcoholic drinks at the city's many taverns, or boiled in hot chocolate or tea. Well water, often briny and hard tasting, was rarely consumed. Lower Manhattan's single freshwater pond, called the Collect, by the early nineteenth century had become such a dumping ground of sewage, excrement, dead animals, and occasional human corpses that the only available wholesome water came from a lone, pumped spring. Known as the Tea Water Pump because its purity favored its use for making tea, it was sold throughout the city in water carts by "tea men" for a handsome premium that only the well-to-do could regularly afford. By the time New York's population began to swell from the Erie Canal's influence, the water system was already at the breaking point.

Catastrophic epidemics and fire finally broke through the political and business intrigues that for decades had stymied the clamoring for water reform. Yellow fever had killed 2,000 in 1798 and various plagues had visited the city periodically before the disastrous global cholera epidemic struck New York in 1832. The horrible dehydration and rupturing capillaries of the bacteria killed 3,500 New Yorkers—almost 2 percent of the population—and sent 100,000 people, or nearly half the city, fleeing to the greener suburbs north of 42nd street and beyond. The lack of a good water system also had left New York vulnerable to the great fire of 1776 and numerous other successive blazes. In December 1835, with residents barely recovered from the cholera epidemic, a terrible fire, fanned by high winds, destroyed some one-third of the city, including much of its commercial and shipping businesses in lower Manhattan. The blaze exhausted the resources of the city's new fire reservoir, while cisterns froze and river water turned to ice in the pipes of volunteer firemen's hand pump carts.

The disasters of the 1830s finally galvanized the city's resolve to build the 42-mile-long Croton Aqueduct system. Its chief engineer, John Jervis, had learned his trade on the Erie Canal. The aqueduct system included a dam, reservoirs, arched bridges, and tunnels and brought water from the Croton River north of the city to a showy distributing reservoir on Murray Hill at 42nd Street and 5th Avenue at the site later conspicuously occupied by New York's central public library. Designed to look like an Egyptian temple, the reservoir's completion in October 1842 was wildly celebrated with fountains being turned on at City Hall and Union Square, cannons firing from the Battery, church bells ringing, the singing of a specially written "Croton Ode," and a seven-mile parade attended by some quarter of a million people. The mood was captured in the journal entry of a future New York mayor, Philip Hone: "Nothing is talked of or thought of in New York but Croton water; fountains, aqueducts, hydrants and hose . . . It is astonishing how popular the introduction of water is among all classes of our citizens, and how cheerfully they acquiesce in the enormous expense which will burden them and their posterity with taxes to the latest generation. Water! water! Is the universal note which is sounded through every part of the city, and infuses joy and exultation into the masses." During the 1850s a sewerage system was begun. Bathrooms became a common feature of New York life. Never again was New York plagued by terrible cholera epidemics and great fires. As in London following its sanitary reforms, childhood mortality plunged and average life spans leaped from historic trend lines.

The Croton water supply had been expected to last for sixty years. Within a decade, however, it became evident that additional supply would be needed to meet the growing population and the surge in per capita consumption that accompanied the sudden availability of freshwater. By 1884 engineers began work on a second Croton aqueduct system with three times the capacity of the original. Even before the New Croton system was completed in 1911, groundbreaking had begun a still larger, new municipal water supply project. The Catskill aqueduct system brought water some 100 miles from the mountains west of the Hudson River to satisfy the thirst of New York's over 3.5 million inhabitants. Completed in 1927, the Catskill system featured a water tunnel 1,114 feet under the Hudson riverbed. Reflective of the era's prevailing Go-Go culture, authorities used high-handed land appropriations to displace entire rural towns and citizens to make way for the system's massive reservoirs.

To carry water throughout the growing metropolis, New York also built its first deep, high-pressure subterranean conduit—City Tunnel Number 1, some 18 miles long—in 1917. Working at perilous atmospheric pressures requiring careful bodily acclimation, workers nicknamed "sandhogs" bore through the solid bedrock under the city and its rivers and harbor at a depth of some 750 feet, or the equivalent of an inverted skyscraper. In 1936 City Tunnel Number 2 added another 20 miles of water distribution capacity far below the city's streets, subways, and buried electrical and gas service pipelines. Work on Tunnel Number 3, one of the world's most monumental engineering projects at the turn of the twenty-first century, started in 1970 and was still being done at the end of the century's first decade.

As the population of New York City climbed toward 8 million in the 1950s, still more water supply was needed. Between 1937 and 1965, the city built its largest aqueduct system with water from the Delaware River that flowed between New York, Pennsylvania, and New Jersey. Construction of the Delaware system proceeded only after a 1931 Supreme Court ruling, written by Justice Oliver Wendell Holmes, rejecting downstream New Jersey's effort to stop the project. Holmes's opinion established the guiding water-sharing principle of "equitable apportionment without quibbling over formulas" because "a river is more than an amenity, it is a treasure. It offers a necessity of life that must be rationed among those who have power over it."

In the early twenty-first century, the water system that keeps New York City running is still an engineering marvel. Although leaky and in a dire race to modernize several critical components ahead of a catastrophic collapse, it delivers, almost entirely by gravity, some 1.3 billion gallons to 9 million people each day—enough for over 140 gallons per person—through three primary aqueducts drawing from 19 reservoirs and three managed lakes across a 2,000-square-mile watershed. Its sewerage counterpart of about 6,500 miles of pipes carries away an equal amount of wastewater each day to 14 treatment plants, 89 wastewater pump stations, and other facilities. History had once again repeated itself: as in ancient Rome and other leading civilizations of the past, the water infrastructure of the foremost center of the greatest power of the age was both a bellwether and a key contributor of its preeminence.

By the late nineteenth century, while New York's water system was being rapidly expanded, America had begun to assert its status as the world's awakening industrial giant. Pivotal to its transformation of the 1900s into the American Century was the nation's response to the challenge of harnessing the untapped resources of its remaining two frontiers—its arid Far West, and its seascape alongside the Earth's two largest oceans. The clarion call west had been sounded on January 24, 1848, when workers erecting Swiss émigré Johann Sutter's new waterwheel-powered sawmill on the south fork of the American River near modern Sacramento, California, found 1.5 ounces of gold dust and nuggets in the millrace. On May 4 news of the gold strike spread wildly through the small town of San Francisco. The following year, word reached the larger world. The California Gold Rush was on. By land and sea, "forty-niners" flocked to California to prospect. By 1853 more than 100,000 had arrived, including 25,000 Frenchmen and 20,000 Chinese. Hundreds of thousands more followed. Within a matter of months, San Francisco swelled into a booming city of over 20,000.

Prospectors panned streams and dug shafts into hillsides. But most effectively, they rigged old Roman hydraulic methods to shear away the rocky skin and foliage of the foothills with high-pressure water jets to expose the underlying gold veins. Water was lifted by wooden waterwheels—called hurdy-gurdy wheels by the miners who improvised them—and funneled through pipelines to fill tanks and dams several hundred feet above the mining site. It was then released in torrents through piping and small diameter metal nozzles to generate pressures up to 30,000 gallons per minute. Hillsides everywhere were stripped bare by the environmental carnage. The residue of smashed rocks and topsoil from neighboring farms washed away toward San Francisco Bay. Farmer pressure finally outlawed the practice in 1884. But by that time the gold had been mostly plundered from the hills and the miners had moved on to the next strike. Johann Sutter, meanwhile, died in Pennsylvania in 1880, at the age of seventy-seven, a man ruined by the unstoppable invasion of squatters and lawless theft of his stores and property.

One of the lasting imprints of the Gold Rush on America's western destiny was the blazing of two main transportation routes linking the East with the nation's western frontier. The parched, overland trails across the Far Western deserts and high mountains drew 300,000 to California by 1860, many of whom later settled down as early farmers and merchants. California's precocious development, in turn, hastened the building of the transcontinental steam railroad. When the Central Pacific and the Union Pacific were linked up in in 1869, travelers could cross the country in comfort from New York to San Francisco in only ten days.

The search for a faster, cheap sea route to California, meanwhile, spurred the epic quest to build a transoceanic canal passage between the Atlantic and Pacific. Until the building of the Panama Canal, the voyage required 15,000 miles and sailing around South America's treacherous Cape Horn. In a single month in 1850, 33 sailing ships arrived in San Francisco after an average 159 days at sea. Speedier, full-rigged clippers later reduced travel time to about 97 days. Before the Gold Rush, in 1840, President James Polk, inspired by America's growing continental ambitions, had already recognized the sovereignty of Colombia over the Isthmus of Panama in exchange for right of American rail passage through it. If the acquisition of California after the 1846–1848 war with Mexico heightened momentum for a canal, the Gold Rush gave it urgency. By 1855 the Panama railway bridging to the oceans was completed with American capital. In the next decade, 400,000 people crossed its tracks, including an army of California-bound miners. To compete with the Panama railway, "Commodore" Cornelius Vanderbilt, notorious baron of Hudson River steamboats (and later railroads), ran steamers across the large lake at Nicaragua that connected to the Pacific coast by mule and schemed of his own interoceanic canal at Nicaragua; in 1855 one of

Vanderbilt's hired associates, William Walker, managed to set himself up as Nicaragua's president. For the rest of the century, powerful interests in the United States and Europe took sides in the mounting high-stakes, multilateral battle to make either Panama or Nicaragua the chosen route for the interoceanic canal that would shape world power in the twentieth century.

CHAPTER TWELVE

The Canal to America's Century

As the 1869 Suez Canal made transparent the supremacy of the steam-and-iron age British Empire, the opening of the Panama Canal in 1914 signaled a reordering of the world's power in favor of the fast-rising leader of the mass production technology era, the United States. The interoceanic canal at the Panamanian isthmus created a highway across the central Atlantic and the Pacific Oceans, integrating Europe, the Americas, and the Far East into a tighter, global web of political, economic, and military relations. No nation was better placed to benefit commercially and strategically from its transformative impact than America, in whose backyard it was and which controlled access to it. Nor was any other nation as capable and so audacious to take on the gargantuan technical, organizational, and political challenges of building it. Its successful completion was a tangible declaration of America's industrial economic superiority and its growing ambition to take its place among the world's great powers. The fast and inexpensive water passage between the two oceans also had a major, catalytic impact upon America's internal growth. At last, the nation became capable of fully exploiting the advantages of its extensive maritime position. By transforming the Caribbean from a dead end into a transportation shortcut across the continent it created fresh synergy from the intermingling of the Far West's potential mineral and agricultural wealth with the prolific industry and markets of the Mississippi Valley, the Great Lakes, and the eastern seaboard. Not least, too, the canal combined America's Atlantic and Pacific fleets into a single, great power upon the high seas.

The very creation of the canal had been closely interconnected with the evolution of American naval power. As a nation surrounded by oceans on three sides, maritime power and commerce had always played a primary role in American history. Although its fledgling navy was overwhelmingly outclassed by the British fleet during the War of Independence, John Paul Jones's heroic sea victories and marauding of England's coast had instilled national pride and the hope that American prowess as naval warriors could augment the natural moat of the Atlantic to defend the young country's vital interests. From 1794, a small, but effective military fleet was built to safeguard American merchant shipping and diplomatic neutrality amid boiling war tensions between France and England. When the three-year Quasi-War with France erupted in 1798, the U.S. Navy earned the respect of all great European powers by prevailing in several Caribbean engagements against top French warships. The navy won further plaudits for liberating America from the onerous tolls and prisoner ransoms charged by the Islamic states of North Africa in the little-remembered Barbary Wars of 1801–1806. By Jefferson's presidency, some $2 million, or one-fifth of America's annual government revenue, had been paid to Algiers, Tripoli, Tunis, and Morocco to allow U.S. merchant ships to sail through the Strait of Gibraltar and trade unaccosted in Mediterranean waters. When Tripoli, greedy for larger tribute and underestimating American naval power, declared war on the United States in 1801, Jefferson sent the navy. The resulting bombardments and audacious marine raids on Tripoli created new naval heroes, swelled patriotic fervor at home and steeled American resolve to pay for defense but never again for tribute or ransom.

It was in the War of 1812 that the navy finally established its permanence as America's indispensable military branch. Despite its small size and poorly maintained state at the start of the badly prepared war, the fleet, led by the USS *Constitution* (aka *Old Ironsides* due to its many protective layers to defend against broadsides), stunned Englishmen and thrilled Americans by winning a series of sea skirmishes against overconfident British warship commanders reveling in England's recent triumphs over French fleets at the Nile and Trafalgar. Most important, U.S. naval freshwater commanders played crucial roles in foiling the concerted British invasion of America in 1814 by winning inland battles on Lake Champlain and Lake Erie. By gaining control of the strategic northern and Great Lakes for America, these victories ultimately convinced England to give up its long-term designs on the Mississippi Valley and instead settle its remaining border disputes with the United States in order to secure the vulnerable borders of British Canada.

After the war, naval power was deployed in America's pursuit of its Manifest Destiny to expand its continental territory to the Pacific Ocean. In 1823, growing confidence in its naval power emboldened the United States to assert its defining Monroe Doctrine, warning European powers not to intervene in the newly independent republics of Latin America because the hemisphere was America's special sphere of influence. In the Mexican War of 1846–1848, American vessels blockaded Mexican ports and in March 1847 gave crucial landing and bombardment support to the U.S. Army's decisive march on Mexico City from Veracruz. With the 1848 treaty transferring nearly half Mexico's territory to the United States—including most of the Southwest, California on the eve of the Gold Rush, and Mexico dropping its claims to Texas—America's continental expansion was nearly complete.

In the 1840s and 1850s, the U.S. Navy also began to assert itself as a Western imperialist steam power in the Pacific. Trading right concessions were extracted by threat of force from China in 1844 paralleling those won by Britain in the Opium War. In 1853 and 1854, Commodore Matthew Perry steamed into Tokyo Bay with an armed squadron of frightening, smokestack-belching "black ships" that showed its overwhelming force through occasional practice gunnery to convince Japan's leaders to open their nation to foreign trade after two centuries of virtual closure.

America's Civil War of 1861–1865 in which the industrialized North applied its decisive naval superiority over the agriculture-based South, provided further evidence that no major modern war, and few rebellions, had been won without the advantage of naval power. Union vessels blockaded Confederate seaports while steam gunboats took command of southern rivers. By 1862, the North controlled the vital points along the Ohio and Mississippi rivers all the way to the port of New Orleans.

Yet by the 1880s, waning investment had caused America's navy to slip far behind the rapid technical advancements in ship speed, and artillery accuracy, distance and power of leading European powers, notably England and rising industrial power, Germany. The defensive buffer provided in the age of sail by America's ocean moats had visibly diminished. American officials grew alarmed when fellow hemispheric nations Peru and Chile in their War of the Pacific (1879–1883) both employed vessels superior to their U.S. Navy counterparts.

America's reaction—the application of its rising mass production industrial might to build a world-class, steel navy—was a decisive turning point that helped tip the balance of power in America's favor and set the stage for the building of the Panama Canal. The gradual buildup of the steel navy from the mid–1880s was paralleled by a transformation in Americans' outlook about the nation's appropriate place in the world and the role of sea power in attaining it. Both, in turn, were spurred by the rapid growth in demand for U.S.-manufactured goods on world export markets. As American economic interests expanded outward, U.S. leaders became convinced that America should behave like a European global power and that a strong navy was a vital component of national prosperity and security.

The most influential intellectual exponent of this view was Captain Alfred Thayer Mahan. His widely celebrated 1890 book, *The Influence of Sea Power Upon History,* shaped the policy framework of a generation of leaders through World War I, not just in the United States but also in England and Germany, including Kaiser Wilhelm II personally. A career U.S. naval officer, historian, and president of the Naval War College in Newport, Rhode Island, Mahan traced the rise and decline of maritime nations, most closely those in Europe from the mid-seventeenth to late eighteenth centuries, and concluded that supremacy at sea held the key to international commercial success, prosperity, and national greatness. To Mahan, a favorable seaborne position, properly exploited, provided a cheap, easy, and safe transport highway that advantaged seafaring states in the key commercial struggle to control traffic on the world's sea-lanes and strategic passages. "The seaboard of a country is one of its frontiers," Mahan wrote. "Numerous and deep harbors are a source of strength and wealth, and doubly so if they are the outlets of navigable streams, which facilitate the concentration in them of a country's internal trade." Conversely, Mahan argued, the failure to fully exploit favorable sea geography resources represented a potential national vulnerability. Mahan's unabashed conclusion was that a great nation needed a strong standing navy, with bases at home and abroad, deployed in the interests of enhancing its sea commerce and worldwide influence. Regarding an interoceanic canal in Central America, Mahan held strongly that if built, "the Caribbean would be changed from a terminus . . . into one of the great highways of the world . . . The position of the United States with reference to this route will resemble that of England to the Channel, and of the Mediterranean countries to the Suez route." He also speculated hopefully that such a canal might excite America's "aggressive impulse" to exert its influence globally.

By focusing his historical study on the mercantilist age of sail, Mahan's study, in hindsight, suffered intellectually from myopic conclusions about the relationship between sea power, commerce, and international standing. In particular he underestimated the enormous national prosperity and military strength that could be derived from industrialized society and free trade. Nevertheless, many of his general observations about the advantages of sea power, from ancient times to the present, were germane, in varying degrees. But the historical significance of his views stemmed from the fact that leaders of the great nations pursued policies based upon them.

Mahan's most important American adherent was Theodore (Teddy) Roosevelt. The future president, who would do more than anyone to turn Mahan's prescriptions into reality, was thirty-one when *Influence* appeared and wrote a glowing review of it for the *Atlantic Monthly.* The two men had been friendly for several years since Roosevelt had lectured at the Naval War College on the topic of one of his own books, the naval history of the War of 1812. Mahan's recommendation helped Roosevelt win appointment as assistant secretary of the Navy when Republican William McKinley won the presidency in 1896.

As assistant secretary, Roosevelt agitated within the McKinley administration for an aggressive expansion of the U.S. naval fleet as well as for the construction of an isthmian canal. Like his mentor, Roosevelt viewed a strong navy as the "big stick" of a new, more assertive American global diplomacy and winning supremacy at sea. As Roosevelt would frame it in speeches as president, "There is a homely adage which runs: 'Speak softly, and carry a big stick; you will go far.'"

Assistant Secretary Roosevelt urged immediate administration of the "big stick" on February 15, 1898, when, amid the Cuban rebellion against Spanish rule, the U.S. battleship *Maine* blew up in Havana harbor from unknown causes, killing 260. In April McKinley yielded to the bellicose clamor of Roosevelt and other young Republicans inflaming public opinion to "Remember the Maine!" by declaring war on Spain and liberating Cuba. Working off a war plan previously developed by the Naval War College, Roosevelt successfully pushed for a strike on Spain's fleet in the Philippines as well as the blockade of Havana. The U.S. Asiatic squadron promptly steamed into Manila Bay and decimated the antiquated Spanish fleet without a single American death. Roosevelt himself soon decamped with U.S. troops for Cuba and attained national war hero status by leading a charge of his personally recruited band of "Rough Riders" up San Juan Hill. In less than three months of fighting, the United States had seized control of Spain's entire remaining American and Pacific empire.

By seeming to validate Mahan's arguments about the benefits of sea power, the Spanish-American War triggered a dramatic escalation in the U.S. buildup of the country's steel navy. Naval investment that totaled 6.9 percent of federal government spending in 1890 soared to 19 percent of a much larger base by 1914. On the eve of World War I and the opening of the Panama Canal, America's navy was the third mightiest on Earth, and was poised to soon overtake those of England and Germany.

The Spanish-American War also infused unstoppable momentum

to build the isthmian canal, which suddenly seemed indispensable to American national security, by unifying the strength of the Atlantic and Pacific fleets. The public case for the canal had been dramatically illustrated during the war by the long delay encountered by the Pacific warship *Oregon* in reaching the Caribbean theater because it had had to steam the additional 8,000 miles around South America's Cape Horn. Two presidential interoceanic canal commissions, one organized in 1897 and a second in 1899, recommended building the passage through Nicaragua. Secretary of State John Hay prepared the diplomatic ground by negotiating an agreement with England to supersede an outstanding 1850 treaty affirming bilateral control over any interoceanic canal: England granted the United States the right to build and operate the canal subject only to the Suez Canal neutrality rules of being equally open in war and peacetime to all vessels. The United States could police the route, but not build fortifications.

It was Roosevelt who led the outraged howls of protests when the treaty was announced in early 1900. He did so from his new platform as Governor of New York, an office to which his war hero status had catapulted him in the 1898 elections. America must not compromise the sea power advantage of an American-built canal by relinquishing the right to fortify it against enemy warships, he insisted stridently. The public was swayed. Hay was forced to renegotiate the treaty. In mid-November 1901 the new Hay-Pauncefote Treaty, omitting the fortification restriction, was signed.

By that time an unlikely set of circumstances had propelled party maverick Teddy Roosevelt himself to the presidency. Irked at Roosevelt's zealous, progressive agenda against corrupt political machines and the big business trusts that dominated the heights of American industry, but eager to capitalize on his popularity, Republican establishment leaders had tried to isolate Roosevelt by persuading him to take the vice presidency in the 1900 elections. Their plans suddenly went awry, however, when McKinley was shot by an assassin at Buffalo, New York, on September 6, 1901.

At forty-three, Roosevelt was America's youngest president ever and unlike any other. Possessed of boundless energy, a determination for action, an outsized vision of what America should be, and, despite personal arrogance and impetuosity, canny political and self-promotional skills, Roosevelt undertook several actions as president that transformed the course of twentieth-century America and water history. Above all, he mobilized the federal government as a strong, proactive agent of policy, whether as a progressive force to countervail big business trusts' distortion of market forces, to undertake large public interest projects that were beyond the resources or risk appetite of private enterprise, to save wilderness areas, and to fulfill his conviction that the march of civilization demanded the artificial remolding and control of Earth's resources, including indispensable water, to mankind's needs.

The isthmian canal was at the very top of Roosevelt's agenda when he assumed the presidency. At the time nearly everyone, including Roosevelt himself, took for granted that the canal would be built at Nicaragua, which was widely viewed as the American route. Men had dreamed about building a Central American canal ever since the Spanish conquistador Vasco Nuñez de Balboa had marched across modern Panama in 1513 and became the first European to set eyes on the "South Sea," as he called the Pacific Ocean. King Charles V, the Holy Roman Emperor and Habsburg monarch, ordered the first canal survey as early as 1534. Modern interest in the canal began in earnest after 1821 with the end of Spanish rule in Latin America. Several locations were considered, including Nicaragua, Mexico, and Panama.

More than anything else, it was the 1869 completion of the world-changing Suez Canal that finally galvanized action on an isthmian canal. Although U.S. president Ulysses S. Grant had sent seven expeditions to Central America starting in 1870 to thoroughly explore various canal routes and in 1876 chose in favor of Nicaragua, it was the renowned French impresario of the Suez Canal, Viscount Ferdinand de Lesseps, who stole the first march to action. In May 1879, de Lesseps unveiled his stealthily laid, private-sector plan to crown his Suez legacy with an isthmian canal at a grand congress in Paris attended by illustrious experts from around the world. The congress's ostensible task was to select the route and technical nature of the canal. In fact, the event was entirely orchestrated, with a predetermined outcome, by de

Lesseps. Although seventy-four years old at the time, he still possessed the charisma, vigor, diplomatic cunning, and grandiose self-confidence that had made him triumphant at Suez. While many engineering plans were presented, the final vote of the all-important technical committee endorsed de Lesseps's favored proposal—a sea-level canal at Panama. In fact, de Lesseps had previously lined up an exclusive contract with Colombia through intermediaries to build the canal in its state of Panama.

By being without locks at sea level, the Panama Canal would be a reprise of the canal engineering approach used at Suez—it was, in effect, Suez II. Yet, in reality, Panama was nothing at all like Suez. Suez had been cut through flat terrain in a hot, dry environment where the main problem was the dearth of water. Panama, by contrast, was a sweltering, tropical environment inundated with too much water, swelling rivers, mudslides, and deadly, disease-carrying mosquitoes. Part of the canal had to be cut through the rocky mountains of the continental divide. Although de Lesseps was a visionary entrepreneur and no engineer, he confidently assured the public that he could assemble the engineers, technology, and finances, and that at 50 miles or just half the length of Suez, the Panama Canal would be easier to build.

Despite adoring press and a soaring share price that excited the whole French nation, the venture soon ran into formidable obstacles on the ground in Panama. Most unanticipated was the outbreak of disease epidemics. Malaria and yellow fever debilitated up to 80 percent of the workforce at any given time with shivering fevers, unquenchable thirst and, in the case of yellow fever, intense headaches, back and leg pain and finally the dark, bloody vomiting that preceded death. An estimated 20,000 workers and managers died of these mosquito-borne tropical diseases whose origins were as yet unknown. Nor was any satisfactory solution devised for containing the wild Chagres River, which could rise 30 feet after a day of torrential rain; the estimated size of the needed dam kept rising to be one of the largest on Earth. Most daunting of all were the incessant mountain mudslides and ever-expanding excavation to cut through the unstable mountain geology of the continental divide; simply disposing of so much dirt proved to be an overwhelming logistical challenge. By late 1886, French engineers realized that the existing excavation technology of the age was simply not up to the task of building a canal according to de Lesseps's sea-level design. For too long, de Lesseps refused to consider any alternative—illustrating the fine line separating inspired vision from disastrous obduracy. By the time he relented to a modified plan, it was too late. The company's financing was exhausted and by mid-1889 work at the isthmus terminated. Some $287 million had been spent—over three times more than the entire Suez Canal—on a glorious but failed dream that in the end proved to be too great for the private sector acting alone with the available technology of the age. With so many individuals and families having lost their life's savings, and with French national pride wounded, government investigations into wrongdoing were launched. In the hunt for scapegoats, de Lesseps was convicted of fraud and maladministration and sentenced to prison. Ailing, broken, and partly senile, he died in 1894, at the age of eighty-nine.

Efforts to revive the French canal project during the 1890s failed. As U.S. government action to build a canal at Nicaragua grew imminent in late 1901 when Roosevelt acceded to the presidency, French shareholders became frantic to salvage something of their Panamanian investments. They replaced company management and signaled the Americans they'd sell their assets on work already done for only $40 million—a 60 percent discount on their previous $109 million asking price. Although the U.S. interoceanic canal commission had recently endorsed Nicaragua, it had decided against Panama not on technical grounds but chiefly on the exorbitant cost of buying the French company's assets. Roosevelt, who to this point had stayed publicly aloof from the debate, summoned each commissioner to the White House individually for a consultation on his private views. He then called a secret meeting of the entire commission in his office. With characteristic boldness, he told the commissioners he wanted a supplemental report favoring Panama—and he wanted it to be unanimous. The combination of factors that caused the new president to challenge the powerful Nicaragua lobby is not definitively known. Clearly, he had become convinced that Panama was indeed the superior technical route and that all the causes of the French failure could be overcome. He may also have found it an opportune means to curry favor with the powerful domestic political interests supporting Panama. Another motivation may have been concern that Germany or another competitive foreign power might buy the French assets if the United States did not.

Roosevelt's intervention reignited the ferocious lobbying clash over the canal route between men of influence at the apex of American society. While most senators and the general public initially supported Nicaragua, the Panamanian route was backed by powerful Wall Street bankers, railroad barons, and Republican senator Mark Hanna, McKinley's key political benefactor and the era's towering national power broker. Although canal commission members were grilled at Senate hearings about why they flipped to endorse Panama after talking to Roosevelt, they insisted that both routes were viable and that the economic value of the work already done by the French made the difference. Ultimately, the tight Senate vote in June 1902 to give preference to Roosevelt's Panamanian route tipped on tremors from a visceral but not technically significant factor—seismic activity in the region. Emotional sensitivities had been heightened by a devastating volcanic eruption that had recently struck the Caribbean island of Martinique. Then just prior to the senate vote, there was a rare, minor eruption in Nicaragua itself. The Nicaraguan government tried to head off any propaganda damage by denying, falsely, that the eruption had occurred at all. But Philippe Bunau-Varilla, a former French canal company engineer and project manager under de Lesseps who had come to America to spearhead the Panamanian lobby, trumped them with a deal-clinching theatrical response: On the eve of the vote he sent each senator an irrefutable, dramatic visual reminder of Nicaragua's seismic peril in the form of that nation's own one centavo stamp—featuring a smoking volcano rising from the middle of Lake Nicaragua. The final vote for Panama was 42 to 34.

In January 1903, after months of stalled negotiations, Colombia's head diplomat in Washington reluctantly yielded to Roosevelt's imperious treaty terms when the U.S. administration threatened to abandon Panama and open negotiations with Nicaragua: the United States would receive a 100-year renewable lease for effective sovereignty over the Canal Zone in exchange for $10 million plus $250,000 in annual rent. However the Colombian Senate, bridling at the loss of sovereignty, and insulted by the fact that the French company would get four times more upfront for its Panamanian assets, rejected the treaty in the summer. Roosevelt was furious. Privately referring to the Colombians as "jackrabbits," "bandits," and extortionists impeding a vital highway of human civilization, Roosevelt tacitly signaled his support for a secret plan for Panama to secede from Colombia and then sign the canal treaty with the United States.

The planned Panamanian revolution was brilliantly stage-managed from New York and Washington by the Frenchman Bunau-Varilla and the U.S. Panama lobby. Panama's secessionist leaders were all prominent professionals employed by the American-owned Panamanian railroad. Bunau-Varilla provided virtually everything they needed—a declaration of independence, a military defense plan, a constitution, a national flag, secret communication codes, a payoff of $100,000 for expenses when the job was done, and most important of all, a promise that the U.S. military would back up the revolution. He even designated November 3 as the date for their revolution. His one condition was that he himself would be appointed by the new Panamanian government as its minister plenipotentiary in Washington to negotiate U.S. recognition and the canal treaty.

Bunau-Varilla had assured himself of Roosevelt's military backing at a personal, informal meeting with the president at the White House on October 10, 1903. Within three weeks of the meeting, three American warships were ordered to sail toward the isthmus; on November 2 the commander of the first to arrive was instructed to prevent Colombian troops from landing in the event of turmoil. Early on November 3, *before* anything had happened on the ground, the State Department in Washington cabled its consul in Panama of reports about an uprising on the isthmus; the consul's office cabled back that the uprising had not yet occurred and was to take place at 6:00 p.m. It did. A fire brigade was commissioned as the new Panamanian army. Marines disembarked from an American steam gunboat and bloodlessly confronted Colombian soldiers. Seeing the superior American vessels, one Colombian gunboat fired a few shells at Panama City—killing a sleeping Chinese shopkeeper and a donkey—and fled.

Independence was declared on November 4. Two days later, the United States formally recognized the Republic of Panama. A ceremony was conducted at the White House, entirely in English, between Roosevelt and the Frenchman Bunau-Varilla acting as Panama's plenipotentiary. Treaty drafting began promptly, interrupted only by Bunau-Varilla's quick visit to banker J. P. Morgan at his Wall Street headquarters to secure a loan for the $100,000 remuneration he'd promised the Panamanian leaders. With a Panamanian delegation rushing to Washington to try to reclaim the extraordinary powers they'd granted to Bunau-Varilla and the press clamoring about America's high-handed involvement in the revolution, Bunau-Varilla and Secretary of State Hay hastily signed the canal treaty on November 18, 1903. The terms were the same as those previously offered to Colombia, with the addition of several clauses—inserted at Bunau-Varilla's initiative—enhancing America's sovereign rights in the Canal Zone and extending its lease into perpetuity. On the centennial of the Louisiana Purchase, another French-American deal had delivered the United States an extraordinary territory that would empower its expansion for much of the ensuing century. As soon as the treaty was ratified by the U.S. Senate in February 1904, and the French company guaranteed the $40 million payment for its canal assets, Bunau-Varilla resigned his Panamanian post and returned to Paris.

Roosevelt's brazen use of naval force and his tortured legalistic justifications for intervening at the isthmus and recognizing Panama's sovereignty ignited a torrent of resentment against U.S. imperialism. "Colombia was hit by the big stick, all Latin America trembled," wrote historian Samuel Eliot Morison. The United States began aggressively policing the Caribbean as if it were an American lake, frequently intervening in basin countries' financial and foreign affairs, and landing marines half a dozen times between 1900 and 1917. America's growing, imperialistic tendencies were also projected by Roosevelt and his successors outside the Western Hemisphere, highlighted by Roosevelt's swaggering display of sea power in sending a great American fleet around the world in 1907. Latin American resentment smoldered for decades. President Wilson tried to soothe Colombia's outrage at the loss of Panama with a $25 million payment; Panamanian restiveness caused the canal treaty to be revised several times and effective control of the canal was surrendered in stages starting in 1979.

Teddy Roosevelt, however, was unapologetic for his actions in Panama. He viewed the canal as an invaluable advancement of civilization and vital to America's defense and prosperity. While absolutely denying any role in making Panama's revolution, retrospectively he wrote that it was "by far the most important action I took in foreign affairs," and in a 1911 speech, declared with characteristic bravado, "I took the Isthmus, started the canal and then left Congress—not to debate the canal, but to debate me." His most famous, instinctive response to his critics was issued shortly after U.S. government engineers took charge of the Canal Zone in May 1904: "Tell them that I am going to make the dirt fly!"

Building the canal was the most monumental, complex engineering challenge of the age, and one of the landmark technological achievements of human history. To complete it required the application of all the qualities underlying America's rise as a great power—prolific industrial production, innovative ingenuity, government financial commitment, tenacity of purpose, and cultural optimism in its ultimate ability to succeed. Although construction lasted through the terms of three presidents, there was no doubt that it was Roosevelt who infused the guiding spirit and was the embodiment of the enterprise. Nothing exemplified this more than the three-day tour through the canal construction sites Roosevelt himself made in November 1906. Like a general inspecting the troops at the front, he trudged in the driving rain through muddy work camps, strode along the ties of the Panamanian railroad, climbed a hill to get a view of the future dam site, peppered everyone with questions, and most memorable of all, spontaneously stopped his touring train during a downpour to clamber into the control seat of one of the huge, workhorse steam shovels that could excavate eight tons of dirt in a single scoop—three times more than the shovels used previously by the French—and unload its contents into a departing railroad car every eight minutes. That no sitting president had ever before traveled abroad magnified the drama of his visit and his subsequent progress report to Congress on America's stupendous endeavor.

After a slow start and a flirtation with reviving de Lesseps' plan to dig a deep trench at sea level across the isthmus, the Americans by 1906 settled on a workable design and methodology for building a lock canal. Dams would be erected to entrap the rain-fed swells of the Chagres River, creating an 85-foot-high artificial lake that bridged much of the isthmus. Ships would ascend the lake through a flight of giant locks at one end and descend through another at the other end. Near the Pacific side, they would pass through a narrow, nine-mile-long canyon that would be excavated through the rocky mountains and rainy forests of the continental divide. Success hinged on managing three main challenges—containing the tropical diseases that debilitated the French workforce, controlling the wildly rising and falling Chagres River, and, hardest of all, cutting through the massive, mudslide-prone mountains. When it was completed in 1914, America's canal construction workforce, which averaged 33,000 to 40,000 annually between 1907 and 1914, had excavated over eight times more earth than their French predecessors.

The eradication of yellow fever and control of malaria attained early on in the Panama project alone merited renown as a notable achievement of the twentieth century. During the French period, the germ theory of disease had been in its infancy and the role of the mosquito in transmitting yellow fever and malaria was only beginning to be perceived. But at the start of the century, American doctors, led by Walter Reed, working in Havana, Cuba, learned how to curb both diseases by attacking the two different mosquito species that carried them. A similar, wider war was waged throughout the jungles and towns of Panama. The silvery, female mosquito transmitting the dreaded yellow fever, which would deposit her eggs only in a container of clean water, depended upon the close proximity to human society for breeding and feeding. By systematically screening all windows and doors, fumigating houses, covering water barrels, oiling cisterns and cesspools, and eliminating standing water everywhere, U.S. Army doctors virtually wiped out yellow fever from Panama by the end of 1905. To attack the more wide-ranging malaria mosquito, sanitary teams also drained hundreds of miles of swamps, installed effective drainage channels, chopped down jungle vegetation, sprayed oil on standing water, and spread mosquito-larvae-eating minnows and mosquito-feeding spiders and lizards near human habitats. While never eradicated, malaria was sufficiently contained so that it didn't interfere with constructing the canal. The success of the disease control program at Panama informed the global wars on yellow fever and malaria that were soon launched in the 1910s and 1920s by the Rockefeller Foundation and other humanitarian organizations.

The Chagres River was controlled by the construction of an immense, wide earthen dam filled with dirt excavated from the mountains. The dam created what was then the largest man-made lake on Earth. The concrete used in the flight of huge, steel-gated locks exceeded the volume of any project until the building of the Hoover Dam in the early 1930s. Some 26 million gallons of fresh lake water—roughly one-quarter of the daily water supply used by New York City in that era—were consumed in filling the innovative flight locks every time a ship was raised or lowered the 85 feet from the lake to the sea. A novel system of towing locomotives guided the ships into and out of the locks. All the tows, valves, culverts, lock gates, and other lock-regulating mechanisms were electrically powered by some 1,500 motors, which were hydropowered by the falling water on site. As a result, the functioning of the canal was centralized and entirely self-contained, requiring no external energy source.

During canal construction, tourists came from around the world to observe what was by far the canal's most Herculean challenge—digging the nine-mile-long, neck-shaped water passage through the mountains of the continental divide. The Culebra Cut that had broken the French was often called the canal's "special wonder." For seven straight years American-managed work crews labored systematically day and night, except Sundays, in the sweltering heat and rain, blasting mountains, hauling away rock and dirt, and relentlessly digging out work equipment that became buried beneath the mountainside avalanches of mud that recurred time and again during the rainy season. The amount of dirt hauled out of the Cut defied imagination. The labor would have been simply overwhelming had not the engineers applied U.S. industrial assembly line methodology and technical ingenuity to the work. The lifeline of the system was the heavy-duty railroad network that operated on precise schedules at different levels within the Cut. Where heavy wheeled vehicles would have become mired in the soft ground, railroads were able to haul in equipment and remove excavated earth at regular intervals. Every day 50 to 60 massive, steam-powered shovels filled 500 trainloads of spoil. Newly invented mechanized excavation equipment that had been unavailable to the French, such as train unloaders and dirt spreaders, accomplished in minutes work that previously had consumed thousands of manual man-hours. Finally, in May 1913, two steam shovels working from opposite directions broke through the last mounds of earth. A few months later, in October 1913, President Woodrow Wilson pushed a signal button in Washington to blow the dike for the final filling of the canal. Ominously, a powerful earthquake had ripped through the Canal Zone some ten days earlier, cracking buildings in Panama City. But the canal passed the final test of nature totally unscathed. It had been finished on time at a total cost of $375 million to the U.S. government. The gala international celebration scheduled for the official opening on August 15, 1914, however, never came off—the start of World War I intervened. Teddy Roosevelt himself never visited the finished canal. He died in early 1919, at age sixty.

By every measure, the canal was a stupendous success. Within a decade, some 5,000 vessels were passing through it every year, as many as through Suez. By 1970, over 15,000 ships made the ten- to twelve-hour journey, paying annual tolls of over $100 million. Periodic enlargements and improvements permitted the passage of large warships and the growing fleet of oil supertankers and giant container ships that formed the backbone of the shipping revolution underpinning the rapid integration of the global economy in the late twentieth century. The canal represented the culmination of the historical transformation of the world's oceans from restrictive boundaries into integrated superhighways that had begun with the European Voyages of Discovery four centuries earlier.

"The fifty miles between the oceans were among the hardest ever won by human effort and ingenuity, and no statistics on tonnage or tolls can begin to convey the grandeur of what was accomplished," summed up David McCullough in his sweeping history of the canal. "Primarily the canal is an expression of that old and noble desire to bridge the divide, to bring people together. It is a work of civilization."

For America, the Panama Canal stood as a beacon of the nation's arrival as a star among world civilizations. It was a national historic turning point when many of the society's dynamic forces coalesced to open a new era. At last, America was in a position to fulfill its continental promise of marrying the maritime resources of its two ocean frontiers. U.S. exports and investment abroad soared after the canal's opening. Overseas markets and raw materials were immediately drawn into the productive circuit of America's prolific industrial economy. By 1929 the United States was producing nearly half the world's total industrial output.

Likewise, the Panama Canal marked the transition between the first and second of American naval history's three eras. It ended the long period when America's navy was focused chiefly on defending the young country's borders and waterways, protecting free trade and market access for its seafaring merchants while opportunistically promoting continental expansion. After Panama, naval orientation projected American power outward to arbitrate affairs within Europe and Asia, while expanding American commercial and military access throughout the world. The United States entered World War I to make the world safe for democracy, as President Woodrow Wilson explained, when German submarines began sinking U.S. and other formally neutral merchant and passenger ships in its effort to break Britain's control of the seas and blockade of its ports. Throughout the third era dating from World War II, the mighty U.S. Navy patrolled the globe's oceans and strategic passages, peerless and rarely challenged, to uphold the international free world order of which America itself was the undisputed leader. American aircraft carriers were the state-of-the-art naval weapon of World War II. Most famously, aircraft carriers played the decisive role in the pivotal Battle of Midway (June 1942) that determined control of the Pacific Ocean. The Japanese navy that had inflicted such a terrible, surprise first strike on the Americans six months earlier at Pearl Harbor never recovered from its Midway defeat; Midway was the first sea battle in history in which the engaging fleets never saw each other and remained from 80 to 170 miles apart throughout. Control of the seas likewise enabled America to carry out the Normandy D-day landings that ultimately reclaimed continental tEurope from Nazi Germany.

America's central position on the world's ocean superhighways, its plethora of good, all-weather seaports, and its naval dominance remained critical advantages in its winning the Cold War against its mostly land-bound Soviet adversary. The USSR's military fleets and supply ships constantly labored against the geographic disadvantages of long distances, bad climates, and confinement imposed by the West's control of key naval passages, such as those in exiting the Black Sea. America's early twenty-first-century capability to project its influence as civilization's unchallenged military superpower continued to hinge upon its worldwide deployment of its dozen world-class, nuclear-powered aircraft carrier task forces. Its dominance was roughly equal to the combined power of the world's next nine leading military nations, and unmatched in Western history since ancient Rome ruled over the Mediterranean world.

By creating an inexpensive, fast water linkage between the Atlantic and the Pacific, the Panama Canal also pointed the direction arrow toward America's next economic boom region—its underdeveloped, arid Far West, whose store of potential mineral and agricultural wealth suddenly came within easier reach of the expanding industries and markets of the East. As a successful federal government-financed and -run enterprise, the canal, moreover, inspired a ready model for undertaking the large, state-run water projects that developed the Far West into America's twentieth-century growth engine. In contrast to the more orthodox, laissez-faire market economy attitudes prevailing in latter-nineteenth-century England and France, the United States had always been governed by a distinctive "American system" that regarded government as an active agent to assist the private development of the nation's resources. The New York State financing of the Erie Canal, the federal government's early policy of promoting internal improvements, and its incentives to the transcontinental railroads and small farmers who homesteaded lands west of the Mississippi after 1862 exemplified how this mixed economic model had worked in the nineteenth century. With the Panama Canal, a fuller, more activist iteration of this mixed system was inaugurated to meet the gargantuan scale of the challenges and opportunities of the industrial age. Teddy Roosevelt himself was the prime mover of the government-led policies that later in the century, under the presidency of his younger distant cousin Franklin Roosevelt, came to fruition in the giant, multipurpose dams that spectacularly transformed America's scantly populated, arid western landscape with cheap irrigation for farming and hydroelectricity for mining and industry. The publicly built dams, in turn, reinforced the general trajectory of the large, state-led governance that characterized so many leading societies of the twentieth century.

CHAPTER THIRTEEN

Giant Dams, Water Abundance, and the Rise of Global Society

Developing the Far West presented a radically unfamiliar set of water challenges to the United States. While the continent as a whole had an abundance of water wealth, its best resources were concentrated in its temperate, river-rich, eastern half, where annual precipitation normally exceeded the minimum 20 annual inches necessary for sustaining small-scale, nonirrigated farming. Moving west from the Mississippi Valley into the semiarid, treeless grasslands of the high plains and prairies of western Kansas, Nebraska, and Texas—America's steppes—rainfall tapered off and became more undependable. Beyond the 99th and 100th meridians, water, not free land, was the main limiting factor in development. Only in wet years did the Great Plains have enough rainfall to maintain cultivation. In 1865 the farming frontier ran roughly along eastern Kansas and Nebraska's 96th meridian. Over the following quarter century successive waves of hardy, yeoman farmers tried settling in wet years across the 100th meridian, only to be pushed back by the dry periods that always ensued. Between 1870 and 1880 the population of Kansas, Nebraska, and Colorado increased by more than 1 million to 1.6 million. But by 1890, in the third year of a ten-year drought and after the terrible winter of 1885–1886, Kansas and Nebraska had depopulated by one-fourth to one-half. West of the Rockies most of the valleys and lowlands were deserts drier than

North Africa; many were scarcely habitable regions with less than seven inches of annual rainfall, such as those of present-day Phoenix and Las Vegas. Western precipitation, including the winter mountain snows that melted into abundant spring runoff, was also highly seasonal and prone to prolonged cycles of drought. Thus even where freshwater existed in sufficient volume for farming, it was often unavailable *when* it was needed. Additionally, most of the Far West's perennial surface water was confined within three large mountain-fed river systems—the Colorado in the Southwest, the Columbia in the Northwest, and the San Joaquin and Sacramento in California's Central Valley—that were often long distances away over rugged terrain from most arable land.

Water scarcity, in short, was the defining geographical condition of America's Far West. As a result, the struggle for water was inseparable from the naked contest for power and wealth. Water rights were the stuff of family blood feuds, such as depicted in the 1958 Hollywood movie *The Big Country*. In the West, as author and humorist Mark Twain wryly put it, "Whiskey is for drinking. Water is for fightin' over." By pioneering the world's first giant, multipurpose dams—the defining water innovation of the twentieth century—America from the mid-1930s successfully converted the West's few wild rivers into dynamic engines of inexpensive irrigation, hydroelectricity, water storage, and flood control. The Far Western deserts were miraculously transformed into the richest irrigated farmland on the planet. The arid western hydrological frontier added potent new impetus to America's rising civilization. Large cities rose in the desert. America's federal government-led development became a standard feature of the national political economy. Very rapidly American dam technology diffused worldwide, spreading the prodigious material benefits deriving from the intensification of man's basic uses of water.

The Far West's water challenge had more in common with the authoritarian, hydraulic societies of ancient Mesopotamia, although over rougher, harsher, and far more expansive landscapes than it did with the rainy, eastern United States, which had helped nurture

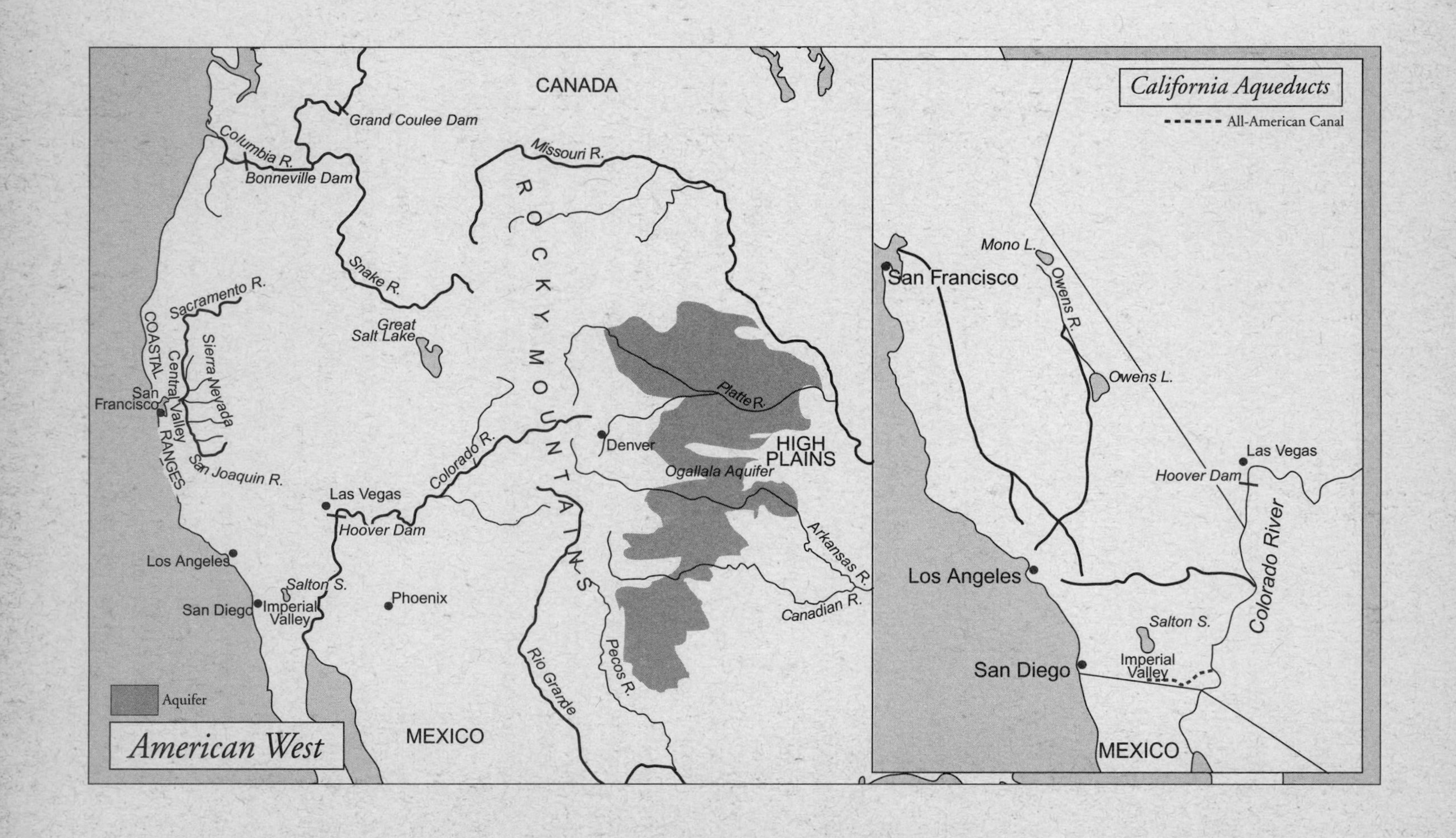
American West
Aquifer
CANADA
Grand Coulee Dam
Columbia R.
Bonneville Dam
Missouri R.
Snake R.
Sacramento R.
Great Salt Lake
COASTAL RANGES
Central Valley
Sierra Nevada
San Francisco
San Joaquin R.
ROCKY MOUNTAINS
Platte R.
Denver
HIGH PLAINS
Ogallala Aquifer
Colorado R.
Las Vegas
Hoover Dam
Arkansas R.
Canadian R.
Los Angeles
Salton S.
San Diego
Imperial Valley
Phoenix
Rio Grande
Pecos R.
MEXICO
California Aqueducts
All-American Canal
Mono L.
San Francisco
Owens R.
Owens L.
Las Vegas
Hoover Dam
Colorado River
Los Angeles
Salton S.
San Diego
Imperial Valley
MEXICO

America's market democracy of independent, yeoman farmers, entrepreneurial industries, and decentralized political power. Indeed, incorporating the arid west into mainstream American civilization, posed political economic and cultural challenges within the purely technical one. These broader challenges were explored by America's great turn-of-the-century historian, Frederick Jackson Turner, whose seminal 1893 essay "The Significance of the Frontier in American History" established the prevailing paradigm that America's uniquely individualistic, democratic, pragmatic, and pluralistic character and institutions had been forged primarily by the frontier experience of continuous westward expansion, rather than by older theories of European values transplanted in the New World or the interplay of conflicting interests between the North and South. In his classic analysis of American history, Turner noted that the effective closure of the free land, farming frontier—which he considered to be virtually complete by 1893—had been accompanied by a gradual trend away from individualism and toward the social tendencies of cooperation, big business combination, and increasing reliance on government assistance. The physical challenges of Far Western settlement, he argued presciently, would inevitably accelerate that trend: "When the arid lands and the mineral resources of the Far West were reached, no conquest was possible by the old individual pioneer methods. Here expensive irrigation works must be constructed, cooperative activity was demanded in utilization of the water supply, capital beyond the reach of the small farmer was required. In short, the physiographic province itself decreed that the destiny of this new frontier should be social rather than individual." Turner predicted hopefully that the frontier spirit of the yeoman farmer might endure by infusing itself in new democratic forms as America faced the historically centralizing, authoritarian tendencies of large states that organized control and distribution of irrigation river water in semiarid landscapes.

By the time Turner delivered his landmark frontier thesis to the American Historical Association in Chicago, it was already apparent that private irrigation alone could not develop the Far West. Irrigation efforts in the region dated back to about 1200 BC, when indigenous southwest natives began digging irrigation canals to cultivate crops. By AD 500 the advanced Hohokam culture, northern neighbors of the Aztecs and Mayas, was well established with an extensive canal network around central Arizona's Salt River, a tributary of the Colorado. Yet by the sixteenth century the Hohokams had vanished, likely victims of one of the prolonged natural droughts to which the region was prone or of ecosystem depletion caused by the soil salinization by-product of intensive irrigation. Some of their canals were reexcavated and reopened in the late 1860s when U.S. settlers moved into the region. Modern western irrigation had begun with the Mormons, who migrated to Utah from 1847. Through centralized organization, religious discipline, and arduous work, the Mormons created numerous small, farm communities growing potatoes, beans, corn, and wheat by diverting small mountain streams into short canals. Between 1850 and 1890, they expanded their irrigated cropland fifteenfold to support a total population of over 200,000. Irrigated farming began in earnest in California after the gold rush with the coming of the transcontinental steam railroad in 1869. Speculative real estate consortiums organized communities of small farmers around water drawn through canals, sometimes with the help of privately built dams, from streams and rivers that flowed through California's fertile Central Valley, including the King, the San Joaquin, the Kern, and the Sacramento. Eastern Colorado, settled from the 1870s by people inspired by utopian community ideals, was another pocket of western irrigated farming. As late as 1909, Colorado had more acres under irrigation than California.

All in all, however, nineteenth-century American irrigators had done hardly much better than their Hohokam predecessors in transforming the Far Western landscape into an agricultural garden. By the mid-1880s, the best irrigation sites on most of the region's small streams already were being tapped. If western farming was to be developed on a meaningful scale, bigger dams on the few large, wild rivers were needed. But enormous risk capital had to be pledged, and complex water rights issues settled, for such an undertaking. With the depredations of the late 1880s drought and the great economic depression of 1893, moreover, private financing all but dried up for large irrigation projects and land values fell. The final blow against private enterprise solutions was struck with the tragic collapse in the spring of 1889 of a privately built eastern dam in Johnstown, Pennsylvania, that unleashed a flood that killed 2,200. Throughout the 1890s, western private sector and elected leaders increasingly beseeched the federal government to take the lead.

The groundwork for federal irrigation had been built over many years thanks to the pioneering efforts of John Wesley Powell. Born in 1834, Powell had explored the Mississippi River in the 1850s. Although he lost his lower right arm as a Union officer at the battle of Shiloh, he intrepidly led a blind expedition in 1869 of nine men on four wooden boats on the first ever explorative run of the wild Colorado River and the Grand Canyon. Parlaying his fame as an explorer into a national platform on western geography and development, Powell in 1874 shocked Congress and the nation in testimony that challenged cherished national myths: He declared that nearly the entire western region of the United States beyond the 99th or 100th meridian was too arid for small-scale, eastern-type agriculture without irrigation, and that even with irrigation the total available supply of water was sufficient to reclaim a much more limited amount of cropland for a much smaller population than irrigation boosters supposed. At the time, many public officials clutched to the comforting but fanciful notion that rain would follow the plow, permitting the unending advancement of small yeoman farmers across the continent. In 1878, Powell expounded his views in greater depth in his influential *Report on the Arid Land of the Arid Region of the United States.* This led to his appointment as head of a new government bureau to scientifically study western lands and later the irrigation potential of western water resources.

In this capacity, Powell became a formidable champion of government-led water storage dams to irrigate America's west. His view was that all the waters of every free flowing river should be commandeered from its natural bed in economic service to the nation. Public lands that naturally stored and accumulated water, such as mountain forests, should be conserved in government hands and not be sold to timber companies or other private interests that would deplete them. Powell did not fret that the United States might repeat the authoritarian history of the ancient hydraulic societies. Indeed, he advocated his own idealistic, technocratic program for development based on political units that would be reorganized around natural watersheds. Powell maintained that his plan would enable 1.25 million small farmers to cultivate 100 million acres of irrigable cropland.

Powell's idiosyncratic views, however, rankled vested establishment interests and disturbed the popular political myth that a federal irrigation program could be based upon a simple adaptation of Homestead-type grants of small, public lots to much-idealized, Jeffersonian yeoman farmers. At the second National Irrigation Congress in Los Angeles in 1893, Powell triggered an uproar by declaring, factually, that large private interests already controlled all the best irrigable lands in the West. But by then the irrigation movement had enough momentum from conventional politicians and powerful private interests to be able to divorce itself from the quirky, water wealth administration schemes of the original champion of the irrigation cause. A year later, Powell resigned from the government. In 1902, he died in obscurity in Maine.

Before dying, however, Powell had the satisfaction of witnessing the birth of a federally run western irrigation effort. The prime mover behind the 1902 Reclamation Act was America's great water president, Teddy Roosevelt, who had recently come to office in September 1901 with McKinley's assassination. Roosevelt was an admirer of Powell. He had lived in the West's South Dakota Badlands, and ardently believed in Powell's prescription of federally supported irrigation to develop its fertile, though dry soil. In his very first formal message to Congress on December 3, 1901—even as he was maneuvering to reopen the debate over the route of the sea canal route through Central America—he declared his determination to open the West through federal water conservation and irrigation. "In the arid region it is water, not land, which measures production," he said, echoing both Powell and Frederick Jackson Turner. "The western half of the United States would sustain a population greater than that of our whole country today if the waters that now run to waste were saved and used for irrigation."

Under the 1902 Reclamation Act, money from the sale of public lands in the West was to fund federal irrigation works administered by a new

Reclamation Service within the Department of Interior. Only family farms of 160 acres or smaller were supposed to benefit from the government's irrigation largesse—a proviso that, in practice, would be routinely breached over the years. Roosevelt also appreciated the intimate interlinkages between water and forests. Forests served as natural reservoirs, conservers of soil, and restrainers of terrible floods. "The forest and water problems are perhaps the most vital internal questions of the United States," Roosevelt stated. Near the end of his presidency, Roosevelt created the celebrated western public parks system, partly to conserve forest watersheds.

Although the Reclamation Service (renamed the Bureau of Reclamation in 1923) eventually became world history's largest government-run water technocracy—a modern democratic version of the ancient priestly elites of the Middle East and China's professional mandarins—the irrigation program started ineffectually. In its first two decades its total projects covered so little acreage as to barely make a noticeable difference in the expansion of western agriculture. Its economic foundations also seemed dubious. Despite the generous water subsidy and extension of payment terms, over half of irrigation project farmers were defaulting on their water loan repayments by 1922. Wealthy land speculators tracked Reclamation engineers like buzzards, swooping in to buy up public homesteads wherever projects seemed likely in order to resell them later at greatly multiplied values to the new, and quickly overindebted, small farmers. Existing private landholders also enjoyed unearned bonanzas from the federal irrigation projects. Then in the early 1920s, the U.S. agricultural sector went into a depression with sharply falling farm prices, one of the contributing factors of the economy-wide Great Depression of the 1930s. Without the remarkable resilience of public bureaucracies to endure over time despite failure and loss of purpose, the Reclamation Bureau and the western irrigation program might well have vanished at that point as a forgettable footnote of one of history's failed policy initiatives.

What changed everything was the Hoover (aka Boulder) Dam. Reclamation engineers had dreamed about building a dam on the Colorado River ever since the bureaucracy's inception. But it wasn't until the late 1910s and 1920s that the combination of political, economic, and technological forces aligned favorably to impel the first serious steps toward undertaking the world's first giant multipurpose dam on the lower Colorado. Another decade of political maneuvering would go by before construction could begin. The completed dam, which started full operations in 1936, was simply stupendous. It dwarfed all previous dams in history by orders of magnitude in scale and novelty. From ancient Roman and Han times through the nineteenth century few dams had surpassed 150 feet in height. The systematic application of the sciences of civil engineering, hydraulics, and fluid mechanics from the mid-nineteenth century, however, enabled the building of much more complex dam structures.

The concrete Hoover Dam would stand 726 feet, over six times taller than the British-built marvel of the first Low Aswan Dam on the Nile even with its final 1929 extension, and more than twice as high as any other dam on Earth. It created the world's largest man-made reservoir, the 110-mile-long Lake Mead, which could store two times the annual flow of the Colorado, or enough to flood the entire state of Pennsylvania under one foot of water. Its world's largest hydroelectric power plant was capable of generating 1.7 million horsepower, upgraded to 2.7 million horsepower in the 1980s. By 2000, some 30 million southwesterners, almost 2 million acres of prized cropland, and metropolises such as Los Angeles, San Diego, Phoenix, and Las Vegas would be dependent upon the Colorado River's water supply. Its wall contained enough concrete to build a highway across the continental United States. Hoover gave mankind, for the first time in history, the technical means for bringing the world's mightiest rivers under almost total control, converting wild, variable currents and unpredictable flooding into tamed pools of carefully regulated flows and allocated levels. Vitally for American water technocrats, too, the Hoover Dam established a viable economic blueprint for dam projects they could emulate again and again to achieve their mission of transforming the arid West.

In addition to its gigantic scale, the key innovation at Hoover was the dam's successful multipurpose design. Throughout history, most dams and their affiliated waterworks had been built for a single purpose only—usually irrigation or flood control, but also improved navigation, drinking supply, or generating waterpower through waterwheels and, since the 1880s, hydroelectric turbines. Divergent purposes presented competing design challenges—for instance, flood control demanded low reservoir levels to catch flood swells while maximum power generation required full reservoirs; navigation presented still other difficulties. The multipurpose approach had been promoted as early as 1908 by Teddy Roosevelt in an effort to jump-start the flagging development of western irrigation. Despite the ingrained skepticism of the water bureaucracy establishment at the old Army Corps of Engineers, Reclamation officials, hungry to find their raison d'être, began to experiment with integrating hydropower to its irrigation dams. Their most celebrated early success was the elegant 280-foot-tall dam on the Salt River in Arizona. Completed in 1911 and named after President Roosevelt, the dam provided a visible boon to the economic life of the Phoenix area by both alleviating irrigation water shortages for farms built around the dredged-out canals of the long-gone Hohokam natives, and by generating electricity. Crucially, the electricity sales added enough revenue to pay for the dam.

Hydroelectric sales underwriting farm irrigation subsidies: This became the working fiscal model for Hoover and the great dam era that it spawned. While American leaders had regularly rejected proposals throughout the 1920s to develop multipurpose dams on the Tennessee River in the eastern United States, since that would make the government a big player in the private electricity business, they were more amenable in the West, where the farming, urban, and railroad lobbying interests from Southern California made a concerted appeal for a giant irrigation, flood control, and hydroelectric dam on the great river that was the lifeblood of the southwest.

Rising in the Rocky Mountains at 14,000 feet and falling toward sea level over its 1,440-mile length through deep canyons—including the Grand Canyon which its torrents had carved over the eons—and deserts, the muddy, turbulent Colorado River flowed within seven states to its delta in Mexico south of the California-Arizona border before exiting through its mouth at the Gulf of California. While its average flow of about 14 million acre-feet per year made it a relatively modest-sized river by volume—comparable in quantity to the eastern Susquehanna, Delaware, Hudson, or Connecticut and only a small fraction of the huge Mississippi and Columbia rivers—every drop was precious because it drained the most arid large basin on the continent. It was the only significant water source within 1,000 miles. For thousands of years its delta of lagoons and earthen mounds, twice the size of Rhode Island, had been a paradise for jaguars, coyote, beaver, a vast assortment of water fowl, fish, and uncountable species of plants.

In its natural, predammed state the velocity of its flow was schizophrenic. When swollen with the springtime snowmelt from the mountains, it sometimes cascaded furiously down from the mountains at 300,000 cubic feet (nearly 2.3 million gallons) per second, tearing away mountainsides and smashing boulders; in the dry season it meandered by at less than 1 percent that speed. The river's annual volume also varied widely, by over 50 percent, depending upon whether the region was in a wet or drought cycle. Most distinctive, its enormous, wild energy made the reddish-brown Colorado one of the siltiest rivers in the world. Sediment from the hillside during its steep descent accumulated in its lower reaches beyond the Grand Canyon, where the river was 17 times siltier than the muddy Mississippi. The river's texture was playfully described by southwesterners as "too thick to drink, too thin to plow." The riverbed was raised by the accumulating deposits year after year until, in a great swell, the river overtopped the sandy bluffs containing it. It flooded violently in new directions, carving fresh channels toward the sea. Before its taming by dams, the Colorado's rampaging floods periodically turned the dry lowlands of Southern California and Mexico into huge swamps that left behind exceptionally fertile soil when the floodwaters evaporated.

Just such a terrible flood event had been a midwife to the Hoover Dam. In the late 1890s, private developers had begun working on an ambitious scheme to clean out one of the Colorado's ancient flood channels, known as the Alamo River, and use it as an irrigation canal to draw river water that would transform the low-lying, silt-enriched soils of the Southern California desert, with less than three inches of annual rainfall, into productive farmland. By 1901, the water started to flow. Its path started with a cut from the Colorado just north of the Mexican border, looped southward some 50 miles through Mexico, and then northward into a Southern California depression known as the Salton Sink. As some 2,000 farmers settled in, the irrigated lowland indeed bloomed forth with crops. Reflecting the new confidence and hope brought by the water, the area's name was changed from the Valley of the Dead to Imperial Valley. By 1904, however, the natural accumulation of Colorado silt began to choke off the diversion channel. While working to dredge it, the irrigation engineers decided to draw river water temporarily from a newly cut bypass. Since it was supposed to be short-lived, the bypass was fitted only with filmsy, wooden control gates. As bad luck had it, the Colorado spring floods arrived two months early in 1905, and ferociously as well. The temporary control gates were washed out—and the full, raging force of the Colorado rushed into the ancient channel. The Salton Sink swelled with water, submerging thousands of acres of prime cropland to become today's inland Salton Sea. Farmers appealed fruitlessly to Roosevelt for government assistance to close the breach. The powerful Southern Pacific railroad with a strong economic interest in the region, feverishly ferried in rock and gravel. But the breach remained open until 1907. Farming in the region revived slowly thereafter. But the shadow of the 1905–1907 floods galvanized a relentless lobbying campaign for a federal flood control dam on the Colorado. The Imperial Valley farmers also campaigned the government to build an accompanying, "all-American" irrigation canal that would run on the U.S. side of the border and thus eliminate Mexico's potential leverage over the vital Colorado flow. By 1920, the political drive for the dam and canal had gathered enough momentum to be seriously proposed in Washington.

When the fast-growing city of Los Angeles stepped forward in 1924 with a proposal to build, at its own expense, an aqueduct to tap Colorado River water and to buy hydroelectricity generated by the proposed dam in order to pump the water 200 miles to L.A. over a rugged escarpment, the economics of what came to be called the Boulder Canyon Project became feasible. Los Angeles desperately needed the water. A dusty, farming town of 13,000 at the edge of the desert at the end of the Civil War, Los Angeles owed its early growth to the region's orange groves and the arrival of the railroads—a Southern Pacific spur line in 1867, and in 1885, a direct link to Kansas City on the Atchison, Topeka and Santa Fe. By 1905 its population had grown to 200,000 and was outstripping the available water supply of accumulated groundwater and flow from the small Los Angeles River—a creek for all but a few weeks of torrential precipitation in winter—whose annual flow was a mere one-fifth of 1 percent of California's total. Los Angeles's population might have crested around such modest levels had not city leaders pulled off one of the most notorious water supply grabs in history. In events that informed the story line of the 1974 film *Chinatown,* Los Angeles municipal water authorities ruthlessly gained command of the flow of the Owens River in one of the valleys formed by the old ice age folds in the Sierra Nevada 250 miles away. Between 1907 and 1913, they constructed an impressive aqueduct that brought Owens water to thirsty Los Angeles.

L.A.'s water boss was the salty-tongued, autocratic William Mulholland, an Irish immigrant who had walked the railroad ties across the Panamanian isthmus en route to California as a young man. A self-taught engineer, he worked his way up from ditch cleaner to become the unchallenged potentate, builder, and personification of the modern Los Angeles water system. In the distant Owens River, Mulholland envisioned a water source, among the city's very few viable options, that could provide L.A.'s hydraulic deliverance for a generation. He and his cronies stopped at nothing to get control of it, including deception, lies, secret agents, spying, and payoffs. With the clandestine collaboration of a paid insider at the federal Reclamation Bureau's local office who was later hired as Mulholland's deputy, and some astute politicking in Washington, they outflanked and killed Reclamation's own farm irrigation plan for the Owens Valley. Posing as cattlemen and as resort developers, they bought farmland that gave Los Angeles the precious associated water rights to the river and gained control of the best site for the future storage reservoir. Adding insult to the injury of Owens Valley farmers, Mulholland rerouted their river water first to suburban Los Angeles's dry San Fernando Valley, where a syndicate of well-connected city insiders, including railroad and trolley kingpins, utility bosses, newspaper barons, land developers, and bankers had been secretly buying up cheap land options. When the aqueduct route became known, San Fernando real estate values shot up, instantly turning millionaires into multimillionaires. San Fernando Valley soon was incorporated into Los Angeles, enriching the city's fiscal base for growth. With the actual arrival of Owens River water in 1913, irrigated acreage in the San Fernando Valley blossomed twenty-five-fold in just five years. In the zero-sum economics of water diversion, Owens Valley withered so that San Fernando Valley could flourish.

Most important, the abundant new supply of freshwater enabled Los Angeles's prosperity boom to continue. The region's population surpassed Mulholland's expectations, reaching 1.1 million in 1920 and climbing to 2.5 million by 1930. By the early 1920s, aggravated by the onset of a new drought cycle in the region, Mulholland realized that once again Los Angeles was facing a water famine unless new water resources could be obtained. That's when he began lobbying vigorously for the Colorado River water and aqueduct. To alleviate the shortages in the meantime, he decided to squeeze every last drop out of the Owens River. Mulholland's imperious combination of monetary inducements and strong-arm tactics to buy up more water rights, however, sparked a violent reaction from irate Owens Valley farmers. Between 1924 and 1927, farmers dynamited parts of the Los Angeles aqueduct and stood off against the armed city agents sent to stop them in one of America's earliest, violent clashes over water between urban dwellers and farmers. But by publicizing the possible specter of a cutoff of Owens River water to Los Angeles, the water war broke the last local opposition to Mulholland's bid for the Colorado River aqueduct. By 1928 the overriding quest for water led to the creation of a new regional political entity, the Metropolitan Water District of Southern California, with taxing power to raise funds to buy Hoover's hydroelectricity to power the aqueduct's water pumps and for other needs. When Colorado River water started to arrive in the mid-1930s, it literally validated the old adage in the arid west that "water flows uphill to money."

Before the dam project could get started, there was one further political hurdle to overcome—the sorting out of water rights on the Colorado River itself. Unlike the eastern United States, which followed the riparian legal tradition of granting water usage rights to landowners abutting a river or a stream, an alterative doctrine had developed in the water-scarce West. Known as "prior appropriation and use," or more colloquially as "use it or lose it," western water doctrine allocated priority water rights to the earliest and continuous users of a water source regardless of their location. In the early 1920s, when the dam project was taking shape, only California had any prospect of using large quantities of Colorado River water. But the other six states in the Colorado basin wanted to protect their rights for future use, lest California claim the entire river flow by using it before they did. As a result, Commerce Secretary Herbert Hoover, an engineer by training, brokered a landmark compromise among the basin states in 1922 that divided the river into an upper and lower basin and assigned equal shares of river water to each—a governance approach that oddly echoed John Wesley Powell's original notion of reorganizing the West's political units around its watersheds. Working with the projection that the river had an average 17.5 million acre-feet per year to share—later proved to be a gross overestimate—some 7.5 million acre-feet were assigned to each basin, 1.5 million set aside for Mexico, and the rest reserved for natural evaporation or storage. The linchpin of the deal was California's agreement to cap its withdrawals; the limit ultimately was set at 4.4 million acre-feet per year. It took over six years more to bring all the major elements, with required approvals, of the Boulder Canyon Project into final alignment. Finally, in 1929, newly elected president Herbert Hoover was able to launch work on the pathbreaking Boulder Dam that in 1947 would be renamed in his honor.

In contrast to its long political gestation, the physical construction of the giant flood control, irrigation, and hydroelectric power-producing dam took only five years to complete. Despite its name, the dam was located not in Boulder, but in Black Canyon about 20 miles downstream and about 150 miles from the Grand Canyon. The misnomer arose from the original authorizing legislation before the final site was selected.

There was no previous engineering model for building such a dam. Like many other pathbreaking civil-engineering feats in history, it was a leap into the unknown. Solutions were improvised to unforeseen problems as they arose in the finest spirit—and one of the last epic expressions—of Yankee ingenuity. No U.S. construction firm was large enough alone to tackle such an immense project, so a consortium of six builders banded together to make the winning bid—thus helping to launch the future destinies of Bechtel, Kaiser, Morrison Knudsen, and other global construction giants. The first stage, which began in 1931 and took nearly eighteen months to complete, was the blasting of four huge tunnels in the canyon walls through which the Colorado was diverted while the dam was built. With the river diverted and the work site secured by a temporary coffer dam, "high-scalers" suspended from long ropes blasted fresh rock from the canyon walls, while other workers excavated 40 feet down below the dry river until they reached the bedrock that would anchor the dam. Since the hot poured concrete filling the immensity of the dam's volume would take a century to cool down naturally, engineers devised an instant refrigeration system by injecting frigid water through one-inch pipes that were inserted for that purpose at measured intervals throughout the structure; within two years, the cooling was accomplished. When no U.S. company could supply plate steel pipes large enough to funnel the falling water through the intake valves on Lake Mead to drive the turbines near the bottom of the dam, the builders constructed their own steel-fabricating plant on-site. Labor was nonstop around the clock. Work conditions in the searing heat were hard, and often deadly. When already low wages were cut in mid-1931, workers, organized by the IWW, or Wobblies, went on strike. But the Great Depression had started and the strike was broken, with the federal government's tacit approval, by the importation of scab labor from nearby Las Vegas.

By 1936 everything was done. Lake Mead began to fill behind the dam. Water began passing through the new turbine-generator units.

The dam's elegant, curved design, art deco flourishes, and 70-story-high grandeur topped off what was instantly recognized as a landmark achievement of civilization. "I came, I saw, and I was conquered," President Franklin D. Roosevelt orated with a paraphrasing twist on Julius Caesar at the dam's dedication in September 1935.

Roosevelt, in fact, already had been won over by Hoover Dam as a model for the public works projects that became a signature centerpiece of his New Deal policy to counteract the Great Depression he'd inherited with the presidency in 1932. Across the country, construction crews were mobilized from the 25 percent unemployed and put to work building scores of new dams that harnessed the untapped water resources of America's rivers. By the mid-1930s, the five largest structures on Earth, all dams, were under construction in the western United States—the Hoover on the Colorado, the Grand Coulee and the Bonneville on the Columbia, the Shasta on California's Sacramento River, and the Fort Peck on the upper Missouri. Indeed, the Hoover Dam inaugurated a seminal turning point in water history—the age of giant, multipurpose dams. Over the three decades from the mid-1930s, America led the way in what became a worldwide frenzy of dam building that intensified production and transformed human society through the delivery of copious new supplies of cheap irrigation water and cheap hydroelectricity, as well as improved flood control and river navigation. By the 1940s, the United States had greater command of its water resources than any other nation. Its innovative leadership in converting the untapped boon of its river water wealth into productive economic and military output played a central role in its visible emergence as the global superpower in the aftermath of World War II.

While the Hoover Dam and its New Deal–born successors did not provide enough stimulus alone to lift America from the staggering economic collapse of the Great Depression, the hope they inspired lent precious legitimacy to a government beleaguered by faltering faith in the efficacy of the country's bedrock political economic system. One lasting impact of the giant, state-built dams everywhere was to help usher in an era of enlarged political and economic centralization with many similar characteristics, albeit in modified forms, to the irrigation-based river states of antiquity—strong government involvement in the economy, policies implemented by a cadre of technocratic engineering overseers and a large workforce of low wage laborers. Control and mass manipulation of river water likewise was a key element of political power, with the chief beneficiaries of the waterworks' wealth, as in ancient societies, both reflective and reinforcing of the society's established power structures—in this case, featuring disproportionate subsidies for politically overrepresented agricultural businesses.

Under the New Deal, all the major river basins of America's Far West came under intensive, multipurpose dam development. Thanks to the exploitation of water, the Far West became the country' most dynamic growth region in the postwar era, exceeding Teddy Roosevelt's vision for it at the turn of the century.

No river in the West was mightier than the Columbia in the Pacific Northwest. It carried nearly 10 times more water than the Colorado and roared out of the glacier-covered mountains and through wide canyons with torrential seasonal force. Its vast potential—especially for hydropower—had long tantalized engineers. Fully exploited, the river could generate enough hydroelectricity for the entire population living west of the Mississippi at the time. One site in particular held colossal potential—the Grand Coulee ("Great Canyon"), a 50-mile-long canyon one to six miles wide within 500- to 600-foot cliffs that were structurally ideal for building a dam.

Yet until the early 1930s the river was completely untamed. Newly elected president Franklin Roosevelt was personally determined to change that. When Congress balked at the extraordinary cost of erecting a high dam at the remote Grand Coulee that would provide far more hydropower and irrigation water than anyone imagined could be profitably used by the 3 million inhabitants of the region, Roosevelt started the project on his own from other relief funds. In the end 36 huge dams would be built on the Columbia and its tributaries between 1933 and 1973—nearly a dam per year. The multipurpose Bonneville in 1938 and Grand Coulee in 1941 were world-class supergiants of their era. Thousands were put to work building them. To help win the public relations battle over the utility of the dam, Roosevelt's staff hired folk balladeer Woody Guthrie as a research assistant. In his folksy midwestern twang, Guthrie communicated the inspiring grandeur of the dam project, in songs like "Roll on, Columbia," as the mightiest object ever built by man.

When finished, the Grand Coulee Dam was indeed the mightiest thing ever built: four-fifths of a mile wide, 550 feet tall, three times the mass of the Hoover Dam, generating half as much hydroelectricity as the entire country at the time, and capable of irrigating over a million acres. The giant concrete plug in the great canyon backed up the artificial Franklin D. Roosevelt Lake some 150 miles to the Canadian border. Downriver, the Bonneville was already generating electricity and taming the wild five-mile rapids with big locks that enabled the passage upriver of large barges transporting farm produce and bauxite to the aluminum smelting industry that developed around the region's rapidly expanding availability of electrical power. By the late 1980s, the Columbia River was providing 40 percent of America's total hydroelectricity.

Like Hoover and the other multipurpose giants of the era, Coulee's hydroelectric sales heavily subsidized the building of the dam and associated irrigation project costs. Yet in the late 1930s, carping critics of Roosevelt were still wondering loudly who would buy so much excess electricity. History, however, repeatedly demonstrated that the development of a useful resource inevitably found unimagined and unforeseen productive applications. Yet no one could have anticipated just how swiftly the great need for the northwest's surplus electricity would arrive. Only five days before the completion of the dam, Japan attacked the American fleet at Pearl Harbor. The United States entered World War II. The war's extraordinary mobilization and economic stimulus caused airplane factories and aluminum smelters to spring up throughout the region. By 1942, 92 percent of Grand Coulee's and Bonneville's electricity output was powering war production. Most of it went to producing thousands of warplanes bound for the aircraft carriers that turned the war in the Pacific in America's favor. An aerospace industry vital to the war effort also arose in Southern California on the electricity available from the Hoover Dam. Japan and Germany had nothing to match it.

It was no exaggeration to say that America's superior industrial productive capacity in general—and its vastly superior and timely availability of hydroelectric power in particular—played a decisive role in the nation's rapid rebound from Pearl Harbor and its ultimate victory in the war. Indeed, rarely in history had exploitation of a water resource so dramatically and immediately affected the military outcome and the rise of a great power. During the war, Coulee's electricity also powered the top-secret Hanford military installation on the Columbia in Washington State to help produce the plutonium-239 that made the United States the preeminent nuclear superpower of the postwar era.

Less than three months after his dedication of the Hoover Dam, Roosevelt signed off on another immense water-moving and irrigation project for the West's third great river basin that ran through California's 450-mile-long, 50-mile-wide Central Valley. Its scale dwarfed the volume of water that Hoover Dam was delivering to the Imperial Valley in Southern California. Set in the basin of the San Joaquin and Sacramento rivers between the Sierra Nevada and Coastal mountain ranges, the Central Valley Project transferred water from California's wetter north to its arid south. In the process it transformed a region as dry as North Africa into America's produce capital and the richest concentration of irrigated farmland in the world. A small boom of private irrigation in the Valley had occurred around World War I when large farmers began to use motorized, centrifugal pumps powered by oil or electricity to tap deeply into the region's store of groundwater. Through the 1920s, some 23,500 well pipes pumped up prodigious amounts to be able to luxuriously irrigate the San Joaquin Valley in the southern Central Valley, and helped California surpass Iowa as the nation's leading agricultural state. But by the early 1930s the uncontrolled groundwater pumping had caused the water levels in its large aquifer to plunge so drastically that thousands of acres had to be retired for want of irrigation water. As the aquifer emptied and drought conditions prevailed on the surface, the big farmers of the Central Valley turned reluctantly to the government for relief.

Inspired by the ambitious water transfers of the Hoover Dam project, they proposed a water transfer plan that would shift resources from watersheds in the north, which were filled by the rainfall and spring snowmelt, to the south through a series of large canals covering hundreds of miles and supplied by two new giant dams. At the heart of the scheme were the Shasta Dam on the Sacramento River and the Friant Dam on the San Joaquin. In essence, the Central Valley Project was a bailout for existing, mainly large farm businesses. What it was not was a reclamation plan intended to create many new small farms as foreseen by the original 1902 legislation. But in the midst of the Depression, Roosevelt approved it anyway. In 1934 the Dust Bowl had hit the midwestern plains, destroying farmland and starting the internal mass migration to California. "The Central Valley Project was without question the most magnificent gift any group of American farmers had ever received; they couldn't have dreamed of building it themselves, and the cheap power and interest exemption constituted a subsidy that would be worth billions over the years," wrote Mark Reisner in his classic work on western water history, *Cadillac Desert*. "It rescued thousands of farms that were already there, including many that were far larger than the law allowed."

The federal waterworks were followed in the early 1960s by another gigantic, entirely state-built California Water Project that moved water even more ambitiously through a network of dams, reservoirs, and hundreds of miles of aqueducts—featuring a section that pumped water, with an extravagant expenditure of energy, *over* an entire mountain range in five stages, including a final, Herculean two-fifths of a mile lift. By the time the water started flowing over the mountains in 1971, California was by far the most intensively water-engineered place on the planet. Every big river flowing from the Sierra Nevada was dammed.

New Deal western waterworks also had their counterparts in the nation's eastern half. Most celebrated was the Tennessee Valley Authority. Launched in 1933, the TVA attempted nothing less than the comprehensive management of the entire Tennessee River basin, three quarters the area of England, with the declared purpose of raising the economic and social well-being of its downtrodden inhabitants.

The TVA's sweeping powers were governed by an independent public body whose precedent was the special authority granted to the Panama Canal Commission. Throughout the 1920s progressives in Congress had stymied presidential plans to privatize the government's big dam at Muscle Shoals, its nitrate factory for munitions, and other assets on the Tennessee River through sale or lease to big businessmen such as Henry Ford. Through the TVA, those assets were converted by the New Deal into the centerpiece of an ambitious, state-managed effort to produce electricity, flood control, irrigation water, improved navigation, and even nitrate and phosphate fertilizers for the region's farmers. No other river in the world had so concentrated an amount of its volume dammed in a staircase of 42 dams and reservoirs, although the 700 miles in the middle reaches of the Missouri River was a close runner-up. The results transformed the Tennessee Valley: rampant spring flooding ceased to afflict Tennessee farmland; improved navigation facilitated freight transport on the river to multiply sixty-seven-fold to 2.2 billion ton-miles in the thirty years leading up to 1963; electricity prices fell by more than half; farm yields multiplied on government-produced fertilizer; endemic malaria was eliminated; even the health of the river valley ecosystem was enhanced by the public reforestation of over a million acres. Tennessee River electricity also powered aluminum and war production factories for World War II, including the Oak Ridge atomic fission center, and brought farmers the first, wondrous benefits of electricity. In the early 1930s, American farmers had been left behind on the dark, Have-Not side of America's electricity divide. Only 10 percent of farms were electrified. By 1950, thanks chiefly to hydropower, 90 percent of U.S. farmers had access to illumination, refrigeration, radio, and the other productive, modern benefits of electricity.

Hundreds of huge dams were erected across the country during the apogee of America's giant dam-building age in the early postwar era. Through all U.S. history some 75,000 dams had been built—about one per *day* from the end of George Washington's presidency to the inauguration of George W. Bush over 200 years later. Most of the 6,600 large ones over 50 feet, and all the multipurpose giants, were built after Hoover. For its seventy-fifth anniversary, the Bureau of Reclamation cataloged its cumulative bureaucratic accomplishments: 345 dams, 322 storage reservoirs, 49 power plants marketing over 50 billion kilowatt-hours, 174 water-pumping plants, 15,000 miles of canals, 930 miles of pipelines, 218 miles of tunnels, more than 15,000 miles of drains, irrigation water for 9.1 million acres, and freshwater for 16 million urban and industrial users. Farming in the arid Far West had not merely been born, but flourished as one of world history's all-time irrigated agricultural gardens. By 1978 the 17 western states had 45.4 million acres under irrigation—10 percent of the world's total.

From the 1940s America was by far the most advanced hydraulic engineering civilization on Earth. As in past eras of history, this leadership was reflected in both robust population growth and an even greater surge of available freshwater supply. American water use for all purposes multiplied tenfold, from 40 billion gallons per day to 393 billion gallons between 1900 and 1975. Population tripled in the same period. The more than tripling of freshwater use per person was a leading indicator and driving factor behind the country's rapid rise in living standards, national economic productivity, and preeminent global influence. America's intensive proficiency in every traditional category of man's use of water, and its pioneering leadership in leveraging innovative water breakthroughs was a major reason why it was the best fed, healthiest, first fully electrified, most industrially productive, most urbanized, most transportation efficient, and most militarily powerful nation on Earth in the postwar decades.

Not all the upsurge in America's freshwater supply in the Far West derived from its innovative dams. From the mid–1940s, the drought-prone western High Plains was transformed from a hellish Dust Bowl into an irrigated cornucopia of grains by a sudden abundance of water drawn from an altogether different source—an immense, heretofore mostly inaccessible, aquifer that lay buried, like a sealed subbasement, deep beneath the near surface groundwater table underlying its semiarid landscape. Ogallala, or High Plains, water accounted for about one-fifth of total U.S. irrigated farming by the late

1970s, and in good years, up to three-quarters of the entire world's wheat crop that was sold on international markets. In addition, 40 percent of American cattle drank Ogallala water and ate Ogallala-watered grain—every ton of which required some 1,000 tons of water to grow.

The hydraulic secret of the arid High Plains was that running far underneath Nebraska, western Kansas, Oklahoma's Panhandle, northwestern Texas, and small portions of South Dakota, Wyoming, Colorado, and New Mexico was a giant honeycomb of up to half a dozen enclosed pockets of freshwater that together were the size of Lake Huron, and contained some 3.3 billion acre-feet, or over 235 years' flow of the Colorado River. The water was wedged between rocks and mixed with silt, sand, and stones. The deepest portion of the aquifer was in the north, so that overall about two-thirds of the water lay under Nebraska and 10 percent each in Texas and Kansas. Ogallala's "fossil water" was the drop-by-drop accumulation from prehistoric ice ages—one of the largest known subterranean reservoirs that existed deep inside the planet's bowels at varying depths. Around the planet there was up to 100 times more freshwater locked away in aquifers than flowed freely and readily accessibly on the surface. Such fossil aquifers existed like nonrenewable, stand-alone reservoirs insulated from the planet's continuous, natural hydrological recycling of surface and shallow groundwater through evaporation and precipitation. They recharged so slowly—Ogallala only half an inch per year from the trickle down from the surface—that they effectively could be used only once before depleting like an empty gas tank.

Due to water's great weight and the technological and cost limitations of pumping up water from aquifers, the High Plains's underground water wealth remained virtually untapped through the 1930s. Waterwheel-powered pumps were useless in an arid land without running streams to drive them, while the cost of transporting coal for steam pumps was prohibitive. Windmills could lift only a few gallons per minute and thus barely skimmed the surface of the Ogallala's deep reserves. The ranchers' prairie-grass-grazing cattle herds of the 1870s and 1880s had disappeared in the droughts and heat of the 1890s; with the return of the rains and demand for grain after World War I, farmers again ventured forth with their plow mules into the water-fragile frontiers beyond the 100th meridian. Then during the prolonged drought years of the 1930s came the man-assisted environmental catastrophe of the Dust Bowl. By clearing their land through cattle grazing and burning harvested wheat stubble, farmers inadvertently turned a fragile ecosystem into an unstable one. Without vegetation to hold the loose topsoil in place, the return of drought, heat, and high, gusting winds kicked up horrific dust storms that devastated farming across the western plains.

Dust storms were created when dry soil was lifted into the air by hot, high, winds; the resulting cloud of swirling, fine particulates grew larger and larger and gathered force as it swept across the open prairies. Eventually it became a gigantic cloud of stinging, shearing dust up to 10,000 feet high and reaching velocities of 60 to 100 miles per hour. The devastation wreaked by the dust storms was of biblical proportions: crops torn up and entire harvests lost, houses shorn of their paint and chickens of their feathers, dirt clogging mechanical farm equipment and water pipes, and millions of tons of fertile topsoil—the precious patrimony of the land—blown far away forever. In the duster that started on May 9, 1934, some 350 million tons disappeared, darkening the skies and dropping dirt residue over Chicago, Buffalo, Washington, D.C., Savannah, and even ships sailing 300 miles into the Atlantic. Between 1935 and 1938 there were an average of more than 60 major, sky-blackening dust storms each year. In the heart of the Dust Bowl—a 400-mile-long by 300-mile-wide area encompassing parts of Oklahoma, Texas, New Mexico, Kansas, and eastern Colorado—the average acre was stripped of 408 tons of fertile topsoil, leaving behind pauperized, sandy earth. Some 3.5 million "Dust Bowl refugees" abandoned the Midwest in search of work by 1940. Many migrated west to pick crops in California, enduring the hardships chronicled in John Steinbeck's classic novel *The Grapes of Wrath.*

Even as the dust storms blew, the High Plains farmers' deliverance was at hand in the form of the centrifugal pump already being used with miraculously transformative effects in California's Central Valley to extract voluminous amounts from its own aquifer. With the postwar recovery and the availability of cheap diesel fuel from the nearby oil patch in Texas and Oklahoma, diesel-powered centrifugal pumps and water wells proliferated. A centrifugal pump could lift 800 gallons of water in only a minute, making widespread irrigation possible on the High Plains for the first time. Oil drilling techniques were also adapted that could raise water even faster. With the postwar invention of the center-pivot irrigation system—a long-tentacled, mobile sprinkler system hooked up to a water well—water pumping and irrigated farming boomed. With some 150,000 pumps extracting huge volumes day and night during the growing seasons, Ogallala annual water use quadrupled between 1950 and 1980, and irrigated acreage septupled to 14 million acres. By the late 1970s, intensified by modern petrochemical fertilizers, pesticides, herbicides, and generous farm subsidies, the 1 percent of American farmers working 6 percent of the nation's cropland that forty years earlier had been a desolate Dust Bowl, were growing 15 percent of that nation's wheat, corn, cotton, and sorghum.

But the boom couldn't last indefinitely. Farmers were drawing water out of the Ogallala 10 times faster than the aquifer network was recharging. Irrigation farmers were living on borrowed time and water. The most profligate drilling was in west Texas and other southern portions of the aquifer. One Kansas region that in 1970 thought it had reserves for 300 years discovered in 1980 that it had only a seventy-year supply left. Water that had accumulated over so many millennia, and acted like an emergency reserve of nature, would be consumed in a one-time irrigation bonanza lasting no more than a century unless it was conserved or used more productively. The prairie would revert to its natural, hardscrabble aridity, its agricultural bounty wither in a new cloud of dust.

Drawing America's groundwater patrimony for unsustainable food exports to foreign countries was a particularly shortsighted policy. From the late 1970s, drawdown allocation agreements and sharply increased pumping costs due to the era's oil price shocks, slowed the rate of depletion and encouraged irrigation efficiencies that got "more crop per drop." Yet the overdrafts—which by 2000 totaled 200 million acre-feet, or 14 Colorado Rivers—were highly concentrated in a few shallower, southern regions. Thus while sustainable equilibrium was being achieved in water-rich Nebraska, Texas and Kansas had already used up some 30 percent and one-sixth of their total shares, respectively, and were still overdrawing at a reckless pace. The looming storm cloud gathering over parts of the prairie was the question how long accessible water from the Ogallala reservoir would last. The day of reckoning for Texas and Kansas was expected to hit between 2020 and 2030.

Recognizing that the future of oil-built Texas depended upon securing enough freshwater, some Texan leaders schemed, and failed, in the late 1960s to steal a march on their regional neighbors by launching an outsized, technically complex, and extremely costly interstate Texas Water Plan to transfer flow from the Mississippi River and pump it across the state to the high plains of west Texas. Robbing one heavily used water ecosystem to replenish another offered no fundamental fix to the depletion challenge. But it did provide a foretaste of the extreme kind of political and resource competition that lay ahead as parts of the Ogallala ran dry.

From the San Joaquin Valley in California's Central Valley and metropolitan Phoenix to El Paso and Houston, Texas, water tables in many arid regions were falling precipitously, causing land subsidence and salt contamination of drinking water and farmland. Despite the respite from California's great water-moving projects, unregulated overpumping in the Central Valley had resumed at such a furious pace that the groundwater tables had plunged up to 400 feet and the land itself had fallen 50 feet in some places. Even the rivers, lakes, wetlands, shallow groundwater, and interrelated water ecosystems of the country's rainy, eastern half were also under growing stress from the intense demands of population and industrial growth. In south central Florida, the straightening, damming, and redirection of streams to benefit the region's large sugar growers had disrupted the fragile Everglades wetlands, which were drying up and shrinking. As clean, fresh surface water became less available across America, groundwater resources were being overdrawn to make up the shortfall. In the thirty years leading up to 1996, total U.S. groundwater usage more than doubled to account for one-fourth of all U.S. water usage.

Although America was one of the world's most water-rich countries, with 8 percent of the world's replenishable freshwater but only 4 percent of its population, shortages of fresh, clean water were starting to impinge upon many regions' patterns of growth, fomenting a new politics of resource competition among neighbors used to plenty. It wasn't that the country didn't have enough *total* water to meet its needs. Rather it was that its profligate use was finally exhausting the productive limits created by the innovative successes of its age of giant dams. The era of cheap, plentiful water was closing. New technologies and more efficient usage were needed. As throughout water history, the success of one era was seeding the defining challenge of the next.

America's age of great dams drew to a close during the 1970s. By then, virtually all the best large dam sites had been exploited. Hardly a major river flowed freely across America's landscape without being interdicted by dams and stored behind reservoirs. While the earliest dams from Hoover onward had returned the largest economic gains for the lowest subsidies, the later ones, by and large, had been built at the more marginal sites, carried the largest subsidies, and had hardly provided any net economic benefit at all. Yet even as the dam-building boom tapered off, demand for more freshwater and hydropower continued to escalate to meet growing populations, intensifying the political struggle among users to control a greater share of the limited, indispensable liquid resource.

The Colorado River told the tale. By 1964 an array of 19 large dams and reservoirs held four times the river's annual flow and gave man total management over the Colorado system. No longer did the river remotely resemble the wildly surging, unpredictably flooding river explored by John Wesley Powell almost a century earlier. Each drop was measured, every release calculated, and every event on the river planned by its central managers. It was the lifeblood of the entire southwestern United States. Every drop was used and reused 17 times before reaching the sea. As demand for its water increased, it also became the most litigated river in the world. By the 1950s Southern California was not only consuming its full 4.4 million acre-feet entitlement under the 1922 water sharing compact, but also was starting to take up to an additional 900,000 acre-feet of unused flow allotted to other states. A Supreme Court ruling in 1963 instigated by fast-growing Arizona, which feared California would claim a permanent right to the water that was otherwise part of its allocation, put a legal hold on California's water overuse—although the political showdown to make it practicable did not occur for another forty years. As the water needs of Arizona and other basin states increased toward their full allocations, something had to give.

The first to feel the squeeze was Mexico. During the 1950s an average annual 4.24 million acre-feet had flowed across the border into Mexico, which used it for irrigation and to replenish the lagoons of the river's lush delta. In the 1960s, the average flow plummeted to the 1.5 million acre-feet minimum entitlement under the 1944 treaty, and the river rarely again reached the sea. Deprived of water and silt, the delta ecosystem shrank into an almost lifeless, salt flat wasteland with a few strips of irrigated cropland. Worse still for Mexico, its 1.5 million acre-feet had become so briny as to be almost worthless for irrigation. The transformation of the Colorado by damming and intensive irrigation had also changed the river's composition as well as its volume. Sediment trapped behind the dams made the river much less silty. Irrigators could partly compensate for the loss of naturally refreshing silt flood deposits by intensive use of artificial fertilizers. But the drainage backflow of used irrigation water contaminated the river with high levels of salts leached from the cropland; by 1972, salinity at the river's halfway point had increased two and a half times over its natural, predam state. Salinity accumulations were highest downriver at the Mexican border. For more than a decade, the United States had rejected Mexico's protests that the 1944 treaty guaranteed it 1.5 million acre-feet of *irrigation-quality* water. Then, in 1973, perhaps mindful of the discovery of large oil fields in offshore Mexico, American diplomats finally agreed to deliver water with an acceptable salt content.

While competition for Colorado water intensified, the river's managers also made the awful discovery that the 1922 Colorado River Compact's baseline estimate of 17.5 million acre-feet per year had been much, much too optimistic. The eighteen-year streamflow data on which it had been measured covered an unusually wet period; by 1965 the Bureau of Reclamation knew that longer-term data suggested an average flow of only about 14 million acre-feet. Subtracting Mexico's 1.5 million and another 1.5 million for evaporation from the giant man-made storage lakes left only 11 million to be divided among states whose irrigation, hydroelectricity, and urban drinking water projects, when built to full capacity, depended upon receiving all the anticipated 15 million acre-feet. The government-brokered compact simply promised more water than it could deliver.

The reckoning day for the Colorado water shortage was postponed by an extremely wet decade from the late 1970s and by reservoir drawdowns from Lake Mead and other storage facilities on the river. The full impact of overallocation finally began to be felt with the long drought in the first decade of the twenty-first century. The river's flow at the Compact's official delivery point from upper to lower basin states at Lee Ferry, Arizona, sank to its lowest level since measurements began in 1922. As Lake Mead, with its 28 million acre-feet capacity, drained to less than half full, water managers scrambled to develop emergency plans in the event it continued to sink below the level of Hoover's intake pipes. A growing body of long-term climate evidence from tree rings, moreover, suggested that the 1900s might have been a relatively moist century. A return to normal climate patterns thus would likely make the southwest even hotter and drier; another megadrought, like the speculated one that may have obliterated native farming civilizations early in the last millennium, was a possibility. Whether man-made or natural, the warmer weather in the Far West over the thirty years to the mid-2000s was already discernibly diminishing Colorado water flows by reducing the winter mountain snowpacks and the replenishing spring runoff it brought when it melted, while also increasing the evaporation loss from reservoirs. The prospect of chronic Colorado River water shortages menaced the basin's 30 million with economic slowdown, possible chronic water crises in large desert cities like Las Vegas and Phoenix, and chaotic political clashes for water among compact states and among metropolitan, industrial, and farm users within them.

The Colorado River shortages signified the dawning of a new Far Western water era marked by supply limitations and ecosystem depletions that demanded fresh responses including alternative technologies, conservation, organizational redeployment of scarce water resources, and new approaches to water management. One of the largest problematic legacies of successful irrigation of the arid west was the extreme economic misallocation caused by the lavish government subsidy for large farm businesses, which consumed over two thirds of the river water and whose runoff by far caused the greatest damage to underlying ecosystems. Such subsidies had served their original purpose in fostering western agricultural development, but long ago had outlived their usefulness. In California, four of every five farms were over 1,000 acres and 75 percent of the state's entire agricultural output came from just 10 percent of the farms. By the late twentieth century, vested agribusinesses had become privileged Water Haves who paid almost nothing for the region's scarce water, while more economically productive and water-efficient industries and cities were taxed by having to pay burdensome premiums of up to 15 to 20 times more to obtain enough. The efficient allocation mechanism of competitive market forces was being grossly distorted, with perverse impacts on economic growth, environmental resources, and basic fairness.

The end of the age of great dams in the United States occurred in the 1970s when an alliance of environmentalist, urban, and recreation industry lobbyists, armed with arguments proving the uneconomic returns of new large dams, united to gradually offset the overrepresentation of irrigation and dam interests in state and federal politics. The breakthrough event came in the late 1960s when the Sierra Club, founded in 1892 by naturalist John Muir and other Californians, rallied a national political effort to defeat proposals to dam the nationally hallowed natural wonder of the Grand Canyon. From then on, the national debate turned increasingly to offsetting the deleterious environmental by-products of dams, such as the drying up of deltas and wetlands, the heavy dependence they promoted on artificial fertilizers, pesticides, herbicides, and monoculture farming, the trapping of soil-replenishing silt, the destruction of river wildlife—the Columbia

River's 15 million wild salmon fishery had collapsed to under 2 million because the fish couldn't surmount the dams to return to their spawning grounds, for instance. By the late twentieth century, the main discussion about dams was their decommissioning and removal—indeed, in the United States decommissioning surpassed new construction by 2000.

America's antidam campaign had gained impetus from the vibrant, grassroots environmentalist movement that had sprung up in reaction to the mounting evidence that mankind was inadvertently poisoning itself with the detritus of industrial growth. Just as the large urban concentrations of the early nineteenth-century Industrial Revolution created foul sanitary conditions that threatened the habitability of large cities and produced the sanitary awakening, rapid industrialization produced unwholesome accumulations of unwanted industrial and agribusiness pollution of society's public waters, air, and soils that was midwife to the modern environmental movement. Over the decades surface freshwater rivers and lakes, seacoasts, and slow-moving, unseen groundwater ecosystems had grown increasingly contaminated. By the mid-twentieth century a new phenomenon—water pollution on a scale and intensity that overwhelmed natural ecosystems' restorative capacities—began to visibly threaten both public health and the long-term environmental sustainability of unfettered economic growth. Before World War II, the overwhelming proportion of pollution emanated from the smokestack technology cluster that burned fossil fuels and produced heavy metals like iron and steel. After World War II hundreds of new plastics, agricultural fertilizers and other synthetic chemicals—many extremely toxic and difficult for natural forces to degrade—became increasingly major pollutants.

For decades chemical companies dumped untreated toxic wastes into local rivers, ponds, and streams, where they leached into groundwater drinking sources and years later brought illness and death to uncounted thousands. By 1980 the United States had more than 50,000 toxic waste dumps. In one infamous incident, residents and schoolchildren in Love

Canal, a neighborhood in Niagara Falls, New York, built on landfill atop a toxic waste dump site, suffered abnormally high rates of cancers and birth defects a generation later. The area was declared a disaster zone and evacuated. Similar horror stories emerged in other countries. Japanese children around Minamata Bay, for instance, showed brain damage after 1956 from eating fish contaminated by the mercury dumped years earlier by a local chemical factory. Islands of toxic waste as long as 18 miles long and three miles wide formed in the Soviet Union over one-mile-deep Lake Baikal, the world's largest freshwater lake. North America's Great Lakes, holding about 20 percent of Earth's fresh surface water, also showed the pollution from the heavy industrial activity around its shores; by the early 1960s much of Lake Erie's fish life had suffocated due to algae blooms run amok from fertilizer runoff and dumping of wastes. Similarly, a large part of the once rich Baltic Sea fishery had become biologically dead from northern Europe's heavy industrial sewage and chemical fertilizer effluents, above all those drained by communist Poland's filthy Vistula River. Acid rain caused by rising sulfur dioxide emissions from industrial smelters and burning fossil fuels contaminated freshwater sources and food chains across national boundaries; the sulfur dioxide emitted in a single decade in the late 1980s from Ontario's giant copper and nickel smelters alone was estimated to have exceeded the entire volume released naturally by all the volcanoes in Earth's history. Nuclear weapons production by the Cold War superpowers also polluted rivers and lakes in America and the U.S.S.R. with deadly radioactive waste.

If the modern environmentalist movement had a specific birth moment it came in 1962 with the publication of a seminal book, *Silent Spring.* Written by Rachel Carson, a former U.S. government aquatic biologist, *Silent Spring* focused the national spotlight on the insidious, water polluting effects of synthetic chemical pesticides such as DDT that were widely applied to kill insects and improve crop yields, and drew attention to the larger ramifications it portended for what man was doing to his habitat. "The pollution entering our waterways comes from many sources: radioactive wastes from reactors, laboratories, and hospitals; fallout from nuclear explosions, domestic wastes from cities and towns; chemical wastes from factories," Carson wrote. "To these is added a new kind of fallout—the chemical sprays applied to croplands and gardens, forests and fields . . . our waters have become almost universally contaminated with insecticides." In vivid prose Carson, who had grown up on the banks of the Allegheny River near Pittsburgh and had witnessed firsthand the effects of industrial pollution from the coal-burning electric plants on the river's ecosystems, synthesized many scientific studies into the bigger picture. "The problem of water pollution by pesticides can be understood only in context, as part of the whole to which it belongs—the pollution of the total environment of mankind."

Observing that for the first time in earthly history, mankind in the twentieth century had gained sufficient power to substantially modify the natural surroundings, Carson worried that it was using it recklessly, polluting air, earth, rivers, and seas in irreversible ways perilous to civilization's own survival. She concluded, "Along with the possibility of the extinction of mankind by nuclear war, the central problem of our age has therefore become the contamination of man's total environment with such substances of incredible potential for harm—substances that accumulate in the tissues of plants and animals and even penetrate the germ cells to shatter or alter the very material of heredity upon which the shape of the future depends."

The publication of *Silent Spring* immediately gave voice to the inchoate, gathering public concern about the environment. Almost overnight, the modern environmentalist movement became a potent political force. Big chemical companies, the U.S. Department of Agriculture and others with perceived vested interests in maintaining the short-term *status quo,* like their counterparts in all eras, mounted a vigorous offensive against *Silent Spring.* Carson's science, her professional credentials, and even her personal traits were assailed. Yet *Silent Spring* resonated deeply within countervailing constituencies of America's pluralistic democracy. President John F. Kennedy took a personal interest. Several expert federal and state studies were duly undertaken and corroborated her allegations.

Before the decade was out, the new environmental movement had gathered unstoppable momentum. Action was further galvanized by a number of high-profile environmental disasters. None was more influential than the spectacular, five story high flames that combusted on Cleveland's Cuyahoga River on June 22, 1969, from the sheets of unregulated, flammable wastes that had been dumped into the river. Within months, the United States took the regulatory lead by enacting comprehensive national environmental legislation and empowering the Environmental Protection Agency to execute it. The 1972 Clean Water Act, and the 1974 Safe Drinking Water Act, were passed to cleanse America's surface and ground waters of pollution. Authorities began to tackle the immense problem of controlling intensive, algae blooms in lakes and coastal seashores. Endangered species were protected. DDT and other harmful chemical pesticides were banned domestically, although not their export to third world countries.

The first annual Earth Day on April 22, 1970, rallied 20 million Americans to support an environmentally healthy planet; twenty years later, 200 million people in 140 countries turned out. Environmentalism went global in the late 1980s. The United Nations played a leading role, starting with the influential 1987 report "Our Common Future," known also as the "Brundtland Report" after its Norwegian chairwoman, that called for examining the relationship between economic growth and environmental sustainability. Thereafter it supported Earth Summits of heads of state every decade since 1992, an ongoing intergovernmental study of climate change from 1988, an influential commission on environmentally sustainable development in 1989, and the first comprehensive, five-year-long assessment of Earth's total ecosystems inaugurated on the occasion of the millennium in 2000 and completed in 2005. International environmental treaties covering environmental problems from air pollution to global warming also were signed by many countries. From the early twenty-first century, water ecosystems received special attention. The U.N. published its first triennial World Water Development Report in 2003 and in 2005 launched the International Decade of Water for Life. Providing clean water and a healthy environment increasingly became a standard measure for domestic legitimacy around the world; horrendous environmental disasters helped undercut the political credibility of the Soviet Union before its collapse and were increasingly becoming focal points of democratic protests in early twenty-first-century China. Giant industrial corporations, such as General Electric, gradually embraced environmentalist agendas and attempted to redefine their images and activities as eco-friendly. Sadly, Rachel Carson never lived to see her handiwork come to fruition. She died of cancer in 1964, at age fifty-six, less than two years after *Silent Spring*'s publication.

The environmental movement represented a turning point in water and world history. For all human history, the governing view was that Earth's freshwater resources were essentially unlimited, naturally self-cleansing, and free to extract from its ecosystem without consequences in any amount of which man was capable. In its place, increasingly, a new recognition was emerging: that in order for industrial civilization, with its prodigious power to alter the natural environment, to continue to thrive it was necessary to establish a sustainable equilibrium between economic growth and its host water ecosystems.

America's pioneering giant, multipurpose dams were the instant envy of the world. Within only a few decades, foreign states everywhere were striving to replicate America's achievement. The result was a dam-building boom of epic proportions on virtually every major river of the planet. The resulting improvements in material well-being helped make communist states credible challengers to the postwar hegemony of liberal Western democracies, and allowed newly independent, poor countries, for the first time in history, to move up the industrial development ladder. Industrialized prosperity spread globally, transforming the world political economy and balances of power. By the end of the century, it had helped bring into existence a multiaxial, interdependent global order that was gradually superseding the long era of western European and U.S. hegemony.

For all countries, the cheap hydroelectricity and freshwater unlocked by large dams was a panacea—irrigation for increased food production, power for industrial factories, healthy drinking water, sanitation services and illumination for large metropolitan centers, and popular hope of betterment in material life. Dams transcended political or economic ideology. Whatever the system, dams meant prosperity, more stable societies and greater governmental legitimacy. American president Herbert Hoover's statement that "Every drop of water that runs to the sea without yielding its full commercial returns to the nation is an economic waste" was virtually interchangeable with Soviet Union leader Joseph Stalin's maxim that "water which is allowed to enter the sea is wasted." Every twentieth-century leader from Teddy Roosevelt to China's Mao Zedong would have concurred. At the dedication of north India's giant Bhakra Dam in 1963, an awestruck Prime Minister Jawaharlal Nehru echoed the rhapsodic Franklin Roosevelt at Hoover as he proudly likened the dam project to "the new temple of resurgent India." Both his sentiment and metaphor were strikingly similar to President Nasser's comparison of Egypt's High Aswan Dam to a pyramid. To each and every leader, water seemed to be a potentially infinite, enriching resource of nature limited only by society's technical virtuosity in extracting ever more of it from the environment.

The centralizing, hydraulic society tendencies of dam-building conformed easily to the model of communist state planning. Marshaling an unpaid army of gulag laborers, Stalin began erecting dams on the Volga River in 1937, and thereafter built them on other great rivers including the Dnieper, Don, and Dniester. All across the huge nation, rivers were rerouted and lakes diverted to the design of Soviet water engineers and state industrial planners. With the help of giant dams, the Soviet Union increased its water use eightfold in the sixty years following the 1917 Bolshevik Revolution and rose to rival America as the world's leading superpower.

Aggressive dam construction and associated water management likewise was a centerpiece of Chairman Mao's effort to reengineer Chinese society to communism in the postwar era. Given Chinese civilization's storied heritage of heroic state waterworks, China's communist mandarins took naturally to the opportunities of dam building on all its rivers, great and small. By the end of the twentieth century, China had some 22,000 large dams—nearly half the world's total and more than three times as many as America—helping to more than double irrigated cropland in the first quarter century of communist rule from 1949. In 2006 it officially opened the world colossus of all dams at Three Gorges on the Yangtze—China's Hoover, and linchpin of its bid for an accelerated economic transformation akin to America's conquest of its western arid lands.

Japan's postwar economic miracle—and the largesse that kept the ruling liberal democratic party in power for so long—rested in part on the intensive exploitation of its limited arable land and its hydropower potential through the construction some 2,700 large dams on its mountain-fed rivers. India's 4,300 large dams ranked it third in the world behind China and America and were vital to its keeping pace in food production for its explosively growing postwar population. Almost every developing nation had its signature giant dam project that was the political and economic centerpiece of its society. As the Aswan Dam transformed the Nile, and with it, all Egypt, Turkey's giant, 1990 Ataturk Dam anchored its immense, region-transforming Southeastern Anatolia Project of 22 dams and 19 hydroelectric plants, while downstream on the Euphrates, the national dreams of Syria and Iraq hinged on there being enough water for their own giant dams. Pakistan's national pride was the huge Tarbela Dam on the Indus. Water-rich South America's stupendous, 1991 Itaipu Dam, on the Parana River on the Brazil-Paraguay border, held the title as the world's largest generator of hydroelectricity—at least until Three Gorges hit full capacity. Central Asia's Tajikistan inherited the world's tallest dam, the Nurek, at 984 feet, when the old Soviet Union broke up.

In all, by the end of the twentieth century mankind had built some 45,000 large dams; during the global peak of dam building in the 1960s, 1970s, and 1980s, some 13 were being erected on average every day. World reservoir capacity quadrupled between 1960 and 2000, so that some three to six times more water than existed in all rivers was stored behind giant dams. World hydropower output doubled, food production multiplied two and half times, and overall economic production grew sixfold.

The international dam boom facilitated one of the most dramatic physical man-made transformations of Earth—the rapid expansion of irrigated cropland, much of it far from natural riverbeds, often through the ferocious conversion of forests and wetlands. Abetted by the extensive mechanization of agriculture, irrigation nearly tripled in the half century after 1950 to cover about 17 percent of the world's arable land and produce 40 percent of its food.

Intensified application of water was a critical linchpin of the world-changing, twentieth-century Green Revolution, which spread from the West to produce surplus yields across the developing world from the 1960s and 1970s. The Green Revolution was based on breeding high-yielding strains of staple crops like corn, wheat, and rice that were highly responsive to intensive inputs of water and chemical fertilizer. One of the pioneering breakthroughs had been in hybrid American corn, starting in the 1930s. By the 1970s, virtually all the corn grown in the United States was hybrid, with yields averaging three to four times more than standard corn of the 1920s. Hybrid dwarf wheat, which carried many more grain seeds in its head than ordinary wheat, triggered its first Green Revolution in Mexico, then spread with spectacular results in the 1960s through the wheat belts of southwest Asia from India's Punjab to Turkey at the head of the ancient Fertile Crescent. Regularly at the brink of mass starvation, staved off only by massive American food donations, India became self-sufficient in food following its 1974 adoption of hybrid wheat. From the late 1960s, hybrid dwarf rice took hold through the world's rice belt, from Bengal to Java to Korea. Between 1970 and 1991, hybrid varietals increased their share from under 15 percent to 75 percent of developing world wheat and rice crop, while yields multiplied by two and three times.

The Green Revolution was akin to other great agricultural revolutions that transformed world history, including the arrival of Champa rice in China in the eleventh century, the introduction of American maize, potatoes, and cassavas to Europe and Asia after the European Voyages of Discovery, and Britain's successive, systematic Agricultural Revolutions from the late seventeenth to early twentieth centuries. Instead of mass starvation and political upheaval from the twentieth century's quadrupling of world population, world living standards per person ruptured from all historical trends and tripled. The global diffusion of wealth creation helped establish a new world economic order marked by an integrated web of fast-moving, cross-border exchanges of communications, capital, goods, ideas, people, environmental impacts as well as buffeting feedback loops. Goods moved around the world on an oceanic superhighway of intermodal container shipping to create a new phenomenon in which demand in any country could be met by supply produced outside its borders as readily as from within. By 2000, some 90 percent of world commerce moved by sea on some 46,000 giant ships amid 3,000 major ports and through a dozen strategically vital straits and sea canals. The stunning abundance of clean, cheap freshwater that became widely accessible through giant dams, motorized drills, pumps, and other advanced industrial technologies highlighted water's indispensability in this remarkable achievement of civilization.

Yet toward the end of the century, the global water cornucopia unlocked by the age of dams began to reach its limits and, as in America, peak out. A similar pattern of ecosystem depletions and limitations, yet on a much larger, planetary scale, was emerging. By 2000, some 60 percent of all larger river systems in the world passed through dams and man-made structures. Most of the best hydropower and irrigation dam sites in the world were already being used. So much freshwater had been redistributed across Earth's landscapes in dams, reservoirs, and canals over the twentieth century as to account "for a small but measurable change in the wobble of the earth as it spins," noted world water expert Peter Gleick. Like the Colorado, great rivers such as the Yellow, Nile, Indus, Ganges, and Euphrates no longer reached the sea much of the year, or did so carrying greatly diminished restorative water flows and sediment to their delta and coastal ecosystems. The deleterious side effects of protracted, intensive irrigation and inadequate drainage on soil pauperization through salinization, waterlogging, and silt erosion were everywhere in increasing evidence. Irrigated cropland that had provided such a spectacular growth in worldwide food production was being retired as fast as new irrigated land was developed—the historic net expansion of irrigated land had ended. As traditional surface resources ran low, more and more regions were mining groundwater for irrigation much faster than nature's water cycle could restore it—some 10 percent of world farming was unsustainable in the long run. Water tables were sinking and desertification was spreading on several continents. Many parts of the world were compounding the problem by poisonously polluting their freshwater supplies, as well as their coastal fisheries, with industrial waste and farm runoff.

At the dawn of the twenty-first century, a new water challenge was rising to the forefront to reshape world civilization, geopolitics, and governing hierarchies between and within societies—an impending famine of freshwater and the depletion of Earth's civilization-sustaining water ecosystems. What was happening was that for the first time in history, mankind's unquenchable thirst, whetted by voracious industrial demand, gargantuan engineering capacities, and sheer multiplication of human population and individual consumption levels, was starting to significantly outstrip many planetary ecosystems' absolute supply of readily accessible and renewable clean, fresh liquid water. Based on current usage trends, practices, and foreseeable technologies, it was doubtful there was enough freshwater returning to the Earth's surface in the natural water cycle of evaporation and precipitation to sustain the economic growth necessary for the developing world's billions to attain anything close to the levels of prosperity and health enjoyed in the West—and for a terrifyingly large percentage of humanity, there wasn't enough clean water to live healthy, natural lives at all. An explosive competition for scarce water loomed. Many of the driest, most heavily populated, and destitute regions, already couldn't feed their populations and had little realistic hope of doing so soon. Even in parts of the world where freshwater was relatively abundant, growing shortages were triggering a new cycle in the age-old struggle to control regional water resources, and with it, new realignments of political and economic power.

The new era of freshwater scarcity was the by-product of the classic historical cycle of resource intensification, population boom, resource depletion, and flattening or falling economic growth until the next round of intensification and growth increased accessible water supply and made more productive uses of existing, available water resources.

In the twentieth century, populations had multiplied based partly on the one-time surge in water supply from the era's great hydraulic innovations. But now that water supply boom was peaking out, leaving behind populations in many parts of the world with greater material needs and expectations than resources to satisfy them. With human population projected to balloon 50 percent by midcentury, the second scissors blade was now ineluctably closing. The planet's supply of accessible freshwater, as presently managed, was insufficient to meet the demands of many of the world's mostly young, restive, and growing multitudes. Fresh, clean—and utterly indispensable—water, in short, was fast becoming a depleted global natural resource and the world's most explosive political economic problem.

PART IV

The Age of Scarcity

CHAPTER FOURTEEN

Water: The New Oil

The challenge of freshwater scarcity and ecosystem depletion is rapidly emerging as one of the defining fulcrums of world politics and human civilization. A century of unprecedented freshwater abundance is being eclipsed by a new age characterized by acute disparities in water wealth, chronic insufficiencies, and deteriorating environmental sustainability across many of the most heavily populated parts of the planet. Just as oil conflicts played a central role in defining the history of the 1900s, the struggle to command increasingly scarce, usable water resources is set to shape the destinies of societies and the world order of the twenty-first century. Water is overtaking oil as the world's scarcest critical natural resource. But water is more than the new oil. Oil, in the end, is substitutable, albeit painfully, by other fuel sources, or in extremis can be done without; but water's uses are pervasive, irreplaceable by any other substance, and utterly indispensable.

The long sweep of history revealed that long enduring civilizations were underpinned by effective water control using the technology and organization methods of its time. Whether it was the irrigation canals of ancient Mesopotamia, the Grand Canal of imperial China, the waterwheels and steam engine of early industrial Europe, or the giant, multipurpose dams of the twentieth century, societies that rose to preeminence responded to the water challenge of their ages by exploiting their water resource potential in ways that invariably were more productive, larger in scale, and unleashed larger usable supplies than their slower-adapting rivals. In contrast, unmet water challenges, a failure to maintain waterworks structure, or simply being overtaken by more productive water management elsewhere was a common factor of many of history's declines and collapses. Likewise, the economic productiveness and political equilibrium of today's advanced societies depends critically upon the robustness, security, and continuous innovative development of an interlinked array of giant dams, electric power plants, aqueducts, reservoirs, pumps, distribution pipes, sanitary sewage systems, wastewater treatment facilities, irrigation canals, drainage systems, and levees, as well as transport waterworks including port facilities, dredgers, bridges, tunnels, and ocean-spanning shipping fleets. In the unfolding reality of the new millennium, water use and infrastructure are also at the heart of the interlinked challenges of food, energy shortages, and climate change dictating the fate of human civilization.

Today, at the beginning of the twenty-first century, there is hardly an accessible freshwater source or a strategically placed waterway on an economically advanced part of the planet that has not been radically, and often monumentally, engineered by man's prodigious industrial power. As world population continues to be propelled toward 9 billion by 2050, and with so many third world inhabitants starting to move up toward consumption and waste-generation levels of the one-fifth living in industrialized nations, demand for more freshwater is continuing to soar. Yet no new innovative breakthrough capable of expanding usable water supply on a large enough scale to meet the demand is anywhere evident on the horizon.

Over the past two centuries, freshwater usage has grown two times faster than population. About half the renewable global runoff accessible to the most populated parts of the planet is being used. Simple math, and the physical limits of nature, dictates that past trends cannot be sustained. Throughout history the ceiling of man's capacity to extract greater water supply from nature had been bounded only by his own technological limitations. Now, however, an additional, external obstacle has arisen to impose the critical constraint—the depletion of the renewable, accessible freshwater ecological systems upon which all human civilization ultimately depends. As a result a new application of water is emerging to join society's traditional four primary uses—the allocation of enough water to watersheds and related natural ecosystems to sustain the vitality of the hydrological environment itself.

The age of water scarcity consequently heralds the potential start of a momentous transition in the trajectory of water and world history: from the traditional paradigm based on centralized, mass-scale infrastructure that extracted, treated, and delivered ever greater, absolute supplies from nature to a new efficiency paradigm built upon more decentralized, scaled-to-task, and environmentally harmonious solutions that make more productive use of *existing* supplies. This transition is fomenting a new politics in the old equations between population sizes and available water resources in societies all over the world. New population-resource equilibriums eventually will be achieved within each society, water-poor and water-rich alike, through breakthroughs in efficiency and organization on the one hand, or stagnation in personal living standards and overall population levels on the other—and very likely some mixture of both. History suggests that it will be a tumultuous process, recasting social orders, domestic economic hierarchies, international balances of power, and everyday lives. Some regions are better placed than others to face the transition. With water demand continuing to outstrip soaring population growth and many planetary ecosystems being taxed beyond sustainable levels, more and more water-fragile nations were already being driven to the brink.

Most prominent, water scarcity is cleaving an explosive fault line between freshwater Haves and Have-Nots across the political, economic, and social global landscapes of the twenty-first century: internationally, among relatively well-watered industrial world citizens and those of water-famished, developing countries; among those upriver who control river flows and their neighbors downstream whose survival depends upon receiving a sufficient amount; and among those nations with enough agricultural water to be self-sufficient in food and those dependent upon foreign imports to feed their teeming populations. Within nations the new freshwater fault line is fomenting a more divisive competition among interest groups and regions for a greater allocation of limited domestic water resources: between heavily subsidized farmers on the one side and industrial and urban users without government assistance on the other; between the well-heeled situated within close proximity of freshwater sources and the rural and urban poor, who, by dint of occupying secondary locations more remote from water sources, endure the added insult of having less piped connectivity and regressively greater expense in obtaining water. The water fault line cut across humanity, between those able to pay the top price for abundant, wholesome drinking water and the water destitute who glean the dregs; between those who dwell in locations with effective pollution regulations, modern wastewater treatment, and sanitation facilities and those on the other side of the sanitary divide, whose daily lives are contaminated by exposure to impure, disease-plagued water. Across geographical habitats, water's fault line contrasts the privileged minority who live in the planet's relatively well-watered and forested temperate zones and the largest part of the human race that live on water-fragile dry lands, oversaturated tropics, or were exposed to the costly unpredictability of extreme precipitation events that cause out of season floods, mudslides, and droughts. Increasingly, the fault line between water Haves and water Have-Nots is being played out on the plane of international policy between traditional economic nationalists trying to manage affairs within the blinkers of domestic boundaries and the growing coalition of enlightened self-interests worried about destabilizing spillovers from the interdependencies of global society and from planetary environmental crises triggered by the regional degradation of water ecosystems.

Every day across the planet, armies of water poor, mainly women and children compelled by thirst to forgo school and productive work, march barefoot two or three hours per day transporting just enough water in heavy plastic containers from the nearest clean source for their barest household survival needs—some 200 pounds per day for a four-person household. The alarming dark side of this humanitarian divide includes over 1.1 billion people—almost one-fifth of all humanity—who lack access to at least a gallon per day of safe water to drink. Some 2.6 billion—two out of every five people on Earth—are sanitary Have-Nots lacking the additional five gallons needed daily for rudimentary sanitation and hygiene. Far fewer still achieve the minimum threshold of 13 gallons per day for basic domestic health and well-being, including water for bathing and cooking. The lives of the most abject of Water Have-Nots, moreover, are chronically afflicted and shortened by diarrhea, dysentery, malaria, dengue fever, schistosomiasis, cholera, and the myriad other illnesses that make waterborne diseases mankind's most prevalent scourge. Half the people in the developing world of Africa, Asia, Latin America, and the Caribbean are estimated to suffer from diseases associated with inadequate freshwater and sanitation. This side of the humanitarian divide includes the 2 billion human beings whose lives are uprooted catastrophically every decade from inadequate public infrastructural protection from water shocks. By contrast, on the Water Have side of the humanitarian divide, industrialized-world citizens use 10 to 30 times more water than their poorest, developing nation counterparts. In the water-wealthy United States, each person uses an average of 150 gallons per day for domestic and municipal purposes, including such extravagances as multiple toilet flushes and lawn watering.

Water rationing is increasingly commonplace in Water Have-Not societies. So, too, are internecine conflicts and violent protests over scarce supplies and high prices. Inadequate water supply commonly manifests itself in the form of insufficient food output, stunted industrial development as critical water inputs are sacrificed to the priority of agriculture, and shortages in energy, whose modern production infrastructure is closely interlinked with copious volumes of water used for cooling, power generation, and other purposes. Chronic water scarcity undercuts the political legitimacy of governments, fomenting social instability, and failed states. Water riots, bombings, many deaths, and other violent warning signs occurred from 1999 to 2005, for example, in various conflicts over water in Karachi, Pakistan, in Gujarat, India, in provinces of arid north China, in Cochabamba, Bolivia, between Kenyan tribes, among Somalia villages, and in the Darfur, Sudan, genocide. In the oddest report of water violence, eight monkeys were killed and 10 Kenyan villagers wounded when the desperate primates descended upon water tankers brought to relieve the drought-stricken village. Cross-border tensions and military threats between nation-states are palpable perils in a growing number of international watersheds in some of the world's most combustible regions. Today, it is a commonplace for statesmen to paraphrase the much publicized 1995 prediction of a former chairman of the World Commission for Water in the 21st Century and senior World Bank official, Egyptian Ismail Serageldin: "Many of the wars this century were about oil, but those of the next century will be over water."

From the early 1990s, a decade marked by the global environmental awakening symbolized by the first Earth Summit in 1992 at Rio de Janeiro, a consensus began to coalesce among attentive world leaders that on existing trajectories and technologies, usable freshwater resources were falling short of what was needed for long-term global economic growth. The consensus helped galvanize in 2001 the first comprehensive, planet-wide assessment of the health of all of Earth's major ecosystems and its effects on human well-being. The headline findings of the landmark Millennium Ecosystem Assessment, launched under U.N. auspices and completed in 2005 with input from over a thousand experts worldwide, was that 15 of the 24 studied Earth ecosystems were being degraded or used unsustainably. Freshwater ecosystems and capture fisheries, in particular, were singled out as "now well beyond levels that can be sustained even at current demands, much less future ones." Up to half the world's wetlands disappeared or were severely damaged in the twentieth century's drive to obtain more arable land and freshwater for agriculture. Worldwide expansion of irrigable farmland is peaking out for the first time in history.

Under demographic and developmental duress, mankind's withdrawal of usable, renewable freshwater from the surface of the planet is expected to rise from half to 70 percent by 2025. Due to heavy overdrafts on slowly replenishing reserves in some water distressed regions, MEA experts estimated that possibly as much as one-quarter of global freshwater use might already be exceeding the accessible, sustainable supply.

In the first decade of the twenty-first century, an increasing number of nations were so critically water stressed that they can no longer grow all the crops they need to feed and clothe their own populations. Growing crops is an astonishingly water intensive enterprise—about three-quarters of mankind's water use worldwide is for farm irrigation. Indeed, food itself is mainly water. To produce a single pound of wheat requires half a ton, or nearly 125 gallons of water; a pound of rice needs between 250 and 650 gallons. Moving up the food chain to livestock for meat and milk multiplies the water intensity since the animals have to be nourished with huge quantities of grain; up to 700 gallons, or nearly three tons of water, for instance, are needed for the feed that produces a single portion of hamburger and some 200 gallons for a glass of cow's milk. In all, a well-nourished person consumes some 800 to 1,000 gallons of water each day in the food he eats. The ordinary cotton T-shirt on his back requires as much as 700 gallons to produce.

As water poor countries fall short of self-sufficiency in producing their daily bread, they are growing increasingly dependent upon importing grain and other foods from water-wealthier farming nations. By 2025 up to 3.6 billion people in some of the driest, most densely populated and poorest parts of the Middle East, Africa, and Asia are projected to live in countries that cannot feed themselves. Due to water scarcity a growing trade in virtual water—food and other finished products imported in substitution for scarce domestic water resources—is redefining the terms of international trade and emerging as a distinctive feature of the changing global order. The growing bifurcation between water-poor food importers and water-rich exporters is often further exacerbated by man-made ruination of cropland from soil erosion and polluting runoff. The prospect of upward spiraling international food prices as the era of cheap water and cheap food comes to an end is already causing experts to warn of grave consequences if there is not a new Green Revolution, perhaps including the development of genetically modified plant hybrids that grow with less water.

The same, finite net, 4/1,000ths of 1 percent of Earth's total water that recycled endlessly and fell over land in the process of evaporation-transpiration and precipitation has sustained every civilization from the start of history to the present. Man's practical access to this renewable freshwater supply remains limited to a maximum of one-third, since about two-thirds quickly disappears in floods and into the ground, recharging surface and ground water ecosystems and ultimately returning to the sea. Even so, that one-third totals enough available renewable water to more than suffice for the planet's 6 billion—*if* it were all distributed evenly. But it is not. A large share runs off unused in lightly inhabited jungle rivers like the Amazon, the Congo, and the Orinoco and across Russia's remote Siberian expanses toward the Arctic in the giant Yenisei and Lena rivers. So the actual total amount of readily available, renewable freshwater per person often averages less—often *far* less—in some regions than the threshold annual 2,000-cubic-meter measure of water sufficiency. And it is declining sharply in inverse relationship to the escalation of world population.

Yet even that does not convey the full measure of the deepening water crisis challenge because the remainder of renewable freshwater that precipitates within the reach of large human society falls in disparate intensities, seasonal patterns and degrees of difficulty in being captured for human use. Hot climates, for instance, suffer much higher losses from evaporation than cool, temperate ones—in Africa only one-fifth of all rainfall transforms into potentially utilizable runoff. The most difficult hydrological environment is not one of extreme aridity, or extreme wetness, however, but where water availability varies widely between seasons and is prone to unpredictable water shocks, such as floods, landslides, droughts and sudden, extreme deviations from usual patterns. Seasonality raises the complexity and the cost of water engineering, while unpredictability defeats even sound waterworks planning, often striking demoralizing setbacks against development. It is not a coincidence that history's poorest societies often have had the most difficult hydrological environments.

As a result, each region's actual water challenges vary enormously by environment, availability, and the population it has to support. Australia is by far the driest continent, with only 5 percent of world runoff. But it has to support by far the smallest human population, a mere 20 million, or less than one-half of 1 percent of world population. Asia, the largest continent, receives the most renewable water, about one-third of the total. Nonetheless, it is the most water-stressed continent because it has to meet the needs of three-fifths of humanity, contains some of the world's most arid expanses, and over three-quarters of its precipitation falls in the form of hard-to-capture, highly variable, concentrated seasonal monsoons. The water richest continent is South America, with 28 percent of the world's renewable water and only 6 percent of its population. On a per person basis, it receives ten times as much freshwater each year as Asia and five times as much as Africa. Yet most of it flows away unused through jungle watersheds, while some high desert regions remain bone dry. North America is water wealthy with 18 percent of the world's runoff and 8 percent of its population. Europe has only about 7 percent of the world's water for its 12 percent share of population, but is comparatively advantaged in its wet, northern and central half because much of it falls year-round, evaporates slowly, and runs off in easily accessible and navigable small rivers.

The continental volumes, of course, mask the all-important disparities among localities and nations that are animating the new water politics. One eye-popping headline of the Millennium Ecosystem Assessment was that the planet's dry lands, encompassing one-third of humanity or over 2 billion people, had only 8 percent of the world's renewable supply of water in its surface streams and fast-recharging groundwater tables. More than 90 percent of the dry-land inhabitants live in developing nations, making water famine one of the key, vexing challenges of international economic development. It is hardly surprising that the vast dry-land belt stretching from North Africa and the Middle East to the Indus valley is also one of the world's most politically volatile regions. At the other end of the spectrum are super Water Have countries such as Brazil, Russia, Canada, Panama, and Nicaragua with far more water than their populations can ever use. The United States and China have large hydrological imbalances with shortages in their far western and northern regions, respectively; while the modestly populated American Far West felt constraints on its rapid growth, the fertile, overpopulated northern plain of China is one of the most severely water-scarce, environmentally challenged regions on Earth. Likewise, India's growing, huge population is outstripping the highly inefficient management of its freshwater resources, forcing farmers, industry, and households to pump groundwater faster and deeper in a proverbial race to the bottom. Western European nations managed successfully because they use their limited water resources more productively, abetted by their higher proportions used for industry and cities, and less for agriculture.

Because water is so heavy and is needed in such vast quantities, chronic shortages cannot be permanently relieved by transporting it over long distances. The challenge of water scarcity, therefore, has to be confronted watershed by watershed, according to local physical and political conditions, and further constrained by the needs of foreign neighbors within the 261 transnational river basins that are home to 40 percent of the world's inhabitants. One of the most reliable indicators of water wealth is the amount of water storage capacity each nation has installed per person to buffer it against natural shocks and to manage its economic needs; almost universally, the storage leaders are the world's wealthiest nations, while the poorest remain most exposed to the natural caprices of water.

Despite its growing scarcity and preciousness to life, ironically, water is also man's most misgoverned, inefficiently allocated and profligately wasted natural resource. Societies' own poor management of water, in other words, is a key component of their water scarcity crises. In market democracies and authoritarian states alike, modern governments still routinely maintain monopolistic control over their nation's supply, pricing, and allocation; commonly, it is distributed as a social good, as political largesse to favored interest groups, and in overweeningly ambitious public projects. Almost universally, governments still treat water as if it were a limitless gift of nature to be freely dispensed by any authority with the power to exploit it. In contrast to oil and nearly every other natural commodity, water is largely exempted from market discipline. Rarely is any inherent value ascribed to the water itself. Only the cost of capturing and distributing it is routinely accounted. Nor is any cost ascribed to the degradation of the water ecosystem from whence it comes and to which, often in a polluted condition, it ultimately returns. By belonging to everyone and being the private responsibility of no one, water for most of history has been consumed greedily and polluted recklessly in a classic case of a "tragedy of the commons."

The result, compounded over time, is a colossal underpricing of water's full economic and environmental worth. This sends an insidious, illusory economic signal that water supply is endlessly plentiful, promoting wasteful use on purposes with low productive returns. The twentieth century's most breathtaking example was the former Soviet Union's inadvertent destruction of central Asia's Aral Sea—its hydraulic Chernobyl—and a symbol of the failure, after less than a century of existence, of its state experiment with communism. What started as a well-intentioned, decades-long effort to transform arid central Asia into a cotton belt that rendered the nation self-sufficient in water-thirsty "white gold" ended as an object lesson in the catastrophic side effects of misguided ecosystem reengineering, and politically, how wretchedly off course unchecked, price-insensitive industrial state planning could go.

In the late 1950s, Soviet engineers began efforts to divert the waters from the two great rivers, the Syr Darya and the Amu Darya, the Jaxartes and Oxus of ancient history, feeding the Aral Sea, the fourth-largest freshwater lake in the world. River flows soon began to decline sharply. By the early 2000s, the Aral Sea had lost fully two-thirds of its volume and had shriveled into two small lakes so saline that its once flourishing fishing industry was decimated. The former lake bed, strewn with abandoned ships and bordered by ghost fishing villages, became a salty dust bowl whose toxic residue was swept up in windstorms over the irrigated cotton fields, crippling yields and corroding the critical infrastructures of production. Worse still, the shrinking of the lake reduced its watery capacity to moderate the local climate, which grew more extreme. Summers were hotter, winters bitterer. Reduced evaporation lessened local precipitation and shrank snowpacks. The volume of water in the two arterial rivers was thus permanently diminished, creating a self-reinforcing pattern of growing desiccation and eroding soil fertility. In the end Soviet planners' stubborn unresponsiveness to environmental signals and misvaluing of water resulted in the loss of everything—drastically reduced cotton output, decimated fishing industry, and a badly depleted environment less habitable by productive society.

A similar fate befell sub-Saharan Africa's immense Lake Chad from the 1970s when uncoordinated dam building, irrigation diversions, and land clearance by bordering countries dried out the lake's nourishing river flows, wetlands, and groundwater. This both accelerated and exaggerated natural climate cycles and resulted in the shocking disappearance of 95 percent of the lake's surface area within only two generations and its replacement by widening desertification. Myriad other locations today are suffering less-pronounced microclimatic changes as a result of upsetting the natural rhythms of their local water ecosystems.

By far, man's most egregious waste of water came from the distortions caused by the chronic underpricing of water for irrigation. Irrigation farmers in Mexico, Indonesia, and Pakistan paid little more than 10 percent of the full cost of their water. Because Islamic tradition held that water should be free, many Muslim countries charged little or nothing except partial delivery costs in some of the driest parts of the world. American government dam water subsidies were grandfathered upon a small number of farmers who cultivated a quarter of the irrigated cropland in the arid lands of the West. Inefficient flood irrigation is still subsidized in many water poor regions, even where sprinkler and drip methods are viable alternatives. These subsidies were so lavish that the farmers grew water-thirsty, low-value crops like alfalfa in the middle of the desert, while more productive, fast-growing industries and municipalities alongside them paid eye-popping premiums to obtain enough water. China's postwar state planners misplaced many water-intensive industries and urban metropolises in the water-short north, where they eventually were forced to compete for water with the region's vital grain farming.

Underpriced water is also a disincentive to urban conservation. Through leaky infrastructure, thirsty Mexico City loses enough water every day—some two-fifths of its total supply—to meet the needs of a city as large as Rome. The world faces a trillion-dollar-plus water infrastructure deficit in the years immediately ahead just to patch the leaks.

Water's peculiar treatment in economic society was famously contemplated in the eighteenth century by Adam Smith. In *The Wealth of Nations,* he pondered, "Nothing is more useful than water; but it will purchase scarce anything; scarce anything can be had in exchange for it." Smith sought an explanation for the "diamond-water paradox," one of the well-known dilemmas so beloved by economists as a means to explore the boundaries of economic theory: Why was water, despite being invaluable to life, so cheap, while diamonds, though relatively useless, so expensive? Smith's answer was that water's ubiquity and the relatively easy labor required to obtain it accounted for its low price. His theory was superseded within mainstream economics in the late nineteenth century by a more refined explanation. Water's price was determined by a sliding scale based upon its availability for its least valued uses, say, for example, watering lawns, filling swimming pools, quenching the thirst of wildlife, or, until the environmental awakening of contemporary times, recharging ecosystems; its premium rose as it became scarce for its most precious uses, reaching its zenith as priceless drinking water. A half century before Smith, Benjamin Franklin, with his characteristic pragmatism, had cut through the theoretical musings to the essence of the water dilemma in his *Poor Richard's Almanac:* "When the well is dry, we learn the worth of water." In the new age of water scarcity, in effect, the global well is starting to go dry. The worth of water is rising to its highest marginal utility value and to reflect Smith's original observation that nothing is more useful.

For the first time in history, the fundamental economic and political rules governing water are starting to be transformed by the power of market forces. Under the duress of scarcity, the iron laws of supply and demand graphically described by Franklin are propelling the market economy's expansive, profit-seeking mechanisms to colonize the realms of water. Beckoning bonanza profit opportunities have set off a worldwide scramble to control water resources and infrastructures, and to commercialize water as an ordinary commodity like oil, wheat, or timber. Bottled water is by far the world's fastest-growing beverage, with global sales of over $100 billion increasing at 10 percent per year and reaping handsome profits for corporate giants Nestlé, Coca-Cola, and Pepsi-Cola; the two latter in the United States sell high-tech filtered and treated common tap water from Queens, New York, and Wichita, Kansas, and elsewhere under the Dasani and Aquafina brand names, respectively, at a 1,700 times markup over public tap costs, more than their famous water-based, sugared soft drinks. Privatized management of water utilities is another huge global sector, as is wastewater services, dominated by corporate multinationals. In total, water is a fast-growing, highly fragmented, competitive, $400 billion per year industry. Specialized water investment funds have been launched on Wall Street. Before its ignominious collapse in 2001, Enron had been promoting a scheme to trade water rights as it traded energy in California. Many cities, such as New York, which had never curtailed water service for nonpayment, have been considering ways to turn off the faucet to force collection of many millions of dollars in delinquent water bills.

Subjecting water to the discipline and productive investment of market forces has enormous capacity to stimulate badly needed efficiency gains and innovations. But water is too precious to human life—and too politically explosive—to be left to the merciless logic of market forces alone. Indeed, warning shots have been fired in high-profile conflicts in India, China, Bolivia, and elsewhere in which international corporations have been compelled to close or make costly modifications to their local operations. Whether the commodification of water ultimately leads to efficiency gains that ease water scarcity or results instead in an unregulated regime of water pricing and allocation that condemns the water poor to choose between desiccated, unhealthy lives and desperate remedies, depends on the terms by which societies choose to inject market forces into the traditional, public realm of water.

The age of water scarcity poses a special threshold challenge for Western liberal democracy: whether such societies can artificially graft a new, effective mechanism that fully prices in the economic costs of maintaining sustainable water and other environmental ecosystems onto the market economy's historically prodigious processes of wealth creation. Adam Smith described how the market's unseen "invisible hand" caused individuals' self-interested, competitive pursuit of profit to simultaneously, as a wholesome by-product, maximize wealth creation for the entire society. Yet the market has glaringly failed to evolve any corresponding invisible green hand to automatically reflect the cost of depleting natural resources and sustaining the total environmental health upon which an orderly, prosperous society ultimately depends. Twice in the twentieth century Western democracies had successfully adjusted to catastrophic market failings through state-led interventions—the trust-busting of Teddy Roosevelt and the progressive movement in the early 1900s, and the New Deal, welfare state response to the Great Depression in the 1930s. Each intervention altered the rules governing the relationship between the private and public realms. In each instance, the market economy's productive power was reinvigorated, helping sustain the West's global leadership. A third adaptation in the unspoken liberal democratic compact between markets and governments is needed for such a new mechanism to thrive.

Every society faces the core question in the age of scarcity of where its increased freshwater supply will come from. Societies have been responding in four general ways, often simultaneously. The first response has been to do little or nothing and await the development of some magic-bullet innovation for extracting more water supply from nature, with the impact of twentieth-century multipurpose dams, and commonly represented by such intriguing processes as seawater desalinization or genetically modified crops that can grow using less water. The second response, most evolved in the mainly water-sufficient industrialized first world, has been to increase effective supply by improving the productivity of existing water use through regulatory and market-oriented methods. The final two responses, while proactive, are mainly expedient postponements by distressed countries of their day of water reckoning. Long-distance water transfer projects that reroute entire rivers and lakes from wet regions to landscapes that are drying up from overuse are prevalent in distressed large countries with severe regional water imbalances. Similarly, many overpump shallow groundwater faster than it naturally replenishes, and if available, drill deeper at great expense and technical difficulty to mine accessible parts of the rocky, geological aquifer reservoirs accumulated by nature over the millennia inside Earth, but that once consumed are gone forever.

The Water Have–Have-Not continuum can be usefully subdivided into four main types of societies. At the abject bottom of mankind's water poor are the masses of destitute souls, mainly in sub-Saharan Africa and Asia, who live without effective infrastructure to buffer them against the tyrannical caprices of water's destructive shocks and without reliable access to adequate clean freshwater to meet their basic domestic and sanitary needs. For the two-fifths of mankind living in such medieval conditions, water represents less of an opportunity for economic development than a daily struggle of life and death. Next are more-modern societies that exist in conditions of such severe scarcity, or water famine, that they typically lack enough freshwater to grow the crops needed to feed themselves, have less than 700 gallons per person each day for all their water needs, and utilize at least one-fifth of their natural runoff. Distressed countries cannot comfortably manage their own food and water needs, average 700 to 1,400 gallons daily per capita, and utilize about 10 to 20 percent of their runoff. Although such borderline nations usually can feed themselves, many are trending toward becoming chronic food importers and face other manifestations of water scarcity as well. Societies that enjoy availability of over 1,400 gallons and have to tap less than 10 percent of their national runoff are typically the world's major food exporters. Their water shortages, in the main, are manageable through relatively modest improvements in existing water productivity alone.

Yet as world population soars by 50 percent and world resource demand increases by a far-greater factor because of those nations transitioning from third world to first world living standards, the entire continuum is lurching sharply toward the Have-Not side of the water spectrum—a massive dry shift—that adds to the stress on everyone. Water famines are worsening in countries already in crisis, and more societies, including some of the world's largest, are joining them. Water scarcity requires nothing less than a comprehensive reevaluation of water's vital importance as the new oil—a precious resource that has to be consciously conserved, efficiently used, and properly accounted for on the balance sheets across the breadth of human activity, great and mundane: from public health, food and energy production to national security, foreign policy and the environmental sustainability of human civilization. In the age of water scarcity, water's always paramount, but its usually discreet role in world history is visibly taking its place at center stage.

CHAPTER FIFTEEN

Thicker Than Blood: The Water-Famished Middle East

One of the front lines of the world's unfolding freshwater crisis is the historically water-fragile Middle East and North Africa—the heartland of Arab Islamic civilization and cradle of the ancient hydraulic irrigation civilizations that arose in the flooding river valleys of the Fertile Crescent. The politically volatile, overpopulated, dry land stretching from Algeria, Libya, Egypt throughout the Arabian Peninsula into Israel, Jordan, Syria, and Iraq, and their regional neighbors, is rife with water tensions, conflicts, and troubled states that hold the potential to combust into a full-fledged water war.

The Middle East is the first major region in modern world history to run out of water. Country after country lacks the freshwater to grow enough crops to feed its population or to provide the basis for long-term rising living standards; per person measures of renewable water availability are widely below the minimum standard measures of scarcity and famine. The desert nations of the Arabian Peninsula and Libya, as well as arid Israel and Palestine, outgrew their internal water resources for sustainable food self-sufficiency in the 1950s. Jordan ran out of water in the 1960s, Egypt in the 1970s, and other regions more recently. The Millennium Ecosystem Assessment reported that in "the Middle East and North Africa, humans use 120% of renewable supplies." They survive by importing growing volumes of food—virtual water—and, where available, by pumping water out of underground aquifers faster than nature can recharge them. Only the bonanza of surging oil income from the early 1970s has staved off a full-blown crisis. Oil wealth paid for the quadrupling of Middle East wheat flour imports to over 40 million tons within a single generation. For most of Middle Eastern and North African history, exploitation of subsurface water deposits had been limited mainly to the excavation of shallow wells and qanats, the horizontal tunnels of antiquity that conveyed water from inside hillsides. Oil opened a new era by facilitating the large-scale subsidization of modern pumping of deep aquifer water for irrigation.

Yet if oil built modern Middle East society, water holds the key to its future. In the end the region cannot escape the same water-fragile geography and stream deficit that shaped its ancient and Islamic civilizations, placed a ceiling on its indigenous, sustainable population size, and ultimately influenced Islam's abrupt decline from glory from the twelfth century. Modern engineering of the region's surface waters began in earnest in the nineteenth century. Irrigation and cheap oil energy metamorphosed the traditional population-resource equation underlying each society. From 1950 to 2008, the population more than quadrupled to 364 million. But country after country soon began to exceed the productive limits of the region's water resources and waterworks capabilities. With population forecast to swell another 63 percent to 600 million by 2050, the Islamic Middle East is becoming a demographic volcano. The region's upsurge in violence, radical religious fundamentalism, and terrorism is likely but a foretaste of what potentially lies ahead as its water famine worsens.

Egypt, the most populous Arab state with 75 million inhabitants in 2006 and projected to reach nearly 100 million within a generation, is being stretched to its breaking point. As it had been since the time of the ancient Pharaohs, the Nile is still by far the paramount factor governing the destiny of Egyptian society. But ever since the completion of the high dam at Aswan in 1971, how the river did so has changed entirely. The giant, multipurpose Aswan Dam utterly

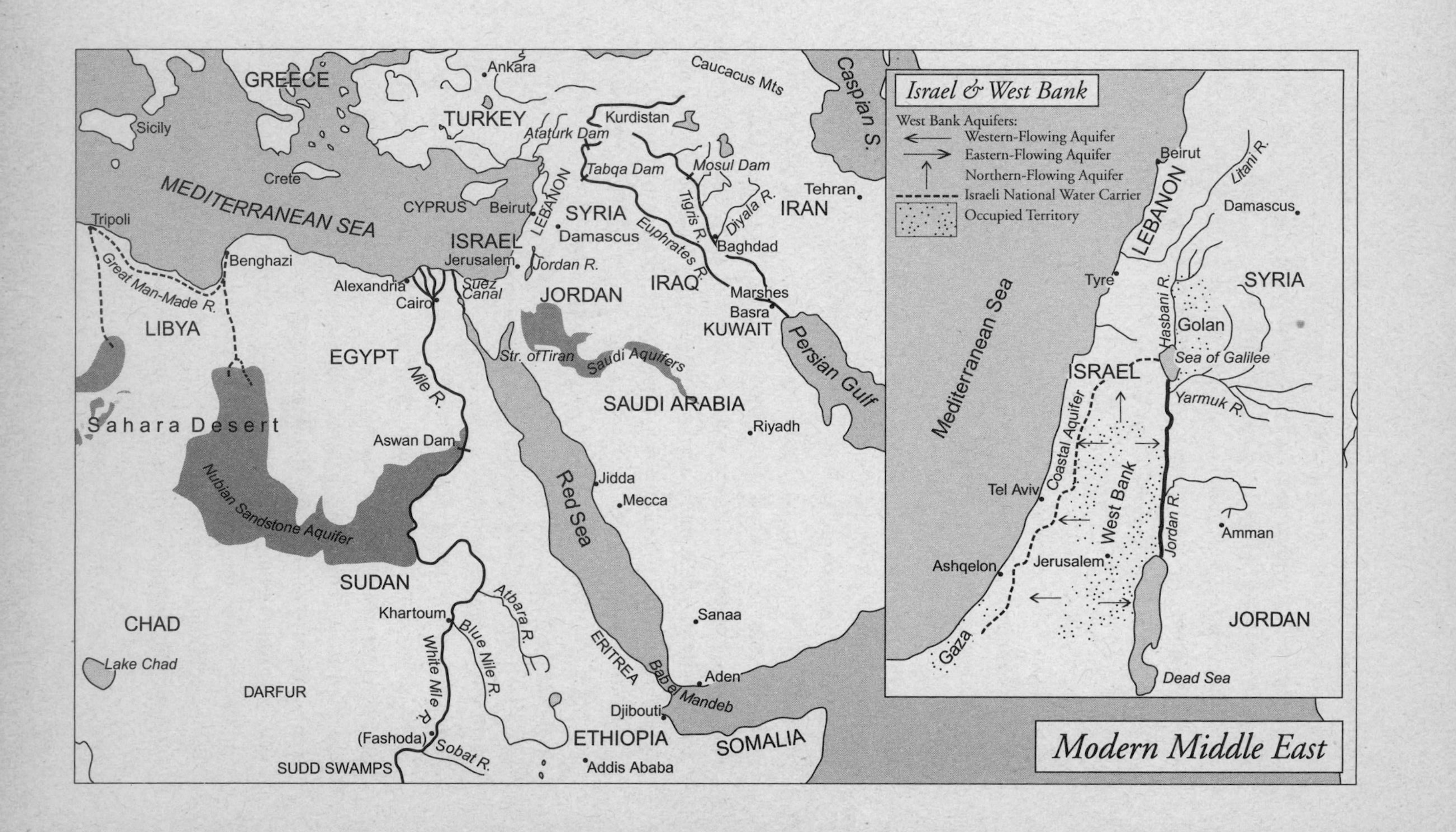

Modern Middle East

transformed the hydrology of the Nile from a miraculous natural phenomenon to a totally managed irrigation channel, and produced copious hydroelectricity for an underpowered nation. The dam fulfilled the dream of 5,000 years by delivering to an Egyptian leader absolute control over the Nile's domestic flow and the power to insulate Egyptians from the dreaded traumas of the river's periodic episodes of extreme droughts and floods. Yet for all its majestic power, the Aswan Dam has not been able to alter one other historical feature of the Nile: nearly every drop of it originates outside Egypt's borders, while the well-being of Egyptian society depends upon consuming a vastly disproportionate share of the Nile basin's water. Beyond upstream Sudan, the countries of equatorial East Africa's great lake plateau provide the sources of the White Nile. By far the biggest contributor of Egypt's water is highland Ethiopia, whose Blue Nile, Atbara, and Sobat rivers supply some 85 percent of the water, and all the silt, that arrives every June at Aswan. Throughout history, impoverished Ethiopia and the White Nile river states have sipped only a tiny fraction of the Nile's water for their own economic development. To alleviate their grinding poverty, they are now determined to use more. In 1989, then Egyptian foreign minister and later U.N. secretary-general Boutros Boutros-Ghali summed up Egypt's geopolitical dilemma to the U.S. Congress: "The national security of Egypt is in the hands of the eight other African countries in the Nile basin."

Paranoid fears about Nile water cutoffs by upstream nations, particularly Ethiopia, have been ingrained in the Egyptian psyche for many centuries, and at times became feverish, one example being when mass famine in 1200 caused by disastrous, low floods killed one-third of Cairo's population. Verdi played upon this angst in *Aïda* by featuring two tragic lovers caught up in a war between Egypt and Ethiopia; in 1875–1876 his story was partly enacted as a bloody reality when Egyptian soldiers were annihilated by 60,000 Ethiopian troops after making several disastrous imperialistic forays into Ethiopian territory. The triumphant achievement of the Aswan Dam, ironically, exacerbated Egypt's national security fears by whetting their poor, upstream neighbors' desire to utilize more Nile waters for giant dams of their own.

Thus, while most of the world viewed Egyptian policy through the lens of the Suez Canal and the Arab-Israelis wars, Egyptian leaders themselves were clear-sightedly focused on their own overriding national security objective—safeguarding their disproportionate consumption of Nile waters—and enlarging the river's overall available flow at Aswan. In May 1978, just prior to making his historic peace treaty with Israel, and with a telescoped eye on Ethiopia, Egyptian president Anwar el-Sadat declared with typical bluntness: "We depend upon the Nile 100 percent in our life, so if anyone, at any moment thinks to deprive us of our life we shall never hesitate to go to war because it is a matter of life or death."

From the dawn of ancient Egyptian civilization, farming along the Nile had operated unchanged as a natural, one crop, seasonal basin agricultural system able to support a peak population of 4 to 5 million. That ceiling had doubled in the nineteenth century with the introduction of barrages and year-round, multicrop irrigation. Under the advances of British hydrologists after 1882, population soared again. It was 25 million on the eve of the Aswan high dam's opening.

At the turn of the nineteenth century, the British-built low Aswan Dam perpetuated the Nile basin's natural, self-sustaining irrigation system by permitting the passage of silt during flood season, while also protecting Egypt for the first time from disastrously large inundations. Yet the dam's reservoir system was too small to store enough water to deliver Egypt from multiyear droughts. In the ensuing decades, British water engineers conceived plans for massive storage dams in highland lake plateaus of equatorial East Africa and at Lake Tana in Ethiopia where evaporation rates are low. They also attempted to augment the Nile's total flow by building a long diversion canal to bypass the huge, stagnant Sudd swamps in British-controlled southern Sudan, where the White Nile loses half its volume to evaporation. But when the era of British hegemony yielded to national independence after World War II, Britain's ambitious Nile plumbing projects were still mostly unfulfilled. With the end of British rule, the basin devolved politically into a fractious cluster of impoverished watershed states unable to undertake cooperative Nile development. The positive sum potential of maximizing the river's productive resources through optimal, nonpoliticized, placement of waterworks vanished with it.

The father of the Aswan Dam, Egyptian president Nasser, had come to power in 1952 with the transcendent dream that a giant dam at Aswan would, at a single stroke, give Egypt economic control of the Nile waters, insulate it from the caprices of nature and the disruptive political machinations of upstream Nile nations, deliver food security and economic modernization, and restore the independent, sovereign glory of Egyptian and Arab civilization. When American secretary of state Dulles withdrew his previous support, Nasser had in 1956 signed a deal with the Soviet Union to build the high dam, which quickly became the incarnate symbol of swelling Egyptian patriotism and a new political phenomenon, pan-Arabism. Although the Soviets desired to make a successful Aswan a beckoning symbol of socialist possibility for the entire third world, the first two years after construction began in 1960 seemed to confirm Dulles's doubt that the Russians had the technical proficiency to meet the challenge. The dam fell behind schedule; despite the availability of a large, cheap Egyptian workforce, less than 10 percent of the rock and sand needed to fill it had been excavated. Nasser got the project on track, however, by breaking his pledge to the Russians and buying superior Western construction equipment to finish the job.

Nasser did not live to see the completion of the Aswan Dam. He died five months before its official opening in January 1971. By 1975 it became fully operational. The high dam itself was a landmark engineering achievement, and influential political symbol of hope for Egypt and newly independent third world nations everywhere. Standing over 360 feet high and sweeping in a great curve for over two miles, it was the world's highest rock-filled dam. If it were ever to burst, the torrential cascade downstream would strike with the destructive fury of a biblical plague, obliterating modern Egyptian civilization in its path. Its immense, 344-mile-long, eight-mile-wide Lake Nasser reservoir, which submerged land and ancient monuments and displaced over 100,000 inhabitants of southern Egypt and Sudan's Nubia as it filled, stored over two times the average annual flow of the Nile. With some 30 times more storage capacity than the low dam it replaced, it protected Egypt for the first time in history against both extremes of drought and flood. Its 12 generators produced half of Egypt's electricity when it opened. The effective gain in controlled Nile flow increased cultivated watered-desert cropland by 20 percent, as well as more extensive double and triple cropping on existing farmland. The ultimate proof of the dam's success was that from the time it opened to 2005, Egypt's population tripled to 74 million.

Critics who warned that it was the wrong dam at the wrong place due to its many technical and environmental drawbacks were drowned out in the triumphal nationalism. Nasser's insistence that it be located in the scorching desert on Egyptian national territory, for instance, caused its giant reservoir to lose a huge amount to evaporation—12 percent of the Nile's estimated average 84 billion cubic meter flow at Aswan. The high dam also blocked the passage of fertilizing silt, transforming the Nile from a natural, self-sustaining irrigation system to an artificially managed river totally dependent on heavy chemical fertilizer and for the first time prone to salinization and waterlogging. Due to the dam, the natural Nile of history died at Aswan. Like America's Colorado, the Egyptian Nile became a glorified irrigation ditch in which every drop was regulated. But when the dam opened with fanfare in the 1970s such problems were but an afterthought. They were left for future generations to contend with.

Indeed, Nasser's monumental legacy at Aswan seemed immediately providential by insulating Egypt from the terrible regional drought of 1979–1988 that resulted from the lowest Niles of the twentieth century. At a time when over a million upriver Ethiopians and an unknown number of Sudanese died of famine, Egypt's growth continued unabated. During the decade-long drought, the average volume of Nile water reaching Aswan plunged 40 percent below normal. By July 1988, Lake Nasser contained so little water that it was within a dozen feet of reaching the total shutoff levels for the dam's hydroelectric turbines, and was able to produce less than one-fifth of Egypt's needs, forcing the country to rely more heavily on costly fossil fuels. Most alarming of all, Egypt was down to its last seven months' reserve of irrigation water.

Then, in August 1988, heavy rains providentially began to pour down in Ethiopia and Sudan. The great drought ended with the highest Nile flood of the century. Over the next few years, the man-made lake behind Aswan began gradually to refill. Egypt was saved.

The great 1980s Nile drought, and the humanitarian tragedy it wreaked on Egypt's southern neighbors, highlighted Egypt's paramount national security priority in securing its near-monopolistic usage of Nile waters and the Aswan Dam's linchpin role in delivering it. At the same time, it painfully exposed the dam's military vulnerability, and the unimaginable devastation that would occur if its towering barrier were breached in an attack. This double-edged geopolitical reality of the dam was pivotal to the historic decision of Nasser's successor, Anwar el-Sadat, to boldly break Arab taboo by traveling to Jerusalem and set the stage for signing the 1979 peace treaty with despised enemy Israel. Egypt had been the key Arab military leader in the wars with Israel in 1948, 1956, 1967, and, under Sadat's presidency, 1973. Yet despite some initial military success, Egypt saw the latter war end with Israel again astride the Suez Canal and holding enough air power superiority to make its rumored readiness since 1967 to bomb the Aswan Dam a palpable threat.

Although it infuriated his Arab brethen, Sadat's strategic decision to make peace with Israel brilliantly secured Egypt's paramount national security interest over the waters of the Nile. At a stroke, it earned Egypt a diplomatic windfall of international goodwill, made it the second largest recipient (after Israel) of U.S. foreign aid, safeguarded the Dam, the Suez Canal and Egyptian territory against Israeli attack, and freed Egypt to redirect its otherwise superior regional military and diplomatic muscle to assert its commanding influence over developments in the vital Nile basin. Upon making peace with Israel, Sadat in 1979 famously highlighted Egypt's shift in national security focus by declaring, "The only matter that could take Egypt to war again is water." He even briefly broached the idea of a peace pipeline to bring a small amount of Nile water to Palestine and Israel, in an effort to ease water tensions and facilitate peace between Palestinians and Israelis.

In his memoirs of the period, Boutros Boutros-Ghali, Sadat's minister for foreign affairs, explicitly confirmed, "Preserving Nile waters for Egypt was not only an economic and hydrological issue but a question of national survival . . . Our security depended on the south more than on the east, in spite of Israel's military power."

Sadat's strategic focus on the Nile's water was also informed by strident declarations of intent to dam the headwaters of the Blue Nile from Ethiopia's new communist military leader, Mengistu Haile Mariam, who had seized power in 1974. Adding to Sadat's unease was that Israel had been sending military support to Ethiopia throughout the 1970s to help it battle internal and neighboring rivals and that the two nations had long-standing affinities through Judaism, history, and a suspicion of Egypt. In the late 1950s, the U.S. Bureau of Reclamation began surveying Ethiopia's immense, untapped, hydraulic potential at the behest of Emperor Haile Selassie. Still stinging from Nasser's alliance with the Soviets on Aswan, America's Cold War leaders had been happy to oblige. The result was an exhaustive, 17-volume bureau report identifying over two dozen irrigation and hydroelectric power projects, the latter with the potential to generate three times the hydropower of Aswan. By capturing and storing Blue Nile and tributary waters in the cool Ethiopian highlands where evaporation loss was only one-third as great as at Aswan, the bureau concluded, Ethiopian projects could vastly boost the region's hydroelectric output and actually increase the overall available net downstream flow to Sudan and Egypt. In theory, it seemed like a win-win situation for all countries. But it put Ethiopia, not Egypt, in ultimate control of the amount of water reaching Aswan—precisely the nightmare that had haunted Egyptians for centuries. Egypt would have none of it. Desperately poor Ethiopia could not finance such ambitious projects itself. Through its far superior international diplomatic political influence, Egypt wielded an effective veto over Ethiopia's multilateral financing and other potential avenues of water development it didn't abide.

Despite the fact that Ethiopia was the source of four-fifths of the waters of the Egyptian Nile, Egypt claimed an historic right of prior use over the lion's share of the river's volume. Upon launching the Aswan Dam project in 1956, Nasser had simultaneously moved to try to secure a Nile water sharing agreement with its southern neighbor, newly independent Sudan, within whose border part of the Aswan reservoir had to be situated. In late 1958, he found an accommodating negotiating partner in a kindred military Islamic leader who had just seized power in Sudan.

The result was the Nile Waters Agreement of 1959. With breathtaking audacity, the accord divided up all the waters of the Nile between Egypt and Sudan: Egypt got three-quarters, or 55.5 billion cubic meters, of the estimated available flow after evaporation; Sudan received one-quarter, or 18.5 billion cubic meters, which, at the time, was far more than it could use. So Egypt effectively got use of the bonanza. The 1959 agreement totally excluded the water claims of Ethiopia and the seven other upriver states—it was, in effect, a Muslim Arab solution involuntarily imposed upon the sub-Saharan Nile basin. Moreover, Egypt and Sudan agreed to move jointly against upstream nations that acted to challenge them. Ethiopia, which used but 1 percent of the Nile basin's water, vociferously rejected the treaty's validity. In 1956 and in 1957 Selassie had obtained public declarations of support for Ethiopia's Nile water rights from American president Eisenhower and vice president Richard Nixon, respectively. Yet in practice Ethiopia was powerless to prevent Egypt's water grab. In the late 1970s, these tensions flared into bellicose exchanges between Sadat and Selassie's communist successor, Mengitsu. Sadat nakedly threatened military reprisals if Ethiopia dared touch the waters of the Nile. The Arabic press in Egypt was soon aflame with anti-Ethiopian rhetoric, including menacing revisionist interpretations of the prophet Muhammad's well-known directive to Muslims to leave Christian Ethiopia alone because its Axumite king had granted refuge to his early followers when they were forced to flee Mecca in AD 615.

Ancient, proud, and never occupied or colonized, Ethiopia's civilization dated back to the days of the Pharaohs. It was to the Ethiopian Land of Punt in the Horn of Africa that Egyptian Queen Hatshepsut in the fifteenth century BC sent her famous Red Sea expedition that brought back myrrh and live frankincense trees. Ethiopian lore told that King Solomon and the Queen of Sheba's son brought the Ark of the Covenant to Axum in northern Ethiopia for safekeeping, where it purportedly remains under guard to this day. The Axumite Empire rose to prominence as an important link in the sea trade that Greek sailors opened around 100 BC between Egypt and India; at its peak its borders reached to southern Egypt and crossed the Red Sea into the Arabian Peninsula. The empire adopted Christianity in the same period of Roman Emperor Constantine's conversion and the Ethiopian Orthodox Church maintained close ties to the Christian Copts in Alexandria until the mid-twentieth century. Despite Muhammad's goodwill, Ethiopia began to decline after the seventh century as Islamic sailors took over more and more of the best trade routes to India and the Orient. From the mid-twelfth to early sixteenth centuries, Ethiopia underwent a golden period of expansion and revival, which put renewed emphasis on its linkages to Jerusalem, King Solomon, and its destiny as the legitimate successor to the Israelities. But by the late twentieth century, it was one of the world's poorest countries with life expectancies of only fifty-three years. The extremely difficult, hydrological conditions on its highland plateaus were one of its most formidable obstacles to economic development. Rains were seasonal and varied unpredictably, while the flash nature of the muddy Blue Nile, which roiled torrentially a hundred feet high in its gorges in wet season and trickled almost uselessly in dry season, made dam control, bridge building, and other waterworks highly complicated, and several times costlier than comparable projects in gentler, temperate, and invariably richer nations.

The confrontation between Ethiopia and Egypt abated after Sadat was assassinated in 1981 by Muslim fundamentalists. The new Egyptian president, Hosni Mubarak, pursued the more conciliatory approach long advocated by a senior Egyptian adviser, Boutros-Ghali, who favored joint, cooperative development of the Nile basin in order to derive its positive sum potential of increased storage capacity, reduced evaporation loss, untapped hydroelectricity, and, above all, more Nile water for irrigation.

Despite Egypt's more diplomatic tone, all the Nile basin plans and technological and financial assistance it offered remained inviolably predicated upon the other river states' acquiescence to the one-sided 1959 Nile Waters Agreement, and gave Egypt an overgenerous share of all new water supply. Political and environmental obstacles also impeded the Nile basin development that was undertaken. Work on the 224-mile-long Egypt-Sudan diversion canal to nearly double White Nile flow by re-rerouting southern Sudan's huge Sudd swamps was abruptly terminated in 1984 with only 70 percent excavated when it was attacked by black, southern civil war rebels who regarded it as a disenfranchising theft of a vital local natural resource and climate regulator for the benefit of Sudan's northern Muslim rulers and their Egyptian ally. In early 1990 Egypt blocked an African Development Bank loan for Ethiopia over concerns that it would consume too much water. Not surprisingly, Boutros-Ghali's Nile diplomacy yielded no major breakthroughs. The decade ended almost as it had begun. When Boutros-Ghali learned that Israeli hydrologists and engineers were doing feasibility studies on a number of dam sites in Ethiopia, with the potential to store as much as half the volume of water reaching Aswan, he summoned Ethiopia's ambassador to the foreign office in Cairo in November 1989 and sternly warned that any damming of the Blue Nile would be taken as an act of war by Egypt unless it had its consent.

Another cycle of water diplomacy began auspiciously in the early 1990s. Egypt and a new democratic government in Ethiopia led by Meles Zenawi agreed in principle that Ethiopia was entitled to an equitable share of Nile waters and to work cooperatively on Nile development. In 1999 Nile countries launched the World Bank–supported Nile Basin Initiative, a model in use in many international river basins around the world. Yet the real motivation behind the diplomacy was Egypt's own ambitious plans to water its desert to ease the explosive population pressures building along its narrow, fertile Nile corridor. In 1997 it had inaugurated the controversial twenty-year New Valley Project, a large water transfer scheme akin to the one that helped transform Southern California in the 1920s and 1930s that required the diversion of an additional 5 billion cubic meters from Lake Nasser through an ancient channel of the Nile—water Egypt did not have and which it needed the cooperation of upriver states to obtain. To entice Ethiopia's cooperation, Egypt offered support for Ethiopian hydropower dams, terracing of Ethiopia's highlands that improved water usage, augmented river flows, and reduced troublesome silt loads arriving at Aswan, and for some small-scale irrigation projects. But any water storage that significantly enlarged Ethiopia's capacity to expand its less than 1 percent of irrigated farmland was still not open to serious negotiation.

By 2005, with one in eight Ethiopians in need of international food relief, Ethiopian prime minister Meles Zenawi was angrily protesting Egypt's monopoly on large-scale Nile irrigation and threatening to unilaterally divert its waters for Ethiopia's benefit. "While Egypt is taking the Nile water to transform the Sahara Desert into something green, we in Ethiopia—who are the source of 85% of that water—are denied the possibility of using it to feed ourselves," he declared. "I think it is an open secret that the Egyptians have troops that are specialized in jungle warfare. Egypt is not known for its jungles . . . From time to time Egyptian presidents have threatened countries with military action if they move . . . If Egypt were to plan to stop Ethiopia from utilizing the Nile waters it would have to occupy Ethiopia and no country on Earth has done that in the past." Ominously, Meles added, "The current regime cannot be sustained. It's being sustained because of the diplomatic clout of Egypt. Now, there will come a time when the people of East Africa and Ethiopia will become too desperate to care about these diplomatic niceties. Then, they are going to act."

It seems unrealistic that Egypt can long maintain its historical hegemony over the waters of the Nile at the expense of widespread poverty, malnutrition, humanitarian crises, and oppressive, dysfunctional government among a fast-growing population of several hundred million Africans upriver. Two of the most horrific genocides in recent times, in Rwanda and Sudan, occurred in Nile basin states. Burundi, like Ethiopia, is one of the three poorest nations in the world, and the Horn of Africa is a war-torn wreckage of failed states and recurring famine. The Nile is these nations' greatest natural asset for development, too. Ethiopia, for instance, has unlocked a mere 3 percent of its hydropower potential. Often heralded as the future breadbasket of Africa, and encompassing over three-fifths of the Nile basin, Sudan is irrigating only 1 percent of its arable land. It defied history that, pressured by dire necessities, Egypt's neighbors wouldn't eventually find means among today's ample global resources to utilize more of the Nile's waters for themselves, with or without Egypt's assent. It is a recurring pattern that civilizations born downstream in the fertile deltas and flooding river valleys saw political power eventually gravitate upstream toward those in the best tactical position to control the river's flow.

In the age of water scarcity, moreover, Egypt's traditional strategy seems shortsightedly disconnected from the new politics of water. It remains rooted in the historical time warp that it could extract an ever greater usable supply of the Nile through continued political dominance on the one hand, and by implementing grandiose engineering schemes for irrigation and new cities on the other. At the same time, its aversion to politically difficult domestic reforms promoting more-efficient use of existing water perpetuates wasteful water practices at a time of growing scarcity. Nile water continues to be given away to irrigation farmers at heavily subsidized prices, amounting to some $5 to $10 billion per year, encouraging profligate flood irrigation techniques that ruinously waterlog valuable cropland. "Among the pervasive beliefs in Egyptian culture is that water, like air, is God-given and free," explains Nile scholar Robert Collins. "Any pricing system and controls on its use are totally unacceptable and almost blasphemous."

Inexorably, Egypt's day of water reckoning is drawing nearer. Its dependence on foreign grain imports—providing up to two-fifths of Egyptians' daily bread—to make up for its freshwater deficit is growing. Simultaneously, the deleterious, long-term environmental impacts of the Aswan Dam are impinging with escalating force. With the river's fertilizing silt being entrapped and building up at the dam, Egypt's farmland is suffering the depletions common to intensively irrigated cropland everywhere. Soil salinization and waterlogging are eroding farm productivity throughout the delta and the Nile Valley. Without the natural silt buffer brought by the predammed river, Mediterranean seawater has intruded as far as 30 miles inland. The fertile delta, home to over 30 million Egyptians and comprising two-thirds of Egypt's agricultural area, is shrinking. The precipitous decline in annual water volume reaching the Mediterranean Sea from 32 to only 2 billion cubic meters following the Aswan Dam's erection is depriving the coastal and marshland fisheries of replenishing nutrients and gradually destroying much of Egypt's once-thriving sardine and shrimp fisheries. The reliance on massive amounts of chemical fertilizers is also exacting a heavy toll in both the Aswan hydroelectricity consumed in its production and from the pollution of the Nile and delta lagoons. Thanks to fertilizer discharge, water hyacinth blooms choke irrigation canals, while infestations of the snail carrying schistosomiasis, the debilitating liver and intestinal disease, has been spreading.

In short, the full cost accounting is just coming due on Nasser and Egypt's fateful mid-twentieth-century decision to submerge forever the Nile's unique identity as world history's only self-sustaining major irrigation system behind the iconic, pyramid-like giant dam in the hot deserts at Aswan. At 75 million people and adding more than a million more each year, Egypt is dangerously outstripping the current productive limits of Aswan and the Nile. A second scissors blade, meanwhile, is closing on Egypt from the swelling demand for Nile water from the population boom going on throughout the Nile basin. In 2006 Egypt, Ethiopia, and Sudan had a combined population of 192 million; they are projected to add nearly 50 percent to reach 275 million by 2025. When all the Nile basin countries are counted, some half a billion people—overwhelmingly young, poor, and bred amid continuous violence—will be struggling to live off the waters of the Nile.

The prognosis darkened further when other related problems of global water scarcity are added to the picture: World food prices, which hit record highs in early 2008, are likely to climb higher in coming years from the 50 percent increase in world population, rising middle class demand for animal protein in China and India, and even possibly from tapering supply if America continues its early rush to corn ethanol biofuels as a gasoline substitute. At the bottom of the food chain are the water indigent, who spend most of the family budget on food and simply have no margin to absorb the higher cost of their daily bread. Climate change forecasts, if they come to pass, add to the potential for cataclysm. Models predict that Nile flows might decline up to 25 percent from altered precipitation and evaporation patterns, while rising sea levels could inundate large tracts of Egyptian delta farmland.

In early 2008, Egypt experienced a possible foretaste of its future when 11 people died in violence linked to lengthening bread lines caused by shortages of the government subsidized, traditional flat round loaves costing one cent (five piastres) arising from the combination of record grain prices and endemic official corruption. Mindful that bread riots could topple governments, President Mubarak called in the army to bake and distribute additional loaves.

In short, Egypt and its basin neighbors are sitting atop a growing demographic and hydrological time bomb, with Nile water scarcity as its detonating fuse. Like many nations facing world water famine, Egypt seems to have but one rational policy response: expand its existing water supply through aggressive efficiency improvements at home, cooperate with fellow basin nations to maximize the absolute volume that can be sustainably extracted from the river, and restructure its economy around the reality of its long-term reliance on the integrated global trading system to import its water-intensive necessities, such as food, until the day when an innovative breakthrough might bring its water resources and population levels into sustainable balance.

It is hard to overstate the immensity of Egypt's political and cultural challenge. Fully embracing such a cooperative approach to the Nile means surrendering its proud self-image as the downriver hegemon that has been ingrained in its national psyche over many millennia. As a thought experiment, it is the symbolic equivalent of Egypt agreeing to tear down the Aswan Dam, and to put its fate as a nation on the goodwill, political reliability, and growth of upriver neighbors with a long history of mutual suspicions and occasional wars, domestic instability, impoverishment, and humanitarian tragedies. This is not a vision that any leader of any nation is likely to embrace welcomingly.

Nevertheless, by the latter part of the first decade of the 2000s, the political economic landscape began to tilt against Egypt. Sudan is planning to erect new dams on the Nile with Chinese assistance. Both Ethiopia and Sudan have begun leasing prospective farmland to grow crops for dry, hungry, rich foreign nations such as Saudi Arabia. Other basin states are launching unilateral projects. A proliferating number of small, earthen dams 10 feet high being built by local Ethiopian and Sudanese farmers on tributaries of the Nile are siphoning a growing volume of Nile water—totaling 3 to 4 percent of the flow at Aswan by the mid-1990s—before it reaches the river's main stem. Ethiopia's diplomatic leverage increased with the discovery of a previously unknown aquifer stocked by Blue Nile runoff that could be pumped for irrigation and the sense that it could outwait an increasingly desperate Egypt for better concessions on the Nile. Against this background Egypt, Ethiopia, and Sudan, with other Nile basin countries, launched a more vigorous new series of meetings to try to reach a comprehensive accord on joint basin investment and development. Throughout the world, nations sharing international waterways have been forming cooperative basin initiatives in response to water scarcity. The Nile Basin Initiative was less advanced than many, and faced challenges graver than most, but offered large potential rewards for all. For the first time, the Egyptians are showing a genuine, pragmatic willingness to seriously entertain, with international monitoring and financing, significant dam storage volumes in the low evaporation rate highlands of Ethiopia. Indeed, through reduced flood loss from dam control, basin experts calculate that Ethiopia will actually be able to provide all the irrigation water envisioned in its own master plan, create much additional hydroelectricity, and still release greater amounts of water downstream to Sudan and Egypt than it does presently. By recouping much of the amounts lost to evaporation in the equatorial Sudd swamps through completion of the long-sidetracked canal, a Nile Basin Agreement, if realized, could yield an extra 10 billion cubic meters or more from the river. Throughout the basin, irrigated food production and hydroelectric generation could expand enormously, the Nile ecosystem become better managed, and, possibly, a cooperative regional sense of shared difficulty rather than competitive enmity could gradually begin to take hold. All the world has a vested interest in helping Egypt and its neighbors muddle through, since any political upheaval in Egypt is likely to transmit instability throughout the water-starved, volatile Middle East and North Africa.

Water had already helped spark an actual hot war in the smallest and driest of the ancient Fertile Crescent cradles of civilization, the Jordan River basin. In one of the world's political hot spots, Israelis, Palestinians, Jordanians, and Syrians contest to control and divide the scarce resources of a region that long ago ran out of enough freshwater for everyone. By 2000, people living in the heart of the basin were withdrawing 3.2 billion cubic meters of water, well in excess of the 2.5 billion cubic meters that recharged each year by natural rainfall. The tiny Jordan River, only 4 percent the size of the Nile, is routinely reduced to a mere trickle south of the freshwater Sea of Galilee, failing to replenish the ever saltier and shrinking Dead Sea into which it empties. Much of the shortfall has been met by overpumping groundwater from the region's main aquifer systems—three on Palestinian lands in the Israeli-occupied West Bank and one in coastal Israel. In all, the more than 12 million inhabitants of the Jordan basin have only one-third as much freshwater as needed for food self-sufficiency; regional stability, therefore, depends upon an uninterrupted flow of virtual water in the form of food imports.

At the time of Israel's creation in 1948, there was enough freshwater for all the Jordan basin's peoples. Shortages started in the 1950s with the doubling of water consumption as Israel's arid landscape was transformed into irrigated cropland by kibbutzim and individual farmers. To head off impending water conflicts, America's Eisenhower administration in the early 1950s sent a special ambassador, Eric Johnston, to try to negotiate a Tennessee Valley Authority–type water-sharing accord that would improve economic, social, and environmental conditions for all basin residents. Remarkably, Johnston forged an agreement among all the representative water professionals. But political polarization, and rising Arab nationalism on the eve of the 1956 Suez Crisis, intervened to doom the landmark water accord when Arab ministers meeting in October 1955 rejected it.

The shortages of the 1950s erupted into violent conflicts in the 1960s. At the start of the decade, Israeli foreign minister Golda Meir had put Israel's Arab neighbors on notice that Israel would regard any effort to divert the northern tributaries of the Jordan River as a vital attack upon Israel itself. When Israel in 1964 built a large pumping station near the Sea of Galilee and began conveying water into its new National Water Carrier network toward coastal Tel Aviv and farms in the southerly Negev Desert, a summit meeting of Arab leaders resolved to stop it. Work began on a mostly Saudi-financed dam in Syria. When Israel responded with its own diversion scheme, Syrians fired upon Israeli works crews. Yasser Arafat's Palestine Liberation Organization's Fatah guerrillas got their baptism with a failed New Year's Day 1965 attack on the water carrier. The cross-border fire culminated in an Israeli tank and aircraft assault that terminated Syria's diversion project and a decision by Arab states to dismantle a dam site on the Yarmuk River. Although a full-blown water war was averted, it added fuel to an escalating chain reaction of violence, which Israeli commander and later prime minister Ariel Sharon called the trigger to the June 1967 war: "In reality, the Six Day War started two and a half years earlier on the day Israel decided to act against the diversion of the Jordan," he wrote. "While the border disputes between Syria and ourselves were of great significance, the matter of water diversion was a stark issue of life and death."

The fateful 1967 war started with Nasser's expulsion in mid-May of the U.N.'s Suez Crisis era buffer force and Egypt's ensuing reblockade of Israel's only shipping route to the Red Sea and Indian Ocean. With united Arab military forces amassing on every Israeli frontier, and Arab popular opinion openly rejoicing at the prospect of Israel's anticipated destruction, Israeli leaders launched a secret first strike on the morning of June 5. While ground forces burst through the Egyptian border into Gaza and Sinai en route to the Suez Canal, Israel's entire small air force, leaving Israel undefended, deployed stealthily for the air bases housing the Arabs' much larger force of Egyptian-led warplanes. Within hours, following furious bombing and antiaircraft fire, the core of the Arab air force lay smoldering in ruins without ever having gotten off the ground. Over the next few days of fighting, Jordanian troops were driven out of the entire West Bank and surrendered the Old City of Jerusalem. Syrian tanks were beaten back up the slopes of the sparsely populated Golan Heights, which they soon abandoned in flight for Damascus.

The stunning results of the Six-Day War transformed the geopolitics of the Middle East. Israel's territorial extent suddenly quadrupled. Of equal significance, but less public notice, the hydrological balance of power between Israel and her neighbors also shifted decisively. Before the war, Israel had controlled less than 10 percent of the Jordan watershed. After the war, it was the basin's dominant water power. Coming under Israel's total control were the West Bank's underground aquifers, including the large, western aquifer that runs north-south along the foothills near the Green Line and flows west toward Israel and the Mediterranean Sea while recharging primarily on the occupied Palestinian territories. By the early 2000s, the West Bank aquifers were supplying one-third of Israel's freshwater. Of even greater strategic importance, the Golan Heights, annexed in 1981, gave Israel geographic possession of a vital catchment for its renewable Sea of Galilee, the source of another third of Israel's water. The Golan also gave greater security over more headwaters and sources of the Upper Jordan River, as well as frontage on the Yarmuk River that fed the Jordan south of Galilee, and was historically famous as the site of the 636 Muslim victory over the entrapped Byzantine army that opened the floodgates of the Levant and Egypt to Islam's early military juggernaut. The conviction that Israel is stealing Arab water that must be recovered has become another of the inflammatory passions and potential hair triggers in the dangerous Arab-Israeli tensions.

Israel exploited its sudden new water bounty to fuel a round of economic growth and modernization. In 1982 it integrated the West Bank's water supplies into its National Water Carrier network. At the same time, it used water as a national political tool by doling it out in disproportionately small drops to West Bank Palestinians through severe restrictions on drilling new or deepening existing wells. In one of the world's starkest Water Have–Have-Not divides, as a result, Palestinians typically had only one-quarter as much water as the Israeli settlers living alongside them. Irrigated Palestinian farmland on the West Bank shrank drastically—from one-quarter to one-twentieth of cropland—as a result. Palestinians were forced to bathe and wash less frequently than their settler neighbors, many of whom enjoyed lawns and swimming pools, while many Palestinians suffered the further burden of having to pay a stiff premium for tanker-transported water to meet their basic drinking, cooking, and hygiene needs; in the villages around Nablus, for instance, some families had to pay as much as 20 to 40 percent of their income for water. On the coast, Israeli dams on the Wadi Gaza diverted for Israeli farms the main recharging source of the shallow and badly overdrawn aquifer that was the sole natural supply of the Palestinian Gaza Strip. Depletion of the Gaza Aquifer grew so bad that that seawater and sewage easily infiltrated, leaving some 1.4 million Gaza Palestinians to drink contaminated water that was often literally sickening and an epidemic health hazard. Water famine and rage over "stolen" Arab water even exacerbated the 1987 intifada, which first erupted in Gaza before spreading across the West Bank.

Given the severe water scarcity and stark inequity in available supply between Israel and its Jordan basin neighbors, water was one of the contentious core issues in the regional peace negotiations that ensued from the famous handshake between PLO leader Yasser Arafat—himself a civil engineer with a professional understanding of water issues—and Israeli prime minister Yitzhak Rabin in September 1993 on the lawn of the White House. Along with territorial borders, settlements, refugees' right of return, and Jerusalem, water was one of the five central issues, and one of Israel's top priorities, in the Israeli-Palestinian Oslo peace process. The September 1995 interim accord affirmed the unequal four to one sharing of the mountain aquifers. Israel formally recognized Palestinians' right to West Bank groundwater, including a small increase to alleviate immediate scarcity and a promise to help Palestinians develop the eastern aquifer—which in fact, the Israelis had unsuccessfully sought to develop themselves for years. The goal of sharing West Bank groundwater on a more equitable basis, however, was put off to the final-status stage; it fell by the wayside when the peace process collapsed in 2000.

The Palestinian-Israeli peace talks provided political cover for the 1994 treaty between Jordan and Israel. The treaty included an Israeli concession to supply an additional 50 million cubic meters of water per year to help meet water-famished Jordan's minimum needs, as well as commitments to cooperate on joint water resource development and to confront the challenge of regional scarcity. The fact that water experts from both nations had been secretly meeting for years since the collapse of the 1950s Johnston Mission along the banks of the Jordan River at regular "Picnic Table Talks" to exchange information and sometimes coordinate water operations facilitated the swift achievement of the peace treaty. Helpful, too, was the pragmatic, unwritten understanding the two nations had forged in 1969–1970 whereby Israel agreed to desist from further destruction of Jordan's vital national aqueduct in exchange for the kingdom's curtailment of PLO raids on Israel across the Jordan River—an understanding that preceded Jordan's bloody expulsion of PLO guerrillas in the battles of Black September 1970. In contrast, progress in negotiating the return of the Golan Heights with Syria, with whom no similar back-channel measure of trust and pragmatic understanding had been built, stumbled in 2000 in large part over Syria's insistence on regaining access to the receded shoreline of Galilee, Israel's main renewable reservoir and a vital linchpin of its national water security.

A near water crisis flared soon afterward, in 2001–2002, when militant southern Lebanese Shiites allied with Syria took advantage of Israel's 2000 unilateral withdrawal from most of south Lebanon after eighteen years of occupation to immediately launch a pipeline to divert a modest amount of water from the Wazzani Springs on the Golan border. The Wazzani fed the Hasbani River, which in turn provided one-quarter of the supply for the Jordan River. With five times more supply per person than Israel, water-sufficient Lebanon could have chosen to divert surplus water from another source, such as the Litani River, that flowed entirely within its borders and would have required shorter transport to the villages the water was allegedly intended to serve. With overtones of events leading to the Six-Day War, Israeli prime minister Ariel Sharon warned that Israel regarded the Wazzani diversion as a deliberate provocation that was a potential cause for war; a timely international diplomatic flurry at the highest government levels of the United States, the U.N., and the European Union in the fall of 2002 averted a violent blowup. Renewed back-channel talks between Syria and Israel, brokered by non-Arab Muslim Turkey, came very close to a breakthrough over the Golan and the Jordan waters, but did not have sufficient American support and was overtaken by new regional Israeli-Palestinian violence in Gaza in the winter of 2008–2009.

Israel's response to its water scarcity challenge was unique among countries in the region. It went beyond simply struggling to secure dominance over as much of the region's natural surface and groundwater supply as possible. It also positively embraced policies that promoted more-productive use of existing supplies and the pioneering use of innovative water technologies. In response to the drought-induced water crisis of 1986, for instance, Israel cut agricultural water consumption by nearly one-third within six years by sharply trimming water subsidies to more fully reflect the total cost of sustainable provision and delivery; subsequent cuts moved Israel closer toward its ultimate national goal of reducing irrigation consumption by 60 percent. Under the stress of a 2008 water emergency, most Israeli farmers agreed in principle to pay full market price to the state water company. Israel redirected its agricultural water savings toward higher economic return uses in industry and high-tech sectors, vital urban water systems, and low-water-intensive, high-value crops. Typical of advanced economies, agriculture's share of Israel's overall economic output was only about 2 percent, even though it consumed three-fifths of the nation's water. The more efficient use of existing water enabled Israel to earn the revenue needed to import food and other virtual water goods it could not itself sustainably produce. Israel's economic restructuring represented a paragon alternative development path for Egypt and other water-famished nations of the Middle East.

Israeli farming's global reputation for efficiency was also burnished by its leadership in using many advanced water technologies. Notable among these was the growing trend toward treating, recycling and then reusing wastewater, for farming and purposes where lower water quality sufficed. Three-quarters of treated sewage from Tel Aviv and other cities, for instance, was pumped to farms in the Negev and other areas to grow crops in the early 2000s. Much of the recycled wastewater was used in high efficiency, drip irrigation systems pioneered by Israeli engineers in the 1960s. Drip irrigation delivered water directly to the plant roots through underground, perforated tubes; modern techniques combined computer monitoring of soil conditions to deliver precisely calculated amounts for optimum crop growing. Through drip techniques, crop yields per unit of water input often doubled and tripled. Only about half the water in traditional flood irrigation, by contrast, ever reaches the plant roots, while much is wasted to evaporation as well. Two-thirds of Israel's agriculture employed such microirrigation methods by the early 2000s; Israeli experts helped transplant the same techniques to neighboring Jordan, which applied it on over half its own farmland. Through the combination of drip irrigation and recycling treated wastewater, Israeli farmers quintupled their water productivity in the thirty years leading up to 2000.

Water efficiency is a necessary, but not a sufficient, response for water-famished regions like the Middle East. New water supplies are also needed. To supplement its water supply, Israel was increasingly turning to state-of-the-art, large-scale desalinization of seawater. Although long employed in some extremely water-scarce, coastal regions where there was no alternative, desalinization had been prohibitively expensive since it required vast amounts of energy to evaporate water or, using more modern, reverse-osmosis techniques, to filter out the salt through a very fine membrane under high pressure. Until recent advances, desal water often cost a hundred times more than natural water supplies. From the 1990s, desal costs at reverse-osmosis plants began to fall sharply—by as much as two-thirds. By the turn of the twenty-first century, water scarcity and falling desal costs induced Israel to launch five large, state-of-the art desalinization plants along its southern Mediterranean coast north of Gaza. The first, at Ashqelon, opened in 2005. It delivers high-quality freshwater at under two times the cost of water pumped up from the Sea of Galilee to Tel Aviv. A drought emergency in 2008 spurred a further commitment to desal—by 2020 Israel expects to produce 750 million cubic meters of desalinated water each year, or more than the amount drawn from West Bank aquifers. Whether economical or not when measured in purely conventional market terms, desalinized water contains a potentially priceless political dividend if it delivers water security and one of the keys to peace with the Palestinians.

To further bolster its long-term security, Israel also began tapping a new, very expensive, but important strategic water source by buying a small amount from the Middle East's rising water superpower, Turkey. In the early twenty-first century, non-Arab, Muslim Turkey looms as an increasingly pivotal regional power not only because it is the frontline between the West and the Islamic world, controls maritime access between the Mediterranean and Black Sea, and is a formidable military power, but also because of its growing leverage as the Middle East's water-wealthiest nation. Turkey's many mountain rivers provide its population with at least 10 times the per capita supply of Israel and triple that of Syria. Of particular strategic significance, its snowcapped, mostly Kurdish-inhabited, southeastern highlands, control the headwaters of both the mighty twin rivers of ancient Mesopotamia, and with it the freshwater lifelines of thirsty modern Syria and Iraq. Some 98 percent of Euphrates water originates in Turkey, before passing through Syria and on to Iraq and the Persian Gulf. Nearly half of Tigris water also comes from Turkey, with most of the rest originating in tributaries from rugged, remote regions of Iran.

For nearly all of history the prime beneficiary of the twin rivers' waters had been the dry-land, downstream region corresponding to modern Iraq, whose fertile swath of soils it brought to life with agricultural abundance. Yet 80 percent of Iraq's waters originate outside the country's boundaries. When Syria acquired the wherewithal to build giant, multipurpose dams and divert a large share of its water in the 1970s, some of the traditional advantage conferred by upstream position migrated northward along the Euphrates. A decade later, the balance of power shifted even more decisively upriver to Turkey when it similarly began to master its abundant natural water resources.

The linchpin of Turkey's ambitious water development program is the Southeast Anatolia Development Project, or GAP, a multidecade plan for 22 dams, 19 hydropower projects, and multiple irrigation schemes. The GAP is intended to transform the poor, politically restless region of 6 million people, more than double national irrigated cropland and electricity output, and turn Turkey from a food importer into a food exporter. Its centerpiece is the monumental rock-and-earth Ataturk Dam—Turkey's Aswan—which was completed in 1990. By itself, the Ataturk reservoir can hold five times the annual flow of the entire Euphrates.

"The twenty-first century will belong to Turkey," proclaimed Turkish leader Turgut Ozal from the ramparts of the 600-foot-high, mile-wide dam at its opening festivities. The prospect of restored Ottoman glory—and Turkish hegemony over the region's water—however, was as alarming to Turkey's downstream Arab neighbors as it was thrilling to the Turkmen. When completed, the GAP is expected to cut Syria's share of the Euphrates' water by as much as 40 percent and degrade its water quality, and leave only 10 percent of its historic flow for parched Iraq. Furthermore, the uncoordinated Euphrates dam and irrigation projects of Turkey, Syria, and Iraq, when aggregated, will consume half again as much water as exists in the entire river—a physically impossible scenario in which upstream Turkey would be the ultimate arbiter of who got how much and when.

In the early, euphoric anticipation of its water windfall, Turkish leaders in 1987 had grandly offered to sell some of its water surplus and deliver it through two 1,000-mile-long Peace Pipelines throughout the Middle East. One pipeline was envisioned to carry water south through Syria and the Jordan Valley—with a spur reaching Israel and Palestine—and on to the holy cities of Medina and Mecca in Saudi Arabia; the second was foreseen to run eastward through Iraq and Kuwait to the Persian Gulf. The price of freshwater transported through these immense pipelines was forecast at only one-third that of desalinized water. Turkey's vision was that its waters would be the tonic of regional cooperation and peace—with Turkey's hand on the strategic and diplomatic control valves.

The Peace Pipelines never got off the engineering drawing boards, however. They were done in by the dry realities of Middle Eastern water politics. Syria and Iraq reacted with bellicose protests when Turkey, in a thinly disguised flexing of its water power, began filling the Ataturk Dam for three weeks in January 1990 and the Euphrates slowed to a trickle. To show its displeasure, militarily inferior Syria broke a 1987 unofficial agreement and began providing greater clandestine support throughout most of the 1990s to the Kurdish separatist rebels of southeastern Turkey, including their most wanted terrorist leader, Abdullah Ocalan. Iraqi dictator Saddam Hussein, meanwhile, moved diplomatically closer to Syria and saber rattled about bombing the Ataturk Dam. Cutting off Euphrates water was evaluated by U.N. members as a countermeasure to try to force Saddam to withdraw his invading Iraqi troops in the run-up to the first Gulf War. Although the plan was not adopted, water became part of the Gulf War battlefield: Iraq's water supply and sanitation infrastructure was intentionally targeted and destroyed, while Iraq itself razed much of Kuwait's desalinization plant as it retreated.

The modern use of water as a tool of diplomacy and warfare on the twin rivers is as familiar as it had been in ancient Mesopotamia. In the mid-1970s, Iraq and Syria had amassed troops on each other's border and narrowly averted war over Syria's constriction of the Euphrates' water as it filled the reservoirs of one of its giant dams—which Saddam also threatened to bomb—while Syria more than once intentionally slowed the flow during the critical sowing season to show its displeasure with Iraqi criticism of its policies. From 1985 to 2000, with special vigor following the Shiite rising against his regime after the first Gulf War, Saddam launched a targeted assault on the fertile, fish-rich marshland ecosystem along the twin rivers' lower reaches north of Basra in historic Sumeria in the vicinity of the putative Garden of Eden that was home to a quarter-million, mostly Shiite, Marsh Arabs. Through massive drainage diversions—complemented by intentional pesticide contamination and high-voltage electrical shocks to the water to kill surviving fish life—the great marshlands shrank to one-tenth of its historical size. Its population fell in close proportion. Reflooding after Saddam's final fall in the second Gulf War was able to restore only 40 percent of the marshes.

The dispute over the waters of the twin rivers flared again in mid-1992 when Turkey's prime minister, Süleyman Demirel, a hydrological engineer, stridently rejected Syrian and Iraqi objections to Turkey's irrigation and hydroelectric projects with an implicit threat of withholding water as retaliation: "We do not say we share their oil resources. They cannot say they share our water resources. This is a matter of sovereignty. We have the right to do anything we like." Throughout the early 1990s, reduced Euphrates flow idled seven of 10 turbines at Syria's giant Tabqa Dam, contributing to Syria's electricity energy crisis. By 1998, troop exercises were being carried out on both sides of the border. War was averted by third party diplomacy and Syria's expulsion from Damascus of Kurdish radical leader Ocalan.

Although some diplomatic progress over water sharing was made in the early 2000s, Turkey pointedly remained one of the three nations in the world—along with China and upstream Nile riparian nation, Burundi—to vote against the 1997 U.N. Watercourses Convention declaring that international waterways should be shared reasonably and equitably, and not cause unnecessary harm to neighbors. With population growing rapidly in all three Euphrates basin countries; soil salinization, pollution, and food shortages worsening in Iraq and Syria; and greater claims being made on the twin rivers than it had total water, the day of reckoning in the Mesopotamian cradle of civilization and location of history's first recorded water war is drawing uncomfortably closer.

In the age of scarcity, Turkey is well positioned to be the pivotal power broker in the great political issues of the Middle East. Iraq's ability to rebuild from the second Gulf War depends much on the quantity and timing of freshwater that Turkey permits to flow downstream. Indeed to help alleviate the 2008 drought, prominent Shiite cleric Ayatollah Ali al-Sistani suggested that Iraq sell its oil at preferential prices to Turkey for more water. Turkey's prime minister, Recep Tayyip Erdogen, then visiting Iraq, demurred by observing that Turkey was already supplying Iraq "with more water than what we had promised, regardless of the high need in our own country."

The degree of Turkey's willingness to open the Euphrates's spigot at the Syrian border is also a significant lever in determining Syria's willingness to compromise on Golan water in its peace negotiations with Israel. Water trade strengthens Turkey and Israel's own military and diplomatic cooperation in an Arab-dominated region, and makes Muslim Turkey a credible intermediary in the conflict between Israel and its Arab neighbors—a role it has been assuming with increasing visibility. Yet Turkey's exercise of water diplomacy risks elevating the domestic leverage—and potential countervailing international support—of its own independent homeland-seeking Kurds, since the Turkish part of Kurdistan is where the Tigris and the Euphrates originates. The Kurds' brethen in neighboring northern Iraq on the Tigris are already using their control of Iraq's largest hydroelectric dam to press their demand for more territory and autonomy in the post-Saddam Iraq as American troops begin to pull out. From a still-larger perspective, Turkey's water supply represents a strategic resource that might, in time, offset some of the economic and political supremacy that oil delivers to the water indigent and ever more overpopulated Arab world.

While the risk of water war in this thirstiest and most politically combustible of regions is high, it is by no means inevitable. The existential threat posed by water scarcity is so palpable that it generates opposing cooperative instincts for mutual survival as well. At the worst moments of the second Palestinian intifada, while Israel's hegemony over West Bank water was being vehemently decried by angry stone throwers, Palestinian and Israeli officials continued to meet quietly and agreed not to damage each other's waterworks. As a religion of the desert, Islam accords water a special esteem that also favors cooperation. All inhabitants in the starkly arid land share an intuitive appreciation of the Turkish proverb: "When one man drinks while another can only watch, doomsday follows." In what might be considered a corollary of the mutually assured destruction doctrine that helped avert direct military conflict in the postwar nuclear age, it is possible that with rare statesmanship and sufficient desperation, a Middle Eastern water famine might lead inexorably not to devastating warfare but to a cooperative model of water détente that helps forge regional peace. It would be ironic, but not impossible, if salvation from the worsening regional water crisis came about through a resurrection of the faded dream of a marriage between state-of-the-art Israeli agricultural know-how and Arab oil investment.

• • •

Meanwhile yet another water shock with the potential to unsettle regional and international balances of power lurks in the torrid, sandy expanse of the Saudi Arabian desert. Geology played a cruel trick on the Saudi kingdom: While blessing it with the world's largest oil reserves, its natural endowment of freshwater is among the planet's most meager. Therefore, its future depends much on how effectively it converts its short-term abundance of oil into a sufficient, sustainable, long-term supply of freshwater. The Saudi landscape holds neither lakes nor rivers. For centuries, almost all its water came from underground—a shallow aquifer that was easily tapped by wells or at oases that recharged with rainfall and was able to support only a modest population at subsistence level. But much farther beneath the surface, about one-quarter of a mile deep, lay other, sizable, nonrenewable fossil aquifers—one-sixth as large America's Midwest Ogallala—containing pristine water from up to 30,000 years ago that had trickled down over bygone ages when the surface climate was wetter.

The oil boom and drilling technology of the 1970s made Arabia's fossil aquifers largely accessible for the first time. Instead of using its new treasure of ancient water prudently, Saudi monarchs went on a water binge as profligate as the industrial West's burning through its oil. They mined it from deep beneath the ground as fast as they could. A spare, arid culture became awash with decorative fountains, modern plumbing, and lush golf courses overnight. Fearing retaliatory Western grain export restrictions in response to their OPEC cartel's 1973 oil embargo, Saudi monarchs sought to achieve food independence by subsidizing the saturation of the desert with free groundwater until it bloomed with grain. In one of economic history's most extravagantly subsidized and hopelessly uneconomic enterprises, the Saudis not only attained self-sufficiency in desert wheat, but from the mid-1980s became one of the world's leading grain exporters. The production costs, however, were staggering—five times greater than the selling price of grain on international markets.

An even more punishing cost was the rapid depletion of the nonrenewable, precious aquifers themselves. Surveys in the 1980s estimated that Saudi Arabia had 400 million acre-feet of groundwater reserves–about thirty years' annual flow of the Colorado River. But by pumping out nearly the equivalent of one Colorado River per year—eight times greater than the volume of renewable water recharge—the Saudis had exhausted about 60 percent of the accessible aquifers by 2005. The pace of depletion was slowed, but not arrested, from the 1990s by the chopping of subsidies that slashed wheat production by 70 percent from the peak in 1992.

Another means of converting Saudi oil into water is to desalinate water from the peninsula's surrounding seas. Yet even with virtually costless oil energy to power desalinization plants on its eastern coast, Saudi production of renewable, desalinized water has been able to substitute for only a fraction of the natural fossil water used by its 25 million people. Early in the twenty-first century, fossil water is still providing half the domestic water for Saudi cities and over 70 percent of the huge volumes consumed by agriculture.

Belatedly, the Saudi monarchy formed a Ministry of Water and began to preach water efficiency. To symbolize its national goal of reducing domestic consumption by half, the de facto ruler, Crown Prince Abdullah, in 2004 replaced the palace's extravagant 10-quart toilets with six-quart models. But it was too little, too late. The reforms lacked strong price incentives or commanding enforcement. Saudi farmers went on planting water thirsty alfalfa, which consumes four times as much water as wheat, as feed for the cattle that satiated the nation's new appetite for dairy products. In the cities, little water is being recycled. Two-thirds of all dwellings are unconnected to sewage treatment plants so that raw sewage pollution is seeping into shallow groundwater, compounding the nation's scarcity crisis. In effect, Saudi Arabia is squandering its one-time opportunity to use its nature-endowed fossil reservoir to restructure its water use and management patterns.

Instead, like so many other water impoverished countries, its fate depends much upon the deliverance from an as yet unforeseeable breakthrough in desalinization or another momentous water technology. By as soon as 2025, the Arabian aquifers may be scraping bottom. In the short run, Saudi Arabia can mitigate the dire impacts by continuing to desalinate with profits from its natural supply of oil for its most indispensable uses and to spend its petrodollar surpluses on importing virtual water as food. Indeed, it was already using some of its oil wealth to buy or lease farmland in nearby Sunni Islamic nations like Sudan and Pakistan, as well as in multireligious Ethiopia just across the Red Sea, in a bid to secure a reliable, future foreign source of food. But reliance on imported food, especially from unstable states dependant on water from their own overstaxed river basins, is a highly uncertain proposition, and slowly consumes the Kingdom's treasury during what looks likely to be the beginning of the end of the golden age of oil. Eventually, its oil reserves will sputter out, too—then modern Saudis may painfully experience the knowledge of their desert forbearers that water, not oil, is man's one truly indispensable resource.

Two stark, alternative futures were playing out on the arid Arabian Peninsula's eastern and western flanks, where the oil had already given out. On the Persian Gulf to the east, Abu Dhabi had astutely invested its oil surpluses to build a low-water-intensive, international services entrepôt of banking and shipping closely integrated into, but reliant upon, the global economy that provided the food it could not grow itself. To the southwest, on the Red Sea, was highland Yemen, one of the world's most dangerous failing states, where plunging water tables were sparking rural violence, mass migration to overcrowded cities, radical Islamic fundamentalism, and international terrorism.

The desert Muslim leader who has been most audacious at turning oil into water is Libya's Muammar el-Qaddafi. Qaddafi bet a large proportion of his country's petrodollar wealth, and his government's own ruling legitimacy, on mankind's grandest underground water transfer scheme—the Great Manmade River. The project mines ancient fossil water from the huge aquifers lying beneath the remote southern Sahara desert, at a depth of an inverted Empire State Building, and delivers it through a 2,000-mile network of subway-sized tunnels buried six feet under the scorching, shifting sands to the Mediterranean coast, where 85 percent of Libya's 6 million people live. Waterless Libya first discovered that below its empty desert forty days' camel trek from the coast lay soft, water-bearing rock structures containing 50 billion acre-feet of water—the largest known fossil water deposit on Earth—when oil-seeking Western drillers began intensive explorations in the mid-twentieth century. Much of the water in the Nubian Sandstone Aquifer originated as rain 25,000 to 75,000 years ago; more filled during a second rainy period, about 4,500 to 10,000 years ago before the Sahara transformed from a wildlife-rich savanna of roaming herds pursued by tribes of hunter-gatherers into today's desert. Enthralled by his vision of a bounteous Libya, Colonel Qaddafi launched his subterranean Manmade River with support from Occidental Petroleum magnate Armand Hammer soon after he seized power in 1969. Despite adversarial political relations with the West over Libya's involvement with state-sponsored terrorism, other American companies, such as Halliburton subsidiary Brown & Root, helped Qaddafi's "new Nile" deliver its first water to the coast in 1991.

Yet the immensity, complexity, and expense of Libya's heroic water transfer project have raised doubts whether it can ever be completed or deliver its intended salvation for Libya. In 1999 pipeline blowouts began to occur, shooting geysers of water 100 feet into the air; repairs were hampered by the difficulty of moving heavy machinery across the desert terrain. Even if fully realized, the Manmade River will likely deliver only enough water—less than the Jordan River system—to meet less than half of the food needs of Libya's small but rapidly growing population. But even that amount has been enough to whet the envious thirst and warnings against Libyan overpumping from Egypt, Chad, and Sudan, under whose territory part of the Nubian Sandstone Aquifer also lies—and exacerbate worries in Egypt that the falling aquifer table might cause intolerable seepage loss of Nile basin groundwater. Like camels smelling water in the desert, Libya's neighbors are homing in on ways to wrest a share of the Sahara's buried water treasure for themselves.

CHAPTER SIXTEEN

From Have to Have-Not: Mounting Water Distress in Asia's Rising Giants

If water famine made the Middle East the world's most explosive Have-Not region, the global scale and depth of the unfolding water crisis is likely to be influenced heavily by what happened in Asia's dynamic, but water-distressed giants at the shifting fulcrum of the twenty-first century world order. Comprising nearly two-fifths of humanity, rising economic stars China and India had mobilized their renewable freshwater through giant dams and other modern waterworks to transform their once chronic hunger into sufficiency through the opportunities of the Green Revolution. Along with India's smaller neighbor on the Indus, Pakistan, they had contributed significantly to the rapid global expansion of irrigated cropland and the spectacular tripling of world grain harvests from 1950 to 2000. Yet worsening water distress from the pressures of demographic and economic growth, and the overexploitation and environmental abuse of their natural ecosystems is causing grain production to peak out in all of them. Unless trends reverse, they may soon lack access to enough clean freshwater to feed themselves. By dint of their sheer size, their domestic responses to their water challenges will powerfully impact economic, environmental, and political conditions worldwide. China and India, together with America, produce half the world's grain—their combined influence on international food markets is like that of OPEC on oil. The looming prospect of India and China becoming major grain importers therefore threatens to dramatically push up global food prices, crowd out the poorest and most water-famished nations, and help trigger humanitarian tragedies and political upheavals around the world.

China with 1.3 billion inhabitants, and India with 1.1 billion, are the world's two largest, and the two fastest economically growing, major countries. Their rapid modernization, and rising average individual consumption and waste generation, puts them at the epicenter of the era's great global tension between breakneck economic development and environmental sustainability. The outcome will help validate or discredit their contrasting and competing development models—India's multicultural democracy driven by private internal demand and China's politically authoritarian, state-guided mix of industrial export champions and domestic entrepreneurship—and strongly influence the nature of the political economic order of coming generations.

When it comes to water, the two giants share many characteristics. With 20 percent of the world's population but only about 7 percent of its freshwater, China's freshwater resources are stretched thin. The case is similar for India, with 17 percent of world inhabitants and 4 percent of its readily accessible freshwater. Both countries must use their freshwater efficiently to satisfy the demands of their large and increasingly protein-hungry populations. Yet paradoxically, as in most of the world, their water management is riddled with wasteful practices, inadequate infrastructure, inefficient allocations, and environmentally unsustainable uses. Both countries have been world-leading dam builders, although with mixed success to show for it. Both are increasingly reliant on overpumped, diminishing groundwater reserves for their irrigated agriculture and metropolitan water supplies. Both are also compounding their natural water limitations with man-made soil erosion and ruination as well as rampant pollution reminiscent of the preenvironmental and presanitary awakenings in the industrialized West. Neither has a coherent, workable, long-term solution for its impending water crisis, and both are hoping to buy time with grandiose, but environmentally dubious national water transfer schemes.

The World Bank warned in 2005 that India was at the cusp of

"an era of severe water scarcity" requiring immediate changes in government management of water. By 2050, water demand is expected to double to a level that exceeded the nation's entire available supply of freshwater. An ongoing series of environmental woes and flagging wheat and rice production are similarly awakening Chinese leaders to their incipient water scarcity crisis. In 1999, Vice Premier Wen Jiabao acknowledged that looming water supply shortages threatened the very "survival of the Chinese nation", and in 2005, as premier, he pledged to deliver "clean water for the people."

India's modern history was marked by a classic cycle of irrigation intensification that created a population boom that in turn demanded further irrigation merely to keep up with living standards and population growth. British irrigation projects during the colonial era had provided the initial impetus, with India's population rising steadily from 175 million in 1850 to nearly 300 million in 1900, then nearly quadrupling to 1.1 billion by 2005. The super crops of the Green Revolution and the postwar burst of large irrigation and hydropower dam building embraced after independence by Prime Minister Jawaharlal Nehru and his daughter Indira Gandhi in the Punjab and elsewhere initially generated an economic flourish that saved India from chronic, near famine and lifted individual living standards. Yet by the 1970s and 1980s, the boom began to fade under the weight of India's ballooning population, the dead hand of an ineffective government bureaucracy, and an overregulated private sector. Farm productivity began to sag as delivery of irrigation water by government bureaucracies became unreliable in timing and quantity, and the second generation of government dam building proved decreasingly effective. India became a global poster child advertising that simply having many dams and large-scale water management projects was insufficient to deliver growth—they had to be appropriately designed and administered effectively. Indians began to bemoan their "Hindu rate of growth."

The multicultural Indian democracy, however, responded with two unexpected developments. A grass roots protest sprung up in the late

1980s that effectively curtailed India's giant dam, state-dominated era. In September 1989, some 60,000 peasants, environmentalists, and human rights advocates rallied in the Narmada River valley in western India against a massive government irrigation project that included 30 major dams that would disrupt local communities while sending most of the economic benefits far away. The Narmada rebellion was fueled by decades of pent-up discontent with state-dictated giant dam and reservoir projects that had displaced an estimated 20 million people and through poor management ruined large portions of India's irrigated lands with salinized and waterlogged soils.

Soon after the protests, the World Bank withdrew its financial support from the Narmada River's centerpiece dam project. Disenchantment with large dams had been growing inside the Bank as well as in the Western-led international environmentalist and development communities. The Bank used the opportunity to support an independent World Dam Commission to assess the efficacy of its entire giant dam program. In late 2000 the commission issued a startlingly harsh condemnation of the development cost-benefits of the $75 billion the Bank had invested in large dam projects in 92 countries: most large dams ended up costing far more, profitably irrigated less cropland, produced less hydroelectric power, and delivered much less water to cities than originally advertised. The overestimated economic returns, moreover, also ignored the social inequities of displacing some 80 million rural people worldwide, the often disproportionate distribution of economic benefits, and the spreading of malaria among the rural poor. Their large impacts on river watersheds and aquatic ecosystems were often deleterious. Indeed, most dam benefits, the commission concluded, could have been achieved without many negative impacts by redirecting funds to decentralized, efficiency-encouraging, often smaller-scale alternatives, including revival of traditional, preindustrial era waterworks. The World Commission on Dams Report was a resounding turning point: The global era of unbridled giant dam building was over. World Bank programs since then demanded demonstrable, broader stakeholder benefits, while nongovernment activists from the industrialized world—where giant dam potentials had long been fully exploited—began to lobby reflexively against almost any giant dam in developing countries whether they had additional capacity for it or not, and for dam removals at home.

In India, the Narmada rebellion contributed to the 1991 economic reform movement led by Manmohan Singh, then finance minister and later prime minister, that liberalized the domestic private sector and transformed the advanced part of India's economy with startling rapidity. India became the world's second fastest growing major economy after China, reaching an average 9 percent per year from 2005 to 2007. World-class technology service companies arose and the size of its professional middle class quadrupled. Toward the end of the first decade of the 2000s, India was on the verge of surpassing Japan as the world's third largest economy. It was emerging as a regional superpower and swing state in the international balance of power and as a potential liberal democratic model for developing countries worldwide.

Yet something was awry. India's rapid economic transformation was precariously lopsided. In effect there were two Indias: a dynamic private upper tier of modern, high-tech service enterprises driven by a rising urban professional class alongside the three quarters of the population that remained locked in desperate poverty and held back by obsolete water infrastructure, weak governing institutions and unproductive practices, particularly in the rural farming countryside. Indeed, one of India's most striking aberrations is the abysmally poor productivity of its surface irrigation grain farming—Indian wheat farmers consume twice as much water per ton as their counterparts in America and China and lag behind those in Egypt, for instance. Untimely and unreliable delivery of government-managed water, poor transportation to get crops to market, shockingly small storage capacity despite the country's plethora of dams, and simple bureaucratic corruption and ineptitude are some of the contributing factors. India's poor, rural farmers are twice dispossessed: first by being located far from prime water sources occupied by the wealthy and powerful, and second by being last to be served due to the higher cost of transporting water to them. Prime Minister Singh himself declared that India needed a second Green Revolution "so that the specter of food shortages is banished from the horizon once again." To do that, it needs to reinvent its wobbly water economy.

For millennia, Indians' lives have been whipsawed year in, year out between the unpredictable cycles of bounty and destruction brought by the torrential monsoons, which concentrates 80 percent of the annual rainfall and runoff into only a few months. A late, small monsoon can mean tragic famine, while a heavy, early one can inundate the countryside with terrible floods and landslides that kill thousands and set armies of climate migrants marching across the landscape. A good and balanced monsoon, on the other hand, waters the crops, replenishes rivers and groundwater and delivers modest prosperity. No single event is more important for India's economy to the present day than the onset of the monsoon, news of which is urgently transmitted to government offices across the country far and wide.

To cope with the capricious tyranny of the monsoon, Indians over the centuries developed small- and medium-scaled local storage technologies that conserved water for the dry season, including simple water tanks and elaborate stone stepwells as deep as several stories that water fetchers descend step by step as the water level falls lower and lower throughout the dry season. In the British colonial era, traditional storage was superseded by mass industrial age water storage methods and partly forgotten. But industrial age water technologies were never effectively enough applied to be able to fully buffer India from the extreme unpredictability of the monsoons. Thus while Aswan gave Egypt a two-year storage buffer, and America's Colorado and Australia's Murray-Darling river system banked two and a half years' river flow against the uncertainties of their drought-prone environments, dam reservoirs on the Ganges, on whose waters half a billion Indians' lives depend, store no more than a couple of months' protection; the Narmada and Krishna rivers stored but four to six months' flow. As a result of India's failure to attain its long sought goal of delinking its annual economy from the monsoon, three-quarters of Indian society still suffers from the persistent poverty of unreliable water management and occasional buffeting from devastating water shocks.

India's failure to provide adequate storage, flood protection, and delivery of clean freshwater is driving the nation to perilously overdraw its national store of groundwater. From the 1980s, when government investment in irrigation virtually ceased, Indian private enterprise had on its own impressively continued to expand domestic water supply. Finally given access to reliable irrigation water, farmers' food production surged—cereals' output more than doubled between 1968 and 1998. But it did so through a ferocious, unregulated, and ultimately unsustainable pumping of groundwater from India's shallow, depleted aquifers and lavish water consumption, both exacerbated by government farm electricity subsidies that made pumping water virtually free. In 1975, before groundwater pumping became significant, India had about 800,000 wells. Most were shallow and dug by hand, with water lifted in the traditional manner by tethered oxen. Only a quarter century later, the nation had an estimated 22 million wells. The majority were small tube wells that had been drilled by power rigs, operating off of government electricity subsidies. They are continuing to increase phenomenally, by about 1 million each year.

Yet each new well has to be drilled ever deeper than the last because water tables throughout India are falling sharply. Through unregulated drilling and pumping, India's private water sector is effectively racing against itself toward each aquifer's bottom. The best financed, deepest wells get the water; the Have-Nots who can afford only shallow wells get the dregs, or go dry. In India's breadbaskets of Punjab and Haryana, for example, the water tables are falling over three feet per year; monitored wells in the western state of Gujarat show a fall in the water table from 50 feet to over 1,300 feet in thirty years. Southern India, an altogether separate geographic zone, is already effectively dry. As a nation, it amounts to a slow-motion act of hydrological suicide.

In the twenty-first century, India relies on groundwater mining for more than half its irrigation water. No other nation in the world pumps nearly as great a volume of groundwater. By some estimates, water is being mined twice as fast as natural recharge. Food produced from depleting groundwater is tantamount to an unsustainable food bubble—it will burst when the waters tap out. One warning occurred in 2006 when, for the first time in many years, India was forced to import large quantities of wheat for its grain stockpile. As the water tables hit bottom, clashes were breaking out between food producers and industrial and domestic users. In 2003 both Coca-Cola and Pepsi bottling plants in southern India were scapegoated and had their licenses revoked on unproven accusations that they were responsible for the region's exhausted groundwater reserves. Elsewhere, textile plants have been forced to shut down, and information technology companies have moved away from Bangalore, over water shortages and undependable supplies.

Geographically, India is blessed with enough arable land, sunshine, and total quantities of water, even for its huge population. Notwithstanding its current plight, it still has significant potential to expand its irrigable cropland, hydropower, storage, and food production. It is a prime example of a nation that with stronger institutions and better water resource management would be in a position to feed itself for years to come, even without a new Green Revolution. Since India accounts for such a large share of global irrigation, even a small productivity improvement would leverage a significant betterment of food conditions on world markets. Yet there is no easy political blueprint for rooting out India's embedded inefficiencies, ineffectual bureaucratic culture and distorted incentives, or for undertaking the many public waterworks that are needed. If the government suddenly eliminates farmers' electricity subsidies to slow the uncontrolled overpumping of groundwater, for example, it is likely that farm failures would soar and tens of millions of Indians face imminent famine.

Yet the subsidies have become so perverse that many farmers no longer bother to grow crops with the precious freshwater they pump up. Instead, they simply sell it to fleets of private water tankers carrying 3,000 gallons apiece that come a dozen times each day to transport it at a tidy profit to India's desperately thirsty and unsanitary cities. A large percentage of urban dwellers are not even connected to India's dilapidated municipal water delivery systems, which commonly lose 40 percent of their supply to leakage, waste, and piracy. For those who are connected, the taps often turn on for only a few hours each day. In New Delhi, with 15 million inhabitants, an Indian government report from 2006 found that one-quarter of households lacked access to piped water, and of those that had it, over one in four got water for less than three hours per day. Nearly 2 million households had no toilets. Upscale neighborhoods typically are the best connected with the best service, while the slums go wanting. Because the charge for municipal water is often less than one-tenth of actual provision costs, the water poor are effectively subsidizing the wealthy with good water connections. Most urban water is unmetered—in some places newly installed meters broke down because they required 24-hour water pressure to function properly—undermining efforts to substitute physical volume rationing for more efficient and equitable regulation by price and measured usage.

To get their daily water, Indian city dwellers—mainly the women—have to be resourceful foragers. To supplement what can be obtained from the public tap, water is purchased from private tankers and drawn from legal and illegal wells drilled in apartment blocks. For Ritu Praser, a middle-aged resident of a middle-class New Delhi neighborhood, every day involves planning to obtain enough water for her family: On the typical day when the municipal service doesn't deliver enough water, she dials the private water tanker on her cell phone, then waits for its arrival. In the meantime she draws what water she can from her apartment block's self-financed, courtyard tube well. The quality of the apartment tube well water is increasingly salty as the aquifer under New Delhi is running dry from overpumping. To get through the day she recycles what water she can, using leftover laundry water to mop the balcony, for instance. In an average month, she gets only 13 gallons from the municipal pipes and 243 gallons from water tankers.

Across India as a whole, three-fifths of households, or over 650 million people, do not have tap water as their primary drinking source. India's sanitation is even more abysmal. Some two-thirds of the population, 700 million people, lack indoor toilets of any kind. Less than 10 percent of urban sewage is treated. Not surprisingly, raw sewage chokes the nation's fabled rivers, from the Ganges to the Yamuna at Agra, with noxious filth and pollution. Such rivers are the drinking source for hundreds of millions of Indians, and the source of immeasurable sickliness and child mortality. An appalling two-thirds of India's total available surface and ground water supply is polluted with agricultural pesticide and fertilizer runoff, industrial discharges, and urban waste. It is a testament to the depth of the public bureaucracy's apathetic ineptitude on pollution cleanup and that a full quarter century after the infamous 1984 poison gas leak from a Union Carbide pesticide plant in Bhopal, the undisposed toxic storage is still seeping into the local groundwater to poison a second generation of residents.

The time bomb of India's mounting water crisis is all the more insidious because it is ticking down unheard and unseen by the public hundreds of feet underground in the falling water tables as well as high atop the fast-melting Himalayan glaciers that are the ultimate source of its great rivers. Due to global warming the Gangotri glacier, an important source of the sacred, but unpredictable and irregular Ganges with perilously little man-made reservoir storage, is shrinking by 120 feet per year and twice as fast as in the 1980s. Glaciers and snowpacks are nature's mountain reservoirs; they accumulate in the cold months and release their precious water to recharge river flows and groundwater tables when they melt in the warm season. By disrupting the glaciers' formation and melt cycles, global warming reduces available water volumes and exacerbates seasonal water mismatches—larger flooding in the wet season and worse drought in the dry months.

A similar, rapid shrinking is taking place on Himalayan glaciers across the entire Tibetan plateau, threatening the fate of Asia's greatest rivers—Indus, Brahmaputra, Ganges, Mekong, Salween, Irrawaddy, Yangtze, and the Yellow—along which live 1.5 billion people and whose waters provide the basic food and energy for most of Asia. The nations most affected hardly understood the dynamics of what was going on in the mountain ranges high above them and hadn't ever discussed the problem together until the World Bank sponsored a round of informal dialogues starting in 2006 in Abu Dhabi. The global climate change impact is predicted to hit especially early and severely on hot India—causing up to a one-third fall in agricultural output by 2080.

The impact of accelerated glacier melt looms perilously over the Indus River, which depends far more than the Ganges on melting snows to supply the waters that make the Punjab one of the world's most intensively irrigated regions of the world and an irreplaceable, stabilizing food lifeline of both nuclear-armed Hindu India and its crowded, and also nuclear-armed Muslim neighbor and rival, Pakistan. The 1,800-mile-long Indus rises in the Himalayan glaciers and gathers flow from tributaries in India and in hotly disputed Kashmir, which has been the object of two of the three wars between India and Pakistan. In Pakistan, the former mainstay of the ancient Indus Valley civilization flows through the Punjab's dense matrix of irrigation canals, and then southward into the Sindh, before fanning out through its delta and emptying into the Arabian Sea. The Indus is not a giant river by streamflow—it is roughly one and a half times the size of the Nile and plays a role of nearly parallel importance to Pakistan as the Nile does for Egypt. Not coincidentally, Pakistan's population at nearly 170 million is twice that of Egypt's, and both countries are suffering from similar water resource scarcity. Indeed, Pakistan's demographic burden has risen even faster than Egypt's, with its population having more than quintupled from 1947 to become the world's sixth largest. Pakistan is projected to have 225 million inhabitants by 2025—and possibly will be in the throes of a full-fledged water crisis.

Like the Nile, the Indus is badly overdrawn through trying to keep pace with rising Pakistani food demand. So much water is diverted upstream that the river carries no freshwater at all for its last 80 miles; its once-fertile, creek-filled delta of rice paddies, fisheries, and wildlife has become a desolate wasteland overrun by salty Arabian Sea water for want of outflow. Despite its growing scarcity, Pakistan's water is poorly managed—decades of underinvestment in basic maintenance, obsolete and leaky infrastructures, chronically dysfunctional and corrupt governance, and an inequitable, feudal-like farm economy have rendered its rich cropland among the world's least productive. The Indus basin's storage buffer is alarmingly precarious—a scant thirty days' capacity protects its crops against the caprice of a drought or flood. Inadequate new storage also contributes to hydropower shortages and regular blackouts that hinder industrial development and fuel urban discontent. Starting in the 1960s with liberal financial aid from the United States, which viewed it as a key strategic Cold War partner, Pakistan began sinking hundreds of thousands of tube wells to pump out fresh irrigation water and lower water tables that threatened to salinize crop roots. But today the private tube wells, which now provide half of all irrigation, are starting to run dry. The result is that Pakistan will likely soon be unable to feed itself.

Water scarcity is also one of the big political forces rending Pakistan apart internally. As water runs short, the southern Sindhis are bitterly complaining that the traditional political and military Punjabi establishment is robbing an unfair share of the scarce water resources for irrigation schemes in the Punjab. Water scarcity and contamination is so acute in the southern port city of Karachi that residents routinely boil their water; water rioting, and deaths, are unexceptional. The water divisions are reflected in the internal ethnic strife and national political party divides, with the aggrieved Sindhi south being the stronghold of the party represented by the late Benazir Bhutto, who was assassinated in December 2007. As Pakistan's available freshwater fails to keep up with the demands of its burgeoning population, it is not at all evident that the tumultuous state will not start to fissure and fail. Indeed, in April 2009 Taliban militants broke out of rugged, northwest Pakistan and overran the pivotal Buner district—putting them less than 25 miles away from gaining control over the giant Tarbela Dam on the Indus, and with it, a strategic chokehold over central Pakistan's electricity and irrigation waters.

As Pakistan's water-scarcity problems intensify, the Indus is again becoming a rising source of potential confrontation with India. The original international boundaries of the postwar divorce between Muslim Pakistan and Hindu India had sliced through the Indus river basin with no regard for its organic hydrological unity—a common problem in many shared international watersheds formerly under European colonial rule. Conflicts over the main tributary rivers of the basin broke out immediately. In 1948, the two states came to the brink of war when India's East Punjab halted the flow of water through two large canals that fed crops on Pakistan's side of the border in a bid to demonstrate its sovereignty over the river. Protracted diplomacy by the World Bank, which then held the key to vital international finance, finally produced the 1960 Indus Water Treaty. Yet the treaty was a minimalist compromise—each side got privileged water rights over three of the six Indus tributaries—rather than the sort of joint basin management agreement that could increase overall water resources through mutual, cooperative development. Although the treaty has held for half a century, including through subsequent Indo-Pakistani wars, a close call in 1999, and ongoing violence in Kashmir, the Indus remains a contentious fuse that can easily ignite again.

With India's population expected to swell by another third to 1.5 billion by 2050, Indian leaders know they need to act swiftly to avert an imminent water crisis. Two main policy paths beckon in the nation's race between economic growth and environmental sustainability: On the one side, India can strive to improve the efficient use of its existing water supply by undertaking the politically painstaking reform of the underdeveloped three-quarters of its economy through gradual elimination of distorting subsidies, rooting out bureaucratic corruption, and introducing new incentives to build small-scale, local storage and other water facilities whose benefits will accrete slowly. Given the political difficulty, and possible explosiveness, of such reforms, and the high uncertainty of success, it is not surprising that many Indian leaders are tempted by the second path of pursuing a grandiose water project that promises to deliver massive new water resources at once. A favored scheme is to build a nationwide plumbing network that interlinks all the nation's main rivers. At the single turn of a few valves by water technocrats all of India's seasonal and regional water disparities, all its extremes between flood and drought, and all the fierce court battles between federal states over water supplies, can be rationally evened out—at least in theory. Environmentalists have decried the idea as another facile, gargantuan, hard-technology folly of oversold benefits and underestimated deleterious ecosystem and human impacts. Yet even if it can be built and does work as advertised, at best it merely buys time for an uncertain, more lasting technological miracle, like Prime Minister Singh's wished-for new Green Revolution, to bail India out from not facing up to the underlying structural causes of its impending water crisis.

India's dream of linking its river systems is already becoming a reality for its booming Asian neighbor, China. In 2001, China launched the first phase of an epic, national river plumbing scheme to try to cope with its own daunting water scarcity challenge. With one-fifth of mankind inhabiting its borders, China faces perhaps the world's most titanic clash between rapid economic modernization and environmental sustainability. Among its many severe environmental degradations, China is running perilously short of available clean freshwater. By about 2030 its supply will run out in vital regions of the country, and its national shortfall will be equal to its total usage in 2008. Increasingly, top leaders are acknowledging that the water scarcity crisis is a primary threat to China's breakneck drive for first world living standards and its capacity to maintain social and political order by fulfilling the elevated expectations of 1.3 billion Chinese.

China's freshwater scarcity is more distressed than its overall population–water resource ratio suggests because much of its water is not easily accessible where or when it is most needed. Overall, China ranks a lowly 122nd among nations in per capita water use; the average Chinaman makes do with only about one-third the world average. Yet this masks China's hydrological mismatch between its wet south and its water-starved north, where Chinese citizens use a scant one-tenth of the world's average and face a steadily worsening water famine. A half century of intensive industrialization and urbanization, moreover, is also depleting the quality of available supply throughout the country with extreme water pollution. While China's economic consumption and waste levels are soaring toward first world levels, in short, its water ecosystem management and waste disposal infrastructure remain irredeemably third world.

After decades of increased agricultural output from the horrific low point of the 1959–1961 famines under state farm ownership in which 35 to 50 million died, China's grain production peaked out in the late

1990s and declined by 10 percent through 2005, forcing China to import large quantities of grain to rebuild its national reserve stocks. Without significantly more clean freshwater and improved land conditions, China is staring at the prospect of running critically short of grain and being compelled to use its growing wealth to outbid poorer, hungry nations for food exports grown in water surplus countries. Making matters worse is that Chinese water demand is soaring to feed the transition of newly prosperous Chinese from subsistence, low-water-consuming, vegetarian-based diets to high-water, meat-enriched ones. In the last quarter century, average individual meat consumption increased two and a half times—along with soaring water consumption to produce it. As Chinese prosperity continues to spread, so will the nation's soaring demand for water.

China's stupendous South-to-North Water Diversion Project to convey rivers of water across the breadth of China's vast landscape, is intended to alleviate northern China's immediate crisis. In effect, it is a modern iteration, writ bolder, of the historic Grand Canal: Yet whereas the Grand Canal of China's medieval golden age had bridged the nation's south-north hydrological divide and sustained water-distressed Beijing and the north with shipments of surplus southern rice—virtual water—the twenty-first-century challenge of meeting the needs of dense urban populations, giant factories, and intensively irrigated agriculture demands the direct delivery of indispensable freshwater itself.

Like the Grand Canal, the South-to-North Water Diversion Project is a large expression of China's traditional Confucian outlook to harness and conquer nature to serve the sovereign vision of the public good. No nation in the postwar era has been as relentless as China in launching immense waterworks projects. Indeed, Mao Zedong's communist government and its market-reformist successors were uniformly super-Confucian in their determination to mold and transform nature with the powerful industrial technologies of the age. The extensive buildings they wrought were reminiscent of the stunning construction bursts of previous Chinese dynastic restorations. In only half a century, they erected 85,000 dams, one-fourth of them giants—over four dams every day—providing irrigation, flood control, hydroelectricity, and seasonal storage capacity across a country that had little. Rivers were pinched by long levees; projects on the Yellow River alone used enough concrete to build 13 Great Walls. Overall water use quintupled; urban water supplies multiplied a hundredfold. Irrigation intensified and spread to many poor, rain-fed regions for the first time. Industrial use likewise expanded at big water-using steel mills, petrochemical plants, smelters, paper factories, and coal mines, and for cooling fossil-fuel-powered electric plants that soon dotted its riversides and lakes. If the human costs seemed high—Chinese officials themselves estimated that 23 million people had been dislocated in the dam-building frenzy, while critics placed the true number at 40 to 60 million—it was culturally consistent with China's forced-labor traditions and facilitated China's remarkable social feat of more than doubling its population while unleashing, especially since its 1978 market-oriented reforms, one of world history's most spectacular bursts of wealth creation and increased standards of living.

Chairman Mao, emulating the role of China's founder, Yu the Great, had instilled the spirit of China's new water age when, upon climbing up a small earthen dam at the Yellow River during first full inspection of the country in 1952, he wondered suggestively how China could better harness the power of the great river for economic development. Within three years the Mother River that gave birth to Chinese civilization was being grandiosely replumbed according to a plan that featured a staircase of dams and 46 hydroelectric power plants. A 60-foot-tall statute of Yu stands approvingly near the giant dam at Sanmenxia in the heart of ancient China, on which is inscribed the old Chinese adage: "When the Yellow River is at peace, China is at peace."

Yet right from the start the silty, unpredictable Yellow River showed it was not going to easily submit its sobriquet "China's Sorrow" to the commands of modern engineers and central planners. The signature project that was to glorify the Maoist "people's victory" over the Yellow was the Sanmenxia Dam at Three Gate gorge, the last gorge in the plateau of soft, loess soil before the river enters the northern plain of China's wheat and millet breadbasket. Yet as soon as its giant reservoir began filling in 1960, the tragic flaw in the dam's design became evident. Thick silt filled it to the brim in only two years, flooding tributary rivers upstream and threatening a catastrophic cascade downstream if the rising waters toppled the dam. Fearing the obliteration of populous cities, and the young communist state's legitimacy with it, Mao indicated his readiness to destroy the dam by aerial bombardment if no other way to solve the siltation problem was found. Perseverant reengineering and a decade of hard reconstruction ultimately saved the dam. But in the end it was only a shadow of its intended magnificence—a reservoir only 5 percent of its original planned size, with corresponding limitations on its capacity to provide electricity and irrigation.

By that time, a new and even more alarming environmental side effect of China's massive hydraulic engineering of the Yellow became visible. The great, fickle river known for its devastating floods and drastic changes of course began to dry up. The phenomenon was first observed in the summer of 1972, when startled staff at a water measuring station near the river's mouth saw a dry, cracking riverbed that no longer carried any water to the Bo Hai gulf. The average length of the dry area grew steadily from about 80 miles in the 1970s to a peak of about 440 miles in 1995. In 1997 the river failed to reach the sea for seven and a half months, much of its last trickles disappearing into the river sand bed near the inland ancient capital of Kaifeng.

The river's failure to reach the important coastal farming province of Shandong during the growing season caused much of the region's wheat crop to shrivel and die. The alarmed Beijing government decided that, henceforth, diversions from the river would be rationed so that some water always flowed to the sea. Like America's Colorado and the Egyptian Nile, the Yellow had become a totally managed river, with electronic maps, real time hydrological readings, and political measurement of every withdrawal. From 1999, the Yellow River never ran dry. But its fundamental problem has not been solved: There is simply not enough water to serve all the competing interests—farms, factories, cities, and natural ecosystems—that depend upon it. The river has effectively tapped out. In 2000, a mini water war erupted in Shandong when thousands of farmers, irate over their inadequate allocation of Yellow River water, illegally tapped reservoir water earmarked for cities.

One policeman died and hundreds of farmers were injured when the authorities moved in to cut off the illicit siphoning.

To compensate for the growing scarcity of Yellow River water, northern Chinese intensified extraction of the only readily available alternative—the large aquifers lying under the north China plain. One aquifer is near the surface and replenishes with rainfall and seasonal runoff; the other lies in a vault of rock and sediment deep beneath it, and is comprised of nonrenewable, ancient fossil water like that in the Sahara and Ogallala aquifers. As Yellow River basin water resources dwindled from overuse, thirsty Chinese on the northern plain began punching through the shallow aquifer and drilling increasingly deeper into the fossil aquifer for water. Extending from the mountains around Beijing in the north, to the Yellow's central loess plateaus, the flat north China plain produces half of China's wheat and a third of its corn and is as vital to the nation's food security as Iowa and Kansas farming is to the United States. Although the plain has little reliable rainfall, and is prone to harsh extremes of heat, cold, drought and winds, its fertile soil yields abundant crops when irrigated. Once rich in surface streams, swamps and springs, with fast replenishing, subsurface water deposits often found only eight feet underground, the north China plain ecosystem is drying out rapidly both from secular climate change and overuse by man. As in India, groundwater overpumping is causing the water tables across the plain to plunge. It is not uncommon for well pumps to have to go 200 feet to strike freshwater. Because contamination from urban, industrial, coal-mining, and farm waste has polluted three-quarters of the region's aquifers, metropolitan areas often have to drill three times deeper than that to obtain enough clean drinking water.

Around water-short Beijing, some wells reach half a mile deep into the fossil aquifer. The city's famous reservoir was declared unfit for drinking in 1997, while a large freshwater lake in the plains to the south shrank by two-fifths between the 1950s and 2000. Beijing is running so alarmingly short of water—the septupling of its population to 14 million in the postwar era has simply outstripped the capacity of its assiduously expanded water supply system—that officials have jocularly suggested the capital will eventually have to move to southern China where water is more plentiful.

In total, roughly half the accessible, nonrenewable water was pumped out of the north's huge aquifer in the second half of the twentieth century. Unless new supplies are found, or radical adjustments made, the bottom will be hit around 2035; some localities could run dry fifteen years before that. With four-fifths of China's wheat crop dependent on irrigation water, the nation's food bubble, and parallel bubbles in urban and industrial expansions that are being unsustainably inflated by overpumping, are in peril of popping.

China's impending water crisis could strike even sooner because north China's microenvironment is becoming gravely parched from other unintended side effects of Yellow River basin engineering. The loss of naturally restorative streamflow due to damming and irrigation diversions, extensive wetlands drainage, deforestation, and grasslands clearance for farming, proliferation of open pit coal mining from the 1990s, and the ravenous groundwater pumping, has combined to create one of the world's most acute crises of soil erosion—itself one of the greatest, though little publicized, water-related environmental challenges of the twenty-first century.

Half the lakes and a third of the grasslands surrounding the Yellow River's source in the Tibetan plateau have vanished. In the severely deforested middle reaches of the Yellow River, some 70 percent of fertile loess plateau soil has eroded away. Desertification is invading north China. Besieging desert sands have replaced the ancient barbarian hordes as the chief menace at the perimeter of China's Great Wall. In a single decade from the mid-1990s some 15 percent of all the region's potential new cropland was destroyed. Mongolian Genghis Khan's memorial tomb, originally emplaced in a beautiful plateau landscape of lake-filled grasslands, now stands nakedly alone amid barren sands. Great dust storms, like those that ravaged America's High Plains in the 1930s, increasingly now choke the skies of Beijing and kill scores of Chinese—China's leaders have been replanting a "green wall" of trees to try to shield the capital. The precious topsoil that China needs to grow the food crops for its next generation is being swept away in whirlwinds and sprinkled eastward over Korea, Japan, and even across the Pacific Ocean on to western Canada. Often the dust mixes with thick clouds of sooty, polluted air that drifts hundreds of miles to drop black blotches on car windshields in Beijing when it rains. The desiccation of northern China, in turn, intensifies regional droughts. The net effect is a significant reduction in total moisture throughout the Yellow River basin and the peaking out of the grain harvest from the late 1990s.

In their massive reengineering of the Yellow River, postwar China's master architects had not fully accounted for their own industrial age power to disturb the complex dynamics and restorative health needs of a total ecosystem. One of the more bizarre mutations of nature they created was that China's Mother River today flows through north China within hundreds of miles of flood dikes and embankments at an altitude of several yards above the surrounding landscape, like some kind of Roman aqueduct or elevated train trestle. Thanks to the ongoing accumulation of silt trapped within its dikes, the bed of the suspended river is rising by about three feet every decade and dikes have to be built higher and higher to keep it in its bed. A vigorous debate rages whether Chinese engineers have struck a Faustian bargain with nature to avert smaller, regular floods today in exchange for a potentially catastrophic, dike-smashing, super cascade caused by a sudden water-and-silt surge in the future.

China's second great river basin is the Yangtze in the rainy south, where the historical problem wasn't scarcity but water excess. It too has been massively reengineered by government central planners—and likewise is suffering serious degradations from the abuse of its ecosystem. As he had with the Yellow, Mao personally spurred the engineering boom when he inspected the river in 1953 and scolded water managers for the timidity of their plans to control the river's infamous floods. The building ultimately produced storage reservoirs with 13 times more capacity than on the Yellow and the world's largest—and most controversial—giant dam at Three Gorges. Building a dam at Three Gorges that could put an end once and for all to the terrible Yangtze floods had been a dream early in the century of China's modern founder, Sun Yat-sen. In his nationally celebrated 1956 poem "Swimming," Mao famously extolled his own vision for a dam that would "hold back Wushan Mountain's clouds and rain, till a smooth lake rises in the narrow gorges." Despite Mao's support, the Three Gorges Dam was much delayed, and in 1984 the project appeared to be shelved forever when a government review recommended against it. And but for China's massacre of prodemocracy Chinese protesters at Tiananmen Square in June 1989, it might have stayed there.

China's stunning transformation into the world's fastest-growing large power had been launched with the successful 1978 reforms of post-Mao leader Deng Xiaoping to enhance administrative efficiency and accelerate economic growth through controlled injection of market forces and some decentralization of political decision making. Many Westerners had hoped that Deng's liberalizing reforms would lead China toward a liberal, Western-style democracy—despite Deng's own disavowal of any such intentions. These hopes were violently crushed at Tiananmen Square. China's hard-line leaders took umbrage at the world's smoldering condemnations. To show their nationalistic defiance and unyielding commitment to China's authoritarian, state-managed market system, they soon resuscitated the Three Gorges project—and imprisoned the dam's domestic critics. Three Gorges was to stand not simply as a dam, but as a crowning showcase of the prowess, wealth, rising world-class status, and unalterable independence of the new China. They hailed it as their civilization's greatest engineering project since the Great Wall.

There had always been little doubt that Three Gorges would transform the Yangtze as thoroughly as the Aswan Dam reshaped the Nile and Egyptian society. Deng's 1978 reforms had done nothing to alter China's traditional attitude toward water management, and when the dam officially opened in 2006 its extraordinary effort to command nature was on full display: it was an impressive 600 feet high and a mile and a half across, with multitiered ship locks and a nearly 400-mile-long reservoir. If 1.4 million people had been involuntarily relocated in its building, its purported greater good to China was that it promised to control floods on the Yangtze, enhance navigation, generate more electricity than any other dam in the world, and serve as the linchpin for a dozen more hydropower megabases upriver that would begin to fulfill China's master plan to triple the nation's hydropower by 2020 and wean it away from its extreme dependency on nonrenewable, dirty coal.

Given the dam's legacy, activists in China's sprouting, broadly based environmental movement were astonished a year later, in September 2007, to hear the senior government official responsible for the dam break a long-standing taboo and not only confess, but publicly warn that Three Gorges posed "hidden dangers" that could cause a "huge disaster . . . if steps are not taken promptly." He added that China cannot win "economic prosperity at the cost of the environment." Speculation that he may have spoken out of turn was erased the next day when the government news agency itself covered the event with the headline "China Warns of Environmental 'Catastrophe' from Three Gorges Dam."

Among the litany of worries voiced by dam critics had been severe water pollution, landslides, riverbank collapses, larger earthquakes in a fragile, fault-prone region, flooding and shipping problems upstream, and crippled hydropower potential from heavy silt buildup in the reservoir. Indeed, the warning signs that things were amiss at the dam had been accruing as the reservoir started to fill. Rising water pressure and seepage had caused scores of landslides upstream and on tributaries, killing dozens of farmers and fishermen in mudslides and the 165-foot-high waves heaved up by the crashing mud. Upstream water quality also had deteriorated because the dam impeded the dispersal of industrial pollutants and urban sewage, contaminating the drinking water of tens of thousands and threatening to turn the dam's reservoir into a giant cesspool. Freshwater shortages turned up in Shanghai at the river's mouth because the decreased flow in the dammed river was no longer able to offset the force of the tidal inflows from the East China Sea; the metropolis's tap water became foul-smelling and yellowish with Yangtze pollution. Two weeks after the government warning about Three Gorges, it announced that an additional 3 to 4 million people would have to be relocated due to the pollution and landside threats. A few months later, dozens of ships became stranded in a stretch of Yangtze waterway as the river recorded its lowest level in a century and a half.

Even before the opening of Three Gorges Dam, China's engineers of the Yangtze had witnessed distressing side effects of their handiwork. Despite the river's reduced streamflow, terrible floods in 1998 had killed thousands. Deforestation, soil erosion and greater siltation upriver, and the draining of water-absorbing wetlands downriver had combined to create a new type of flood risk on the river. Their dreaded nightmare was a major earthquake in the active fault zone around Three Gorges—possibly made catastrophic by the sheer pressure from the water's weight in its own reservoir. The tragic, 7.9 magnitude quake of May 2008 in Sichuan province near Dujiangyan, site of Li Bing's famous third century BC Min River diversion and irrigation works, that killed 80,000, extensively damaged 400 dams and compelled the draining of the giant, 50-story-tall Zipingpu dam reservoir, only 3.5 miles from the quake's epicenter, might have been a catastrophe beyond imagining had it struck instead 350 miles east at Three Gorges. Indeed, many scientists contended that the anomalously extreme size of the 2008 quake itself may have been caused by geological pressure from the 320 million tons of water in the Zipingpu reservoir—a charge strenuously denied by the government, which also blocked websites suggesting that ongoing giant reservoir-building in the region might be putting inhabitants in jeopardy.

The government's public warning about Three Gorges reflected a deepening concern among China's post-Tiananmen leaders of the severity of the environmental danger imperiling China's future—and their own credibility to govern as public anger boiled with each deadly ecological disaster. Just a few months earlier, in June 2007, some 10,000 middle-class environmental protesters had taken to the streets against the construction of a new chemical plant in the coastal city of Xiamen. This followed the angry national headlines in May that the nation's third largest lake and famous national beauty spot, Lake Tai, on the lower Yangtze delta near a branch of the Grand Canal, had suddenly erupted with fetid, fluorescent green toxic cynobacteria—pond scum—depriving more than 2 million local residents of potable drinking and cooking water.

The pollution outbreak at Lake Tai had been building for decades as irrigation and flood works reduced the lake's circulation of cleansing, oxygenating freshwater. From the 1980s, some 2,800 chemical plants also proliferated along the transport canals around the lake, which provided both the large volumes of water they needed for processing and discharge and for shipping the end products to the industrial port of Shanghai downstream. Local officials had encouraged the chemical plants to locate around the lake because their taxes provided four-fifths of local government revenue. Although reports of the extensive pollution they were causing had reached China's top leaders as early as 2001, local political resistance and chemical company cover-ups had kept the national inspectors at bay. A lone, dogged private environmental whistle-blower lost his job and in 2006, after further agitation, was arrested on dubious charges. He was still in prison—and became an instant national hero—when the toxic combination of chemical waste, untreated sewage, fertilizer runoff, and lack of rainfall finally exploded in the lake with oxygen-choking cynobacteria bloom. Within six months the central government enacted antipollution measures and promised to restore China's major lakes to their original pristine states by 2030.

By the early twenty-first century, pollution has reached epidemic proportions throughout China, and is seriously exacerbating the nation's natural water shortages. Over half the freshwater in the nation's major river systems and lakes, and a third of its groundwater, is unfit for human consumption. Two in three major cities suffer serious water shortages. Only one-fifth of wastewater is treated compared to about four-fifths in first world nations. Electricity generation at power plants is sometimes curtailed for want of adequate river volumes, which likewise forces temporary factory production halts at big water users like petrochemical plants, smelters, and paper mills. To keep up with demand, reliance on groundwater has doubled since 1970 to constitute one-fifth of the national supply. By the government's own admission, one-third of its land is severely degraded due to water loss, soil erosion, salinization, and desertification. In 2007, the World Bank concluded that some 750,000 Chinese were dying prematurely each year from the nation's water and air pollution, but acceded to Chinese official requests to excise that finding from the final report for fear that it might stir domestic unrest.

Upon ascending to power after 2000, in fact, China's post-Tiananmen leadership has tried to nudge the country's economic system toward a more environmentally sustainable path. One public initiative, launched in 2004 by President Hu Jintao, attempted to modify China's obsessive growth culture with a new Green GDP calculation that imputed the negative growth costs of environmental degradation in each province. Hu's Green GDP report, however, was issued just once. It met strong political resistance from provincial leaders, who had been empowered under the 1978 reforms and who resented both the conclusions that much of their province's celebrated economic achievement was being canceled by environmental damage and being called to account by central party leaders. Green GDP calculations continued to be made by others, however. The World Bank found that nearly 6 percent—over half—of China's national GDP growth should be canceled from air and water pollution damages to sustainable ecosystems and human health. The deputy minister of China's own, weak state environmental protection agency went further. He estimated the annual cost of environmental loss at 8 to 13 percent of GDP—negating *all* China's vaunted economic growth.

China's environmental challenge is reminiscent of the unsanitary, overcrowded conditions that hallmarked Britain's cities in the early Industrial Revolution—writ immensely larger and intensified by modern-scale technologies and much more concentrated, rapid development. Somewhere between 2025 and 2035, the clean freshwater may run out in its water-famished north and its promise to clean up its polluted lakes and rivers will fall due to a public that is increasingly restive about environmental hazards. China's conundrum is that it can ill afford to disappoint the soaring material expectations of its 1.5 billion citizens with remedies that might significantly hamper its dazzling economic growth; but its long-term growth may become unsustainable—and possibly suffer an abrupt, destabilizing environmental shock—if it doesn't move fast enough to reverse the systematic overexploitation of its freshwater resources. Its governing dictum thus far remains frozen: Growth first, clean up later.

As the failure of the Green GDP initiative illustrated, changing an entrenched political economic culture is difficult, even for authoritarian China. In the absence of a clear and present emergency, Chinese leaders are mainly sticking to traditional, Confucian approaches that have prevailed since the Han era. Although incremental new pollution regulations and reforestation programs have been issued, only modest steps have been made to encourage more efficient use of existing water supply through pricing that more fully reflects its total cost. As a result, even in the face of widespread scarcity, the price of water in cities, industries and agriculture continues to be closely politically controlled and heavily subsidized. Chinese farmers consequently are still irrigating water-guzzling crops in dry regions in competition with cities and factories that treat and recycle far less water than is common in the West.

Chinese industry generally uses three to 10 times more water than its counterparts in the West—a significant, long-term competitive disadvantage in the global marketplace when the subsidies or the water itself give out. Clean freshwater shortages also impose a ceiling on China's future competitiveness in water-intensive, high-tech industries such as biotechnology, semiconductors, and pharmaceuticals. Other hidden competitive disadvantages stem from the interdependencies between water, energy, and food: China's heavy reliance on coal-based ammonia production for its fertilizer and textiles manufacturing, for instance, consumes 42 times more water than the West's cleaner natural-gas-based ammonia production methods. Inefficient flood irrigation and heavy artificial fertilizer use, moreover, continue to pauperize soils, and add to the pollution loads diminishing China's long-term ability to feed itself.

Notwithstanding their warning of possible environmental catastrophe at Three Gorges, China's leaders show no sign of wavering from their tenacious determination to harness and conquer nature through many more giant infrastructure-based water schemes in active fault zones. In addition to the country's master growth plan of a dozen more hydropower bases on the upper Yangtze, China is launching immense dams on the upper basins of the Mekong and the Brahmaputra that have the potential to alter the flow patterns and water quality of those great rivers before they exit China to bring life to the Asian nations founded around their middle and lower reaches.

Not just China's own Yellow and Yangtze, but most of the great rivers of Asia originate in the Tibetan Plateau. Indeed, China's aggressive international stance toward its domination of Tibet is as much about pragmatic control of its own and Asia's regional water and hydropower resources as it is about nationalist politics. Given the vital importance of those rivers to societies downstream, it is troubling that China stands uncooperatively apart—as it generally does on any international agreement that might possibly constrain its freedom to pursue its overriding national growth goals—from all but two other countries in the world in voting against the 1997 U.N. Watercourses Convention recognizing the need to fairly share international waterways in ways that don't significantly harm other river states and are equitably shared among them.

Yet all China's grandest hydraulic plans are dwarfed in scale and ambition by its heroic South-to-North Water Diversion Project. The inspirational spirit, once again, had been Mao Zedong, who during his 1952 inspection of China's water resources, noted: "Southern China has too much water and the north has too little. We should try to borrow some from the south to help the north." With water famine looming in the north and desirous of making the 2008 Beijing Olympics an international showcase for the new China, Chinese leaders in 2001 launched the transnational civil engineering water transfer scheme of uncertain technical feasibility and environmental side effects to redirect rivers of water—two and a half to three times the volume of the Colorado River or 25 times more than Libya's subterranean Manmade River—northward from the Yangtze basin. Three separate channels, totaling 2,200 miles in length, were designed to carry the water across mountains, canyons, waterways, railways, and other arduous natural and man-made landscapes, to deliver parched north China from its dire thirst.

Work on the eastern and central routes began on an accelerated timetable to be ready to deliver water for the Olympics; the more complex, western project is programmed to start after 2010. The eastern route diverts water from the mouth of the Yangtze and, with the help of 13 pumping stations, lifts it to channels that carry it along the coast through the north China plain and on to the cities Tianjin and Beijing. Large sections run through still-functioning portions of the Grand Canal. The central route is to enlarge the huge artificial reservoir on the Han River, a main tributary of the Yangtze, and build a new 200-foot-wide water canal and aqueduct the length of France across the heavily populated north China plain toward Beijing and Tianjin. Supplemented by water from a Yangtze aqueduct near the Three Gorges Dam, it is designed to travel by tunnel under the Yellow River, cross 500 roads and 120 railway lines, and displace a quarter million people in its building. The goal is to relieve pressure on the Yellow by supplying water to the thirsty regions around it. The final western route is envisioned to reroute water from the Yangtze headwaters in the glacial, Tibetan plateau directly into the Yellow River—the only one of the three routes to replenish the Yellow River directly.

The doubts of environmentalists and water engineers about world history's largest water transfer project were brushed aside in the urgency to deliver water to the north. The main worry about the eastern route has been that it might simply spread the extreme pollution from the Grand Canal and contaminate regional river basins. Following the great flood of 1855, large stretches of the Grand Canal had been left in a state of disrepair and left dry. The hundreds of miles still in use, which transported over 100,000 cargo ships annually, had become a filthy, malodorous, lifeless, black-colored cesspool of factory effluent and urban sewage; even touching its waters was treacherous to one's health. The diversion project requires building hundreds of new sewage treatment plants, closing the endless rows of dirty factories along its banks, massive dredging, and other major cleanups.

Along the central route, a chief concern has been that taking too much water from the Han tributary would upset the region's ecological balance and worsen pollution farther downstream. Pumping water across the mountains from the Three Gorges Dam would, in turn, lower that dam's hydroelectric output by at least 6 percent. The western route would be by far the most technically challenging, having to cut through mountains and gorges through tunnels up to 65 miles long in an earthquake-prone zone.

The South-to-North Water Diversion scheme is China's most ambitious hydraulic undertaking to harness and conquer nature since the Grand Canal itself. Like the Grand Canal, it represents a potential new landmark chapter in water and world history—the opening of a new era of nationwide plumbing networks that consolidate all accessible surface and ground water into a single supply that, if successful, likely will be imitated by other water-distressed nations. Such long-distance, gigantic-scale water-moving approaches are increasingly disfavored in the United States and other industrialized, liberal democracies as environmental anathemas whose benefits can be provided with fewer negative side effects in more ecosystem-sustainable ways. Critics often liken it to the Soviet Union's disastrous alteration of the Aral Sea and regional climate from its replumbing of central Asia.

On one level, the international debate over water management is a recasting of the ancient Chinese Taoist-Confucian philosophical argument over the degree to which man should bend to the natural order to live in modest harmony with it or strive to command and harness it artificially to his will. In modern parlance, the debate is framed in terms of soft-path versus hard-path solutions. Tao-like soft-path advocates, who have been gaining influence internationally with the spread of the environmentalist movement, emphasize improved efficiencies from existing water supplies and "right-scaled" solutions tailored to users' needs that give preference to smaller, more decentralized technologies and administration, and operate in closer harmony with the flow of nature to achieve systemic environmental balances. Hard-path proponents, who have dominated engineering thinking for most of human history and reached their zenith of achievement in the twentieth-century age of dams, continue to favor technologies and centralized infrastructures that strive to remold Nature's ecosystems and water resources on a grand scale. In the twenty-first century, China is the unapologetic, leading state representative of the hard path. Yet on a simpler, everyday level, the Chinese predilection for outsized waterworks projects simply reflects the nation's desperate thirst with few viable short-term alternatives.

If successful, the South-to-North Water Diversion Project could vault China beyond the immediate peril posed by its water-scarcity crisis. But it is unlikely to solve China's longer-term crisis for an obvious reason—the Yangtze itself is already becoming overtaxed and doesn't have enough surplus to send north to keep pace for long with China's rapid modernization. It postpones, but does not constitute a direct response to, the fundamental problem of China's freshwater security and ecosystem depletion.

Meanwhile, global warming lurks as a potential environmental Hiroshima over China's total water supply. Glaciers in the Tibetan plateau that are the source of its major rivers are rapidly melting, as they are across the Himalayas. All of China's giant dam and water transfer schemes could be transformed overnight into an epic boondoggle if disappearing glaciers render them all wrong sized for the new, more extreme seasonal climate patterns. China is not alone in facing the global warming threat, but the sheer scale of its gamble on the success of its large dams and water transfer schemes means that it has more at stake than everyone else. No one knows for certain how much time China has before it feels the full brunt of its climate change reckoning. But the Intergovernmental Panel on Climate Change estimated that over time, global warming is likely to result in the melting away of enough of the glaciers to cause the nation's freshwater supply to fall as much as a third below its farming needs—even while China struggles to cope with the climaxing of its other water environmental crises.

China's future hinges heavily upon how it meets its water and environmental crises. The twentieth-century restoration of its illustrious, old civilization and its rising superpower status both depend upon it. One possible political outgrowth is that the nation's grassroots environmentalist movement may emerge as an enduring domestic force that nudges the government in a more liberal democratic, responsive direction. On the other hand, it is equally possible that the environmentalist pressures might provoke an authoritarian backlash that collides violently with the inexorable, immediate need to provide vast new supplies of water and other resources to meet the material expectations of a billion and a half people. Whether China succeeds or fails in meeting its water challenges, the outcome will be felt internationally and leave indelible impressions on the history of the twenty-first century.

While water scarcity and ecosystem depletion is a vulnerability for fast-growing, water-stressed giants China and India, it is simultaneously delivering to the relatively water-wealthy, liberal industrial democracies of the West a renewed strategic opportunity to revive their own waning leadership status in the changing world order. In an age of scarcity in which freshwater is becoming the new oil, the industrial democracies enjoy an enormous comparative resource advantage that they have yet to fully recognize or exploit.

CHAPTER SEVENTEEN

Opportunity from Scarcity: The New Politics of Water in the Industrial Democracies

The headline scarcity crises among the world's demographically stressed, water poor overshadows one tantalizing, emerging trend in the relatively water-wealthy, industrial democracies—an unprecedented, sharp productivity gain in the use of existing freshwater supplies. This new development is being driven by the growing engagement of market forces as fresh, clean water resources run short and pollution regulations firm up. It offers an alternative, beacon path for alleviating the water crisis—and a pathway for the Western-led market democracies to relaunch their global leadership. Generations of water resource underpricing and inefficient political management have led to colossal waste in every society's use of water—and therefore created correspondingly huge opportunities to increase effective total water supply by using those current resources more productively. For example, each North American uses two and a half times more renewable freshwater than the world average—and could unleash a proportionately prodigious new supply to productive uses simply by adopting readily available, high-efficiency practices and technologies. Tapping such already available supplies, moreover, comes at a lower environmental cost than any incremental new supply that can be extracted from nature or reallocated among river basins.

The democracies born in ancient Greece and Western traditions enjoy greater leeway to pursue improved efficiency solutions to their water shortages because, in the main, they have more favorable water profiles, competent governing mechanisms, and many fewer demographic burdens on their resources than the world's water Have-Nots. Most have renewable water supplies that are ample, available year-round on a predictable basis, and fairly easily accessible. Rain-fed agriculture is widespread and provides a reliable natural food base. While America's groundwater use is large and growing in some major regions, the nation is not excessively dependent upon it for irrigation to feed itself as are India, Pakistan, China, and countries in the Middle East. Water infrastructures, while in many cases obsolete, leaky, and in need of overhaul, are comprehensive and functional. Industrial and urban water pollution is regulated and monitored. Although the industrial democracies' overall relative population size is shrinking to only one-ninth of humanity, their comparative hydrological resource advantages help put them in a good position to make the breakthrough innovations to meet the era's defining challenge of augmenting the productive supply of freshwater in an environmentally sustainable and economically vibrant manner. As with other water breakthroughs in history, doing so would leverage their wealth and influence in the new century's global order.

Indeed, by aggressively reallocating its current supplies using existing technologies, America is well positioned to not only remain the world's leading food exporter, but also to free up water resources to boost its energy output, accelerate industrial production, and maintain robust growth in its services and urban economies. The comparative impact upon a world economy and political order constrained by water resource scarcity is potentially akin to the advantages gained from the early, large discoveries and production of oil in the twentieth century.

The remarkable increase in water productivity under way in the advanced industrial democracies represents a startling historic break in the correlation between absolute water withdrawals and economic and population growth. After three centuries of increasing twice as fast as world population, average water withdrawals per person are declining in many advanced democracies without any slowdown in economic growth. American water withdrawals peaked in 1980, and declined about 10 percent by 2000; in the same period, the nation's population expanded by 25 percent and the economy continued on its long-term growth trajectory. From 1900 to 1970, U.S. water productivity per cubic meter withdrawn had remained relatively constant at about $6.50 of gross domestic product; by 2000 it had soared toward $15. Japan's economic productivity per unit of water increased fourfold between 1965 and 1989. The pattern was similar in much of Europe and Australia.

The sudden upsurge in water productivity is a market response to the economic incentives created by the combination of growing water shortages and water pollution regulations that came into force from the 1970s with the environmental movement. The environmental golden rule of thumb is that users have to discharge water to the ecosystem in the same pristine condition as they have taken it from nature. Led by thermal electric power plants, industry, and cities, large water users soon realized that they could save money on pollution cleanup by using less water overall through more efficient conservation and recycling technologies.

Gradually, the first generation of government environmental rules are being refined into a subtler, soft-path efficiency approach more attuned to ecosystem needs and services. In messy, pluralistic Western democratic style, government officials, market participants, and environmentalists are often working together as constituent representatives in devising solutions tailored to specific user needs and conditions, including appropriate scaling. Small-scale, ecosystem-friendly solutions are preferred when feasible. The European Union's Framework Directive on water policy (2000), for instance, expressly discouraged new dams where economically and environmentally viable alternatives existed; dams are also starting to be removed and replaced by wetlands restoration and reforestation in America. Legislatures and courts for the first time are granting ecosystems a legal entitlement to a sustainable share of contested water supplies. Creative concepts of valuing provision of ecosystem services are also being devised so that environmental regulations can be fulfilled and exchanged in a more flexible, market-oriented manner. Greater attention is being focused on suppler efficiency measures, such as water consumption—that is, water that is used and lost for other purposes such as irrigation water—instead of simpler, gross withdrawals, which fail to capture the productivity of recycling water for multiple uses or treated releases that are reused again by others downstream. Intermediation mechanisms are being launched, often supported by governments, to help users sell their annual water rights at premiums to other, presumably more efficient, users.

As water grows scarce and soft-path regulatory approaches take form, a market price and a marketplace for water services are coming into existence. A few businesses are pioneering measurements of their water-use "footprints," a parallel of the carbon footprint tool gaining traction to help each entity reduce its contribution to global warming. Business enterprises are making major investments in water to compete for market share and profits. In vague outline is the embryonic contour of an artificially grafted, long-missing Invisible Green Hand mechanism that fully prices in the cost of utilizing and restoring water resources and that can enlist the historically prodigious forces of private market wealth creation in the provision of a sustainable environment. It is still early days. Large, crucial sectors, notably agriculture, remain heavily subsidized, lightly regulated for pollution, and insulated from market forces. Development is occurring locally, and sporadically, in response to needs as they arise. The change faces strong entrenched and ideological opposition on all sides. Hard-path approaches to moving, storing, draining, and cleaning water still prevail overall within the slow-changing governing water bureaucracy. At the same time, traditional environmentalists remain suspicious of any treatment of water as an economic good. They fear it may lead to its being governed solely as a profane commodity, according to dictates of market forces with inequitable outcomes and without regard to its inherently priceless and sacred value to nature and human life. But between these two poles, something novel is taking hold.

• • •

One place where the combination of water shortage, ecosystem protection, and market responses has been catalyzing a more productive use of existing freshwater is in America's arid Southwest. At the dawn of each new era of water history, societies face the classic transitional problem of how to reallocate water resources from old uses to newer, more productive ones. By the end of the twentieth century, America's Southwest water productivity gap had become enormous between its privileged farming businesses that were still guzzling from the trough of socialized irrigation water from the bygone age of government dams and the modern West of dynamic cities and high-tech industries. The same volume of water—250 million gallons per year—could support 10 agricultural workers or 100,000 high-tech jobs; California's agribusinesses were using 80 percent of the state's scarce available freshwater but producing just 3 percent of its economic output. Within agriculture, too, water was being inefficiently consumed in one region by water-thirsty, low-value crops like rice and alfalfa, while high-value fruit and nut trees were being cut back in another place for lack of water. The essential problem, even in the arid Southwest, is not that absolute water availability is too scarce to sustain robust economic growth, but rather that regulated water is both too cheap and vested in less efficient users and thus impedes the simple market price incentive mechanism that would otherwise reallocate it toward more productive uses.

In few places was the disparity greater than among the 400 farm agribusiness descendants of Southern California's Imperial Valley water district who consumed 70 percent of California's 4.4 million acre-foot allocation of Colorado River water under the 1922 compact at the delivery cost of only about $15 an acre-foot, and the 17 million thirsty Southern California coastal dwellers who were paying 15 to 20 times more to meet their much more modest needs. Imperial Valley's water surfeit, furthermore, encouraged particularly profligate farming practices, including the desert planting of water-thirsty crops, and the consumption of twice as much water per acre as other California farmers.

Such an enormous disparity between the farm and city price of water and the formation of a political alliance of city, industrial, and environmentalist interests with the clout to offset the farmers' historical dominance within California's hardball water politics, did not escape the notice of the billionaire Bass brothers, scion investors of one of Texas's oil empires. In the early 1990s, the Basses invested about $80 million of their $7 billion fortune to buy up 40,000 acres of Imperial Valley farmland—and the water rights that were conferred with it. Reminiscent of William Mulholland's deception in purchasing Owens Valley land to gain access to its river water in the infamous Los Angeles water grab earlier in the century, they avowed to farmers wary of losing their water rights that they wanted the land merely to raise cattle and not to speculate on water.

Soon enough, though, the Basses managed to persuade the cooperatively owned Imperial Valley water district that its best interest lay in selling San Diego 200,000 acre-feet of its 3.1 million acre-feet entitlement each year starting at $233 per acre-foot—a markup of nearly 20 times its own subsidized effective cost—for an immense, cumulative profit of more than $3 billion over seventy-five years. The plan, moreover, called for Imperial Valley to invest a slice of these profits in water efficiency improvements intended to save at least as much water as it was selling to the city, so that in practice it wouldn't lose any of its precious Colorado River water at all. Despite the farmers' exorbitant profit, San Diego liked the deal because it provided an independent water source and a one-third savings over what it was then being forced to pay to the powerful, Los Angeles–dominated Southern California water authority.

Although applauded by federal and state regulators, environmentalists, and most nonfarm participants because it finally began the transfer of water from agriculture to the cities, the deal got bogged down in the internecine battles between California water authorities and other interests. As the millennium approached, the proposed water sale became engulfed entirely in the larger regional crisis on the Colorado: Diminished by drought, full allocation draws by fast-growing Arizona and Nevada, and the fact that the river's average annual volume was less than the 1922 compact had estimated, the Colorado basin was fast running out of enough total water to supply everyone's existing needs. Storage in Lake Mead was dwindling to alarmingly low levels. Without major modifications, the iconic Hoover Dam and other Colorado water infrastructure, would have sufficed for less than a single century.

In late 1999, U.S. secretary of the interior Bruce Babbitt, backed by the other Colorado basin states, issued the first-ever ultimatum to California to end its decades of river overdrafts, then running about 800,000 acre-feet per year, and to live within its compact limit of 4.4 million acre-feet. California was given until year-end of 2002 to come up with a plan to wean itself off its Colorado overdraw by 2016. The regulators also insisted that the plan include the transfer of water from Imperial Valley to the coastal cities and to protect existing water ecosystems. Failure to devise an acceptable program, the interior secretary warned, would lead to the immediate cutoff of the excess flows.

Imperial Valley agribusinessmen furiously resisted being forced to accept the terms of a deal that they correctly foresaw would be the start of a slippery—even if lavishly gilded—slope to losing control of the virtually free irrigation water, which American taxpayers had granted their forbearers for settling the barren desert long ago. In particular, they bridled at the demands that they allocate some of their water to preserve the ecosystem health of the Salton Sea, the inland lake that had formed when the Colorado River flooded its levees in 1905, and that was replenished only by the runoff from the valley's 82-mile All-American Canal and 1,700 miles of irrigation ditches.

When the December 31, 2002, deadline passed without an acceptable plan, the unthinkable happened: At 8:00 a.m. on New Year's Day 2003, the new interior secretary, Gale Norton, stood by the pledge of her predecessor from the opposition Democratic administration, and switched off three of the eight pumps controlling the flow from the Colorado into the 242-mile-long aqueduct to Southern California. At the turn of the spigot, Imperial Valley lost as much water as it was to have sold to the cities—without any compensation. Still the farmers did not buckle. Then, in August 2003, the Interior Department's Bureau of Reclamation increased the pressure by releasing a study that concluded that in response to the drought the government could cut off water to Imperial Valley because farmers were using it wastefully—the whispered amount of the waste was 30 percent. Playing "good cop" to the Fed's "bad cop," the California state government stepped forward and offered to share some of the water and infrastructure cost burden to preserve the Salton Sea.

Within two months, in October 2003, the agribusinessmen of Imperial Valley capitulated. The ceremonial signing of the landmark agreement to transfer Colorado River water to San Diego and other cities was held at the Hoover Dam. In all, 500,000 acre-feet per year, or one-sixth of Imperial Valley's water, would be reallocated. An estimated 30 million acre-feet—some two years' annual flow of the Colorado River—would move from primarily agricultural to urban uses over seventy-five years. No one doubted that further calls on agricultural water lay ahead as the New West continued to rise.

Despite the multibillions of dollars of profit they would receive for selling a fraction of their taxpayer-subsidized water, some bitter Imperial Valley farmers felt cheated. "They should pay $800 an acre-foot versus $250," complained farmer Mike Morgan. "The greatest water heist ever is going on right under your feet." Others, however, promptly got over their grievances and moved forward to recoup much of the water lost to the sale by improving existing water productivity through investments in repairing leaky irrigation networks and in new technologies, such as high-tech satellite sensors to monitor crop and soil moisture and activate precision watering devices. In fact, the only big losers from the Imperial Valley water deal were Mexican farmers, who for decades had been pumping up groundwater that had leaked from the irrigation ditches on the American side of the border. They now suddenly found their wells running dry due to the Californians' more-efficient irrigation methods. Ever lucky, Imperial Valley soon stumbled upon another potential bonanza under the Salton Sea's southwest corner—a large geothermal field that could significantly boost California's renewable electricity production.

By breaking the political logjam over the Colorado, the landmark agreement with Imperial Valley paved the way in late 2007 for a second breakthrough accord—an emergency plan among the Colorado River compact states about how to allocate scarce water among themselves should the river flow fall below the 7.5 million acre-feet promised to the lower basin. Given the downward revision in the Colorado's long-term average flow to only 14 million acre-feet per year, and with Lake Mead only half full because of a severe drought, such an emergency was likely to occur; moreover, with climate change models forecasting a 20 percent decline in rainfall compared to most of the twentieth century, and shrinking annual mountain snowpacks exacerbating summer shortfalls, the emergency seemed likely to strike sooner rather than later. The shared sense of impending crisis spurred unusual, proactive efforts to improve the productive use of existing water supplies before the crisis threshold was crossed.

The 2007 emergency accord included innovative market and ecosystem-management arrangements that stimulated interstate water trades and consumption reductions by allowing users to bank their savings in Lake Mead or aquifers for later use. Fast-growing, desert-bound Las Vegas, for instance, offered to pay for new water storage facilities or desalinization plants in California in exchange for an extra draw on California's Colorado River allotment. Las Vegas was already one of the most conservation efficient cities. Every drop of its sewage wastewater is treated and released into Lake Mead, where it is further purified by dilution and pumped back to the city's taps. Even with its growing population, water use has fallen from its peak in 2002 thanks to various conservation methods, including promotion of low-flow toilets and appliances, paying residents to replace water-thirsty lawns with natural desert flora, and higher consumer prices. In the eastern Rockies, the city of Aurora, Colorado, was creating a recycling loop even more elaborate than Las Vegas's. It was buying agricultural land downstream along the South Platte River so it could draw water that had been naturally filtered by the sandy riverbanks through a series of adjacent wells. The water was then to be pumped to the city in a 34-mile-long pipeline, purified, used, treated, discharged back into the river, and then recaptured in the riverbank wells to start a new circuit. Each round-trip took forty-five to sixty days and recycled half of every drop.

Spurred by continuing drought, California introduced a state water bank that allowed northern California farmers to sell their seasonal water rights on fallowed land to farmers using more efficient farming techniques and growing more valuable crops. In 2009, the state-set price for watering parts of California's fertile, but naturally arid and badly overpumped Central Valley, was $500 an acre-foot—nearly three times the price of 2008, but still far below what the free-market price would likely have fetched. Coastal Southern California has also been turning to wastewater recycling to supplement drinking supplies because all available local and aqueduct-delivered natural water supplies had been exhausted. The Colorado River had maxed out. Mountain snowmelt and reservoir levels were diminishing with drought and global warming. Even long-distance freshwater piped from northern California was being curtailed, as court rulings and federal restoration for the Central Valley elevated the priority of conserving water to improve the health of the fish and wildlife of the depleted ecosystem of the San Joaquin–Sacramento River delta estuary and San Francisco Bay. With population growth forecasts still rising, Los Angeles and San Diego, as a last resort, were turning to large-scale recycling and purification of sewer wastewater—long used for irrigation and lawn watering—to augment urban drinking supplies.

The disparaging name critics label such projects—"toilet to tap"—is a misnomer. Not only is the wastewater intensively cleansed to a level that can be purer than naturally derived tap water; it does not go straight to the tap, either. Instead it is injected into the ground to be further filtered by natural aquifers before being drawn into public drinking supplies. There is little novelty in the concept. For decades, cities across the United States routinely discharged their treated sewage, or effluent, into local rivers such as the Colorado, the Mississippi, and the Potomac, where in diluted form it was taken into the drinking supply of cities downstream. The same principle had been followed by London in building its sanitary system in response to the mid-nineteenth-century Great Stink. The Southern California recycling projects differ in using slow-moving groundwater instead of surface river flows to do the additional natural filtering. The pioneering prototype was the facility that opened in January 2008 in Orange County, California, which has a capacity of 70 million gallons per day. Its labyrinth of tubes and tanks take in dark brown treated sewage water, then remove solids with microfilters and smaller residue through high-pressure reverse osmosis before a final cleansing with peroxide and ultraviolet light. The final product, as pure as distilled water, is injected into the aquifer for natural filtering before entering the public drinking supply. Water managers in water-stressed southern Florida, Texas, and San Jose, California, have been contemplating similar projects to help meet their future needs. Only one major city in the world—Windhoek, in Africa's arid Namibia—actually recycles water on a large scale from treatment plant directly to the drinking tap. Yet aside from the revolting idea of the source of such water, there is no technological, or cost-efficiency obstacle to believe that such truly closed, recycled infrastructure loops will not become more commonplace as the age of water scarcity advances.

Water shortage is also propelling Southern California's leadership in a modest global movement toward state-of-the-art desalinization technologies. Desal costs in California had fallen from $1.60 to 63 cents per cubic meter between 1990 and 2002, putting it on par with large, efficient reverse osmosis plants built in water-starved Israel, Cyprus, and Singapore. By 2006, there were enough proposals for new seawater desal plants to increase California's capacity a hundredfold, and supply up to 7 percent of the entire state's urban water use. The first major test of desal's mass production capabilities in California was joined in 2009 with a decision to build a giant, reverse osmosis plant near San Diego that was projected to produce 50 million gallons of drinking water daily from the ocean by 2011—10 percent of northern San Diego's requirements. While total desalinization capacity is still very small, California's sheer size and its special, water trendsetting status makes it a potential catalytic tipping point—especially if coupled with breakthroughs whereby solar or wind power can substitute for nonreplenishing and polluting fossil fuel energy—for the long hoped for takeoff of water desalinization.

A half century earlier, President John Kennedy had expressed mankind's age-old dream of desalination. "If we could ever competitively—at a cheap rate—get fresh water from salt water," he mused, "that would be in the long-range interest of humanity, and would really dwarf any other scientific accomplishment." Ever since man first took to the seven seas, sailors had dreamed of desalting seawater. Long-distance European mariners in the Age of Discovery pioneered the installation of primitive desalting equipment for emergencies. Crude, large-scale water desalting was enabled by advancements in the distillation process made in the mid-nineteenth century by the sugar refining industry. Modern desalinization, however, was brought to fruition by the U.S. Navy, which developed it during World War II to provide water to American soldiers fighting on desolate, South Pacific islands. By the 1950s, a thermal-desalinization process based on steam-pressure-induced evaporation was developed; although very expensive, it was adopted on a fairly large scale in Saudi Arabia and other oil rich, waterless coastal nations of the Middle East. Also in the 1950s, the American government supported university research for a better desalinization technique—the reverse osmosis process was invented during Kennedy's presidency and was put into action on a small scale using brackish water in 1965. With the development of a much-improved membrane in the late 1970s, reverse osmosis desalinization plants for seawater became possible. Since they required enormous amounts of energy and the water they produced was so costly compared to water obtained by other means, it was unsurprising that the first big city desal plant was opened in Jedda, Saudi Arabia, in 1980, where energy was cheap and water pricelessly scarce.

Major improvements in energy recovery techniques and membrane technologies occurred with such speed from the 1990s that by 2003 desal costs had fallen by two-thirds, and desal was becoming a viable component of the diverse portfolio of water supply solutions being adopted in water-famished, coastal regions where supply was abundant and expensive long-distance water pumping unnecessary. Perth, Australia, for instance, got nearly one-fifth of its water from desalinization. Israel's desal share was poised to rise rapidly and desal offered hope of quenching some of the mounting thirst in the Muslim Middle East and North Africa. Reverse osmosis membrane technologies at the heart of desalinization were also being applied in recycling wastewater in Orange County's pioneering plant and in Singapore, where it helped replenish local reservoirs. With growth stirring in desal, major corporations were gearing up to win market share in order to earn large profits as the market developed. Projections of market growth in the decade to 2015 ranged widely, from a trebling to a septupling of the $4 billion spent in 2005.

On its most optimistic projections, however, desalinization cannot be the panacea technology to solve the world's water crisis in the short term. Installed desal capacity is simply too tiny—a mere 3/1,000ths of 1 percent of the world's total freshwater use. Even if costs plunged, there are unsolved environmental problems about how to dispose of the briny waste; inland regions cannot be reached without expensive pumping and building long aqueducts. In the most likely, best case scenario, desal will become one of a portfolio of freshwater supply techniques that help countries muddle through their scarcity crises.

In the rainy, temperate eastern half of America, New York City, the nation's urban trendsetter in long-distance water storage and delivery systems, is also in the vanguard of the new soft-path movement. One of its most closely watched experiments is to exploit the natural, cleansing services of forested watersheds to improve the wholesomeness of its drinking water—and simultaneously save billions of dollars for the region's 9 million inhabitants. Ever since its gravity-fed Croton water system opened in 1842, New York had routinely extended its aqueducts and reservoirs farther and farther upstate into the Catskill Mountains and the upper reaches of the Delaware River to obtain more clean freshwater. By the 1990s New York City's water network featured three distinct water systems with one and a quarter year's storage capacity that delivered 1.2 billion gallons per day from 18 collecting reservoirs and three lakes in upstate New York. But a serious problem of deteriorating water quality had been building as the pristine rural, forested countryside surrounding the reservoirs degraded with modern development and farming. As a result, half the city's reservoirs were chronically choked with poisonous phosphate and nitrogen runoff from dairy farm pastures and over 100 sewage treatment plants that depleted oxygen levels and produced foul, algae blooms, as in China's Lake Tai, that killed cleansing biological life. When U.S. fresh drinking water standards were toughened in the late 1980s, New York City faced an ultimatum: build a state-of-the-art filtration plant—at a staggering cost of $6 to $8 billion, exclusive of the huge operating expenses of the energy-intensive filtration facility—or devise an alternative method to protect the quality of the city's water.

New York's innovative response was a $1 billion plan to improve the upstate forests and soils surrounding the reservoirs so that they conserved more water and filtered out more of the pollutants in a natural way—in effect, New York was enhancing the natural watershed ecosystem and putting a market value on its antipollution services in place of far more expensive, traditional, artificial cleansing infrastructures. Also remarkable was that New York's ecoservices project was forged by a new, politically inclusive consensus among city and state officials, environmentalists, and rural community representatives. Their multiyear negotiation was formalized in a 1,500-page, three-volume agreement signed in January 1997.

At the heart of the plan, New York City would spend $260 million to purchase some 355,000 acres—nearly twice the geographic area of the city itself—of water sensitive land from voluntary sellers to buffer the reservoirs. Some of the new city-owned land would be open to the public for recreational fishing, hunting and boating, and leased to private interests for environmentally controlled commercial activities such as growing hay, logging and production of maple syrup. Up to $35 million more would be spent to clean up and modernize several hundred dairy farms—including reducing their water consumption in milk production by up to 80 percent—to help them compete against the encroachment of concrete road polluting and waste-producing subdivisions. To mollify local communities still resentful of the city's imperious, historical use of compulsory sales to acquire watershed land for its reservoirs, the city agreed to spend another $70 million for sundry infrastructure repairs and environmentally friendly economic development. A new environmental division was created within the city's century-old watershed police force; armed with chemistry kits and looking for leaky septic tanks and rivulets of frothy, toxic discharges, they patrolled the countryside and subdivisions to protect the reservoirs. In effect, New York City has created a market price for the ecosystem services provided by its watershed. A decade later, it took another step toward marrying ecosystem sustainability and market economics by negotiating a complicated land swap with a big resort developer whereby a public forest acquired watershed-protective mountainside real estate in exchange for a smaller resort project on a less environmentally sensitive side of the mountain. The developer also agreed not to build on runoff-prone steep slopes or use chemical fertilizers on its golf courses.

The early results of New York City's watershed experiment are auspicious. Environmentalist watchdogs gave New York City good grades for drinking-water quality in 2008—a year after the city had won a further conditional ten-year filtration plant exemption from the U.S. Environmental Protection Agency. In economic terms the program had saved the city up to $7 billion in unnecessary construction, expanded recreational revenues, and augmented the long-term sustainability of New York's water supply. With continued success, it offered a potential template for the next generation of urban water development. Indeed, other American cities, and some abroad including Cape Town, South Africa; Colombo, Sri Lanka; and Quito, Ecuador, also adopted variants of New York–style ecosystem service valuations to help solve their local challenges.

With echoes of both New York and Southern California, Florida's governor Charlie Crist launched in 2008 a novel initiative to revive a moribund restoration plan for the state's famous wetlands, the dying Everglades. For nearly a decade, a joint federal-state plan had been held hostage by the political grip of the state's water-guzzling, phosphate-polluting, and price-subsidized big sugarcane farmers. Deprived of clean water, half the Everglades had already dried up. By spending $1.34 billion in state funds to buy 181,000 acres of land from the giant U.S. Sugar Corporation, Crist opened the way for a land swap with other agribusinesses that would open channels to renew the historic flow of fresh, clean water to the Everglades from Lake Okeechobee.

In addition to enhancing its upstate watersheds to improve the quality of the water entering its reservoirs and aqueducts, New York City also embarked on a showcase water conservation program in the early 1990s aimed at trimming the system's total demand, thus reducing the absolute volume that had to be supplied and subjected to expensive purification and wastewater treatment. First, water and sewerage rates were raised sharply toward market levels to discourage wasteful use. A highly publicized, $250 million toilet rebate program for poorer families was also launched to jump-start a citywide trade-in of old 5- and 6-gallon toilets for newer toilets that consumed only 1.5 gallons per flush. Toilets are by far the biggest single water consumer in the household—accounting for about a third of consumption—and in 1992 the government mandated a gradual national conversion to low-flow models. By 1997 the toilet replacements, higher prices, and other measures, including comprehensive metering and leak detection, helped New York's daily water consumption to plunge dramatically to 164 gallons per person from nearly 204 gallons in 1988—a 20 percent savings, or 273 million gallons per day. As a result, New York officials projected that the city would not need any additional water supply for another half century, while incalculable millions of dollars were saved on sewage treatment and pumping. The replication of New York's conservation methods by cities across the United States has been one of the driving forces behind the unprecedented increase in America's water productivity since the 1980s.

New York City faced one other gargantuan challenge, however, for which it had no low-cost, water-productivity-enhancing alternative to old-fashioned, large government expenditure—the decrepit, leaky and potentially failing state of vital components of its aging water infrastructure. Significant leaks had sprung below ground in the original aqueducts that conveyed water from its upstate reservoirs, beneath the Hudson, to a final storage reservoir on the city's outskirts in Yonkers. More threatening still, New York's two, leaky urban distribution tunnels, completed in 1917 and 1936, respectively, which conducted water from the Yonkers reservoir throughout the city, hadn't been shut down for inspection for over half a century for fear they might fail catastrophically—forcing the evacuation of large portions of New York. From

1970, New York tunnel crews had been laboriously drilling through the solid bedrock 600 feet underground—some 15 times below the depth of the subway—to construct a modern, third tunnel that would enable the original two to be shut down and rehabilitated. The $6 billion Tunnel Number 3 was the largest construction work in New York City's history and one of the most monumental, although invisible and virtually unknown, civil engineering feats of the era—a subterranean descendant of the Brooklyn Bridge and Panama Canal. Until the day it was finished and ready for service, around 2012, New York would continue to live in a slow-motion race against time and potential disaster.

The status of any society's waterworks network is both a bellwether and a foundational element of its economic and cultural dynamism. Many metropolises in America and Europe that industrialized early face the formidable challenge of modernizing their original domestic water systems. Although the quest for an era defining water innovation captures the headlines, maintaining good infrastructures for all the four main historical uses of water—domestic, economic, power generation, and transportation—is also a necessary condition of the industrial West's ability to fully exploit its comparative, global freshwater advantage. The failure to do so imperceptibly erodes efficiency and resiliency, and makes society more prone to shocks, such as the levee failures and flooding of New Orleans during Hurricane Katrina in the summer of 2005. Yet the engineering complexities, and the low political reward of supporting costly repairs, pose enormous obstacles. Often the work involves difficult, subterranean construction, amid intense atmospheric pressures and large, fast-moving volumes of water that cannot be shut off, and in systems that had not been designed with future renovations prominently in mind.

Following revelations from Riverkeeper, the private Hudson River environmental watchdog group and a major player in the watershed program, New York authorities in 2000 admitted publicly for the first time that a branch of the Delaware Aqueduct, the city's largest, had been leaking significantly for a decade. When first detected in the early 1990s, the tunnel's leakage had been about 15 to 20 million gallons per day; by the early 2000s, the leakage had swelled to 35 million gallons.

While that totaled only 4 percent of the aqueduct's overall capacity, the leaks had to be fixed before they got worse and eventually the tunnel's structure gave way. The last inspection in 1958 had been done by driving through the drained, 13-foot-diameter tunnel in a modified jeep. But with all the cracks the tunnel could no longer be shut down for fear of structural damage from the change in water pressure. So in 2003, in an unprecedented action, the city sent an unmanned, remote-controlled, torpedo-shaped, minisubmersible, with protruding, catfish-whisker-like titanium probes that had been specially designed by the sea experts at the Woods Hole Oceanographic Institution, on a 16-hour data gathering mission through the dark, watery 45-mile-long tunnel. After studying the results for four years, the city decided upon the first phase of the complicated repair, which would cost $239 million. A team of deep-sea repair divers, working round the clock for nearly a month in a sealed, pressurized environment, were lowered 700 feet to perform the preparatory inspections and measurements amid the tunnel's currents in winter 2008.

New York's struggle to plug its twenty-year-old aqueduct leaks paled in degree of difficulty and urgency, however, to completing Tunnel Number 3. The project's genesis went back to 1954 when New York engineers descended a city shaft several hundred feet to the main control site for Tunnel Number 1 to prepare a long-overdue inspection. Their intent was to shut off the water flow so cracks could be found and repaired by welders from inside the tunnel. But when they began to yank on the old, rotating wheel and long bronze stem at the bottom of the shaft that controlled the six foot diameter open and shut gate inside the tunnel, it began to quake from intense pressure. Terrified that the brittle handle might break—or worse that the inside gate might shut permanently in the closed position and cut off all the water flowing to lower Manhattan, downtown Brooklyn, and part of the Bronx—they dared not continue. They returned to the surface. From that day onward, New Yorkers had lived in ignorant bliss that no one could repair the two badly leaking, antiquated distribution tunnels providing all the water for their homes, hospitals, fire hydrants, and 6,000 miles of sewage pipes—or even know if a structural weakness was building to a critical threshold that would cause the tunnel to rupture and collapse in a sudden apocalypse. Some believed that only the outward force of water pressure was maintaining the tunnels' integrity. "Look, if one of those tunnels goes, this city will be completely shut down," said James Ryan, a veteran tunnel worker. "In some places there won't be water for anything . . . It would make September 11 look like nothing."

It took sixteen years before city officials were able to break ground on the elaborately planned remedy. Tunnel Number 3 was to be a redundant, citywide water network with many branches and a state-of-the-art central control facility. Once operative, it would allow flows to be easily turned off and repairs made anywhere in the city. The project's problems were time, immense cost—in its early years the project was delayed by New York's 1970s financial crisis—and the arduous, dangerous work of blasting and drilling through bedrock in tunnels that were as deep as some of New York's tallest skyscrapers. The work was done by a specialized, grizzled, close community of urban miners, known as sandhogs. Sandhogs had built virtually every notable New York tunnel system from subways to utility shafts; in the 1870s they worked inside high atmospheric pressure caissons, excavating the foundations of the Brooklyn Bridge, where they were the first workers to encounter the agonizing chest pains, nose bleeds, and other symptoms of the bends. Many were killed. Two dozen had died digging Tunnel Number 3 alone. Because of the danger, they were well paid. Sandhog jobs tended to be passed down from father to son; many sandhogs were of Irish and West Indian descent.

The excavation work on Tunnel Number 3 was all the more difficult because the sandhogs knew they were digging against doomsday if Tunnels Number 1 or 2 collapsed before they finished. Usually they could advance no more than 25 to 40 feet per day, chiseling, dynamiting, removing endless tons of rubble. Their methods were modern-age equivalents of the fire and water rock-cracking technique used by ancient Rome's aqueduct builders and Li Bing's Chinese tunnelers along the Min River. Progress accelerated when a new mayor, Michael Bloomberg, set a high priority on improving water facilities citywide and invested an additional $4 billion toward finishing Tunnel Number

3. The excavation rate more than doubled with the introduction of a new 70-foot-long boring machine—called the mole—with 27 rotating steel cutters, each weighing 350 pounds. Donning a hard hat in August 2006, Mayor Bloomberg descended into the tunnels and took a seat at the mole's controls to bore through the final foot of rock to complete excavation of the second, and most crucial, of Tunnel Number 3's four stages. The work, however, was not finished. At least six more years of work lay ahead to line the tunnel with concrete, fit it with instruments, and sterilize it so it could carry water. By then, it would be linked up with the water system's space age, electronically regulated, new central command center—featuring 34 precision stainless steel control valves, specially fabricated in Japan under constant, two-year vigil of New York city engineers, housed inside 17 giant cylinders weighing 35 tons. The control chamber itself was 25 stories beneath the Bronx's Van Cortlandt Park in a domed vault three stories tall and the length of two football fields. Nothing aboveground, save a small guard tower and door leading into the grassy hillside indicated that it was the entrance to one of New York's most critical infrastructures.

Throughout the industrialized democracies, localities are facing infrastructure challenges similar in kind, if usually smaller in scale, to New York's. Estimates for upgrading America's 700,000 miles of aging water pipes and wastewater, filtration, and other facilities at the core of its domestic water systems range from $275 billion to $1 trillion over the next two decades. Global water infrastructure needs are several quantum orders of magnitude greater. Many major world cities have notorious leaks; possibly up to half of drinking water entering cities worldwide is lost before reaching residents.

Regions that fail to improve their efficient use of existing water resources are more prone to water shocks, slower economic growth, and to become enmeshed in political clashes over water with neighbors. The state of Georgia's unwillingness to invest to upgrade fast-growing Atlanta's water supply system, for example, caught up with it in 2007 when a prolonged drought caused the city's water reserves to dwindle to only four months. The governor's only immediate recourse was to impose emergency measures and to try to wrest a greater share of water from the Apalachicola-Chattahoochee-Flint river system away from downriver neighbors Alabama and Florida, which depended upon the flow to keep its own electric power plants and factories running, and to sustain the Gulf coast ecosystem for its shellfish industry. Implementing simple efficiency measures, Georgia reckoned retrospectively, could have alleviated its water crisis by reducing water demand by 30 percent.

Relentless regional freshwater demand and diminished ice cover due to warming temperatures is also taking a costly toll in the north by lowering normal water levels in the immense Great Lakes. Every inch of lost water depth forces the lakes' fleet of 63 transport ships to lighten their annual cargo load by 8,000 tons to avoid grounding mishaps. This adds another cost to the global competitiveness burdens already faced by America's aging industrial belt of steelmakers and heavy manufacturers situated on the lakes' edges for its cheap transport and industrial water. Seaports that don't keep pace with the modifications required by the new generation of giant, ocean cargo supercontainers, some as long as a 70-story skyscraper and traveling halfway around the world between ports of call, likewise risk losing out on global shipping business. Extensive port restructuring helped New York recover some of its historical greatness as a harbor with renewed Asian trade following a prolonged loss of business in the second half of the twentieth century to more modern ports in America's southern and western coasts. With Great Lakes states ever fearful of schemes to siphon their water to dry parts of America, the U.S. Congress in 2008 passed a new legal compact governing lake water that provided strict conservation measures and banned the export of the lakes' water out of their basins.

The Great Lakes conservation measure was disappointing news to some in Texas, which had designs on its water dating back many decades. Although oil had built Texas, the state's future prospects—its economic prosperity and its outsized leverage on American national politics—rested chiefly upon whether it could rationalize its water use to sustain its large cities and industries. In the absence of a comprehensive program that increased effective water supply through efficiency and conservation, Texas seemed set to live through an accelerated reprise of Southern California's history of water grabs and speculations.

Billionaire water speculators, including oil magnate T. Boone Pickens and Qwest Communications cofounder Philip Anschutz, for years had been exploiting a Texas law to acquire unrestricted water rights through land purchases and lobbying government officials to fulfill their ambitious plans to pump and sell nonrenewable Ogallala Aquifer water through multibillion-dollar pipelines hundreds of miles to thirsty cities such as Dallas, San Antonio, and El Paso. At $1,000 an acre-foot, their profit potential was spectacular and Texas's good fortunes could be extended for a while—until the Ogallala fossil water itself gave out. Yet even as certain regions declined, the industrial democracies enjoyed an enormous advantage in the water infrastructure-building challenge facing the world, thanks to the existence of a competitive industry of large and small companies seeking to profit from the growing thirst and capable of expeditiously delivering solutions.

While cities are learning to use their existing water more efficiently, industry has been the largest single contributor to the unprecedented surge in water productivity. Across the industrial spectrum, water is a major input of production. Alone, five giant global food and beverage corporations—Nestlé, Danone, Unilever, Anheuser-Busch, and Coca-Cola—consume enough water to meet the daily domestic needs of every person on the planet.

Superior water productivity is one of Western industry's competitive advantages in the global economy, helping to offset the low wages and laxer environmental standards of industries based in poorer nations. American companies began to treat water as an economic good with both a market price for acquisition and a cost of cleanup before discharge in response to federal pollution control legislation in the 1970s. With characteristic business responsiveness wherever operating rules were clear and predictable, they sought ways to do more with the water they had and to innovate in their industrial processes so that they needed to use less overall. The results were startlingly instructive of the enormous, untapped productive potential in conservation.

No industrial sector uses more water than thermoelectric power plants. Huge amounts—two-fifths of all U.S. water withdrawals—are sucked out of rivers and other water sources as coolant, even though overall net consumption is low because the water is returned to its source a few minutes later. Galvanized by federal regulations requiring that the quality of the discharged water be as pure and cool as it was when withdrawn, the power plants increased recycling and converted their once-through systems to more efficient cooling technologies. By 2000 some 60 percent of all thermoelectric power capacity was using modern systems; the amount of water needed to produce one kilowatt-hour had plunged to only 21 gallons from 63 gallons in 1950.

Manufacturers, likewise, responded impressively to the water pollution regulations. Chemicals and pharmaceutical companies, primary metals and petroleum producers, automakers, pulp and paper mills, textile firms, food processors, canners, brewers, and other large water users increased recycling and adopted water-saving processes. In just the fifteen years from 1985 to 2000, American industry's total withdrawals were trimmed by a quarter. Pre–World War II American steel mills that needed 60 to 100 tons of water for every steel ton produced were superseded by modern mills using only six tons by the turn of the twenty-first century. Similarly, water-intensive semiconductor silicon wafer makers reduced their intake of ultrapure freshwater by three-quarters between 1997 and 2003, and recycled much of the discharge for use in irrigation. In the decade from 1995, Dow Chemical cut its water usage per ton produced by over a third. Europe's Nestlé nearly doubled its food production while consuming 29 percent less water from 1997 to 2006. In a scheme reminiscent of New York City's landmark ecosystem services plan, bottled water company Perrier Vittel invested in reforesting some heavily farmed watersheds, and paid farmers to adopt more modern methods, in order to protect the quality of its mineral water sources.

For years water had scarcely commanded a line item in corporate budgets or more than cursory attention from top planning executives. In the age of scarcity, more and more water-conscious companies were treating water as a key strategic economic input, like oil, with clearly reported accounting and future target goals. The most forward-looking and global-minded analyzed water risks facing their key suppliers around the world, and helping insulate the vulnerable by helping them adopt conservation and ecologically sustainable practices. Unilever's technical and economic support, for example, enabled its Brazilian tomato farmers to adopt drip irrigation that trimmed water use by 30 percent and reduced water-contaminating pesticide and fungicide runoff. Brewer Anheuser-Busch became acutely aware of the importance of its water supply chain when it was whipsawed by a drought in America's Pacific Northwest. Water shortages for crops pushed up the price of a key beer-making ingredient, barley, while diminished dam flows elevated hydroelectric prices and with it the cost of producing aluminum beer cans. Environmentalists, too, have been getting on board with collaborative efforts: for instance, the Nature Conservancy has been developing a plan to award good standing certificates to companies who use water efficiently.

Improved industrial water productivity not only enhances competitiveness directly. It also creates economic benefits by freeing water and lowering its cost for other productive uses. Yet the potential scale of its benefit pales next to the boon that can accrue from water productivity breakthroughs in the least efficient, most subsidized, and heaviest polluting sector of society—agriculture. That is because agriculture is still by far the greatest user of freshwater, often consuming over three-quarters of usage. As much as half of all irrigation water is simply lost due to inefficient flood techniques without ever reaching the crop's roots. Cutting irrigation consumption by one-quarter roughly doubled the water availability for all other productive activities in the region, including industry, power generation, urban use, or recharging groundwater and wetlands. Moreover, proven technologies to multiply agricultural productivity already existed. Microirrigation systems, such as drip and microsprinklers, and laser levels of fields to cause water to distribute more uniformly, were widely successful in reducing water consumption by 30 to 70 percent and increasing yields by 20 to 90 percent in venues around the world, including Israel, India, Jordan, Spain, and America. In the long run these and other methods are necessary elements to meeting the growing challenge of global food shortages. The problem, at bottom, is political—how to promote rapid adoption and how to level the subsidized playing field so that the most efficient farmers reap a proportionate bounty of the market profits they deserve.

American irrigation agribusinesses—led by those in water-poor California—have slowly been making investments to migrate from flooding fields to sprinklers and microirrigation systems. Yet still mostly protected from the discipline of full-market costs by price supports, tariffs, and exemptions from cleaning up all the pollution runoff they caused, politically entrenched agribusinesses lack sufficient incentives to move faster. The result is more than a missed opportunity for the United States to boost its overall economic growth and competitiveness through more efficient allocation of water. There are increasing negative economic, environmental, and equity costs, too. Inevitably, American irrigators are becoming more and more reliant on mining groundwater aquifers beyond replenishable rates to produce America's crops. Over two-fifths of all U.S. irrigation came from groundwater by 2000, nearly twice as much as a half century earlier.

Both from irrigated and rain-fed farmland, vital water ecosystems are also being damaged from the runoff of artificial fertilizers and pesticides. Since it is hard to pinpoint the runoff to a single source, American farm pollution still is not adequately regulated. The pollutants that seep into slow-moving groundwater, wetlands, and rivers are poisoning drinking water and coastal fisheries near and far away. The Mississippi River carries so much nitrogen-rich nutrients from fertilizer runoff that an expanding, biological dead zone without fish life as large as the state of Massachusetts now rings its mouth in the Gulf of Mexico. Similar dead zones around the world have doubled in size since the 1960s and are a major contributor to the alarming collapse of ocean fisheries. It is a classic tragedy of the unmanaged commons, where the producer of an environmental problem is exempted from bearing the full responsibility of its costs and thus of any incentive to rectify it—and, in the age of water scarcity, as well, one of the growing, hidden inequities between water Haves and Have-Nots.

The most intriguing models of improved agricultural water productivity, however, are developing far from America in smaller, water scarce industrial democracies, like Israel and Australia, where necessity is again acting as the mother of innovation. Australia faces the industrial world's harshest hydrological environment: The continent-nation suffers acute aridity, erratic rainfall patterns, exceptionally nutrient-poor, aged soils, and lacks long internal waterway transport routes across its vast expanses. As a result, its population of only 20 million, on a land as large as the lower 48 states of America, is concentrated in the river basin of the southeastern Murray-Darling, which also produces 85 percent of the nation's irrigation, and two-fifths of its food.

Australia developed along an economic model with many similarities to the American West—dammed rivers, subsidized irrigation, and profligate water use by farmers. By the early 1990s, the damage to river ecosystems became too great to ignore. Over three-quarters of the Murray-Darling's average annual flow was consumed by human activity. As on other overused rivers, the mouth was silting up. Water in the lower reaches became so saline that it was poisoning the municipal water supply of downriver Adelaide. Fertilizer runoff was triggering deadly algae blooms along a languid 625-mile stretch of the Darling.

The government's response to the Murray-Darling's ecosystem crisis was to radically restructure its water policies by emphasizing market pricing and trading, and ecological sustainability. The new governing principles ended irrigation subsidies, required farmers to pay for maintaining dams and canals, and, of critical importance, established a scientist-calculated baseline of how much water had to be left in the river to ensure the health of its ecosystem. To facilitate independent water trading, water rights were clearly separated from private property. Governance was managed by a new basin commission.

In little more than a decade, water trading between farmers, farmers and cities, and across state lines, had taken off. There were two computerized water exchanges; farmers were even accustomed to trading over mobile phones. A kindred scheme, akin to America's cap and trade in greenhouse gas emissions, enabled irrigation farmers, who added salt to the soil and into the river basin, to buy "transpiration credits" from owners of forests, whose trees removed salinity by sucking water up through their roots.

Just as its architects had hoped, Australia's water reforms are facilitating the transfer of irrigation water from salty soil to more fertile regions, from use on lower value to higher value crops, and generally from less to more productive methods. Soil salinization has fallen sharply. River fish populations are reviving. Overall water productivity is soaring. Australia's water reforms were implemented none too soon. In the early 2000s, the continent was enduring its worst drought in a century, reviving internecine political rivalries between states and vested interests that could have torn the democracy apart without a preexisting plan. Sheep farms in the arid outback are now being bought by the government to conserve the water the animals had consumed in order to replenish the basin. Water is being more tightly rationed and the government is stepping in to pay the highest price to obtain sufficient water for the priority need of recharging wetlands and safeguarding other components of ecosystem health. Climate change, too, stalks the political struggle over Australia's freshwater—scientists predict a decline in the Murray's flow by 5 percent to 15 percent in coming decades.

As Americans feel about their own bygone, settler frontier, Australians are nostalgic, uneasy, and sometimes despairing at the prospective decline of its individualistic family farm homesteads and livestock and sheep ranches, which alone consume half the nation's agriculture water. But the reality of water scarcity imposes tough, new choices upon modern societies about how to most productively allocate its precious resources. The hard truth is that less than 1 percent of Australia's agricultural land produces 80 percent of its agricultural profits—the vast majority of the rest are marginal enterprises that lived off resource-depleting farm subsidies. In effect, they are cultural relics, worthy perhaps of preservation for social and political reasons but carried along at the expense of some of Australia's competitiveness in the twenty-first-century global economy.

America and other leading industrial democracies have not yet fully awakened to the era's defining water challenge—or to their own strategic advantages in a world order being recast by water scarcity and ecosystem depletion. While the soft-path response emphasizing improved existing water productivity has been gaining ground, it has been doing so only fitfully. No coherent, national policy is helping nurture its embryonic development into an automatic invisible green hand mechanism with the potential to marshal water's full catalytic potency and possibly deliver a transformational, era-defining breakthrough.

Inertia and long-rooted institutional forces are formidable impediments to innovative change at any given moment of history. So it is today. Powerful water bureaucracies cling unimaginatively to approaches forged in previous eras; the U.S. Army Corps of Engineers, for example, is still scoping plans for giant, river interbasin transfers between the Colorado and the Mississippi. Farm subsidies and protective tariffs are so firmly entrenched in the political landscape that Congress has been concentrating on how to extend them to biofuels like corn ethanol, even though doing so will divert water from food production and add to greenhouse gas emissions and global warming. Despite the success of thirty-five years of clean water legislation in improving water quality and stimulating dramatic water productivity gains among private enterprises, the Bush administration's Environmental Protection Agency unsettled the regulatory environment and reopened the door to special interest lobbying by reflexively dropping 400 cases against illegal industrial discharges after a split 2006 Supreme Court decision muddied the terms under which seasonal or remote wetlands and streams deserved 1972 Clean Water Act protections. Similarly, most environmental groups continued to view the world through the original regulatory prism of simple top-down government prohibitions and remain highly suspicious of any market-oriented, soft-path innovations. In short, the jury is still out on whether the water sufficient industrial democracies will fully grasp their leadership opportunity to achieve the water breakthroughs that could trigger another dynamic cycle of creative destruction within market economies or whether its trend toward improved water productivity will merely become a modest way to slim down from an abundant water diet without seriously confronting the underlying, politically entrenched and outdated practices.

Momentous innovations in water history only become clear in

hindsight, after they have meandered and permeated through society's many layers, catalyzing chain reactions in technologies, organizations, and spirit that sometimes combine in new alignments to foment changes transformational enough to alter the trajectory and destinies of societies and civilizations. The way James Watt's steam engine, for instance, interacted with the nascent factory system, canal craze, coal mining and iron casting boom, Britain's growing imperial reach and the nation's new capital accumulation and entrepreneurship-friendly political economic atmosphere, to help launch the Industrial Revolution would have defied prediction at the time. Yet at times it is possible to foresee at least some of the channels through which a great water breakthrough might multiply its effects.

One such channel visible on today's horizon is through water's interaction with three other global challenges—food shortages, energy shortages, and climate change—that together are likely to profoundly influence the outcome of civilization's overarching challenge of learning how to sustainably manage the planet's total environment. While not always perceived as such, the four are so inextricably interdependent that a profound change in any one alters the fundamental conditions and prospects of the others. Irrigation, for example, depends not just on water to nourish crops but also on prodigious energy to pump water from underground aquifers, transport it long distances over hilly landscapes, and drive the sprinklers and other methods that deliver it to plant roots. Artificial fertilizer, too, a mainstay of large-scale irrigated agriculture, requires great energy to produce, and its runoff from cropland has significant impacts on water quality and nourishing ecosystems. Clearing grasslands, rain forests, and wetlands for agriculture, meanwhile, worsens global warming on at least two counts—by adding greenhouse gasses to the atmosphere directly through burning and plowing, and by removing nature's sponges that absorb carbon emissions. A zero-sum conundrum of using water either to grow fuel or food to meet shortages is inherent in the decision over biofuels like corn ethanol. The growing, interoceanic shipping trade in virtual water crops vital to alleviating impending food famines depends upon burning expensive, fossil fuel to power the world's supercontainer fleets.

Near the end of the production chain, processing and canning food products are both extremely water and energy intensive processes.

Ever since the age of waterwheels, water and energy have been coupled in power generation. Today, they are wed on a mass scale through hydroelectricity and in the cooling process of fossil fuel thermoelectric plants; indeed, one of the main constraints on adding more power plants is insufficient volumes of river water to cool them. Filtering, treating, and pumping water for cities also consumes vast amounts of energy. To gauge some idea of the scale of the water-energy nexus, nearly 20 percent of all California's electricity and 30 percent of its natural gas are used by its water infrastructure alone.

Energy crises often became water crises, and vice versa. During the great northeastern U.S. power failure of August 2003, Cleveland mayor Jane Campbell soon discovered she had an even bigger crisis than darkness and a flustered White House wanting her to reassure the public that the cause was a local power grid failure and not international terrorism, when four electric water pumping stations shut down, and threatened to contaminate the city's drinking water with sewage; to stave off a public health catastrophe, she had to launch a second emergency action to warn citizens to boil their water, a practice that continued for two days after the lights returned. The causality of crisis transmission also frequently works in reverse, with drought-induced electrical power shortages diminishing drinking water supplies, irrigation, industrial operations, and shipping. With the river Po 24 feet below its normal level during Italy's severe drought in 2003, power stations shut down from lack of water to cool turbines, and electricity was curtailed to homes and factories. Likewise, hydroelectricity output was halved and shipping reduced on the Tennessee River when it shrank to record levels during America's 2007 southeastern drought.

High energy costs are also one of the major constraints on many approaches to easing water scarcity. A third to a half of desalinization costs are energy, mainly fossil fuels—indeed, any large-scale takeoff of desal seems to be contingent upon a cost breakthrough in some renewable energy source. Likewise, the amount of weighty water that can be lifted from deep aquifers, or transported great distances through interriver basin pipelines like China's South-to-North Water Diversion Project is limited chiefly by the expenditure of energy for pumping such a heavy, hard to manage liquid.

Energy generated from fossil fuels, of course, worsens the mounting global warming crisis. When James Watt invented his steam engine in the late eighteenth century, carbon dioxide in the atmosphere was 280 parts per million; after two centuries of industrialization, the levels had risen by a third to over 380 parts—the highest level in 420,000 years and rapidly approaching the catastrophic threshold of 400 to 500 parts per million that scientists calculate could trigger the irreversible disintegration of the Antarctic or Greenland ice sheets.

The main feedback loops of warming-induced climate change are, in fact, also water related—an increase in what forecasting scientists call "extreme precipitation events": more prolonged droughts and evaporation, heavier flooding and landslides in wet seasons, more intense storms like hurricanes that need minimum temperatures to form, melting polar ice caps and rising sea levels, and, most widely felt of all, a disruptive alteration in historical seasonal precipitation patterns. Due to global warming more spring precipitation is falling as rain instead of snow, intensifying spring flooding and mudslides, and diminishing summertime mountain snowpack melt that normally arrives just in time to replenish dry cropland. Since the world's dam and water storage infrastructure had been designed to accommodate traditional patterns, climate change is rendering that infrastructure increasingly "wrong-sized"—dam reservoirs can no longer capture and store all the available spring precipitation runoff, while its summertime irrigation and hydropower turbine output dwindles from reduced snowmelt. Food and energy output suffers, potentially tipping fragile, water scarce conditions to full-blown water famine. At the very least, a massive rebuilding of infrastructure looms to accommodate the change in climate.

Leading the way is one of history's stellar water engineering nations, Holland, whose society's very physical and democratic political foundations derive from extensive, ongoing water and land reclamation management in a low-lying, heavily flood-prone region. Following a giant 1916 flood, the Dutch accomplished one of the great engineering feats of the first half of the twentieth century. By closing off the Zuider Zee inlet from the North Sea with a giant dike, they created a Los Angeles–sized, artificial freshwater lake and a new water supply source near Amsterdam, known as the Ijsselmeer, or IJ. More recently, Dutch water engineers created a sophisticated combination of water pumps in winter and the natural phenomenon of planting trees—each of whose roots can suck up to 80 gallons a day—to help maintain drainage on reclaimed lowlands. But as rainfall and sea levels have been rising with early climate change, the Dutch have begun to pioneer what may become a new trend in the struggle to sustainably manage water ecosystems—the government is buying reclaimed land so that it can be flooded, thus diverting the rising water from cities and other invaluable societal infrastructure. Among those seeking to learn from the Dutch experience are state leaders from low-lying Louisiana, which is still recovering from the devastating floods of Hurricane Katrina.

In water poor, monsoonal, subsistence countries that lack modern infrastructure buffers from water's destructive extremes, however, the impacts are likely to be reckoned by increased deadliness: Traditional, hand-built mud dams that aren't washed away in the intensified flooding often run dry of their precious, captured, seasonal flow during the prolonged drought that follows, withering crops and killing livestock. For the hundreds of millions who live daily in this precarious, impoverished condition, the consequences are often famine, disease, misery, and death. Worse lies ahead: Climate models predict that the harshest effects of global warming are likely to fall disproportionately on regions with the scarcest water; the temperate zones, inhabited by mostly water-wealthy nations, are expected to suffer the mildest initial effects. Yet in the end, no one will be spared if, as some models predict, the alarmingly rapid melting polar ice caps raise sea levels by 15 to 35 feet and inundate shorelines, and ultimately change the salinity and temperature mix of the North Atlantic enough to halt the interoceanic conveyor belt to bring a frosty, ice age ending to human civilization's brief reign during Earth's unusual 12,000-year stable and warm interlude.

More optimistically, the same relationships work in converse—any important innovation that alleviates water scarcity is likely to multiply the upside benefits to help societies meet their food, energy, and climate change challenges. Genetically modified crops that require less water, or breakthroughs in diffusing microirrigation and remote-sensing systems, would help feed the world's soon-to-be 9 billion and save fossil fuel burning energy now used to overpump groundwater for irrigation. Breakthroughs in desalinization could help provide water for crops and cities in coastal areas. Free standing, small water turbines, another promising innovation, could generate renewable electricity in fast-running streams and rivers around the world, producing inexpensive local electrical power, facilitating the removal of ecosystem-injuring dams and providing a clean alternative for communities, possibly augmenting their autonomy over the means to produce wealth and with it, their democratic voice in society. Much-ballyhooed fuel cells, which might get their hydrogen from water and yield water vapor as a by-product, could provide widely available clean renewable energy that liberates resources for food, water, and ecosystem health. But at least as important as any extraordinary new technologies—indeed, likely much more so—is the gradual, humdrum accumulation of low-tech and organizational advancements in the productive use of water supply already available to man in the form of more efficient existing waterworks, increased small-scale, decentralized capture and storage of existing precipitation, and smarter exploitation of nature's own cleansing and ecosystem renewal cycles. By one estimate, statewide application of existing efficiency techniques could reduce California's total municipal water consumption—with commensurately reduced energy costs—by one third. Water savings in profligate agriculture would be far greater.

With no technological panacea in view comparable to the giant dams and Green Revolution in the last century, the winning responses to the world's water crisis are most likely to emerge fitfully out of a messy, muddling-through process of competitive winnowing and trial and error experimentation with diverse technologies, scales and modes of organization, as each locality and nation seeks to find solutions tailored to meet its particular conditions. Uncertainty, multiplicity, and fluidity are likely to characterize the landscape until clear trends emerge. Historically, Western democracies' market economies have excelled at innovating and creating growth in just this sort of environment—indeed it is one of their main claims to fame. Centrally managed economies and authoritarian states, on the other hand, have tended to do best where technological trends are clear and the main challenge has been to apply them effectively. Thus the Western model enjoys a built-in organizational, as well as water resource, advantage in the unfolding global competition to find the most effective responses to the novel challenges of water scarcity.

Yet history also bears witness that the West's great water advances have been often brought forth by special leadership at key moments. Teddy Roosevelt's visionary commitment at the turn of the twentieth century to exploit the undeveloped potential of America's Far West by launching a new federal institution to promote irrigation and by building the Panama Canal stood out. Similarly, so did Franklin Roosevelt's Depression-era commitment to swiftly multiply the benefits of the Hoover Dam by erecting similar government-built giant, multipurpose dams elsewhere in the country, and De Witt Clinton's use of New York State financing for the Erie Canal early in the nation's history to fulfill the founding fathers' vision of opening a route through the Appalachians to the Mississippi Valley. By creating in each case a coherent environment with clear goals and reliable rules, these leaders inspired confidence among individuals and private enterprises whose participation was necessary for the achievement of their purpose. It is precisely such galvanizing, visionary leadership and reliable commitment to principles that is yet to arise today. Albeit, given the awareness and means in today's world to resist the social and economic displacements often attenuating to such bold, society-changing projects, doing so is comparatively harder. Nevertheless, until it does, the full potential of the organizational innovation of enlisting market forces in the delivery of a sustainable environment—an invisible green hand mechanism that improves water productivity, allocation and ecosystem health through an automatic market price signal for water that reflects the full cost of water supply, delivery, cleansing and ecosystem maintenance—is likely to be impeded by embedded vested interests, incomplete frameworks, and rules of the game that are too uncertain to fully engage private market participants.

Without any imminent solutions to the deepening global water scarcity crisis, water rich nations are likely to be buffeted by a growing number of unfamiliar foreign water shocks, much as they had been from oil in the latter twentieth century. Diplomatic standoffs, water violence, and possibly even water wars are likely to occur in overpopulated regions of extreme scarcity, such as the Middle East. Soaring world food prices, famines, and environmental spillover from the global quantum jump in resource consumption and waste generated by fast-growing Asian giants like China and India threatens to destabilize poor countries dependent upon food imports. When grain prices were spiking in the spring of 2008, World Bank president Robert Zoellick warned that without a new Green Revolution some 33 countries faced social unrest.

The smooth functioning of the integrated global economy and the critical trade in oil and food also depends upon some nation, or group of nations, stepping forward to commit their navies to guarantee unimpeded supercontainer sea passage through nearly a dozen strategic straits and canals that are potential choke points if closed. Feasible threats include terrorists or pirates sinking an oil supertanker in the narrow, pirate-infested Strait of Malacca, a war that closes oil flows through the Strait of Hormuz at the mouth of the Persian Gulf, or a blockage of the Red Sea's southern strait at Bab el Mandeb.

Foreign policies are likely to be realigned and influenced by water-driven alliances, just as they were in the last century by oil. Saudi leasing of cropland in friendly nearby states; a similar, but ultimately unsuccessful effort by South Korea to secure the fruits of Madagascar's potential farmland; and China's provision of work crews and dams, bridges, and other water infrastructure to resource-rich African nations are possible harbingers of the formation of new virtual water and other resource-security and diplomatic blocs within the larger world order that could prove more bonding and outflank the defense umbrellas currently provided by the West. Indeed, water-based alliances could emerge as one of the new international paradigms of the post–Cold War order. New, nontraditional foreign policy thinking is required. Strategic alliances with other regional water Haves, for example, could offer many avenues for exerting increased leverage in many parts of the world. Turkey was already exerting its influence as the Middle East's water superpower to act as broker—and presumptive water enforcer—of peace talks between Syria and Israel. Over four-fifths of fresh river water flowing to oil-rich Arab lands originates in non-Arab states. Under more dire and polarized political conditions as water grows scarcer, it is conceivable as a thought experiment—however highly unlikely in practice—to imagine the formation of a water bloc among Ethiopia on the headwaters of the Nile, Turkey on the Tigris-Euphrates, and Israel on the tiny Jordan, perhaps in league with a cartel among international exporters of food—virtual water—as a diplomatic countermeasure should Middle Eastern oil suppliers turn extremist and try to take excessive advantage of their disproportionate oil power. Similar considerations could apply in central Asia, where the currently dysfunctional state of Tajikistan has potential control over 40 percent of the region's water sources and, through a program of giant dam-building, could deliver badly needed hydropower to nearby Afghanistan and Pakistan. Forward-looking Western foreign policy makers also have to be cognizant of the enormous leverage China's control of Tibet gives it over the mountain sources of the great rivers, and therefore the economic and political fate, of Southeast Asia.

Endless foreign policy challenges are also likely to emanate from the world's abject water poor, roughly calculated as the one-fifth of humanity without access to enough clean water for their basic domestic needs of drinking, cooking and cleaning, the two in five without adequate sanitation, including simple pit latrines, and the 2 billion more whose lives are devastated every decade by their exposure to recurring water shocks like floods, landslides, and droughts. For the most part they live in Africa and Asia, both in failing states and poor, usually rural regions of developing ones. For them, progress is not primarily measured in terms of harnessing hydrological resources to enhance their productive society but in terms of brutal survival against the natural ravages of unmanaged water and the prevention of catastrophes stemming from the collapse of aging and often poorly built waterworks. As world population soars, so too will the absolute number of abject water poor and international spillover to the richer parts of the world. From India to Africa, hundreds of thousands of climate migrants are already on the march from unbuffered water shocks, shortages and infrastructure failures—there is no reason to expect that they will politely stop at their national or regional borders to quench their driving thirst for survival.

On the hopeful side, a Western breakthrough in exportable techniques that dramatically increases existing water use productivity, improves sustainable water ecosystems, and enhances international food export supplies, of course, would quickly become a powerful lever to helping other nations and individual communities cope with their water scarcity challenges. Abundant production of internationally traded food could help strengthen the existing world political economic order by reassuring water-poor countries that their best interests lay in relying upon the liberal, free-trade region to provide, at fair prices, the food they need to import. They could yield extensive diplomatic goodwill for Western interests and promote indigenous democratic development in other parts of the world as well.

But any such water-driven democratic development would likely require imaginative, flexible, and conditional solutions beyond solely large-scale, national government-ministry-directed projects of the twentieth-century variety, including a willingness to build upon and help revive traditional, small-scale water management practices from the precolonial era. In rural parts of India and central Asia where British colonialism did not penetrate with its centralized, modern water techniques, for example, some such traditional methods and local governing mechanisms have remained intact. Village built and managed water tanks in India offer small, local, partial, but helpful solutions to the nation's great water storage shortages. In rural Afghanistan and eastern Iran, highly respected village *mirabs,* or water foremen, are still selected annually among local orchard growers and farmers who share a water source to set watering schedules and amounts and to settle disputes so that wellhead and upstream farmers do not consume more than their fair share before it flows to users at the bottom. The *mirab* system is remarkably reminiscent of the Dutch water parliaments that became a prototype for the founders of the Dutch Republic's democracy, as well as of democratically functioning local institutions like Valencia's public water court. It does not require too great a leap of thinking to imagine how expanding the power base of such long-established, local water institutions and practices might become one of the building blocks to rebuilding failed, or never fully formed, states that otherwise menace the world order.

Although the water crisis of the world's poorest has been on the international agenda and the subject of numerous, high-level meetings among serious-minded people since the 1970s, and the U.N. Millennium Development Goals, endorsed by world leaders at the second Earth Summit at Johannesburg in 2002, included a specific target of halving the proportion of people without access to clean water and basic sanitation by 2015, the truth is that the legions of the world's water disenfranchised are continuing to swell. The familiar dynamics of ruthless indifference among those far away and diffused political power are at perpetual play. Moreover, one perverse, unintentional effect of the multilateral campaign for clean drinking and sanitary water has been to divert increased investment away from also badly needed food production infrastructure. Without a pressing crisis to rivet all world leaders' serious attention, there is not nearly enough financial commitment from rich countries, nor even sufficient political will from government leaders of many suffering, water poor ones. In a changing global order without a single dominating world power to set the agenda, the task of rallying action is chiefly being left to an amorphous international process led by weak, multilateral institutions and diverse nongovernmental entities. If only a small fraction of the debate and study they have committed over the years had been translated into concrete action, the water crisis might have been solved many times over.

Several promising principles have been enunciated. These include striking a balance between the "3 E's": Environmentally sustainable use of water; Equitable access by the world's poor to fulfill their basic water needs and for communities to share in the benefits of local water resources with the poor; Efficient use of existing resources, including recognition of water's value as an economic good. Yet no galvanizing consensus has emerged on how to practically realize these or other principles. As a result, the small army of jet-setting, water conference-goers often resemble the proverbial endless talking shop, issuing declarations of broad good intentions but disagreeing too much to get on board with concrete paths proposed to achieve them. This was illustrated at the third triennial World Water Forum held in Japan's historic capital of Kyoto in 2003, impressively attended by 24,000. Conference-goers became embroiled in a furor over a report of a high-profile committee headed by former IMF managing director Michel Camdessus that proposed specific financial means to achieve the Millennium Development Goals for water. Citing the staggering investment sums needed—on the order of $180 billion globally per year—for water infrastructure, and recognizing the paltry commitments industrialized governments were willing to make, the Camdessus report strongly endorsed private sector participation; adding fuel to a controversial suggestion, it cited large-scale, centralized waterworks like dams as potential targets for private financing that are an anathema to activists who had fought against them on the World Commission on Dams. Protests erupted at the session where the Camdessus report was launched. Angry anti-private-market water activists, NGO representatives, and union members marched through the venue, and unfurled a banner that read, "Water for People, Not for Profits."

On current dynamics and trajectories, not only will the U.N.'s self-declared International Decade for Action "Water for Life" (2005–2015) likely expire without achieving the Millennium targets, but the massive dry shift in the global water continuum of Haves and Have-Nots will continue to lurch toward deepening scarcity. Countries with scarcity are likely to veer toward famine; countries already in water famine face greater human catastrophes and political upheavals. Overtaxed water ecosystems are likely to grow more and more depleted and less and less capable of sustaining their societies. As the gulf between those with sufficient water and those without deepens as a source of grievance, inequity and conflict, the new politics of scarcity in mankind's most indispensable resource is becoming an increasingly pivotal fulcrum in shaping the history and environmental destiny of the twenty-first century.

EPILOGUE

Looking back over time brings into relief the close association between breakthrough water innovations and many of the turning points of world history. From about 5,000 to 5,500 years ago, following several millennia of experimentation and development, large-scale irrigated agriculture in the arid, flooding river valleys of the Middle East's Fertile Crescent and the Indus River, and along the Yellow River's soft loess plateaus, provided the technological and social organizational basis for the start of modern human civilization. During the same period, man began transporting large cargoes on rivers and along seashores in reed and wooden sailing vessels, eventually aided by a steering rudder. Sailing in turn, nurtured the rise of international sea trade and Mediterranean civilizations where indigenous agricultural conditions were relatively poor. Civilization's slow march through rain-watered, cultivatable lands began in earnest a little under 4,000 years ago with the spread of plow agriculture that allowed more intensive farming over a greater expanse of cropland through the application of animal power.

Mastery of the art of quenching red hot iron in water to make steel weapons and tools about 3,000 years ago made possible construction of qanats and aqueducts, which reliably conveyed enough freshwater to sustain the rise of the great cities that anchored every civilization. The inland expansion of civilization was facilitated by the innovation of transport canals that connected natural waterways, starting in China 2,500 years ago and replicated everywhere with great impact over the centuries from southern France's seventeenth-century Canal du Midi to America's nineteenth-century Erie Canal. Some 500 years ago, global distance barriers were defeated by Europeans' momentous discovery of how to sail back and forth across the open oceans; from the mid-nineteenth century, interoceanic sailing times were compressed by the cutting of great sea canals for new, speedy steamships and gunboats that forged the world order of the colonial age.

Just prior to start of the Christian Era 2,000 years ago the seminal invention of the waterwheel captured the power of flowing water to turn mills to grind man's daily bread; a thousand years later water-power was applied with more complex gearing to a widening array of industrial applications and ultimately, a quarter of a millennium ago, to power the first factories. The waterpower barrier was finally shattered by the steam engine in the late eighteenth century—arguably the greatest invention of the last millennium which catalyzed the defining innovations of the Industrial Revolution—and was transcended yet again by hydroelectric power in the late nineteenth century and a panoply of water-assisted power generation inventions in the twentieth century. The sanitary revolution helped foment transformations in human health, demography, and clean drinking water that sustained massive modern industrial urban concentrations. Less than a century ago, 5,000 years after the original big dams of antiquity, history's first giant, multipurpose dams began harnessing the planet's great rivers to deliver electricity, irrigation water, and flood control on a massive scale that remade landscapes at a stroke and was vital to launching the worldwide Green Revolution that nourished humanity's stunning population surge. Modern industrial technologies also permitted man to mine the earth of water from its deep underground reservoirs as he had drilled oil, and to pump the water unprecedented distances over and beyond mountains in long-distance aqueducts. By the end of the twentieth century, an ocean fleet of intermodal supercontainers speedily delivering goods ordered from foreign factories from a nearly real-time information web to local markets across the planet served as the transport backbone of the new, integrated global economy.

With each major breakthrough, civilization had been transformed by the conversion of a key water obstacle into a source of greater economic power and political control; invariably its accessible water resources became more productively utilized and more voluminous in absolute supply. Time and again, the world order of the age was recast, elevating societies to preeminence that proved most adept at harnessing the new form of water's catalytic potency and pushing the laggards toward decline. Today, man has arrived at the threshold of yet a new age. His technological prowess has reached the point that he possesses the power, literally, to alter nature's resources on a planetary scale, while soaring demand from swelling world population and individual levels of consumption among the newly prospering urgently impel him to use that prowess to extract as much water as he can. The alarming, early result is a worsening depletion of many of Earth's life-sustaining water ecosystems that, nonetheless, are not keeping pace with the growing global scarcity.

Until now, all history's water breakthroughs have fallen into four traditional categories of use—domestic needs, economic production, power generation, and transport or strategic advantage. At the dawn of the twenty-first century, civilization faces an imperative fifth category that defines the era's new water challenge: how to innovate new governing organizations and technical applications that make available sufficient supplies of freshwater for man's essential purposes in an environmentally sustainable manner and relieves the scarcity of an increasingly thirsty planet. No technological panacea that extracts more renewable water from nature is available or on the near-term horizon to answer the call. Some societies may borrow time by mining Earth's underground reservoirs or transferring freshwater from river basin to river basin until their total water reserves give out. For others, comprising many hundreds of millions of people, the day of reckoning has already arrived. For everyone sharing the planet, the destiny of human civilization as we know it hinges on the responses to this challenge. History suggests those societies that make big breakthroughs that maximize productive use of their renewable water resources and possibly usher in a turning point in practices and applications are the likeliest to be rewarded with rising economic wealth and international power.

The most obvious, environmentally sustainable large source of freshwater at hand to alleviate the crisis is simply to use the current supplies more efficiently. Tapping them, however, is more difficult than it seems at first glance. For starters, it requires major organizational changes in the way water is managed, politically and economically. Enormous inefficiencies, waste, and political favoritism have been built up in the government command systems that controlled water use in almost every society through the centuries—the true paradox of water is that despite its scarcity, it nearly everywhere remains the most shortsightedly and poorly governed critical resource. Reform can come in one of two main ways: by foresightful, effective, top-down political leadership that uproots its own embedded systems and then makes wise choices about the governing technologies and methods to replace them; or by turning loose the proven reorganizing power of impersonal market forces within a properly regulated, governing framework to winnow out the inefficiencies and redeploy the existing water resources from less to more productive hands.

It is, of course, conceivable that uncommon leadership might arise within a handful of governments around the world to implement the necessary internal reforms. Yet judging from history, it seems highly imprudent, even fanciful, to bet that such exceptional leadership will arise across many continents at one time. Better—more pragmatic—odds of success almost surely lie with greater reliance upon the self-interested, profit motive of individuals organized by the politically indifferent market anchored in a pricing mechanism for valuing water that reflects both the full cost of sustaining ecosystems through externally imposed environmental standards and a social fairness guarantee for everyone to receive at affordable cost the minimum amounts necessary for their basic needs. Those uneasy with the market system's history of yielding widely unequal wealth distribution patterns should be partially heartened by the fact that competitive, free markets' singular devotion to lucre has on its side the considerable merit of being one of history's most subversive and undiscriminating enemies of unfairly entrenched privilege and deserves credit as a prodigious creator of the wealth that necessarily precedes any debate about how to make its distribution more equitable.

A second obstacle is that the precondition for any effective organizational innovation, either market-based or government-imposed, is adequate water infrastructure and control for basic delivery, protection against shocks, waste removal, and measurement of use. In vast swathes of the world this precondition is in shocking deficit. The dearth of infrastructure is central, for example, to the deplorable failure to achieve the most elementary, universally sought goal of providing at least 13 gallons, or 50 liters, to meet the minimum basic daily domestic and sanitary needs for each individual. This is a minuscule drop—the equivalent of eight low-flow toilet flushes—that even the water poorest societies have enough supply to provide. Any legitimate government would readily strive to do so. Moreover, many nongovernmental and official international institutions have been trying to assist countries to achieve it and other very basic water needs. Prominent water experts are campaigning for this tiny amount to be recognized as a universal human right to water. Yet it is unachieved for two-fifths of mankind for one overriding, simple reason—the deficit of existing infrastructure and competent, institutional governance.

Finally, there is no one-size-fits-all remedy for the global crisis of water scarcity. Each society's hydrological reality and challenges, like its political, economic and social conditions, are unique. Some societies have to cope with monsoonal seasonality, others with perennial rainfall, and some with almost none at all. Some entire regions, such as Africa, have scarcely tapped their hydroelectric power development and irrigation water storage potential; while in America and Europe, additional giant damming has mostly yielded environmentally counterproductive and diminishing economic returns. Investing local, mostly poor stakeholders who have historically been dispossessed by large waterworks in the success of a new water project is a paramount challenge in many developing countries but almost nonexistent in leading industrial democracies with responsive governing structures. Some nations' most urgent need is to resurrect and expand traditional small-scale, low-tech methods for water storage and terracing, while for others it is to apply modern water technologies on a large scale as rapidly as possible. Pragmatism, not universality or bias of principle, is what is called for: It is, quite frankly, hypocritical and even morally obscene, to witness activists and officials from water-Have nations whose material benefits—albeit often gained with ugly social, economic, and environmental side effects—have been so visibly aggrandized by giant dams to use their international clout to reflexively oppose virtually all similar development in water poor ones. In short, the world water crisis is a multidimensional crisis. It requires myriad responses targeted at each specific layer and situation, much trial and error adaptation of what works elsewhere, vast capital investment in infrastructure, relentless hard work governed by a pragmatic intelligence and a few, flexible guiding principles. The world has no previous model or institutional framework for coping with it. Everything has to be invented on the fly.

Every society in the age of scarcity faces its own particular version of the era's defining water challenge. How each copes with its challenges, and which societies make the most dynamic breakthroughs, will partly dictate the winners and losers in a century where water's role is of increasingly paramount importance. History is agnostic as to whether a water rich society is likeliest to seize upon its opportunity to exploit its initial water resource advantage in a dynamic new way or whether its relative comfort instead will make it a complacent onlooker while some water indigent society, driven to innovation by the dire necessity of survival, makes the pathbreaking innovations that unlock a new, hidden aspect of water's extraordinary, catalytic properties and transforms the obstacle of scarcity into a propellant of expansion toward wealth and possible global leadership. Whether in the end it is a Western liberal democracy, China's authoritarian, state-directed market system, a resurgent totalitarian, command economy state like antiquity's hydraulic societies and the industrialized twentieth century's Nazi Germany or Soviet Union, or a nation rising on some other new model, which proves most adept at making the breakthrough responses, will influence the type of governing model that prevails in this round of history's endlessly shifting contest between political economies.

Throughout history water has been a great uniter and a great divider, a barrier and a conveyance, but always a great transformer of civilization. As history's most critical natural resource, vital in virtually every aspect of human society, and one that interactively leverages food, energy, climate change, and other grave problems facing a world rising toward 9 billion souls, all striving for first world material standards, water also represents an early proxy test for human civilization's impending survival challenge of learning how to sustainably manage Earth's total planetary environment. Geographer Jared Diamond has grimly concluded that, on current trajectories, there are simply not enough planetary environmental resources, including accessible freshwater, to even come close to satisfying the aspirations of several billions to move up the development ladder to industrial-world levels of consumption and waste. As in previous eras, human population and available environmental resources are again widely out of balance. Famines, genocides, wars, disease, mass migrations, ecological disasters, and untold miseries are history's remorseless mechanisms for reequilibration. In the end all nations will be buffeted, if not engulfed, by the myriad feedback channels of water crises that originate elsewhere. How much tumult and suffering lies ahead depends in significant measure upon how well mankind manages the total global freshwater crisis on our shared planet. Looking farther ahead, the extraordinary, unique substance that gave life to man and shaped the destiny of human civilizations is still the indispensable, prerequisite stepping-stone to some day transplanting our species beyond Earth's sphere to colonize other orbs in the solar system.

There is one more special attribute about water that must inform any study of its role in history: The inextricable affinity between water and our own essential humanity—not merely with human life, but with a dignified human life. My visit to Kenya in the summer of 2004 set off a personal alarm of just how dehumanizing and economically crippling the lack of water for basic needs could be. It drove home the mind-numbing inequity that a majority of humanity still struggles to extract its meager material surplus from nature using obsolete and even ancient water technologies. In the semiarid, rural Chyulu Hills in southeast Kenya on the edge of the Great African Rift Valley congeries of otherwise vibrant, culturally robust communities live in literally dirt-poor subsistence for one overriding reason—insufficient freshwater.

It shocked my sense of common humanity to see the small group of men and women work so tenaciously with hand tools such as picks, shovels, and sisal sacks to perform the backbreaking manual labor of digging and carrying the reddish dirt week after week to reinforce the earthen dam they'd built nineteen years earlier—precisely like those built in ancient times–to trap the seasonal monsoonal rainwater through the dry season so that their cattle can survive, when they and I knew that one-day access to a simple bulldozer could do the job of a whole season, and a few days with a cement mixer could alleviate the task for years. In the nearby Machacos Hills, where low-tech terracing has improved water management and agricultural production, Kenyan farmers step up and down for hours each day on a treadle water pump—much as Chinese rice farmers did using bamboo tubes centuries ago and modern Westerners do at the gym on their exercise StairMasters—to lift water from a muddy creek up the hillside in plastic tubes to fill cans they use to hand water their crops.

More striking still is the ubiquitous sight of large numbers of women and children acting with their feet by marching two to three hours or more per day on dusty roads to fetch clean water from wells or other sources in large, yellow, plastic "jerry" cans, which they carry on their heads, on the ends of poles laid across their shoulders, and packed on bicycles or donkeys. A family of four needs to transport around 200 pounds of water each and every day to meet its most minimal drinking, cooking, and cleaning needs. To manage such an impossible weight, two trips to the well each day by mother and children are not uncommon. Carrying water for basic subsistence devours school time for children and places a dispiriting burden on the enterprising will of parents to struggle out of their material privation. That the water carrying falls traditionally on women adds the insult of gender inequity to the tragedy. There was genuine rejoicing when the two miles of piping our small, humanitarian group of American volunteers had financed, for a pittance in Western terms, was connected to the well pump and began to deliver water directly to a simple, plastic water tank located in one of the villages.

I will never forget the sense of disempowered injustice we felt when we met a thoughtful young man as enterprising, vivacious, and worthy of a fair opportunity as anyone his age in industrial America, Europe, or Asia who was studying on his own every night, in a home without electric lighting, for a high school equivalency exam on the remote chance that he could qualify for a special scholarship to attend the University of Nairobi; his family was too poor to pay the couple of hundred dollars for his formal high school education, and I knew that had the water pipes we financed arrived years earlier and been put to productive economic use for modest, gravity-fed irrigation as well as drinking and cleaning, this young man might well have gotten the professional chance he deserved and which my own daughters take for granted. His country would also have gained important human capital in its struggle for development. In Ethiopia, where my wife, a high school teacher, traveled in the summer of 2008, the situation was similar, and the poverty even more desperate. When she arrived in the beautiful, remote mountain highlands that provide the headwaters of the Blue Nile, she felt as if she had been dropped back into medieval times as she saw farmers scratching out meager livelihoods with oxen-pulled wooden plows.

As recently as the 1950s in early postwar France, my Bretagne mother-in-law was still washing clothes with river water and carrying upstairs water buckets of captured rainwater with which to bathe her children and cook the family's food. It further illustrates how much water history was everywhere a layered history: Ancient, medieval, and modern methods always coexist; yet, crucially, it is an *unevenly* layered history, imparting enormous—and easily overlooked—advantages to the comfortable water Haves and crippling disadvantages, starting with a life handicapped by malnutrition, ill health, and sacrifice of education to the daily search for water, to the world's water Have-Nots. The need for water trumps every human principle, social bond, and ideology. It is literally indispensable. With extreme water scarcity showing through as a root cause of many of the world's famines, genocides, diseases, and failing states, I am inclined to believe that if there can be a meaningful human right to any material thing, surely it starts with access to minimum clean freshwater.

At the end of the day, how each member of the world community ultimately acts in response to the global freshwater crisis is not just a matter of economic and political history, but a judgment on our own humanity—and the ultimate fate of human civilization. As one scientist succinctly put it: "After all, *we* are water."

ACKNOWLEDGMENTS

In writing the history of water, I have had the intellectual pleasure of having been able to stand upon the broad, high shoulders of many first-rate thinkers and scholars who have written insightfully about the subject from the perspective of their own disciplines and times. I salute them, and the civilized enterprise of accumulating knowledge that hopefully helps human society to inch forward with better understanding and management of our shared world.

Many exceptional personal contributions have also informed my work. I have learned an immense amount from David Grey, who heads the World Bank's water group. David not only has an amazingly profound and broad understanding of the complexities of today's water issues, but he brings inspiring passion, energy, intelligence, and an encyclopedic knowledge of water history to his work. At the outset of the project, Dr. Allan Hoffman of the U.S. Department of Energy and senior adviser to Winrock International's Clean Energy Group, impressed upon me the inextricable interconnectedness of water, energy, and climate change issues, and pointed me in fruitful directions. Many of my conceptual frameworks developed from stimulating conversation with Peter H. Gleick, president of Pacific Institute, a fantastically useful, research-based NGO specializing in water issues, and with Professor J. A. "Tony" Allan of the School of Oriental and Asian Studies, Kings College London, who imparted his important idea of thinking of food as "virtual water" as we tackle the world's interrelated food and water problems.

Others who generously shared their time and minds were Jim McMahon at the Lawrence Berkeley National Laboratory, Philip Duffy and Andy Thompson of the Lawrence Livermore Laboratory, and Ambassador John McDonald of the Institute for Multi-Track Diplomacy. They provided me with an educated start in understanding energy, climate change, hydrology, and global water diplomacy, respectively. One delightful experience was discussing the waters of Rome with Katherine Wentworth Rinne, whose interactive cartographic project, "Aquae Urbis Romae," at the University of Virginia tracing the evolution of the Eternal City's water development is exploring new boundaries in the use of online technology to study history. Professor Peter Aicher of the University of Southern Maine enthusiastically shared his extensive knowledge about Rome's aqueducts and water management. Veteran journalist Bill Kelly was a veritable Virgil in guiding me through the intricate underworld of California and Colorado River water politics and ever graciously answered my many follow-up questions. I'd also like to thank Bob Walsh of the Bureau of Reclamation office at Boulder City near the Hoover Dam for a warm and illuminating welcome to an unannounced walk-in.

My parents, Ruth and Lee Solomon, deserve a special "shout-out" in more ways than can be expressed for their unwavering, lifelong encouragement and comfort whenever adverse winds buffeted. My father's insightful critique of each chapter as it was written provided invaluable feedback that enriched the final text of *Water*. I've been privileged to have him as a best friend and intellectual companion as well as a father.

Jean Michel Arechaga and Nicole Macé were tireless and resourceful field detectives in investigating water mills, canal locks, and weirs with me in northwestern France. The Monagan Writers' Group was a tolerant foil for my testing out of myriad stories and ideas about water during lively give-and-takes at our regular biweekly luncheons at the Women's Democratic Club; I regret only that John Monagan did not live long enough to see the final publication of the work. My long involvement with the environmentalists and local community activists of Washington, D.C.'s Klingle Valley Park Association deepened my appreciation of the profound ways water interacts with urban ecosystems and infrastructures, as well as the obdurate difficulty of overcoming entrenched, reflexive political opposition even when all objective analyses argue overwhelmingly for environmentally sustainable, economically less-expensive, and democratically more-equitable alternatives to the status quo.

Nola Solomon did a magnificent job of preparing the endnotes and the bibliography, and correcting some of the text along the way. Brittany Watson was a stellar blend of artistic creativity, flexibility, and perseverance in creating the maps. Stephanie Morris deserves thanks for her generous guidance on the artwork. Cordelia Solomon provided valuable assistance in marketing research, and, along with Brittany Wilbon, helped me organize the research material at an early stage. Aurelia Solomon made a valued contribution on publicity research, as well as stimulated fruitful discussion on the environmental aspects of water.

Tim Duggan of HarperCollins has been a paragon of what an editor should be: patient, encouraging, considerate, always with the big picture in view, ready with sensible suggestions, and possessing a knack for knowing just the right time and degree to apply pressure. The foresight, efficiency, positive spirit, and all-around intelligent beneficence of Allison Lorentzen, Tim's wonderful assistant, facilitated the project from start to finish.

As always, my agent, Melanie Jackson, has been outstanding in all facets and phases—a great collaboration.

This book also could not have been completed without some timely and superlative medical intervention. Above all, I owe unrepayable debts of gratitude to neurosurgeon Fraser Henderson and infectious disease expert Dr. Mark Abbruzzese, as well as to doctors William Lauerman, Kevin McGrail, Gil Eisner, and James Ramey, and the remarkable team of nurses at Georgetown Hospital's concentrated care unit.

The most special acknowledgment of all, however, is reserved for Claudine Macé, my comrade in passion and life's adventure for nearly three decades over many continents and conditions. A dedicated high school teacher in Washington, D.C., Claudie organized a service learning trip to the Rift Valley of Africa a few years ago to lay water pipes for waterless villages in rural Kenya that became a transforming voyage of discovery about the surpassing importance of water to human life for all who participated in it. I aspire to fulfill her unflagging expectation that the best is yet to come.

Finally, I'd like to thank the unnamed, many tens of thousands who are out in the field working each and every day in all kinds of conditions doing the good work to alleviate, and hopefully one day solve, the local and global water challenges facing us all.

NOTES

Prologue

4. ninefold increase in the twentieth century: Paul Kennedy, foreword to McNeill, *Something New Under the Sun,* xvi.

Chapter One: The Indispensable Resource

10. water in the soil: Water's high specific heat capacity, which allows it to maintain its liquid form over an extremely wide range of temperatures and pressures, is essential to Earth's having maintained its moderate climate despite the fact that the Sun had grown about 33 percent hotter over the past 4 billion years.
10. planet's infancy: Most scientists now believe that instead of being a hellish fireball 4.2 billion years ago, Earth was fairly settled geologically, with both land and oceans, with parts of the surface covered in ice due to the 30 percent lower heat output of the young Sun.
10. climate change cycles: Short cycles covering centuries of warm, wet climate commonly alternated with long cold, dry, windy periods; sometimes climates fluctuated unstably between extremes within a single year. Over the past 700,000 years, these short cycles have been dominated by dramatic swings between very long, severe, dry ice ages and warm, wet interludes.
11. favorable climatic conditions: Alley, 3, 14. The stability of the current warm period is the longest in the 110,000 years of ice core data. Alley notes that the fluctuations that marked Earth's past "were absent during the few critical millennia when humans developed agriculture and industry."
11. atmospheric water vapor: Water vapor was the planet's most prolific, heat-trapping "greenhouse gas."
12. warm Atlantic Gulf Stream: Water temperature variations also help drive the

oceanic wind systems, including both the sinking, weak doldrums near the equatorial horse latitudes loathed by mariners in the age of sail as well as the Atlantic's favorable, wet trade wind system, which, when at last decoded, became the ocean-crossing expressways for European explorers' world-transforming Voyages of Discovery.

12. conveyor belt: Too much extra cold freshwater introduced into the North Atlantic by the melting of polar glaciers—say, from global warming—might trigger a new shutdown of the conveyor, setting off an abrupt return to ice age conditions. Past shutdowns and slowdowns appear to have been quite abrupt, as short as fifty years. Once shut down, the conveyor was difficult to get moving again.
12. fresh liquid water: Water stock data is primarily from Shiklomanov and Rodda, 13, And Gleick, *World's Water, 2000–2001,* 19–37. The total amount of water on Earth is 1.386 billion cubic kilometers, of which 96.5 percent is in the oceans and only 2.5 percent (or 35 million cubic kilometers) is fresh.
13. three lake systems: Shiklomanov and Rodda, 8, 9.
13. constantly being replenished: Transpiration from plants also adds to water vapor. Much of the precipitation never reaches land because it evaporates en route. To give some sense of proportion, it takes about 3,100 years for a volume equal to all the world's oceans to recycle through the water cycle.
13. lost in floods: Some 15 percent of falls occur in the Amazon rain forests, which have less than one-half of 1 percent of the world's population; water-short Asia receives 80 percent of its rain as hard-to-capture monsoons that fall during only five months (from May to October).
14. "Every day the sea": Durant and Durant, 14.

Chapter Two: Water and the Start of Civilization

15. Arnold Toynbee: Toynbee, *Study of History,* chap. 5, "Challenge and Response," 60–79.
17. biological cycles: During a day of normal activities, approximately 0.3 quarts were exhaled, 0.5 quarts sweated out, and the excess expelled as waste.
17. death struck: Swanson, 9. As the body dehydrated, the blood thickened and the heart had to pump harder as circulation became less efficient.
17. "Almost every mythology": Campbell, *Hero's Journey,* 10.
17. four primary terrestrial elements: Ball, 3, 4, 117–120. Water, Earth, Fire, and Air were the Greek foursome; Chinese philosophers, from about 350 BC, agreed with the first three but replaced Air with Wood and Metal. The Mesopotamian cosmology concurred on Water and Earth, but substituted Sun for Fire and Sky for Air, and added its unique fifth, Storms.
19. mini ice age: Alley, 3, 4, 14; Kenneth Chang, "Scientists Link Diamonds to Quick Cooling Eons Ago," *New York Times,* Janurary 2, 2009. The well-documented, millennium-long paleoclimatic episode, called the Younger

Dryas event (after a tundra-loving plant), was probably triggered by the collapse of a huge melting ice sheet or lake in North America that sent a torrent of cold freshwater draining through the St. Lawrence Seaway into the North Atlantic, slowing the oceanic conveyor belt and temporarily reversing the retreat of the ice age. What caused the water surge is much debated, with some hypothesizing a meteor strike. The event was incomparably more extreme than Europe's Little Ice Age that ended in the mid-nineteenth century and triggered significant lifestyle adaptations around the continent.

20. Jericho's location: Braudel, *Memory and the Mediterranean,* 40–45. Control of trade routes and the watery sources of salt, so prized over the centuries that it was accepted as money and traded for gold, was a source of power and wealth until modern times. Jericho's founding goes back to about 9500 BC. Two other important original cities were Jarmo, on the edge of a deep wadi in the Zagros Mountains that fed the Tigris River, and Catalhüyük in mountainous Anatolia, which was advantaged by its virtual monopoly in the trading of the highly prized, hard-edged volcanic stone, obsidian.
20. farmers to relocate: Some paleoclimatologists believe that the proximate force driving the advent of irrigation farming in the Near East may have been an increasing regional aridity exacerbated by a 200-year cold drought period between 6400 and 6200 BC, which caused farm hilltop settlements to be abandoned across the Levant and northern Mesopotamia.
21. independent, smaller communities: McNeill, *World History,* 46.
22. barbarian waves: The four great barbarian waves were (1) the Bronze Age charioteers, circa 1700–1400 BC; (2) the Iron Age invaders from around 1400–1200 BC; (3) the Hsiung-nu from 200 BC and then in the fourth century AD the Juan-juan confederations of the eastern steppes; and (4) the great Turkish-Mongol invasions from the 700s arguably to the fall of Constantinople in 1453.
23. world population: Ponting, 37.

Chapter Three: Rivers, Irrigation, and the Earliest Empires

25. "creates a technical task": Wittfogel, 15.
25. fast-growing maize: Braudel, *Structures of Everyday Life,* 161. Maize was a miraculous plant due to three attributes: (1) it was fast growing, (2) it was edible even before it was ripe, and (3) it grew with little effort—requiring less than fifty days of total farming work. Potatoes thrived at high altitudes.
26. giant dams built in the twentieth century: The pioneering Hoover and Grand Coulee dams were built by New Deal America; major Russian and European giant dam building coincided with the rebuilding after World War II; and Communist China, along with many newly independent developing countries, erected dams as foundations of their new regimes.
27. three great kingdoms: Kingdom date estimates vary by source. Those used

here combine the Thinnite period with the Old Kingdom, and follow Grimal, 389–395.

28. nilometers: Collins, 13–14. The earliest existing nilometer readings, covering the period up to 2480 BC, are from Memphis; although the nilometer itself has disappeared, its data were carved on the stela fragment known as the Palermo Stone.

28. total water volume: Based on renewable water resources per year. Shiklomanov and Rodda, 365.

28. fertile black silt: Ancient Egyptians called this flooded, silt-laden plain the "black land," or *kmt,* which was also their name for Nile Valley Egypt itself. The barren soils untouched by the floodwaters were known as the "red land."

30. Menes: Grimal, 37–38; Shaw, 61.

31. reservoir dam: Smith, *History of Dams,* 1–4. It is believed this dam failed from overflow shortly after its construction.

31. peasant's duty: Egyptian frescoes and bas-reliefs depicted the dreary, duty-bound daily life of peasants performing their routine farm toil in the fields, carrying grain to the granary, drawing fishing nets, unloading boats, and brewing beer, all under the stern watch of an armed supervisor.

34. a transformative innovation: The world's earliest surviving water clock also dates from the New Kingdom.

34. secure precious, high-quality timber: Braudel, *Memory and the Mediterranean,* 59–60. Owing to the dearth of useful tree species, both Egypt and Mesopotamia traded and sometimes waged war to secure vital timber from Levantine forests. Egypt's only hardwood trees were the sycamore and the acacia.

36. Neko's canal: Neko's canal may have tracked a possible previous canal effort obscured to history by the filling in of the desert sands.

36. 120,000 died: Herodotus, *Histories,* 193.

36. sultan and the Christian king: Lewis, *Muslim Discovery of Europe,* 34, 38.

38. "A society dependent": McNeill, *Rise of the West,* 32; Unlike on the Nile, upriver transport on the twin rivers required laborious oar power and portage.

39. Mesopotamia: Van De Mieroop, 13.

41. "the first efficient means": Mumford, 71.

42. flood myth: Archaeologists have uncovered evidence of frequent, huge inundations. The flood that submerged the Sumerian city of Shuruppak in 3100 BC may have inspired the Bible's great flood story.

42. easier to control: Campbell-Green.

42. "Why . . . if Sumer": Leonard Woolley, *Ur of the Chaldees* (1929), quoted in Ponting, 69–70.

43. "black fields becoming white": Cited in Pearce, 186.

43. 1700 BC almost no wheat: Ponting, 71. See also McNeill, *Rise of the West,* 48.

44. water war: Van De Mieroop, 48–49. See also Gleick, *World's Water, 1998–1999,* 125; Reade, 40–41; and Pearce, 186.

44. under modern Baghdad: Van De Mieroop, 64.

45. earthworms had perished: Kolbert, 95, 97. The original research was done by Yale archaeologist Harvey Weiss, who led the excavation of ancient Tell Leilan in modern Syria near the Iraq border.
45. "provider of abundant waters": Harris, 123.
46. "If anyone be too lazy": Hammurabi, Law 53.
47. Hard iron weapons: Refined, harder steels with much-sharper edges were produced in the ensuing centuries, starting in India and China. For centuries, Western smithies vainly tried to reproduce "watered steel" (as it was known in Persia) or "Damascus," or "damask," steel (as it was known in Europe). Success came only with the application of waterpower in the early nineteenth century—the birth of modern metallurgy.
47. "gleaming in purple": George Gordon, Lord Byron, *The Life and Work of Lord Byron*, "The Destruction of Sennacherib" (1815), http://englishhistory.net/byron/poems/destruction.html.
48. stone aqueduct: Smith, *History of Dams,* 9–12; Smith, *Man and Water,* 76–78. The aqueduct is known as the Jerwan aqueduct bridge.
49. Tehran's water supply: Smith, *Man and Water,* 70–71.
49. tried almost every water supply technique: Ibid., 79.
49. King David discovered: Johnson, 56, 72–73; Smith, *Man and Water,* 77.
51. "only deep enough": Herodotus, *Histories,* 113–118. Herodotus also relates that a previous ruler had rechanneled the Euphrates from its previously straight path into a winding course in order to slow its current through Babylon and to impede any direct approach by enemy vessels.
51. "No Persian king": Herodotus, Ibid., 117. The river was the Choaspes.
54. contacts with Mesopotamia: McNeill, *A World History,* 34.
54. Great Bath: Keay, 12–14.
55. rivers that had radically changed course: Some are referred to in the Rig Veda. Rivers that dried up included an eastern tributary of the Indus and the Ravi, upon which Harappa had been located.
55. decline and emigrate: One possibility is that some Indus people migrated to southern India and Sri Lanka. Indus writing has some earmarks of being a proto-Dravidian language, which is among that region's tongues. The ingenious, huge artificial reservoirs and canal networks that before the third century BC irrigated Sri Lanka's golden age might also hint at the possible knowledge of the lost Indus descendants.
55. irrigation canals: Pacey, 59.
56. drought cycle: Diamond, *Collapse,* 157–176. Regional Mayan collapses in 810, 860, and 910 coincided with severe intracycle drought peaks. The rise of classic Mayan civilization started during a wet period, which had followed a 125-year drought (after AD 125) that brought about the demise of the preclassic Mayan era. See also Harris, 87–92, and Pacey, 58–61.
57. monsoon's start date: As late as the 1970s, the arrival of clouds in the southern state of Kerala, where the monsoon first appeared, would trigger an urgent

message to the prime minister's office in New Delhi heralding the start of the monsoons. Economic growth could fall to zero if monsoon rainfall was poor; even in India's more advanced twenty-first-century economy, deficits in precipitation could reduce economic growth by up to four percentage points.

57. in the aftermath: Keay, 83.
58. Sabaeans from the Arabian Peninsula: Smith, *History of Dams,* 15; Gunter, 2–19, 104–113. The Sabaeans were also famed pioneer irrigators; their huge dam at Marib—by far the largest city in ancient Arabia—on the Wadi Dhana was enlarged several times from its first 1,800-foot-wide earthen iteration in about 750 BC, and intercepted the wadi's periodic floodwaters to intensively irrigate over 4,000 acres.

Chapter Four: Seafaring, Trade, and the Making of the Mediterranean World

63. Bronze had first appeared: Braudel, *Memory and the Mediterranean,* 60. Copper smelting began in the fifth millennium, but it took a long while before it was discovered that adding tin could strengthen it as bronze.
64. their civilization: According to Greek myth, under Minos's palace at Knossos lay a labyrinth inhabited by a sacrificial-maiden-devouring Minotaur, the monstrous offspring of Minos's wife and a bull sent by the sea god Poseidon that ultimately was slain by the Greek hero Theseus.
66. mariners from Miletus: Cary and Warmington, 37.
66. manifestation of water: Jones, *History of Western Philosophy,* 32–34.
68. trireme lay low: Casson, 85.
69. great cajoling: To entice Xerxes—as well as to prevent his vacillating allies from changing their minds at the last moment—Themistocles devised one of history's most famous deceptions. Pretending to turn traitor, he sent an informant to Xerxes' headquarters with the credible news that the Greeks were preparing to slip away and disperse rather than fight a single big battle against long numerical odds. Xerxes took the bait. He ordered his patrols to row all night to prevent a Greek breakout.
69. "they gathered the grass": Herodotus, *Persian Wars,* 642–643.
69. asymmetrical advantages: Athens's surrounding seas and rugged landscape provided a further defensive buffer against land armies—a distinguishing advantage lacked by both Phoenicia and Miletus.
71. more representative: Athens's laws and magistrates were decided by a majority vote of the citizens' assembly, normally in accordance with a representative advisory council. In time voting rights were extended to the poor as the growing wealth of the state came to depend on the large naval manpower needed to pull the galley oars. Even naval commanders, such as Themistocles, were elected by popular vote.
73. Alexander turned it into an opportunity: Cary and Worthington, 179–180; Foreman, 188–189.

74. 700,000 items: Daniel J. Wakin, "Successor to Ancient Alexandria Library Dedicated," *New York Times,* October 17, 2002. Government officials boarded ships in Alexandria's harbor, seized whatever scrolls were on board, and then had them copied. The originals were returned to their owners; the copies were added to the library's collection.
75. body lay in state: After his death Alexander's body had been intercepted en route from Babylon to its final resting place in Macedon by Ptolemy I, his trusted general and boyhood friend, to bolster the legitimacy of the Egyptian dynasty he founded and which would rule Egypt until Rome incorporated the country, and its agricultural bounty, into its empire. The site of Alexander's tomb was lost in the riots of the third century AD.

76–77. consolidated slowly: Rome's expansion progressed slowly through military victories, regional political alliances, and the granting of citizenship to absorbed Italic tribes; plebeian classes that served in the army gradually gained greater political representation in government.

77. 100 quinqueremes: Casson, 145.

78. "by the sea": Mahan, 15.

79. Carthage's surrender: The brief, one-sided Third Punic War, initiated by Rome on flimsy pretexts, ended in 146 BC with the destruction and plowing under of the city of Carthage itself.

79. influence indirectly: Where force was required against a hardened enemy, such as Macedon, it deployed its army as a first resort. Only when absolutely required by military exigencies did it exercise its naval might directly.

80. 1,000 ships: Casson, 180.

80. ruling triumvirate: Norwich, *Middle Sea,* 34.

81. civil war: In all some 1,000 ships and tens of thousands of Roman mariners were lost throughout Rome's civil wars.

81. help of the catapult grapnel: Reinhold, 29–34, 161. Agrippa also built Rome's first naval port to support the sea war.

84. thirty- to sixty-day voyage: Casson, 206–207.

84. position vertical to the water: Braudel, *Structures of Everyday Life,* 355.

84. grind 10 tons: Williams, 55–56.

85. hydraulicking: Bernstein, *Power of Gold,* 14. Hydraulicking's horrendous environmental impacts, including the denuding of hillsides, topsoil erosion that destroyed farmland, and the silting up of rivers and harbors, finally caused Californians in 1884 to rise up and have it outlawed.

85. concrete was derived: Braudel, *Memory and the Mediterranean,* 30; "Secrets of Lost Empires: Roman Bath." Heating common limestone to high temperatures for a prolonged period produced a very light derivative, quicklime. Adding water caused the hot quicklime to sizzle, steam, swell, and ultimately transmute into a new material: a very fine powder, or "hydrated lime." Adding more water to the lime powder created a putty adhesive strong enough to bind sand, stone, and crushed tile chips, which were the coarse components of Roman

concrete; later, where possible, Romans used volcanic ash. When hardened, the substance became miraculously waterproof.

86. Aqua Appia: Evans, *Water Distribution in Ancient Rome,* 65–74.
86. Hellenist water engineering: Aicher, 2–3.
87. total aqueduct water: Evans, 140–141.
88. only six cities: McNeill, *Something New Under the Sun,* 282.
88. 150 to 200 gallons: Peter Aicher, cited in "Secrets of Lost Empires: Roman Bath."
88. best water quality: Rome's suburban hills had fresh springs and deep volcanic lakes, while the porous travertine bedrock of its surrounding valleys acted like a natural purifying filter for underlying aquifers. Romans tried to use the best-quality water for human consumption and route brinier and poorer-tasting water for tasks like irrigation, street cleaning, and filling theater basins for mock sea battles.
88. "have laid hands upon the conduits": Frontinus, 128.
89. Waterworks were the centerpiece: Evans, 137–138; see also Reinhold, 47–51; Shipley, 20–25.
90. "sheltered place": Mumford, 225, 226; "Secrets of Lost Empires: Roman Bath."
91. periods of aqueduct building: Smith, *Man and Water,* 84; Evans, 6.
91. Emperor Claudius: Claudius added the Aqua Claudia and Anio Novus in AD 52. Trajan's Aqua Traiana was the first to serve the Trastevere quarter across the Tiber.
91. emperor's baths: The Aqua Alexandriana was built for the baths of Alexander Severus to replace Nero's baths.
91. pirates, Goths: Casson, 213.
92. The Huns: McNeill, *A World History,* 195–197. The fleeing Huns displaced the Ostrogoths from southern Russia in 372 and caused their weaker Visigoth neighbors to enter Roman frontiers.
93. floating water mills: Procopius of Caesarea, 5, 191–193.
94. "Rome's decay": Hibbert, 74.
95. Martin V: Karmon, 1–13.
95. "Water Popes": Nicholas V (founder of the Vatican Library, who hired Leon Battista Alberti to work on the aqueducts) added a simple terminal fountain in 1453 that in the prosperous eighteenth century was transformed into the elaborate Trevi Fountain. Gregory XIII built the conduits that give its name to Via Condotti as well as many fountains. Sixtus V, born Felice Peretti, in the late sixteenth century rebuilt the last aqueduct, Aqua Alexandriana, and renamed it Aqua Felice, after himself; he also added many underground pipes, 27 fountains, and some bridges across the Tiber. Paul V, who became pope in 1605, outdid him with monumental fountains, some by Bernini, supplied by rebuilding Trajan's aqueduct, now called Aqua Paola, after himself.

Chapter Five: The Grand Canal and the Flourishing of Chinese Civilization

96. "The Chinese people": Needham, vol. 4, pt. 3, 212.
97. 33rd parallel: Fairbank and Goldman, 5.
98. 15 times more water: Shiklomanov and Rodda, 365.
100. "mastered the waters": Yu the Great, quoted in Fernández-Armesto, 217.
100. humble water's yielding: Lao-tzu wrote, "Water flows humbly to the lowest level. Nothing is weaker than water. Yet for overcoming what is hard and strong, nothing surpasses it." Cited in "Sacred Space: Rivers of Insight," *Times of India*, http://timesofindia.indiatimes.com/articleshow/msid-3423508,prtpage-1.cms.
102. millet noodle: Among other revelations, the noodle put an end to the centuries-long canard that Marco Polo had introduced the pasta noodle to China during his famous late thirteenth-century trading expeditions.
103. Li Bing: Kurlansky, 23–25; China Heritage Project, "Taming the Floodwaters: The High Heritage Price of Massive Hydraulic Projects," *China Heritage Newsletter* 1 (March 2005), China Heritage Project, Australian National University, http://www.chinaheritagequarterly.org/features.php?searchterm=001_water.inc&issue=001.
104. population of 5 million: Needham, 288.
104. thousands of waterwheels: Ibid., 296.
104. bamboo tubes with leather flap valves: Kurlansky, 26–28.
105. Treadle chain pumps: Temple, 56–57.
106. government monopolies: Elvin, 29.
106. government controlled: Fairbank and Goldman, 59.
106. malleable cast iron: Temple, 42–43.
107. noted Chinese engineer Tu Shih: The device had reciprocating action. Ibid., 55–56.
107. vertical waterwheels: Gies and Gies, 88–89.
107. same essential design: The machines, lacking only the steam engine's crankshaft, operated on the reciprocating action of a rod-driven piston attached to a waterwheel-powered crank. Temple, 64.
107. one pound of raw silk: Fairbank and Goldman, 32.
108. Emperor Tiberius: Edwards, 20.
108. silk industry: Persia, India, and Japan each developed silk culture independently. By some accounts Alexander the Great brought silkworm cocoons back with him from India, but the art of cultivating them was lost by the time of the Romans.
110. web of international exchange: McNeill, *Global Condition,* 92, 96–99.
110. barbarian raiders: The Han's main tormentors had been the Hsiung-nu, but by 350 a new powerful Mongolian confederation, known to the Chinese as the Juan-juan, had arisen. It was their westward irruption that put to flight the fearsome Huns, who displaced the Ostrogoths from southern Russia

in 372 and caused their Visigoth neighbors to enter Roman frontiers. The Juan-juan confederacy was finally destroyed in 552 by an alliance of Chinese armies with Turkish tribes, who quickly established a formidable steppe empire of their own.

110. "there was insufficiency": *Record of the Three Kingdoms,* quoted in Elvin, 37.
111. Grand Canal: Needham, 307–310; Elvin, 54–55.
112. one-third less: Elvin, 138.
113. pound lock: Temple, 196–197.
113. Yangtze salt and iron fleet: Elvin, 136. Each of the 2,000 boats built had a capacity of 110,000 pounds.
114. "the amount of shipping": Polo 209.
114. rice-farming revolution: Braudel, *Structures of Everyday Life,* 146–155.
115. Champa rice: McNeill, *Rise of the West,* 527; Pacey, 5; Elvin, 121.
115. 120 million: Fairbank and Goldman, 89.
116. technological leader: Pacey, 7.
116. coke-burning blast furnaces: A similar coal-for-wood substitution was the coking process developed by England's Abraham Darby—one of the watershed events of England's Industrial Revolution—but only in 1709.
116. 114,000 tons of pig iron: Fairbank and Goldman, 89.
116. water-powered spinning machines: Elvin, 194–195; Pacey, 24–28, 103.
117. water clock: Boorstin, 60–61, 76. Imperial ladies of the highest rank were bedded by the "Son of Heaven" nearest to the full moon when their female yin influence was strongest and best able to balance his yang, or male, aspect, and thus ensure the favorable virtues for offspring then conceived. The "Heavenly Clockwork," invented by government official Su Song, also corrected an astronomical error that had corrupted the accuracy of China's official calendar. Driven by a noria wheel mounted with 36 water-lifting buckets that made exactly 100 revolutions each day, Su Song's water escapement ingeniously exploited the fluid properties of water.
117. river- and canal-fighting vessels: McNeill, *Pursuit of Power,* 42.
119. gorge at Chü-tang: Elvin, 93–94.
120. Cheng Ho's 27,000 man fleet: McNeill, *Rise of the West,* 526; Fairbank and Goldman, 137–139; Boorstin, 192.
121. ruler in Ceylon: McNeill, *Pursuit of Power,* 44.
122. Heaven Well Lock: Elvin, 104.
122. "With the re-construction": Ibid., 220.
122. rely exclusively: Some Ming officials worried about exclusive reliance on the inland waterway network. Likening the summit passage portion of the Grand Canal (the Hui-t'ung Canal) to a man's throat that if choked off for even a single day would result in death, they argued for maintaining the sea transport network. In the event, save for periods of Yellow River flooding in 1571 and 1572, the Grand Canal passage remained uncut until the end of the Ming dynasty in the mid-seventeenth century. Ibid., 105.

123. China's inner dynamism: Ibid., 203.
123. labor-saving technologies: The labor-intensive bias of China's state-directed economy was notably evident in its prodigious iron industry, in which, despite waterpower's demonstrated superiority, use of manually powered bellows remained predominant. Pacey, 113.
124. opium imports: British India's opium exports rose from 400 chests in 1750 to 5,000 in 1821 and 40,000 in 1839. McAleavy, 44.
124. free trade: Britain's policy shift from mercantilism to espousal of "free trade" principles coincided with the rise of its world-class industrial factories, which enjoyed unrivaled competitive advantage. This was not the first, nor would it be the last, instance in world history that self-serving economic advantage informed the adoption of grand economic principles.
125. worst flood: McAleavy, 59.

Chapter Six: Islam, Deserts, and the Destiny of History's Most Water-Fragile Civilization

129. water was always highly esteemed: By Islamic custom visitors are always offered free water. Water is central to daily purification rituals at prayer. Paradise is described as a shaded garden with cooling fountains. And the ritual Muslim pilgrimage, or hajj, to the Ka'bah at Mecca includes racing seven times between two nearby hills to commemorate Hagar's frantic quest for water after Abraham had expelled her and Ishmael from her tent.
129. reputable but weaker clan: The clan was the Hashemites, whose descendants include today's royal family of Jordan.
130. armed struggle: Hourani, 18.
131. 2.5 million: Collins, 20–21.
132. ships loaned: The Christians had their own religious and political divisions. The Byzantines were rivals of the Visigoths, who in 589 had adopted the *Filioque* interpretation to the Nicene Creed that was vigorously rejected by Constantinople and would become a factor in the Great Schism between Latin and Eastern Christendom in the eleventh century.
133. caliphate's revenue: Braudel, *History of Civilizations,* 73.
133. Its agriculture was confined: Hourani, 100.
134. built on slopes: Braudel, *Structures of Everyday Life,* 507.
134. "Not being well endowed": Braudel, *History of Civilizations,* 62.
134. camels: Saharan camels could carry half the weight of their heavier, cold desert Bactrian cousins. Camels originated in North America and were close relatives of the South American llama and alpaca. They were domesticated for food in the Middle East by 2000 BC. By 1000 BC they were commonly used as transport animals.
134. deserts: Fernández-Armesto, 67.
136. seasonally reversing wind system: Ibid., 384, 389. Reliable sailing conditions

and a relatively safe way home were the major reasons for the Indian Ocean's precocious development as mankind's richest, earliest long-distance trade highway.

136–37. In Mesopotamia goods: Hourani, 44.

138. "Greek fire": White, *Medieval Technology and Social Change,* 96. Greek fire seems to have been invented, fortuitously for Constantinople and the West, just prior to 673 by a refugee architect from Syria named Kallinikos. Its spectacular effects in repelling Muslim forces ignited the history of the search for combustible weaponry, which ultimately produced the seminal invention of gunpowder and cannons.

139. long-distance aqueduct: Valens, in the fourth century, and Justinian, in the sixth century, were the major aqueduct and cistern builders, respectively.

140. Famine and disease: Norwich, *Short History of Byzantium,* 110.

141. lifted the siege: Davis, *100 Decisive Battles,* 102.

141. the First Crusade: Ironically, the most immediate effect of the halt of Islam's expansion at Constantinople's seawalls in 718 was to sow discord within Christianity itself. Soon after the victory, Leo III decided to forbid the use of religious icons, following Muslim and Jewish practice. But iconoclasm was an anathema to the pope in Rome. Although Constantinople ultimately renounced it just over a century later, the rivalry between Eastern and Latin Christianity endured for centuries.

143. Abbasid engineers: Pacey, 10.; Smith, *Man and Water,* 16, 18.

144. "in Cairo": Ibn Battutah, 15. By way of comparison, Paris in the glory years of the eighteenth century employed 20,000 carriers of Seine River water.

145. paper pulp mill: Gies and Gies, 42.

145. over a hundred bookshops: Pacey, 41; Public libraries were opened, too. Caliph al-Hakam of Cordoba in the latter part of the tenth century reportedly had a library of 400,000 manuscripts—by comparison, the library of France's mid-fourteenth-century king, Charles V, had only 900. Braudel, *History of Civilizations,* 72.

147. Mesopotamia's irrigation system: Smith, *History of Dams,* 81; Pacey, 20; McNeill, *Rise of the West,* 497.

147. Nahrwan transport and irrigation: Smith, *Man and Water,* 18; Temple, 181.

148. cannibalism, plague, and decaying waterworks: Collins, 21; Smith, *Man and Water,* 16.

148. water court at Valencia: All the elected members of the weekly Tribunal de las Aguas sit at a round table and in full public view discuss and settle farmers' disputes over use and maintenance of water and infrastructure. Judgments are based on common sense and no written records are kept.

149. parity with Islam: Pacey, 44.

152. under Sultan Süleyman the Magnificent: Lewis, *Muslim Discovery of Europe,* 32.

153. 30,000 men died: Howarth, 18–21.

154. engage on equal terms: Africa was slow to learn about the development of the wheel and the plow, for instance. Moreover, Africa may also have been impeded by its southerly latitude to the main Eurasian belt; biota seemingly adapts best in similar latitude bands, which may have added the benefit of scale to Eurasia's other comparative advantages.

Chapter Seven: Waterwheel, Plow, Cargo Ship, and the Awakening of Europe

161. The key technical breakthrough: White, *Medieval Technology,* 43. The plow had three main functioning parts. The coulter, or heavy knife, was attached to the pole of the plow and cut into the earth. Set at a right angle to the coulter was a flat plowshare that dug into the turf horizontally. The moldboard turned the unearthed clods to the side. After the stiff, nonchoking horse collar was introduced into western Europe before the tenth century, horses increasingly replaced oxen as the favored plow animals.
162. drier and milder climate: Gimpel, 29–30, 205–206. The advance and retreat of the Fernau glacier over 3,000 years suggested that the first millennium BC was a cold period, followed by a warming trend in late Roman times. The medieval warm period lasted from about AD 750 to 1215, followed by a brief cold spell until 1350, and may have contributed to the conditions that produced the Black Death. The Little Ice Age in Europe from 1550 to 1850 was followed by a century-long warming trend.
162. south of the Loire River and Alps: Ibid., 44.
163. new cog: The cogs were clinker-built, meaning that the planks overlapped, like tiles on a roof. Originally, the cogs had been flat-bottomed for easy landing on natural shorelines, but as they grew larger they became harder to control and inadequate for use in the growing number of improved ports.
163. Cologne, situated at the juncture: Braudel, *Structures of Everyday Life,* 51.
164. 85 percent of commercial traffic: Gies and Gies, 221. Weirs are small obstructions that block part of a waterway, often for multiple purposes such as to maintain current flow speed for waterwheels and sufficient depth for navigation.
164. "Commerce between": Lopez, 86–87.
165. built by monastic orders: Interestingly, the relationship of monks with bridges had an Eastern parallel with Buddhist monks, who built and maintained many of the suspension bridges across Himalayan passes as part of their duties.
167. several times more powerful: Estimates of waterwheel power vary greatly. Wheel size, the construction material, the angle and timing of water entry to its blade, and streamflow rate all affect output. Gies and Gies, 34–36, 115; Braudel, *Structures of Everyday Life,* 371; Smith, *Man and Water,* 143, 145; Williams, 54–55.
167. Leonardo da Vinci: Smith, *Man and Water,* 147; Gies and Gies, 258, 265. Da Vinci rejected the popular and incorrect view of contemporaries that waterpower offered a key to perpetual motion, and understood the basic physics that

water's work potential depended much upon its fall minus the wheel's frictional resistance and that of the machinery it powered. He understood that efficiency depended upon the angle of the water's impact with the wheel's blades. His theory that the overshot wheel was the most efficient form was not supported by any mathematical quantifications; that was left to John Smeaton, the father of modern civil engineering, in his experiments in the mid-eighteenth century. Leonardo's drawings offered one of the earliest models of the highly efficient breast wheel, in which the water strikes blades positioned at ten o'clock and two o'clock.

168. ocean-tide-powered mills: White, *Medieval Technology,* 84, 85. In the eleventh century, there were tide-powered mills near Venice on the Adriatic and at the mouth of the port of Dover in England.

168. king hastened the surrender: The king was Philip Augustus; the town, Gournay (near Beauvais); and the author, William the Breton. Smith, *History of Dams,* 144.

168. half a million water mills: Braudel, *Structures of Everyday Life,* 358.

169. description of a contemporary: Mumford, 258–259; see also Gies and Gies, 114–116, and Gimpel, 66–68.

170. one silk mill: Gies and Gies, 178–179; Lopez, 133–135; White, *Medieval Technology,* 44. The earliest referenced water-powered fulling mills in Europe date to 983 in Tuscany, 1108 in a Milan monastery, 1010 in Germany, between 1040 and 1050 in Grenoble, and 1080 in Rouen.

170. huge iron church bells: Lopez, 145. On casting, Gimpel, 66–68.

171. on parity with: Pacey, 44, White, *Medieval Technology,* 82.

172. "where there is no Nile or Indus": Harris, 167, 169.

174. Benedetto Zaccaria: Lopez, 139–141; see also Norwich, *History of Venice,* 202. For the history of the control of Gibraltar, see Casson, 65; Cary and Warmington, 45–47, 60.

175. Genoese republic: To give an idea of Genoa's power, by 1293 its sea trade alone was three times greater than all the revenue of the French kingdom. Lopez, 94.

176. Dante Alighieri's special embassy: Norwich, *History of Venice,* 204.

177. naval help: McNeill, *Rise of the West,* 514, 515.

177. sack Constantinople: Villehardouin, in Joinville and Villehardouin, *Chronicles of the Crusades;* Norwich, *History of Venice,* 122–143.

177. three-eighths of Constantinople: Norwich, *History of Venice,* 141.

178. 1280 to 1330: McNeill, *Pursuit of Power,* 70.

179. until after 1480: Ibid., 70.

Chapter Eight: The Voyages of Discovery and the Launch of the Oceanic Era

180. "the two greatest": Smith, *Wealth of Nations,* 281.

183. African slaves: Boorstin, 167–168. After 1445, some 25 caravels per year voyaged to West Africa to carry out commercial trade in slaves, gold, and ivory.

183. circumnavigation and coastal exploration of Africa: Cason, 118, 120-123; Cary and Warmington, 62, 128, 131, 229-230.
186. "Considered as a whole": Fernández-Armesto, 406. There were various exceptions to the Atlantic wind system. For instance, inside the Gulf of Guinea was a wind system that blew straight into Africa's large bulge, creating, in effect, a treacherous lee shore and helping explain why West African civilizations in that region were so disadvantaged at seafaring. In the far north, the Vikings, in their explorations of Iceland, Greenland, and North America, were able to take advantage of a clockwise current system that moved west from Scandinavia.
188. European diseases: Europeans, having been exposed to many diseases through Old World trade, had an overwhelming immunity advantage in the contest with the "virgin" Amerindians.
188. "Get gold": Timothy Green, *The World of Gold: The Inside Story of Who Mines, Who Markets, Who Buys Gold* (London: Rosendale Press, 1993), 11, quoted in Bernstein, 121.
188. Water-powered mills: Pacey, 70.
189. Treaty of Tordesillas: The new line gave Portugal claim to Brazil when it was discovered in 1500 by Pedro Cabral in his southwesterly arc through the Atlantic to catch the winds for Portugal's second Indian Ocean expedition.
189. out of sight of land: McNeill, *Rise of the West,* 570.
190. "Christians and spices": Quoted in Lewis, *Muslim Discovery of Europe,* 33.
192. cost of his voyage sixtyfold: Clough, 188.
192. use of crossbows: McNeill, *Pursuit of Power,* 100.
192. sea artillery: Braudel, *Structures of Everyday Life,* 388–389.
192. "There is no doubt": Kennedy, *Rise and Fall of the Great Powers,* 26.
194. price of pepper: Boorstin, 178.
194. reopen Pharaoh Neko's old "Suez": Lewis, *What Went Wrong?* 13.
194. Venice's desperate offer: Cameron, 121.
195. boiled hot drinks: Braudel, *Structures of Everyday Life,* 227. Chocolate and coffee were both considered medicinal when introduced to Europe, most probably because they were served hot. Boiled water was commonly sold on the streets in China.
195. yellow and putrid: Cited in Boorstin, 265.
196. Spanish Main: The Spanish Main was an area in the Caribbean enclosed by ports from Cartagena, Colombia, to Nombre de Dios, Panama, to Trujillo, Honduras, to Veracruz, Mexico.
196. interdicting the pay: Trevelyan, 238.
197. "difference of social character": Ibid., 233.
198. more than 10 tons: Bernstein, *Power of Gold,* 138.
199. Spanish Armada: Howarth, 24–33; Davis, *100 Decisive Battles,* 199–204.
200. shift of European power: Braudel, *Afterthoughts,* 84–86, 98. Historian Fernand Braudel reckons that the center of gravity of the European economy was anchored in Italy for several centuries until 1500, when it moved to Antwerp, then from

1550 to 1600 back to the Mediterranean in favor of Genoa (due to the wars in the north), and then definitively back north between 1590 and 1610 to Amsterdam, where it remained until the late eighteenth century, when it moved to London. In 1914 the center of the world economy crossed the Atlantic to New York.

200. contracted dysentery and died: Bernstein, *Power of Gold,* 138.
201. continued to be a leader: Smith, *Man and Water,* 28–33; Kolbert, 123–127.
202. half of the shipping: Cameron, 121–122.
202. closed Lisbon harbor: Spain had taken control of Portugal in 1580.
202. sea passages to the Spice Islands: Braudel, *History of Civilizations,* 263–264.
204. superiority at sea: French fleets, whose sailors were weakened by food and water shortages and disease caused by the insanitary conditions aboard ship at Brest, were slow to press their advantage.
204. Britain's navy reigned supreme: Lambert, 104.
206. had kept their powder dry: Davis, *100 Decisive Battles,* 241; Lambert, 122; Keay, 381–393.
206. winning command of the sea: Kennedy, *Rise and Fall of the Great Powers,* 124.
207. low water supplies: Davis, *100 Decisive Battles,* 275.
209. Nelson himself, shot: Howarth, 75.

Chapter Nine: Steam Power, Industry, and the Age of the British Empire

211. King George III: George III ascended to the throne when his grandfather died suddenly of a burst blood vessel while in the royal water closet.
212. Little Ice Age: Ponting, 99–101. The Thames froze over 20 times between 1564 and 1814. France's Rhone froze three times in the thirteen years between 1590 and 1603, and even the Guadalquiver at Seville in Spain froze in the winter of 1602–1603. By contrast, in a dramatic illustration of the large effects small temperature changes can have, the warm climatic period that ended about 1200 had fostered vineyards in England to the Severn in the north, arable farmland over large parts of southern Scotland's uplands, and even habitable climates on the southern coast of Greenland.
213. converting coal into coke: Coke, an almost pure form of carbon, was produced from coal in a method similar to the way wood was converted into charcoal—it was heated in a closed vessel to burn off impurities, leaving behind a residue that was coke.
213. new shipbuilding was being outsourced: Pacey, 114.
214. "many domestic hearths cold": Trevelyan, 430.
214. price of his coal: Bernstein, *Wedding of the Waters,* 40–45.
214. Canal du Midi: Ibid., 38–40. The driving force behind the Canal du Midi was a self-made tax collector of King Louis XIV's, Baron Pierre-Paul de Riquet de Bonrepos, who was close to the king's influential finance minister, Colbert, and spent his entire fortune in building it.

217. canal frenzy added 3,000 miles: Cameron, 174.
217. growing financial markets: The Glorious Revolution (1688–1689) played a critical role in creating the political and economic atmosphere favorable to private capital accumulation and investment, which was so essential to stirring the entrepreneurship and innovations of the Industrial Revolution.
218. Thomas Savery: Bronowski and Mazlish, 314; Cameron, 177–178; White, *Medieval Technology,* 89–93.
218. less than a hundred: Pacey, 113.
219. pumping water from a coal mine: Lira.
219. the watt: One watt is equal to 1/746 horsepower. Ironically, Watt invented the term *horsepower* by imagining the amount of coal a horse could lift from a mine in a defined period of time. He calculated that one horse could lift 33,000 pounds one foot in one minute.
221. "I sell here, Sir": Matthew Boulton, quoted in *English Merchants,* by H. R. Fox Bourne (London: R. Bentley, 1866), cited in Heilbroner, *Making of Economic Society,* 119.
221. "The people in London": Matthew Boulton, "Document 14, 21 June 1781: Matthew Boulton to James Watt," in Tann, 54–55.
222. Darby silk-stocking factory: Pacey, 103, 107. The original silk-stocking factory was opened in 1702, but failed. Subsequent owners made a success of it after secretly copying the designs of an Italian silk-stocking plant.
222. spinning mule: The mule got its name by merging aspects of Arkwright's water frame with James Hargreaves's non-water-powered spinning jenny (1764). Crompton never earned the fruits of his invention; although mules were in use everywhere, he himself remained indigent.
223. had 52 two decades later: Cameron, 181.
223. produce goods less expensively: McNeill, *World History,* 368.
223. accelerated twelvefold: Heilbroner, *Making of Economic Society,* 81.
223. 1 percent to 4 percent per year: Simmons, 201.
224. generated about 25 horsepower: Tann, 6–7. The British government, trying to preserve the country's industrial leadership, limited the sale of larger Watt engines abroad.
224. nearly 500: Ibid., 6–7.
224. fountains and gardens at Versailles: The three-level waterworks was known as the Marly machine. Smith, *Man and Water,* 100–106. See also Braudel, *Structures of Everyday Life,* 227–231.
225. Paris's 20,000 omnipresent water carriers: Braudel, *Structures of Everyday Life,* 230.
226. nearly 1.4 million tons: Heilbroner, *Making of Economic Society,* 81.
227. Mastodon Mill: Ponting, 276.
228. 1.7 percent per *century:* Per capita economic growth figures are derived from McNeill, *Something New Under the Sun,* 6–7.

228. freshwater use grew: Ibid., 120–121.
229. 400 miles per day: McNeill, *Rise of the West,* 766–767.
230. communications cable: Gordon, 212.
230. age of the ocean steamer: Cameron, 208.
232. traumatic, long-term challenges: Of the pattern of asserting of Western hegemony in the age of steam and iron, historian Fernand Braudel observes, "It is only a step from market to colony. The exploited have only to cheat, or to protest, and conquest immediately follows. . . . When civilizations clash the consequences are dramatic." Braudel, *Structures of Everyday Life,* 102.
232. invention of the torpedo: Williams, 136.
233. Torpedo ranges multiplied: McNeill, *Pursuit of Power,* 284.
233. cut Germany's five transatlantic cables: Gordon, 212–213.
234. one-fourth of world commerce: Cameron, 224.
234. travel time to India: Karsh and Karsh, 43.
235. ruler, Muhammad Ali: Ibid., 27–29.
235. De Lesseps finally got his chance: Ibid., 42–44.
236. no technical background: McCullough, 49.
236. funding from the Rothschild banking family: Ferguson, 231.
238. Fashoda Incident: Collins, 57–59.
238. 1,300 liters of claret, 50 bottles of Pernod: Barnes, n.p.
240. Nasser himself likened it to a modern pyramid: "In antiquity we built pyramids for the dead," Nasser said in 1964. "Now we build pyramids for the living." Gamal Abdel Nasser, speech, May 14, 1964, quoted in Waterbury, 98.
241. "Well, as you have the money": Fineman, 46–47, 48.
241. prearranged code word: "An Affair to Remember," *Economist,* July 29, 2006, 23; Fineman, 40. Dulles should not have been so shocked by the canal nationalization because he had been warned of that potential consequence by the French ambassador.
241. "have his thumb on our windpipe": Anthony Eden, quoted in Fineman, 62.
241. colluding to seize back the canal: Some 70 percent of the traffic in the canal was British; France was perturbed because it was at war in Algeria to put down a rebellion that Nasser was supporting.
242. "Anthony, have you gone out": Dwight D. Eisenhower, quoted in Urquhart, 33. The Americans' sense of betrayal was, in part, based on poor communication—from both sides. The Americans were not totally explicit about their unwillingness to support any military action in the Suez Affair, while the allies, knowing the Americans' predilections, were not eager to ask for permission before acting, and miscalculated the Americans' readiness to back them up once they had acted.
243. hydroelectric power station: The first hydroelectric plant was in Appleton, Wisconsin, on the Fox River in 1882.
244. pay up to eight times: McNeill, *Something New Under the Sun,* 175.

Chapter Ten: The Sanitary Revolution

250. infant mortality that claimed some 15 of every 100 children: Pacey, 187.
250. "What a pity": *Times* (London), June, 18, 1858, cited in Halliday, ix.
251. clean freshwater daily: Peter H. Gleick, Elizabeth L. Chalecki, and Arlene K. Wong, "Measuring Water Well-Being: Water Indicators and Indices" in Gleick, *World's Water, 2002–2003,* 101.
252. early spring rainwater: Braudel, *Structures of Everyday Life,* 230.
252. artesian wells: One famous artesian well gave a much-needed boost to Paris's water supply in 1841 when a large water deposit was struck at a depth of about 1,800 feet after eight laborious years of boring. The well, to public fascination, jetted 100 feet above the ground and was soon enclosed in a tall tower. Smith, *Man and Water,* 108.
252. distilled spirits: Braudel, *Structures of Everyday Life,* 241–242, 248.
253. fleet of water boats ferried freshwater: Ibid., 228. The boatmen, much like ubiquitous water carriers throughout European cities, even formed their own trade guild.
254. three times each week when dyers dumped: Ibid., 229.
254. "Whole quarters were sometimes without water": Mumford, 463.
254–255. 30 gallons of wholesome springwater: Smith, *Man and Water,* 111.
255. private water carriers: The water supply expanded significantly in the thirteenth century, especially after a wealthy individual gave a grant to the city in 1237 of all the springs on his estate issuing from the Tyburn, a tributary of the Thames, near today's Marble Arch.
255. three times per week: Chelsea Waterworks used water piped from Hertfordshire into Islington in north-central London through the 36-mile artificial New River to deliver its pledge. At the time of the Great Stink, the New River still supplied the largest volume of London's water. Halliday, 21.
256. "charged with the contents": John Wright, "The Dolphin or Grand Junction Nuisance," published March 15, 1827, quoted in Smith, *Man and Water,* 112–113.
256. "Going down to my cellar": Pepys, "Entry: Saturday 20 October 1660."
257. guano: Halliday, 41.
257. "a certain flush with every pull": Ibid., 42.
258. created a central board of health: McNeill, *Plagues and Peoples,* 240.
259. Death came from collapse: Biddle, 41.
259. first pandemic spread in Asia: McNeill, *Plagues and Peoples,* 232–233.
259. murdering victims in order to dissect: Karlen, 133–139.
261. three river embankments were constructed: The narrowing of the river caused by the embankments speeded the Thames's flow, with the salutary benefit of helping whisk away the waste that had eluded the intercepting sewers.
262. Typhoid fever: Milk pasteurization and vaccines against tetanus, diphtheria, and tuberculosis bacilli were among the other major antibacterial successes that inspired the medical conquest of many viruses.

262. human longevity to leap: McNeill, *Something New Under the Sun,* 199–200. U.S. life expectancies for white males rose from 56 to 75 between 1920 and 1990, up from a mere 30 to 40 years before the sanitary awakening, when infant mortality was so high. Worldwide average life spans leaped from 36 years in 1900 to over 65 in 1995.

262. Infant mortality plunged, falling to half of 1 percent: Cameron, 328; Economist staff, *Pocket World in Figures, 2009 Edition,* 83. Japan achieved the most spectacular improvement of any advanced nation, with a more than thirtyfold drop in infant mortality to the world's lowest absolute levels.

263. 20-mile-long sewage storage tunnel: "My Sewer Runneth Over," *Economist,* March 22, 2007.

263. municipality-run water supply: Smith, *Man and Water,* 127. In heavily wood-constructed U.S. cities, firefighting was another important motivating factor in the early evolution of public water systems. New York City launched a board of health in 1866 directly modeled on the British prototype and driven by the same cholera fears.

264. waterborne disease fell sharply in America: McNeill, *Something New Under the Sun,* 196.

264. cleaner than the water in the Thames: Halliday, 107. Leftover liquids from the sludge were aerated to promote microbacterial activity that eliminated further impurities.

265. Moscow River received untreated nearly all the sewage: Ponting, 356.

Chapter Eleven: Water Frontiers and the Emergence of the United States

271. two froze to death marching: Morison, 243–244.

271. take control of the strategic Hudson waterway: They were to rendezvous at Albany. Both sides considered that the critical strategic spot for controlling the unbridged Hudson was West Point, south of Albany, because the river was wide enough for sailing ships to navigate up to that point, but not beyond, without the help of rowed tugs. To defend West Point, the colonials built a ring of forts buttressed by a chain they laid nearby across the Hudson.

271. personal entourage some three miles long: Wood, 33.

273. King George III's bid to reassert monarchal authority: Trevelyan, 389–390.

274. Only seven rivers carried a greater volume of water: McNeill, *Something New Under the Sun,* 183.

275. encompassed over two-fifths of the continental United States: Barry, 21.

275. unusual feature of the lower Mississippi: Ibid., 38–39.

276. 17 million acres of surrounding wetlands: Clarke and King, 70.

276. inveigled, to obtain U.S. domain over the lands: At the crucial moment, the Americans infuriated their French ally by contravening the spirit, though not the letter, of the Franco-American alliance by secretly negotiating separately with England to preempt the possibility that France and Spain might try to

secure Gibraltar in exchange for England's right to the lands west of the Appalachians.

277. in 1794 signed a controversial treaty: Despite losing the Revolutionary War, Great Britain did not give up on on its hope of winning the Mississippi for itself and British Canada until after the War of 1812. Its strategy was to try to hem in the United States to the east by creating native Indian buffer states west of the Appalachians.

277. feared ulterior American and British designs: Spain had good reason to worry. Alexander Hamilton was lobbying in Washington to personally lead an invasion force to seize Louisiana and Florida from militarily vulnerable Spain by arms.

278. "The day that France takes possession of New Orleans": Thomas Jefferson to Robert Livingston, April 18, 1802, quoted in Tindall, 338.

279. "What would you give for the *whole*": Talleyrand, quoted in Morison, 366.

280. he raised capital from private investors: Bernstein, *Wedding of the Waters,* 70–71; Achenbach, 19–20.

280. one-third of Britain's fleet: Heilbroner and Singer, 43. See also Pacey, 114.

281. producing more total pig and bar iron than England: Heilbroner and Singer, 63–64.

281. Britain vigorously enforced sanctions: Some U.S. states offered bounties to anyone who smuggled out the sanctioned technology.

283. 1,200 automated factories: Groner, 60.

283. interchangeable parts: In 1801, to demonstrate the effectiveness of his innovation, Whitney famously produced 10 muskets that he disassembled, put into piles, and then reassembled before the eyes of President John Adams and Vice President Thomas Jefferson.

284. 1,200 factories with 2.25 million spindles: Morison, 483. A parallel celebrated woolen manufacturing city evolved, somewhat more slowly, on the same river in Lawrence.

285. turbines capable of 190 horsepower: Smith, *Man and Water,* 179. This was Uriah Boyden's turbine for the Appleton Company of Lowell, starting in 1844. Pioneering breakthroughs in turbine design had been made in the 1820s by French engineers Jean-Victor Poncelet and Benoit Fourneyron.

285. Francis turbine: Ibid., 179–180, 185.

285. electricity could be produced: Man's awareness of electricity dated at least to the sixth century BC to the father of Greek philosophy, Thales of Miletus, who observed static electricity's effects after rubbing amber on light objects.

286. generating hydroelectricity from 5,500 horsepower Francis turbines: Smith, *Man and Water,* 185, 187.

286. consuming more electricity: Heilbroner and Singer, 262.

288. John Fitch: Williams, 100. See also Groner, 87.

289. western river steamboats were carrying freight: Groner, 88; Heilbroner and Singer, 97.

290. "when the United States shall be bound together": Robert Fulton, "Mr. Fulton's Communication." Fulton made a similar point in a much-earlier letter (February 5, 1797) to President George Washington, who had just received a copy of Fulton's *Treatise on the Improvement of Canal Navigation* (1796). Advocating the benefits of canals over investments in land or river transport in general, and a business proposal for a canal between Philadelphia and Lake Erie in particular, Fulton wrote that such canals "would penetrate the Interior Country And bind the Whole In the bonds of Social Intercourse." Fulton, "Letter from Robert Fulton to President George Washington."

290. "It is little short of madness": Thomas Jefferson, quoted in "Claims of Joshua Forman," in Hosack, *Memoir of De Witt Clinton,* Appendix Note U. Reminded years later of his comment in a letter from DeWitt Clinton, Jefferson mused in his late 1822 reply upon what marvelous qualities they were that enabled the state to execute such a great enterprise that "anticipated, by a full century, the ordinary progress of improvement."

291. New York state limestone that acted like waterproof Roman cement: Chittenango cement, as it was called, was found near Syracuse.

291. foreigners held over half: Bernstein, *Wedding of the Waters,* 235.

292. through seven miles of solid rock face: Ibid., 280–284. Lockport, as the nearby town was called, later used the canal's surplus water as an important electricity producer. The normal locks were eight feet, four inches.

292. symbolic wedding of the waters: Ibid., 319. This ritual wedding of the waters was reminiscent of how Venetians tossed rings into their city's canal to symbolize its marriage to the sea.

292. slashed freight transportation costs by 90 percent: Heilbroner and Singer, 94.

292. cheapest route to Pittsburgh: Morison, 478.

293. more than 3,000 miles of canals: Cameron, 230.

294. economy expanded on average about 2.8 percent per year: Bernstein, *Wedding of the Waters,* 347.

295. 100 gallons of water per day: Koeppel, 287. Other main sources used in this section are Galusha and Grann.

297. specially written "Croton Ode": Koeppel, 280–283.

297. "Nothing is talked of or thought of": "Croton Water: October 12, 1842," in Hone, 130–131.

297. surge in per capita consumption: Galusha, 35. Daily consumption rose from 12 million gallons per day to 40 million in the eight years from 1842 to 1850.

297. authorities used high-handed land appropriations: In the same period, Los Angeles was constructing its aqueduct (completed 1913) from ruthlessly acquired water rights to the Owens River 250 miles away.

298. first deep, high-pressure subterranean conduit: Galusha, 113; Grann, 93.

298. "equitable apportionment without quibbling": Oliver Wendell Holmes, Supreme Court of the United States, No. 16, *State of New Jersey v. State of New York and City of New York,* May 4, 1931, cited in Galusha, 113.

298. 1.3 billion gallons to 9 million people: Galusha, 265. About 50 percent of the water came from the Delaware Aqueduct, 40 percent from the Catskills, and 10 percent from the nineteenth-century Croton system. In addition to the central water tunnels, the system included 6,200 miles of water mains that helped distribute water to end users.
298. Its sewerage counterpart: Chicago's water system also featured one of engineering history's innovative and culturally indicative early twentieth-century marvels. In contrast to New York, Chicago drew its freshwater from the huge natural reservoir at its doorstep, Lake Michigan. In the nineteenth century, the lake also was the sewage dump of the Chicago River. Disease plagued the city until 1867, when it built a drinking-water intake tunnel two miles out into the lake. But population growth overtook it. In 1885 a heavy storm flushed the increased volume of sewage discharge out beyond the intake valves. The epidemics returned. Chicago responded with an innovative, ambitious civil-engineering project—the reversal of the flow of the Chicago River. By 1900 the 28-mile-long Chicago Sanitary and Ship Canal diverted the river southward to dilute and drain into the Mississippi watershed instead of into Lake Michigan. Not everyone hailed the largest earthmoving civil-engineering project until the Panama Canal, however. The state of Missouri, complaining about increased pollution on the Mississippi at St. Louis, pursued litigation. The earthmoving technology used on the Chicago River project was soon applied in building the monumental Panama Canal.
299. Sutter's new waterwheel-powered sawmill: Bernstein, *Power of Gold,* 223–225.
299. San Francisco swelled into a booming city: Morison, 569.
299. hurdy-gurdy wheels: Smith, *Man and Water,* 182. Bernstein, *The Power of Gold,* 14.
300. drew 300,000 to California by 1860: Worster, 65.
300. speedier, full-rigged clippers: Morison, 583.
300. the Panama railway: McCullough, 36.
301. set himself up as Nicaragua's president: Morison, 580–581.

Chapter Twelve: The Canal to America's Century

303. John Paul Jones's heroic sea victories: Love, 1:22–24.
303. U.S. Navy earned the respect: Morison, 350–351.
303. one-fifth of America's annual government revenue: Morison, 363–364.
304. give up its long-term designs on the Mississippi: Napoléon's abdication in April 1814 allowed England to concentrate on invading the United States, which it planned to do in three places in succession—Niagara, Lake Champlain, and New Orleans—while raiding the Chesapeake. While the Chesapeake raid led to the torching of the White House and the bombardment of Baltimore that inspired Francis Scott Key to write "The Star-Spangled Banner," the other battles

were determinative. The dramatic naval victories were Captain Oliver Hazard Perry's victories on Lake Erie and the U.S. victory at Niagara Falls, and, even more dramatically, Captain Thomas Macdonough's victory at Plattsburg on Lake Champlain, which halted the British plan to take the Hudson and sever the United States, as it had tried to do in the War of Independence. The navy assisted in the defense of New Orleans, where Andrew Jackson made his fame.

304. America's special sphere of influence: In the 1830s U.S. military ships also began around-the-world explorative expeditions.

305. rapid growth in demand for U.S. manufactured goods: Heilbroner and Singer, 180–181, note that U.S. exports tripled from 1870 to 1900 and that manufacturing's share doubled from 15 percent to 32 percent. Kennedy, *Rise and Fall,* 245, writes that from 1860 to 1914 U.S. exports grew sevenfold while imports rose only fivefold.

306. "The seaboard of a country is one of its frontiers": Mahan, 35.

306. "the Caribbean would be changed from a terminus": Ibid., 33.

306. excite America's "aggressive impulse": Ibid., 26.

306. wrote a glowing review of it: McCullough, 252.

307. "There is a homely adage": Theodore Roosevelt, quoted in Morison, 823.

307. "Remember the Maine!": Love, 1:388–389; Morison, 800–801.

307. Naval investment that totaled 6.9 percent: Kennedy, *Rise and Fall,* 247. In absolute dollars, naval spending rose almost sevenfold, from $22 million in 1890 to $139 million in 1914.

309. Several locations were considered: De Witt Clinton, impresario of the Erie Canal, gave his blessing to a canal at Nicaragua; celebrated British engineer Thomas Telford, who had designed Scotland's pioneering, lock-based Caledonian Canal, also studied a water passage near Darien in southern Panama.

310. excited the whole French nation: De Lesseps got off to a dazzling start. When the major French and international financial institutions eschewed his company's initial public offering, he broke new ground in French capitalism by launching the venture with funds raised from the savings of 80,000 small investors, most purchasing one to five shares each.

310. 20,000 workers and managers died: McCullough, 235.

311. de Lesseps was convicted of fraud: De Lesseps was sentenced to five years, but due to age was excused from prison. His son, Charles, who had overseen the day-to-day operation, was convicted, too, and served jail time.

311. U.S. interoceanic canal commission: McCullough, 264–265.

311. Panama was indeed the superior technical route: Ibid., 326–327.

312. backed by powerful Wall Street bankers: The backroom Panama lobby had been influential enough to get McKinley to appoint a second interoceanic commission with several new members after the first had ruled in favor of Nicaragua, but not enough to get it to change its recommendation.

312. nation's own one centavo stamp: McCullough, 323–324.

313. Roosevelt tacitly signaled his support: Ibid., 338, 340, 382; Morison, 824–825.

313. the uprising had not yet occurred: Morison, 825; McCullough, 364–367.
314. "Colombia was hit by the big stick": Morison, 826.
315. "by far the most important action I took": Roosevelt, *Autobiography,* 512.
315. "I took the Isthmus": Roosevelt, "Charter Day Address," *Theodore Roosevelt Cyclopedia,* 407.
315. "Tell them that I am going to make the dirt fly!": Roosevelt, quoted in *Nation,* November 23, 1905, cited in McCullough, 408. See also Morison, 826.
316. 33,000 to 40,000 annually: Panama Canal Authority—Canal History, "Panama Canal History—workforce," www.pancanal.com/eng/history/index.html.
317. 26 million gallons of fresh lake water: Cornelia Dean, "To Save Its Canal, Panama Fights for Its Forests," *New York Times,* May 24, 2005.
318. By 1970, over 15,000 ships: McCullough, 611–612. In 1955 Suez had 14,555 ships. Morison, 1,093.
319. shipping revolution: The revolution had transformed the world's ports. No longer were cargo ships unloaded at docks. Instead, intermodal containers were lifted directly onto waiting trains and trucks to be transported directly to their final destinations.
319. "The fifty miles between the oceans": McCullough, 613–614.
319. American naval history's three eras: Love, 1:xiii.
319. United States entered World War I: The March 1917 sinking of three U.S. merchantmen, with heavy loss of life, as well as the interception of the Zimmermann telegram suggesting a German-Mexican alliance that could threaten U.S. security, were proximate causes.
320. Midway was the first sea battle: Howarth, 152–163. No U.S. aircraft carriers had been destroyed at Pearl. The intelligence breakthrough that tipped the Americans off to Japan's secret intention to attack the Midway atolls occurred when U.S. radio signalers purposely sent out a bogus, uncoded message that the water distillation plant on American-controlled Midway had broken down, and then later intercepted Japanese radio operators relaying the message, in Japanese code, that Midway was without water.
320. combined power of the world's next nine leading military nations: Kennedy, "Eagle Has Landed," I, III; Kennedy, "Has the U.S. Lost Its Way?" Some estimates have the United States spending as much on its armed forces as the world's next nine biggest military powers combined.

Chapter Thirteen: Giant Dams, Water Abundance, and the Rise of Global Society

322. Kansas, Nebraska, and Colorado increased by more than 1 million: Smith, *Virgin Land,* 174, 184.
322. depopulated by one-fourth to one-half: Reisner, *Cadillac Desert,* 107.
325. "When the arid lands": Turner, 258. Turner also wrote, "No longer is it a question

of how to avoid or cross the Great Plains and the arid desert. It is a question of how to conquer those rejected lands . . . It is a problem of how to bring precious rills of water to the alkali and sage brush." Ibid., 294.

326. expanded their irrigated cropland fifteenfold: Worster, 77.

327. unleashed a flood that killed 2,200: Reisner, *Cadillac Desert,* 107–108.

328. 1.25 million small farmers to cultivate 100 million acres: Worster, 132–139; Smith, *Virgin Land,* 196–198; Reisner, *Cadillac Desert,* 45–50. By careful management of water rights, Powell argued, small farms of only 80 acres—half the Homestead Act size of 160 acres for dry farms—could be viable.

328. "In the arid region it is water": T. Roosevelt, "State of the Union Message, December 3, 1901," http://www.theodore-roosevelt.com/sotu1.html.

329. "The forest and water problems": Ibid.

329. over half of irrigation project farmers were defaulting: Reisner, *Cadillac Desert,* 116.

329. 1920s, the U.S. agricultural sector: Two of the main causes of the agricultural depression were a decline in foreign export demand from war-recovering Europe and a fall in commodity prices due to increased productivity from farm mechanization; as a result, four out of 10 of the 7 million U.S. farmers were tenants, not freeholders, in 1929.

329. might well have vanished at that point: Instead, in 1923 the Reclamation Service was purged, its leader replaced, and renamed the Bureau of Reclamation.

330. few dams had surpassed 150 feet: A Roman dam at Subiaco was about 130 feet, and was hardly surpassed for 1,500 years. In Persia, the Mongols of the thirteenth and fourteenth centuries built the 190-foot Kurit Dam, which was the tallest on Earth for 500 years. Smith, *History of Dams,* 32, 235, 236; Billington et al., 50.

330. Hoover Dam: Hoover is a concrete, arched-gravity dam. U.S. Department of the Interior, Bureau of Reclamation, Lower Colorado Region, 30–36.

331. ingrained skepticism of the water bureaucracy establishment: Billington et al., 90–91. The multipurpose approach was especially controversial inside the U.S. Army Corps of Engineers, whose main mission was navigation.

331. government a big player in the private electricity business: In the 1930s private power plants in the West generated 3.5 million horsepower versus only 50,000 by the government in 1920. Hoover's original 1.7 million horsepower, therefore, dramatically altered the political economy of electricity in the West.

332. 14 million acre-feet per year: An acre-foot measured the volume that would cover one acre with one foot of water. It was equal to 325,851 gallons, or 1,233.5 cubic meters.

332. its flow was schizophrenic: Billington et al., 136. The Colorado's intensity ranged from 2,500 to more than 300,000 cubic feet per second.

332. 17 times siltier than the muddy Mississippi: McNeil, *Something New Under the Sun,* 178.

332. "too thick to drink, too thin to plow": Michael Cohen, "Managing across

Boundaries: The Case of the Colorado River Delta," in Gleick, *The World's Water, 2002–2003,* 134.

333. name was changed from the Valley of the Dead: Reisner, *Cadillac Desert,* 122–123.

333. Salton Sink swelled with water: The years 1905 to 1907 were some of the wettest in the Colorado basin's history. Since then, the Salton Sea, which initially acted like a reservoir, has continually shrunk from natural evaporation and is now very salty.

333. Los Angeles stepped forward in 1924 with a proposal: Billington et al., 160–161.

334. small Los Angeles River: Reisner, *Cadillac Desert,* 53, 60, 73.

334. outflanked and killed Reclamation's own farm irrigation plan: At a critical moment, Teddy Roosevelt threw his support to Los Angeles and arranged for his Forest Service to kill the Reclamation Service's claims by declaring that much of Owens Valley would henceforth be national parkland.

334. Posing as cattlemen and as resort developers: Reisner, *Cadillac Desert,* 68–69.

335. secretly buying up cheap land options: Ibid., 75–76. The key reason Mulholland wanted to route the water through San Fernando was that the unused portion could be stored there. This allowed him to use all of Los Angeles's share of the Owens River, which was essential to maintain the city's claim under the western water law of appropriation rights, popularly known as "use it or lose it." As a direct result of the Owens River water, the San Fernando Valley was soon incorporated into Los Angeles. Among the insiders were Harrison Gray Otis and Harry Chandler of the *Los Angeles Times,* railroadmen Edward Harriman and Henry Huntington, and bankers Joseph Artori of Security Trust and Savings Bank and L. C. Brand of the Title Guarantee and Trust Company.

335. population surpassed Mulholland's expectations: Billington et al., 161.

335. violent reaction from irate Owens Valley farmers: Reisner, *Cadillac Desert,* 92–95.

335. broke the last local opposition to Mulholland's bid: A chief opponent was the powerful *Los Angeles Times* publisher Harry Chandler, who placed greater importance on the near-term hit to the value of his large acreage in Mexico than on the long-term growth of Los Angeles. The Owens Valley problem also expedited the creation of regional water districts with taxing powers in order to raise funds to purchase the dam's hydroelectricity to pump its water up the escarpment through the aqueduct and across the Mojave Desert.

336. divided the river into an upper and lower basin: The upper-basin states of Colorado, Wyoming, Utah, and New Mexico supplied over 90 percent of the Colorado's water.

336. 7.5 million acre-feet were assigned to each basin: Billington et al., 158–159; U.S. Department of the Interior, Bureau of Reclamation, Lower Colorado Region, 10; Reisner, *Cadillac Desert,* 262–263.

337. builders constructed their own steel-fabricating plant: Reisner, *Cadillac Desert,*

128–129; U.S. Department of Interior, Bureau of Reclamaton, Lower Colorado Region, 15–23.

337. the strike was broken, with the federal government's tacit approval: Billington et al., 174–175. The Bureau of Reclamation also declared the construction site to be federal land to circumvent Nevada law prohibiting the underground use of internal combustion engines on health safety grounds.

338. "I came, I saw, and I was conquered": Franklin D. Roosevelt, quoted in Billington et al., 179.

338. five largest structures on Earth, all dams: Reisner, "Age of Dams and Its Legacy."

339. hydroelectricity for the entire population living west: Reisner, *Cadillac Desert,* 155.

339. the Grand Coulee: Billington et al., 206.

339. Roosevelt started the project on his own: Reisner, *Cadillac Desert,* 156–157.

339. 36 huge dams would be built on the Columbia: Ibid., 165.

340. Grand Coulee Dam: Worster, 271.

340. providing 40 percent of America's total hydroelectricity: Billington et al., 191.

340. hydroelectricity sales heavily subsidized the building of the dam: Worster, 271. Ninety percent of the costs were covered with hydroelectricity sales; in the absence of a strong agribusiness lobby, the government made a concerted effort to limit existing users to the same water subsidies that were to be provided only to small 160-acre farms under the 1902 Reclamation legislation.

340. 92 percent of Grand Coulee's and Bonneville's electricity output: Reisner, *Cadillac Desert,* 162, 164. By the middle of the war, half of total U.S. aluminum production—which requires electrical power—was located in the Pacific Northwest. The United States produced some 60,000 warplanes in four years of war.

341. 23,500 well pipes pumped up prodigious amounts: Ibid., 151, 335.

342. "The Central Valley Project": Reisner, *Cadillac Desert,* 336–337. Among the big landowners receiving subsidized waters were food giant DiGiorgio Corporation, the Southern Pacific Railroad, and Standard Oil.

342. most intensively water-engineered place on the planet: Reisner, "Age of Dams and Its Legacy."

342. Tennessee River basin: Morison, 960–964. The river, 652 miles long, rises in the Appalachians of North Carolina and Virginia and flows west, where it empties in the Ohio River near Paducah, Kentucky.

343. staircase of 42 dams and reservoirs: Specter, 68; Reisner, *Cadillac Desert,* 167.

343. results transformed the Tennessee Valley: Morison, 963. Electricity prices fell from 2.4 cents to 1 cent per kilowatt-hour.

343. 75,000 dams had been built: Specter, 68.

343. 6,600 large ones over 50 feet: Peet, 9; Sandra Postel, "Hydro Dynamics," 62.

343–344. Bureau of Reclamation cataloged: Worster, 277.

344. 17 western states had 45.4 million acres under irrigation: Ibid., 276–277.

344. American water use for all purposes multiplied tenfold: Ibid., 312. Water use rose from 40 billion gallons per day to 393 billion between 1900 and 1975. U.S. Census figures show that population rose from 76 to 216 million in the same period.
345. 40 percent of American cattle: McNeill, *Something New Under the Sun,* 154; McGuire, "Water-Level Changes in the High Plains Aquifer, 1980–1999"; Pearce 59.
345. Ogallala only half an inch per year: Reisner, *Cadillac Desert,* 438.
346. gigantic cloud of stinging, shearing dust: Ibid., 452. See also Evans, *American Century,* 232–233.
346. 60 major, sky-blackening dust storms each year: Evans, *American Century,* 232. There were 40 major dust storms in 1935, 68 in 1936, 72 in 1937, and 61 in 1938.
346. 3.5 million "Dust Bowl refugees": Ibid., 234.
347. centrifugal pump could lift 800 gallons: Reisner, *Cadillac Desert,* 436.
347. could raise water even faster: Glennon, *Water Follies,* 26. The new techniques were capable of pumping 1,200 gallons per minute.
347. Ogallala annual water use quadrupled: McNeill, *Something New Under the Sun,* 154; McGuire, "Water-Level Changes in the High Plains Aquifer, 1980–1999."
347. growing 15 percent of that nation's wheat, corn, cotton: Reisner, *Cadillac Desert,* 437, 448–449.
347. drawing water out of the Ogallala 10 times faster: McNeill, *Something New Under the Sun,* 154.
348. Ogallala reservoir would last: Ibid. Irrigation peaked in northern Texas in the mid-1970s and began contracting across the High Plains as a whole in 1983. On Central Valley overpumping, see Felicity Barringer, "As Aquifers Fall, Calls to Regulate the Use of Groundwater Rise," *New York Times,* May 14, 2009.
348. U.S. groundwater usage more than doubled: Robert Glennon, "Bottling a Birthright," in McDonald and Jehl, 17. Over that thirty years, groundwater usage increased from 8 to 18.5 billion gallons per day—65 gallons per person.
349. 19 large dams and reservoirs held four times: The last dam on the river was the hydroelectric giant Glen Canyon, completed in the mid-1960s.
349. Every drop was used and reused 17 times: Cohen, 134.
350. starting to take up to an additional 900,000 acre-feet: Reisner, *Cadillac Desert,* 260–261.
350. salinity at the river's halfway point: Worster, 321–322.
351. scrambled to develop emergency plans: Gertner.
352. 75 percent of the state's entire agricultural output: Bureau of Reclamation, "Central Valley Project—General Overview," www.usbr.gov/dataweb/html/cvp.html.
352. water-efficient industries and cities: For example, 1,000 acre-feet of water used to produce semiconductors and other high-tech applications created some

16,000 jobs, while the same water on pasture farms added only eight jobs; Las Vegas and Reno used 10 percent of Nevada's water but accounted for 95 percent of its economy—while marginal alfalfa farmers who consumed most of the rest couldn't survive without the water subsidy.

353. United States decommissioning surpassed new construction by 2000: Clarke and King, 44. As it often did, the change in American domestic attitudes within the leading world power helped condition opinions at world institutions. In 2000 the U.N.'s World Commission on Dams reported that the negative effects of many large dam projects outweighed the benefits and urged nations to explore alternative approaches to satisfying their water resource needs.

353. the United States had more than 50,000 toxic waste dumps: McNeill, *Something New Under the Sun,* 29.

354. released naturally by all the volcanoes in Earth's history: Ponting, 366.

354. deadly radioactive waste: Nuclear waste afflicted both America's Columbia River and the Soviet Union's upper Ob River basin in western Siberia, which became the most radioactive place on Earth. In 1967, when a prolonged drought dried the bed of Lake Karachay, into which the Soviets had disposed nuclear waste, lake dust carrying 3,000 times the radioactivity of the bomb dropped on Hiroshima was scattered by the winds over half a million people in central Asia; the area remained so radioactive twenty years later that anyone visiting the lakeshore for an hour risked death from the radiation.

354. "The pollution entering our waterways": Carson, 39, 41.

355. "The problem of water pollution by pesticides": Ibid., 39.

355. "Along with the possibility of the extinction of mankind": Ibid., 8.

356. combusted on Cleveland's Cuyahoga River: Specter, 69. Similar fires on India's Ganges and Russia's Volga rivers in the same period attested to the universality of the environmental problem.

356. Earth Summits of heads of state: Earth Summits were held at Rio de Janeiro (1992) and Johannesburg (2002). The Intergovernmental Panel on Climate Change issues reports every five or six years (1990, 1995, 2001, 2007). The Millennium Ecosystem Assessment, inaugurated by Kofi Annan, was published in 2005.

358. "Every drop of water that runs to the sea without yielding": Herbert Hoover, quoted in Glennon, *Water Follies,* 13; Joseph Stalin, quoted in Peet, 11.

358. "the new temple of resurgent India": Jawaharlal Nehru, quoted in Specter, 68.

358. Soviet Union increased its water use eightfold: McNeill, *Something New Under the Sun,* 163.

358–359. double irrigated cropland in the first quarter century: Ibid., 179, 278. Water use, meanwhile, quintupled during the same period; see also Jim Yardley, "Under China's Booming North, the Future Is Drying Up," *New York Times,* September 28, 2007.

359. India's 4,300 large dams ranked it third: Peet, 9.

359. 13 were being erected on average every day: Ibid., 9–10.

359. World reservoir capacity quadrupled: Millennium Ecosystem Assessment, 26.
359. World hydropower output doubled: Ibid., 5. World population doubled from 3 to 6 billion from 1960 to 2000, while economic output sextupled.
360. irrigation nearly tripled in the half century: Hans Schreier, "Mountain Wise and Water Smart," in McDonald and Jehl, 90.
360. all the corn grown in the United States was hybrid: McNeill, *Something New Under the Sun,* 220.
360. Hybrid dwarf wheat: Dwarf wheat started in the 1920s with Norin 10, a semi-dwarf variety developed in Japan that crossed Japanese and U.S. varieties, then was further crossbred in Mexico in the 1950s by pathbreaking plant breeder Norman Borlaug, who won a Nobel Prize in 1970 for his work.
360. hybrid varietals increased their share: McNeill, *Something New Under the Sun,* 222.
361. 60 percent of all larger river systems in the world: Millennium Ecosystem Assessment, 32.
361. best hydropower and irrigation dam sites: This was not true in Africa, which had for the most part been bypassed by the Green Revolution and still had good, untapped hydropower potential.
361. "for a small but measurable change in the wobble of the earth": Gleick, "Making Every Drop Count," 42.
361. retired as fast as new irrigated land was developed: Simmons, 258.
362. 10 percent of world farming was unsustainable: Postel, "Growing More Food with Less Water," 46–47.

Chapter Fourteen: Water: The New Oil

368. half the renewable global runoff: Millennium Ecosystem Assessment, 106.
370. 1.1 billion people: United Nations Millennium Project Task Force on Water and Sanitation, 4.
371. lives are uprooted catastrophically: Millennium Ecosystem Assessment, 13; United Nations Millennium Project Task Force on Water and Sanitation, 17.
371. occurred from 1999 to 2005: Peter H. Gleick, "Environment and Security: Water Conflict Chronology," in Gleick, *World's Water, 2006–2007,* 207–212. Yemen, Jordan, Namibia, Sicily, and Algeria were among the multitude of places where water was rationed. Fierce, perennial litigation over water rights was normative in the United States and other countries governed by credible rules of law. The annals between 1999 and 2005 provided an illustrative sample of the increasingly commonplace violent protests and clashes within countries. Chinese farmers from Hebei and Henan provinces fired mortars and bombs at one another in a battle over limited water resources; a year later small-scale water wars and riots broke out, leading to several deaths, in Shandong province along the Yellow River when the government tried to stop thousands of farmers illegally diverting water from a reservoir earmarked to supply China's

drying northern cities. In Cochabamba, Bolivia's third-largest city, one person died when 30,000 protesters clashed with police for several days in a fury over the government's privatization of the municipal water delivery system, which pushed prices up to one-quarter of many residents' wages. Karachi, Pakistan, was shaken by four bombings and riots from demonstrators chanting "Give us water" during a period of prolonged drought. Neighboring India had several incidents and deaths in different parts of the country, including riots in Gujarat when water trucks regularly failed to provide enough water. More than 20 were killed in tribal violence in northwestern Kenya following charges by Masai herdsmen that a local Kikuyu politician had diverted a river to irrigate his farm. Somalia's "War of the Well" claimed 250 dead as villagers clashed in the extensive violence that accompanied the three-year drought and dysfunctional central government. Water wells in Darfur were intentionally destroyed and contaminated as part of the campaign of genocidal ethnic cleansing.

372. "Many of the wars": Ismail Serageldin, quoted in "Of Water and Wars."

372. "now well beyond levels that can be sustained": Millennium Ecosystem Assessment, 6.

372. rise from half to 70 percent by 2025: Sterling, 30.

372–373. one-quarter of global freshwater use: Millennium Ecosystem Assessment, 6, 106–107. Fifteen percent to 35 percent of withdrawals for irrigated crops were drawn from depleting resources.

373. 265 gallons for a single glass: Pearce, 3–4.

373. ordinary cotton T-shirt: Sterling, 31.

373. By 2025 up to 3.6 billion people: Postel, *Last Oasis,* xvi.

373. virtual water: J. A. Allan, professor at Kings College London and at the School of Oriental and African Studies, University of London, won the 2008 Stockholm Water Prize for his pioneering work on the concept of virtual water in the early 1990s.

374. evaporation-transpiration: Transpiration is the process of water vapor emission from organic matter such as plants and humans.

374. that one-third totals enough: McNeill, *Something New Under the Sun,* 119; Postel, *Last Oasis*, 28; Pearce, 28.

374. large share runs off unused: The Amazon watershed alone accounted for 15 percent of the runoff, while only four-tenths of 1 percent of the world's population lived there.

374. in Africa only one-fifth of all rainfall: Clarke, *Water: The International Crisis*, 10.

374. history's poorest societies often had: Grey and Sadoff, 545.

375. 90 percent of the dry-land inhabitants: Millennium Ecosystem Assessment, 13.

376. governments still routinely maintain monopolistic control: In the United States, water was the one surviving great state monopoly, which had previously included electricity and telecommunications.

377. "tragedy of the commons": As used here, the term *tragedy of the commons* refers to the social trap caused by combined individual exploitation of a shared resource that is injurious to the greater public good. The concept has a long history, but was coined in modern times in a famous 1968 essay in *Science* by biologist Garrett Hardin.

377. Aral Sea: McNeill, *Something New Under the Sun,* 163–164.

378. Lake Chad: Pearce, 85. Lake Chad had oscillated in size from natural forces since at least the Middle Ages. A natural peak was reached in 1962 with a drainage zone comparable in size to continental western Europe. About half to one-third of the shrinkage from the 1960s to 2004 was estimated to have come from man-made irrigation water diversions. The major irrigation dam projects were in Nigeria and Cameroon.

378. paid little more than 10 percent: Postel, *Last Oasis,* 166–167.

378–379. Mexico City loses enough water every day: Sterling, 32; Gleick, "Making Every Drop Count," 43.

379. "Nothing is more useful than water": Smith, *Wealth of Nations,* 174.

379. "When the well is dry": Benjamin Franklin, *Poor Richard's Almanac, 1733,* cited in "Water Fact Sheet Looks at Threats, Trends, Solutions," Pacific Institute, www.pacinst.org/reports/water_fact_sheet.

380. 1,700 times markup: Lavelle and Kurlantzick.

380. $400 billion per year industry: Peter H. Gleick and Jason Morrison, "Water Risks That Face Business and Industry," in Gleick, *World's Water, 2006–2007,* 158–165. Water utilities were dominated by two French and one German company: Veolia Environnement, Suez S.A., and RWE Thames Water. GE had made a $3.2 billion investment in the $140 billion per year wastewater services sector. The fragmentation of the water business made reliable estimates of comprehensive industry size hard to come by. Some $85 billion per year was spent on private industrial water treatment to supply purified water to water-intensive industries, such as semiconductors, pharmaceuticals, certain chemical processing, pulp and paper, and food and petrochemicals. Drinking-water purification, desalinization, and water distribution infrastructure were other large segments.

382. aquifer reservoirs accumulated by nature: A good deal of this deep fossil water was inaccessible even with modern drilling technologies.

382. have less than 700 gallons per person: Postel, *Last Oasis,* 28–29. Less than 1,000 cubic meters (2,740 liters) daily per capita defined water scarcity, 1,000 to 2,000 per day defined water stress, and more than 2,000 per day defined water sufficiency. Clarke, *Water,* 12, cites more than 20 percent of runoff used as a sign of water scarcity; 10 to 20 percent usage as a serious water problem, and less than 5 percent usage as water sufficiency.

382. world resource demand increases: Diamond, *Collapse,* 495. Diamond argues that by far the greatest impact comes from the 80 percent who live in the third world, including the rising Chinese and Indian populations, increasing their

meager consumption of water and other resources to the prodigious levels of Western industrial societies.

Chapter Fifteen: Thicker Than Blood: The Water-Famished Middle East

384. outgrew their internal water resources: Allan, 6.
384. "the Middle East and North Africa": Millennium Ecosystem Assessment, 33.
385. quadrupling of Middle East wheat flour imports: Allan, 8.
385. shallow wells and qanats: Much of the drinking water of modern Tehran was still supplied by qanats.
385. forecast to swell another 63 percent to 600 million: Andrew Martin, "Mideast Facing Difficult Choice, Crops or Water," *New York Times,* July 21, 2008.
385. 75 million inhabitants: Economist staff, *Pocket World in Figures, 2009,* 16.
385. completion of the high dam at Aswan: Some 96 percent of Egyptians lived on the crowded "ribbon of land along the Nile's banks," which comprised a mere 4 percent of the country's total area. Elhance, 6.
387. "The national security of Egypt": Boutros Boutros-Ghali, quoted in "Water Scarcity, Quality in Africa Aggravated by Augmented Population Growth, *International Environmental Reporter,* October 1989, cited in Postel, *Last Oasis,* 73.
388. "We depend upon the Nile 100 percent": Anwar el-Sadat, quoted in Collins, 213.
388. 25 million: See Smith, *Man and Water,* 205; Collins, 140.
390. within a dozen feet of reaching the total shutoff levels: Collins, 225–226.
391. "The only matter that could take Egypt to war again": Anwar el-Sadat, quoted in Gleick, *World's Water, 2006–2007,* 202. Senior Egyptian officials have made the same point repeatedly since then, including then Egyptian foreign minister and future U.N. secretary-general Boutros Boutros-Ghali, who said in 1988: "The next war in our region will be over the waters of the Nile, not politics."
392. "Preserving Nile waters for Egypt": Boutros-Ghali, 322.
392. Mengistu Haile Mariam: Collins, 214–215. Mengistu overthrew Emperor Haile Selassie, whose longtime dream had been a dam at Lake Tana, in 1974. By 1978 he began pressing for Ethiopia's internal development of its water resources with a dam. Mengistu aggravated his bad relations with Sadat by conjuring up historic Ethiopian fears of Egyptian-Islamic territorial ambitions in the Horn of Africa, accusing Egypt of stirring up trouble in his country's backyard by arming Somalis in the Ogaden and supporting breakaway rebels in Eritrea.
392. the bureau concluded: Collins, 171.
393. found an accommodating negotiating partner: At first Nasser had tried and failed to bully Sudan's leaders into acquiescence by pressing Egyptian territorial claims on Sudan's ancient Nubia.
393. agreed to move jointly against upstream nations: Erlich, 6.

393. Selassie had obtained public declarations of support: Collins, 170. The United States conditioned its backing for Nile waters development for the Aswan Dam upon the cooperation of all Nile states.
394. several times costlier: Grey and Sadoff, 545–571. World Bank water experts David Grey and Claudia Sadoff note that water-shock-prone countries are typically among the world's poorest, and that such countries often face more difficult hydrological patrimonies than industrialized nations did during their earlier phases of economic takeoff.
395. Nile Waters Agreement: Egypt's Master Water Plan of 1981, which envisioned increasing Nile water yields by up to one-quarter through new projects situated upstream, also rather fancifully ignored the region's rampant political instability and any deleterious environmental side effects.
395. abruptly terminated in 1984: The Darfur genocide in western Sudan, likewise, was supported by Sudan's northern-led Muslim government and included assaults on the water supplies of indigenous, mostly black non-Muslim residents.
395. Egypt blocked an African Development Bank loan: Alan Cowell, "Cairo Journal: Now, a Little Steam. Later, Maybe a Water War," *New York Times,* February 7, 1990.
395. Israeli . . . engineers were doing feasibility studies: Darwish; Ward, 197.
395. diversion of an additional 5 billion cubic meters: Allan, 67–68, 152–153. The New Valley Project was intended to transform Egypt's desolate desert northeast of Aswan into an agricultural and industrial oasis for some 7 million people relocated from Egypt's overcrowded Nile corridor.
396. "While Egypt is taking the Nile water": Meles Zenawi, quoted in Mike Thomson, "Nile Restrictions Anger Ethiopia," *BBC News,* February 3, 2005, http://news.bbc.co.uk/2/hi/africa/4232107.stm. While steadfastly denying that they blocked international financing for other countries' irrigation projects, Egyptian leaders argued that they were compelled by Egypt's lack of natural rainfall and domestic demographic trends to expand water diversion for ambitious new desert developments. In 2005, Dia El Quosy, a senior adviser to the Egyptian Ministry of Water Resources and Irrigation, told the BBC, "It's not only the production of food. It's also about the generation of employment. Some 40% of our manpower are farmers and if these people are not given opportunities and jobs they will immediately move to the cities and you can see how crowded Cairo is already." When asked in 2003 if Egypt would respond with force if Ethiopia or an independent southern Sudan region cut Nile flows north, Boutros-Ghali replied, "I don't believe that any country will dare to cut the water because . . . the national security of Egypt is based on water, on the sources of the Nile." "Talking Point."
397. heavily subsidized prices: "Of Water and Wars"; Elhance, 60. In the late 1990s, subsidies for energy, some for pumping and moving water, amounted to another $4 to $6 billion.

397. "Among the pervasive beliefs in Egyptian culture": Collins, 218.
397. imports—providing up to two-fifths: Brown, "Grain Harvest Growth Slowing."
397. 30 miles inland: McNeill, *Something New Under the Sun,* 170–171.
398. from 32 to only 2 billion cubic meters: Lester Brown, "The Effect of Emerging Water Shortages on the World's Food," in McDonald and Jehl, 85; McNeill, *Something New Under the Sun,* 170–171.
398. they are projected to add nearly 50 percent: Economist staff, *Pocket World in Figures, 2009,* 16, 17.
399. might decline up to 25 percent: Elhance, 58.
399. Mubarak called in the army: "Not by Bread Alone," *Economist,* April 12, 2008, 55.
400. Nile Basin Initiative: Sadoff and Grey, "Beyond the River." Sadoff and Grey posit four potential sources of gain from cooperation: (1) better management of ecosystems supporting the basin; (2) higher yields from the rivers; (3) reduction in costs from competition and tensions; and (4) benefits, such as increased trade, between nations arising from their amity in river cooperation.
400. 10 billion cubic meters: Interview with Nile Basin Initiative participant.
401. withdrawing 3.2 billion cubic meters: Sher, 36. See also Allan, 74–77. The total of 3.2 billion cubic meters (2.6 million acre-feet) includes Israel, Palestine, and Jordan, but excludes Syria on the basin's periphery.
401. one-third as much freshwater as needed: Allan, 76.
401. Eric Johnston: Ibid., 78; Postel, "Sharing the River out of Eden," 61.
401. doom the landmark water accord: Elhance, 113. The Johnston plan operated on the principle that surface water should be allocated by reasonable proximity to irrigable land fed by natural gravity, which placed useful needs above territorial land rights to water. Although the plan was not enacted, the proportions it allocated remained the baseline for water-sharing negotiations over the next half a century.
401. Golda Meir had put Israel's Arab neighbors on notice: Postel, "Sharing the River out of Eden," 62. On the Skirmish over the Jordan, see also Darwish.
402. "In reality the Six Day War": Sharon, 167.
402. Arab air force lay smoldering: Goldschmidt, 326. The Suez Canal would be closed for eight years. It remained the violent front line between the two enemies, marked by occasional firing across its waters and air skirmishes above it. The Yom Kippur War (1973) began with a simultaneous surprise assault by Syria on the Golan Heights and an amphibious thrust by Egypt across the canal to establish a bridgehead that allowed its troops to temporarily recover much of the Sinai. But when General Ariel Sharon and a small tank force snuck behind Egyptian lines nine days later and managed to cut off the Egyptian expeditionary force from Egypt proper, the Suez Canal water boundary line was reestablished for another two years.
403. another third of Israel's water: Allan, 82. The three headwater tributaries of the

Jordan—the Banias, the Hasbani, and the Dan—all originated in springs fed by an underground aquifer on the slopes of Mount Hermon in the Golan. Another one-fifth of Israel's water supply was recycled, and desalinization plants had capacity for another one-third and were replacing the water in the depleting coastal aquifer. Israel wasted little time in augmenting the flow from the Golan by opening new springs in the Huleh Valley to channel more floodwaters to the Sea of Galilee.

404. 20 to 40 percent of their income for water: Pearce, 160–161. On the decline in West Bank Palestinian irrigated cropland see Darwish.

404. Gaza Aquifer: Postel, "Sharing the River out of Eden," 63. Gaza water was below the minimal drinking water standards of the World Health Organization. The international community financed, with Israeli approval, a state-of-the-art sewage treatment plant in Gaza to try to mitigate the problem.

404. 1987 intifada: Aaron T. Wolf, "'Water Wars' and Other Tales of Hydromythology," in McDonald and Jehl, 116–117.

404. put off to the final-status stage: Postel, *Last Oasis,* xxiv, xxv. By 2000 Israel was drawing half to three-quarters more water than envisioned by the original Johnston plan that the Arabs had rejected.

405. secretly meeting for years: Elhance, 107, 113. Jordan depended heavily on the water from the Yarmuk-Jordan since its only other major water source was the nonrenewable Qa Disi aquifer on its southeastern border with Saudi Arabia, which the Saudis were rapidly drawing down toward exhaustion at the prodigious rate of up to 250 million cubic meters per year.

405. Wazzani fed the Hasbani: Tensions were further heightened because Israel's water reserves were at their lowest historical level at that moment.

405. international diplomatic flurry: "Israel Hardens Stance on Water," *BBC News,* September 17, 2002, http://www.bbc.co.uk/ 2/hi/middle_east/22265139.stm; Luft; Stefan Deconinck, "Jordan River Basin: The Wazzani-Incident in the Summer of 2002—a Phony War?" Waternet (July 2006), http://www.waternet.be/jordan_river/wazzani.htm.

406. very close to a breakthrough: Working through the offices of Turkish prime minister Recep Tayyip Erdogan, who was trusted by the Israelis and had developed good relations with Syria's leadership, the two sides reportedly had worked out virtually all the major issues of a Golan water deal between them and were close to being ready to move the final negotiations up to the official level. But the opportunity was reportedly killed by the Bush administration, which refused to give the comfort sought by Syrian leaders that they could count on American support as Syria moved out of its orbit of relations with Iran, Hezbollah, and Hamas. Israel's attack on Gaza over Hamas's resumption of missile launches at Israel in late 2008 reinflamed discord, which scuttled any chance of an early settlement.

406. cut agricultural water consumption by nearly one-third: Allan, 96–97; Elhance, 96.

406. pay full market price: "Don't Make the Desert Bloom," *Economist,* June 7,

2008, 60. The actual water price farmers paid in 2008 was still only about half the market rate due to hidden subsidies, but the agreement signaled the direction of things.

406. agriculture's share: Postel, "Sharing the River out of Eden," 64.
407. microirrigation methods: Ibid., 43, 64.
407. quintupled their water productivity: Pearce, 300.
407. fall sharply—by as much as two-thirds: Ibid., 254; Economist staff, "Tapping the Oceans," *Economist Technology Quarterly,* June 7, 2008, 27. The main improvements were in energy recapture and membrane technology.
407. Israel to launch five: From the 1970s Israel had been studying elaborate schemes to pipe water from the Mediterranean or Red sea to the Dead Sea, exploiting the decline in altitude to generate the great amount of electricity needed to power desalinization.
407. Ashqelon, opened in 2005: Ashqelon water costs were about 55 U.S. cents versus about 30 cents per cubic meter from Galilee.
407. by 2020 Israel expects: "Don't Make the Desert Bloom," 60; Postel, "Sharing the River out of Eden," 64.
408. 10 times the per capita supply of Israel: Sher, 36.
408. control the headwaters: Turkey's inclusion in NATO was strongly influenced by its strategic control of the strait controlling access to the Mediterranean from the Black Sea. During the Cold War, it helped deny the Soviet Union's navy easy access and influence in the Mediterranean and the Middle East, stretched Soviet supply lines in aiding distant allies like North Vietnam, and in general added to the burdens that helped lead to the Soviet Union's collapse. The strait today remains a strategic choke point of key shipping lanes, including those for oil from the large new fields and pipelines of central Asia to the West.
408. upriver to Turkey: From the early 1970s to 2002, Turkey built some 700 dams and had plans to build 500 more. Douglas Jehl, "In Race to Tap the Euphrates, the Upper Hand Is Upstream," *New York Times,* August 25, 2002.
408–409. double national irrigated cropland and electricity: Elhance, 148–149.
409. Ataturk reservoir: "One-third of Paradise," *Economist,* February 26, 2005, 78.
409. "The twenty-first century will belong": Turgut Ozal, quoted in Ward, 192.
409. cut Syria's share of the Euphrates' water: Ibid.
409. consume half again as much water as exists: Jehl, "In Race to Tap the Euphrates, the Upper Hand Is Upstream." The Euphrates held about 35 billion cubic meters of water.
409. Turkey's vision: Elhance, 150–151; Sher, 35–37. The first pipeline, which drew water from the little-used Seyhan and Ceyhan rivers, was expected to carry 1.28 billion cubic meters per year, and the second, drawing from the Tigris, 0.9 billion cubic meters annually.
409–410. Euphrates slowed to a trickle: Elhance, 144.
410. clandestine support: Allan, 73.

410. force Saddam to withdraw: Gleick, *World's Water, 2006–2007,* 204.

410. Syria's constriction of the Euphrates' water: Elhance, 142–143. Syria slowed flows in the spring of 1974 to show its anger with Iraqi criticism of its policy toward Israel and again in 1975 after Iraq signed an accord with Iran. Saudi Arabian and Soviet diplomacy resolved the 1975 crisis by getting Syria to release additional water downstream from the Tabqa Dam. Ironically, the greatest danger from a catastrophic dam break in the region was probably in Sadaam's own Iraq. The large Mosul dam on the Tigris near ancient Nineveh had been so poorly constructed in 1984 that by 2007 the U.S. Army Corps of Engineers was warning occupying American military commanders that it was in imminent danger of collapse—threatening to kill hundreds of thousands of people between Mosul and Baghdad and deliver a devastating setback to the American effort to build a stable, pluralistic state in post-Saddam Iraq. Patrick Cockburn, "Iraqi Dam Burst Would Drown 500,000," *Independent,* October 31, 2007, www.independent.co.uk/news/world/Middle-east/Iraq's-dam-burst-would-drown–500,000–398364.html.

410. restore only 40 percent of the marshes: Alwash, 56–58; "One-third of Paradise," 77–78; Edward Wong, "Marshes a Vengeful Hussein Drained Stir Again," *New York Times,* February 21, 2004; Marc Santora, "Marsh Arabs Cling to Memories of a Culture Nearly Crushed by Hussein," *New York Times,* April 28, 2003. More water than Iraq had available was needed to flush out salts and other toxins to restore a larger area.

411. "We do not say we share their oil": Süleyman Demirel, quoted in "The Euphrates Fracas: Damascus Woos (and) Warns Ankara," *Mideast Mirror,* July 30, 1992, cited in Elhance, 144.

411. idled seven of 10 turbines: Whitaker.

411. "with more water than": Recep Tayyip Erdogen, quoted in Sally Buzbee, "Drought Threatens Iraq's Crops and Water Supply," Associated Press wire on Yahoo!News, July 10, 2008, AP20080710.

412. control of Iraq's largest hydroelectric dam: Daniel Williams, "Kurds Seize Iraq Land Past Borders in Blow to U.S. Pullout Plan," March 5, 2009, Bloomberg, http://www.bloomberg.com/apps/news?pid=20601087&sid=aeLL5Yyjul18&refer=home.

412. Palestinian and Israeli officials continued to meet: Postel, "Sharing the River out of Eden," 64.

412. "When one man drinks": Cited in Elhance, 122.

413. costs, however, were staggering: Craig A. Smith, "Saudis Worry as They Waste Their Scarce Water," *New York Times,* January 26, 2003. Allan, 85.

414. exhausted about 60 percent: Pearce, 61.

414. slashed wheat production: Brown, "Aquifer Depletion."

414. 10-quart toilets: Smith, "Saudis Worry as They Waste Their Scarce Water." See also Pearce, 61.

414–415. Arabian aquifer may be scraping bottom: Patrick E. Tyler, "Libya's Vast Pipe Dream Taps into Desert's Ice Age Water," *New York Times,* March 2, 2004.

415. Yemen: Yemen, ancient home of the Sabaean kingdom and source of precious myrrh and frankincense, was in danger of becoming a failed state amok with religious jihadists, political insurgencies, and anarchic social conflicts over scarce freshwater that had left dozens dead in recent years. The groundwater tables supplying Yemen's life-giving wells were plunging by six feet a year in the countryside and by 15 feet a year in its major cities; its capital, Sanaa, was expected by the World Bank to run dry by 2010, with no solution in sight. Meanwhile, Yemen's 22 million mostly poor, restive citizens were expected to double within a generation—making the country a constant source of potential regional and international destabilization.

415. subway-sized tunnels buried six feet: Tyler, "Libya's Vast Pipe Dream Taps into Desert's Ice Age Water"; McNeill, *Something New Under the Sun,* 155. See also Pearce, 45–48. Like Saudi Arabia and Yemen, Libya was an effectively waterless land with scant rainfall and no surface rivers or lakes that withdrew seven times more freshwater from groundwater sources than its total renewable supply.

416. largest known fossil water deposit: Earth's biggest aquifers are the Sahara's Nubian sandstone aquifer, with 50 billion acre-feet under Libya, Egypt, Chad, and Sudan; South America's Guarani aquifer, with 40 billion acre-feet lying beneath 400,000 square miles of Brazil, Argentina, Paraguay, and Uruguay; the Ogallala, in the United States; and the North China Plain.

416. Occidental Petroleum magnate Armand Hammer: McNeill, *Something New Under the Sun,* 155.

416. pipeline blowouts: Tyler, "Libya's Vast Pipe Dream Taps into Desert's Ice Age Water."

416. less than half of the food needs: Pearce, 45–48.

Chapter Sixteen: From Have to Have-Not: Mounting Water Distress in Asia's Rising Giants

418. diminishing groundwater reserves for their irrigated agriculture: India, Pakistan, and China together accounted for 45 percent of global groundwater use; the other leading groundwater user was the United States, but only a small portion of its agriculture depended upon it.

419. "an era of severe water scarcity": Quoted in Somini Sengupta, "In Teeming India, Water Crisis Means Dry Pipes and Foul Sludge," *New York Times,* September 29, 2006.

419. "survival of the Chinese nation": Wen Jiabao, quoted in "Drying Up," *Economist,* May 19, 2005, 46.

419. "Hindu rate of growth": Das, 4.

420. Narmada River valley: McNeill, *Something New Under the Sun*, 161–162; Postel, *Last Oasis*, 55–56; Specter, 68.

420. the commission concluded: Pearce, 134–135; Katherine Kao Cushing, "The World Commission on Dams Report: What Next?" in Gleick, *World's Water, 2002–2003*, 152.

421–422. "so that the specter of food shortages": Manmohan Singh, quoted in Somini Sengupta, "In Fertile India, Growth Outstrips Agriculture," *New York Times*, June 22, 2008. On Indian wheat farmers' water use, see *Economist*, "Awash in Waste," April 11, 2009.

422. store no more than a couple of months' protection: World Bank.

423. small tube wells: Marcus Moench, "Groundwater: The Challenge of Monitoring and Management," in Gleick, *World's Water, 2004–2005*, 88; Pearce, 36–37.

423. India relies on groundwater mining: Pakistan, however, relied more on groundwater as a percentage of its total water use.

423. being mined twice as fast: Brown, "The Effect of Emerging Water Shortages on the World's Food," in McDonald and Jehl, 82.

424. Coca-Cola and Pepsi: Peter H. Gleick and Jason Morrison, "Water Risks That Face Business and Industry," in Gleick, *World's Water, 2006–2007*, 146; Saritha Rai, "Protests in India Deplore Soda Makers' Water Use," *New York Times*, May 21, 2003.

424. Indian government report: Sengupta, "In Teeming India, Water Crisis Means Dry Pipes and Foul Sludge." New Delhi had 5,600 miles of water pipes, which lost an estimated 25 percent to 40 percent to leaks.

425. poor are effectively subsidizing: Peet, 8; Specter, 63.

425. newly installed meters broke down: Gleick and Morrison, 148.

425. Ritu Praser: Sengupta, "In Teeming India, Water Crisis Means Dry Pipes and Foul Sludge."

425. India's sanitation: Gleick and Morrison, 148.

425. surface and ground water supply is polluted: Meena Palaniappan, Emily Lee, and Andrea Samulon, "Environmental Justice and Water," in Gleick, *World's Water, 2006–2007*, 128.

426. pesticide plant in Bhopal: Somini Sengupta, "Decades Later, Toxic Sludge Torments Bhopal," *New York Times*, September 29, 2006.

426. Gangotri glacier: Emily Wax, "A Sacred River Endangered by Global Warming," *Washington Post*, June 17, 2007; "Melting Asia," *Economist*, June 7, 2008, 29. Similarly, the Kashmir valley's sole year-round water source, the Kolahoi glacier, had shrunk by half a mile in the twenty years since 1985. "How Green Was My Valley?" *Economist*, October 23, 2008. On a positive note, in early 2009 India broke its domestic political logjam and agreed to move forward cooperatively on the Ganges with Nepal, the mountainous upriver state where half its waters originated but which itself had only one-twentieth of the basin's population.

426. round of informal dialogues: The diplomatically quiet Abu Dhabi Dialogue, under World Bank auspices, brought together the often-rival neighbors for three meetings between 2006 and mid-2008.
426. one-third fall in agricultural output: *Economist,* "Melting Asia," 29.
427. Indus is not a giant river: McNeill, *Something New Under the Sun,* 159.
427. Indus is badly overdrawn: Erik Eckholm, "A River Diverted, the Sea Rushes In," *New York Times,* April 22, 2003.
427. scant thirty days' capacity: World Bank.
427. Rich cropland: Moench, 88. Today, groundwater pumping is an indispensable source of the country's heavily irrigation-dependent agriculture; indeed, on a per person basis, no major nation in the world outside the Middle East was more addicted to its depleting groundwater for its survival.
428. Sindhis are bitterly complaining: Eckholm, "River Diverted, the Sea Rushes In"; Erik Eckholm, "A Province Is Dying of Thirst, and Cries Robbery," *New York Times,* March 17, 2003.
428. residents routinely boil: Michael Wines, "For a Sickening Encounter, Just Turn On the Tap," *New York Times,* October 31, 2002.
428. overran the pivotal Buner district: Pakistan's semiautonomous mountainous northwestern Pashtun-tribe-dominated provinces were already host to the Muslim fundamentalist extremists like Afghanistan's Taliban and Osama bin Laden's al-Qaeda. Their breakout into Pakistan proper posed perilous potential repercussions for India and the world. Carlotta Gall and Eric Schmidt, "U.S. Questions Pakistan's Will to Stop Taliban," *New York Times,* April 24, 2009.
428. brink of war: Postel, *Last Oasis,* 85; Elhance, 167, 174–175.
430. shortfall will be equal to its total usage: John Pomfret, "A Long Wait at the Gate to Greatness," *Washington Post,* July 27, 2008.
430–431. peaked out in the late 1990s: Brown, "Aquifer Depletion." On famine numbers, see Mirsky, 39.
431. meat consumption increased two and a half times: "Sin Aqua Non," *Economist,* April 11, 2009.
432. projects on the Yellow River: Ma, *China's Water Crisis,* ix, 39. The large-scale waterworks were common to both the Communists and their Nationalist predecessors. In 1934 the Nationalist government had dredged and almost entirely rebuilt the span of the Grand Canal between the Yangtze and the Huai rivers and installed ship locks for medium-sized steamers. Between 1958 and 1964, Mao's Communist government did even more extensive work so it could handle larger ships.
432. water use quintupled: Jim Yardley, "Under China's Booming North, the Future Is Drying Up," *New York Times,* September 28, 2007.
432. If the human costs seemed high: "China's Growing Pains," *Economist,* August 21, 2004, 11. See also Jim Yardley, "At China's Dams, Problems Rise with Water," *New York Times,* November 9, 2007. In the quarter century after 1978,

per capita living standards rose about sevenfold; some 400 million were lifted out of poverty and a huge middle class was born. The 23 million dislocations come from Premier Wen's 2007 work report to the National People's Congress; Palaniappan, Lee, and Samulon, 134, cite the critics' estimates of 40 to 60 million displacements.

432. staircase of dams and 46 hydroelectric power plants: Ma, 8–11, 39.

432. "When the Yellow River is at peace": Quoted in Gifford, 105.

433. shadow of its intended magnificence: Ma, 10.

433. dry area grew steadily: Ibid., 11, 12.

433. river would be rationed: Pearce, 108, 112.

434. have to drill three times deeper: Yardley, "Under China's Booming North, the Future Is Drying Up."

434. reservoir was declared unfit for drinking: Marq De Villiers, "Three Rivers," in McDonald and Jehl, 47.

435. capital will eventually have to move: Ma, 136.

435. bottom will be hit around 2035: Yardley, "Under China's Booming North, the Future Is Drying Up"; Ma, viii. The northern plain originally had 60 billion cubic meters of nonrenewable groundwater. Reliance on groundwater was increasing across China as a whole, reaching one-fifth of the nation's water supply.

435. Half the lakes: Pearce, 109.

435. potential new cropland was destroyed: Diamond, *Collapse,* 364, 365. Erosion pauperized the soil for agriculture, clogged irrigation canals and navigable river channels, and increased the risks of major flooding. Some one-fifth of all China's land, north and south, suffered major soil erosion.

435. Genghis Khan's memorial tomb: De Villiers, 49; Ma, 31.

435. replanting a "green wall" of trees: Diamond, *Collapse,* 368, 369. In the 2,000 years leading up to 1950, major dusters occurred on average every thirty-one years. From 1950 to 1990, they hit once every two years; from 1990, they struck almost every year. A big one in May 1993 killed a hundred people. The green wall project was budgeted at $6 billion.

436. dust mixes with thick clouds of sooty, polluted air: Jim Yardley, "China's Path to Modernity, Mirrored in a Troubled River," *New York Times,* November 19, 2006.

436. desiccation of northern China: Ma, 19. The headwaters of the Yellow had dried up, and had reduced water flows, just like the lower reaches from the mid- to late 1980s.

436–437. "Swimming": Mao Zedong (1956), quoted in Ma, 57.

438. hydropower megabases: China has exploited only about one-fourth of its hydropower potential.

438. "hidden dangers": Wang Xiaofeng, speaking at the September 25 forum at Wuhan, composite quotes cited in Lin Yang, "China's Three Gorges Dam under Fire," *Time,* October 12, 2007, http://www.time.com/time/world/

article/0,8599,1671000,00.html; Yardley, "At China's Dams, Problems Rise with Water"; Jane Macartney, "Three Gorges Dam Is a Disaster in the Making, China Admits," *Times* (London), September 27, 2007, http://www.timesonline.co.uk/tol/news/world/article2537279.ece.

438. 3 to 4 million people would have to be relocated: Howard W. French, "Dam Project to Displace Millions More in China," *New York Times,* October 2, 2007.

439. water in the Zipingpu reservoir: Sharon LaFraniere, "Scientists Point to Possible Link between Dam and China Quake," *New York Times,* February 6, 2009.

440. dogged private environmental whistle-blower: Joseph Kahn, "In China, a Lake's Champion Imperils Himself," *New York Times,* October 14, 2007.

440. promised to restore China's major lakes: Keith Bradsher, "China Offers Plan to Clean Up Its Polluted Lakes," *New York Times,* January 23, 2008.

440. unfit for human consumption: Data from 2005 Chinese Ministry of Water Resources, cited in Gleick and Morrison, 147.

440. one-fifth of wastewater is treated: Diamond, *Collapse*, 364.

440. curtailed for want of adequate river volumes: "Drying Up," *Economist,* May 19, 2005. In the northwest, some factories were permanently closed due to water shortages.

440. reliance on groundwater has doubled: Yardley, "Under China's Booming North, the Future Is Drying Up."

440. one-third of its land is severely degraded: De Villiers, 48.

441. Chinese official requests to excise: "Don't Drink the Water and Don't Breathe Air," *Economist,* January 24, 2008. In 2006 China had recorded 60,000 pollution-related domestic disturbances.

441. Hu's Green GDP report: Joseph Kahn and Jim Yardley, "As China Roars, Pollution Reaches Deadly Extremes," *New York Times,* August 26, 2007.

441. cost of environmental loss: Economist staff, "A Ravenous Dragon: Special report on China's quest for resources," *Economist,* March 5, 2007, 18; David Barboza, "China Reportedly Urged Omitting Pollution-Death Estimates," *New York Times*, July 5, 2007.

442. uses three to 10 times more water: Yardley, "Under China's Booming North, the Future Is Drying Up."

442. consume 42 times more water: Diamond, *Collapse*, 362.

442–443. immense dams on the upper basins of the Mekong and the Salween: Jim Yardley, "Seeking a Public Voice on China's 'Angry River,'" *New York Times,* December 26, 2005; Seth Mydans, "Where a Lake Is Life Itself, Dam Is a Dire Word," *New York Times,* April 28, 2003; Ma, x. The plans included 13 dams on the heretofore undammed Nu River—called the Salween in Myanmar—including one of the world's biggest that would produce more hydroelectricity than Three Gorges. The dams on the Mekong and its tributaries would threaten the unusually powerful, oscillating flow of Tonle Sap—causing the tidal lake's size to expand and contract fourfold—which was vital to Cambodia's livelihood, as well as the volume and quality of the river reaching Vietnam.

443. 1997 U.N. Watercourses Convention: Turkey and Burundi, both upriver riparians, were the other two treaty rejecters. Many other countries abstained and the treaty was never ratified. However, it became part of the growing body of customary principles governing international water issues. Its two main principles, evolved over three decades, were that all riparians were entitled to equitable utilization of the watercourse's resources and that countries would not behave in ways that significantly harmed other river states. A third, less-well-established principle held that countries would not act in any way that foreclosed another riparian's future use of the river's resources—a placeholder principle aimed at protecting late-developing, poor countries against overexploitation by early users.

443. "Southern China has too much water": Mao Zedong, quoted in Ma, 143.

443. eastern and central routes began: Erik Eckholm, "Chinese Will Move Waters to Quench Thirst of Cities," *New York Times,* August 27, 2002; Ma, 136–137, 143–144; Kathy Chen, "China Approves Large Project to Divert Water to Dry North," *Wall Street Journal,* November 26, 2002.

444. travel by tunnel under the Yellow River: Pearce, 219–221.

444. The diversion project: David Lague, "On an Ancient Canal, Grunge Gives Way to Grandeur," *New York Times,* July 24, 2007; Eckholm, "Chinese Will Move Waters to Quench Thirst of Cities." By 2007 progress was being made—some of the stench had cleared, small fish life had returned, and urban renewal was visible along rehabilitated stretches—but many experts remained incredulous that it could be restored to an environmentally healthy state.

444. Pumping water across the mountains: Eckholm, "Chinese Will Move Waters to Quench Thirst of Cities"; Ma, 144.

446. fall as much as a third below its farming needs: *Economist,* "Ravenous Dragon," 18.

Chapter Seventeen: Opportunity from Scarcity: The New Politics of Water in the Industrial Democracies

448. each North American uses: *Economist,* "Sin Aqua Non," April 11, 2009.

449. three centuries of increasing twice as fast: Millennium Ecosystem Assessment, 107.

450. American water withdrawals peaked in 1980: U.S. Geological Survey, "Estimated Use of Water in the United States in 2000."

450. U.S. water productivity: Gary H. Wolff and Peter H. Gleick, "The Soft Path for Water," in Gleick, *World's Water, 2002–2003,* 19. All figures are in constant 1996 dollars.

450. Japan's economic productivity per unit of water: Specter, 70. Japan's water use per $1 million of water fell from 50 to 13 million liters between 1965 and 1989.

450. soft-path efficiency approach: The soft path concept was originally proposed by the influential Amory Lovins of the Rocky Mountain Institute in the mid-1970s in response to the oil crisis. He argued that the West's chief response should be to reduce energy demand through greater efficiency, thereby lowering supply needs and breaking the long-standing correlation between growth and the absolute level of energy consumption. The soft path to water, based on similar reasoning, was elaborated by Peter Gleick of the Pacific Institute, with acknowledgment of his intellectual debt to Lovins.

452. 10 agricultural workers or 100,000 high-tech jobs: Gleick, "Making Every Drop Count," 45.

452. profligate farming practices: Peter H. Gleick, cited in Timothy Egan, "Near Vast Bodies of Water, the Land Still Thirsts," *New York Times,* August 12, 2001; "Pipe Dreams," *Economist,* January 9, 2003; Douglas Jehl, "Thirsty Cities of Southern California Covet the Full Glass Held by Farmers," *New York Times,* September 24, 2002.

453. Bass brothers: Charles McCoy and G. Pascal Zachary, "A Bass Play in Water May Presage Big Shift in Its Distribution," *Wall Street Journal,* July 11, 1997; "Flowing Gold," *Economist,* October 10, 1998; Brian Alexander, "Between Two West Coast Cities, a Duel to the Last Drop," *New York Times,* December 8, 1998.

453. internecine battles: The first battle was with the Metropolitan Water District of Southern California, which controlled almost all of San Diego's access to water and didn't want to lose its largest client.

454. ecosystem health of the Salton Sea: Kelly, n.p.

454. Imperial Valley lost: Dean E. Murphy, "In a First, U.S. Officials Put Limits on California's Thirst," *New York Times,* January 5, 2003. Los Angeles's water authority also lost water; California was forced to draw more from its reservoirs to make up the critical shortfalls.

455. 30 million acre-feet: San Diego County Water Authority, Water Mangement, "Quantification Settlement Agreement," www.sdcwa.org/manage/mwd-QSA.phtml#overview. See also Imperial Irrigation District, "News Archive 2003," November 10, 2003, http://www.iid.com/sub.php?build=view&idr=1264&page2=1&pid=761.

455. "They should pay $800": Mike Morgan, quoted in Kelly.

455. large geothermal field: "Something Smells a Bit Fishy," *Economist,* April 10, 2008. One caveat to exploiting the geothermal field was that existing environmental preservation plans for the Salton Sea had to be modified so its geothermal corner could be drained and exploited.

456. desalinization plants in California in exchange for an extra draw: Gertner.

456. Las Vegas: Las Vegas was also pursuing traditional hard infrastructure, such as controversial multibillion-dollar long-distance pipelines to carry groundwater pumped from land purchased in east-central Nevada, and building a new, deeper intake valve in Lake Mead.

457. $500 an acre-foot: "Dust to Dust," *Economist,* March 7, 2009, 39.
457. court rulings and federal restoration: Under the groundbreaking federal Central Valley Project Improvement Act (1992), wildlife and ecosystem uses were given equal priority with long-favored irrigation; many farmers' water rates had increased tenfold in the 1990s as a result.
457. Orange County, California: Randal C. Archibold, "From Sewage, Added Water for Drinking," *New York Times,* November 27, 2007.
458. Desal costs in California had fallen: Peter H. Gleick, Heather Cooley, and Gary H. Wolff, "With a Grain of Salt: An Update on Seawater Desalinization," in Gleick, *World's Water, 2006–2007,* 68. Coastal Florida, where groundwater supplies were badly overdrawn and plentiful brackish estuaries provided low-salt-content water that was cheaper to purify to drinking quality levels, and ever-thirsty California had long been America's leading laboratories for desalinization experiments. Texas was also another leading player.
458. 7 percent of the entire state's urban water use: Ibid., 65. That figure is based on usage in 2000. On the San Diego plant, see Felicity Barringer, "In California, Desalinization of Seawater as a Test Case," *New York Times,* May 15, 2009.
458. "If we could ever competitively": John F. Kennedy, quoted in Economist staff, *Economist Technology Quarterly,* 24.
460. Projections of market growth: Peter H. Gleick and Jason Morrison, "Water Risks That Face Business and Industry," in Gleick, *World's Water, 2006–2007,* 161.
460. New York City's water network: Galusha, 265.
460. reservoirs were chronically choked: Andrew C. Revkin, "A Billion-Dollar Plan to Clean the City's Water at Its Source," *New York Times,* August 31, 1997.
461. 1,500-page, three-volume agreement: Galusha, 258–259.
461. New York City would spend $260 million: Winnie Hu, "To Protect Water Supply, City Acts as a Land Baron," *New York Times,* August 9, 2004; U.S. Environmental Protection Agency, Region 2, "Watershed Protection Programs," U.S. Environmental Protection Agency, http://www.epa.gov/region02/water/nycshed/protprs.htm. Some $200 million was allocated for upgrading treatment plants.
461. $70 million for sundry infrastructure repairs: About $200 million in improvements to treatment of wastewater entering the reservoirs from sewage plants was also approved.
462. complicated land swap: "A Watershed Agreement," editorial, *New York Times,* September 10, 2007.
462. Water to the Everglades: Damien Cave, "Everglades Deal Shrinks to Sale of Land, Not Assets," *New York Times,* November 12, 2008.
463. city would not need any additional water supply: Galusha, 229.
465. last inspection in 1958: Andrew C. Revkin, "What's That Swimming in the Water Supply? Robot Sub Inspects 45 Miles of a Leaky New York Aqueduct," *New York Times,* June 7, 2003.

465. team of deep-sea repair divers: New York City Department of Environmental Protection, "Preparation Underway to Fix Leak in Delaware Aqueduct," press release, August 4, 2008. The high-pressure diving operation was not New York's first underwater repair experience. During a weeklong exercise in December 2000, a team of divers was lowered by crane inside a diving bell into another portion of aqueduct and worked to seal off a coin-sized hole in an old bronze valve from which water was spewing at 80 miles per hour. The great nightmare of engineers would be a tunnel leak under the Hudson River, which would be very hard and perilous for divers to get to and repair.

466. "Look, if one of those tunnels goes": James Ryan, quoted in Grann, 91, 96, 102.

467. took a seat at the mole's controls: Sewell Chan, "Tunnelers Hit Something Big: A Milestone," *New York Times,* August 10, 2006. The mole had been used in drilling the innovative Channel Tunnel linking England and France.

467. one of New York's most critical infrastructures: Grann, 97.

467. America's 700,000 miles of aging water pipes: Lavelle and Kurlantzick, 24.

467. Global water infrastructure needs: Pearce, 304.

469. T. Boone Pickens: In Pickens's case, the 250-mile pipeline from the Panhandle to Dallas was being developed imaginatively in conjunction with electrical power generated by the world's largest wind farm.

469. five giant global food and beverage corporations: J. P. Morgan calculated the amount to be 575 billion liters per year; cited in "Running Dry," *Economist,* August 23, 2008, 53.

470. water needed to produce one kilowatt-hour had plunged: U.S. Geological Survey, "Estimated Use of Water in the United States in 2000."

470. modern mills using only six tons: Gleick, "Making Every Drop Count," 44.

470. Perrier Vittel: "Are You Being Served?" *Economist,* April 23, 2005, 77.

470. water-conscious companies: Among those reporting corporate water use and setting future targets were Intel, IBM, and Sony in high tech/electronics, pharmaceutical/biotech producer Abbott, Nippon Steel, automotive giants Volkswagen, Toyota, and General Motors, forestry products maker Kimberly-Clark, and food and beverage companies Unilever, Nestlé, and Coca-Cola. Gleick and Morrison, 154–155.

471. Brazilian tomato farmers: Ibid., 149.

471. Anheuser-Busch: Among the companies engaging their supply chains were Anheuser-Busch, Coca-Cola, McDonald's, Unilever, Nestlé, Gap, Johnson & Johnson, and oil refiner Chevron. Coca-Cola experienced a bitter foretaste of water's potential political risk when it was accused of abusing scarce groundwater resources in India. Although Coke was later exonerated in court, the negative publicity posed a reputational threat to its priceless brand name, as well as harming local market sales. To publicize its green commitment to treating all its wastewater by 2010, Coke began putting schools of fish in tanks filled with treated wastewater at its bottling plants across the world.

471. by one-quarter roughly doubled: Wolff and Gleick, "Soft Path for Water," 19. The calculation is based on 80 percent agricultural water use in the area.
472. to sprinklers and microirrigation systems: According to the U.S. Geological Survey, irrigated acreage under sprinklers or microirrigation rose from 40 percent in 1985 to 52 percent in 2000. McGuire, "Water-Level Changes in the High Plains Aquifer, Predevelopment to 2002, 1980 to 2002, and 2001 to 2002."
472. American farm pollution: Europe's farm pollution regulations were more muscular.
472. biological dead zone without fish life: Bina Venkataraman, "Rapid Growth Found in Oxygen-Starved Ocean 'Dead Zones,'" *New York Times,* August 15, 2008.
473. Australia faces the industrialized world's: Diamond, *Collapse,* 379–380, 384, 387, 409.
473. southeastern Murray-Darling: "The Big Dry," *Economist,* April 28, 2007, 81.
473. facilitate independent water trading: Peet, 13–14.
473. "transpiration credits": "Are You Being Served?" Prices adjusted to higher-volume use and seasonal availability, and including wastewater treatment in calculating water's final price, lay ahead.
474. decline in the Murray's flow: "Big Dry," 84.
474. Australia's agricultural land: Diamond, *Collapse,* 413.
475. Bush administration's Environmental Protection Agency: "Clearer Rules, Cleaner Waters," editorial, *New York Times,* August 18, 2008.
476. inextricably interdependent: Elizabeth Rosenthal, "Biofuels Deemed a Greenhouse Threat," *New York Times,* February 8, 2008.
477. 20 percent of all California's electricity: Wilshire, Nielson, and Hazlett, 252. Data is from a 2005 California Energy Commission report. See also Meena Palaniappan, Emily Lee, and Andrea Samulon, "Environmental Justice and Water," in Gleick, *World's Water: 2006–2007,* 151.
477. northeastern U.S. power failure: Jane Campbell, interview with author, March 17, 2008.
477. Italy's severe drought in 2003: "Emergency Threat in Dry Italy," *BBC News,* July 14, 2003, news.bbc.co.uk/go/pr/fr/-/2/hi/Europe/3065977.stm; "The Parched Country," *Economist,* October 26, 2007.
478. carbon dioxide in the atmosphere: Kolbert, 201–203. In 2007, the U.N.'s Intergovernmental Panel on Climate Change concluded that with almost total certainty planetary warming was man-made. Andrew Revkin, "On Climate Issue, Industry Ignored Its Scientists," *New York Times,* April 24, 2009.
479. Dutch have begun to pioneer: Smith, *Man and Water,* 28-33; Kolbert, 123–127.
480. reduce California's total municipal water consumption: Wilshire, Nielson, and Hazlett, 252.
482. potential choke points: Simply trying to keep bands of Somali pirates from hi-

jacking vessels off the lawless Horn of Africa enlisted the navies of more than a dozen nations—including China, India, Italy, Russia, Saudi Arabia, Malaysia, Turkey, England, France, and the United States—in 2008, without notable success.

483. world's abject water poor: United Nations Millennium Project Task Force on Water and Sanitation, 13, 17.

485. U.N. Millennium Development Goals: A more ambitious target of providing every person with access to safe, clean water and sanitation by 1990 had failed to be achieved as part of the U.N.'s International Drinking Water Supply and Sanitation Decade (1981–1990; the new downscaled targets were set to coincide with the conclusion of the U.N.'s new, aspirational International Decade for Action "Water for Life" (2005–2015).

486. Camdessus report: Nicholas L. Cain, "3rd World Water Forum in Kyoto Disappointment and Possibility," in Gleick, *World's Water 2004–2005,* 189–196.

Epilogue

493. not enough planetary environmental resources: Diamond, *Collapse,* 487–494, 495. Diamond estimates that the average Westerner consumes 32 times more resources than low-impact third world citizens and that the per capita effect of everyone attaining a high environmental-impact lifestyle would increase world resource consumption twelvefold—an unsustainable environmental burden on planetary resources based on today's technologies and practices. Water strongly influenced almost every one of the 12 great problems Diamond concludes have to be solved for twenty-first-century civilization to adjust without great trauma. These include deforestation, collapse of fisheries, loss of biodiversity, soil erosion, energy shortages, freshwater depletion, photosynthetic capacity, toxic chemical pollution, invasions by alien species, climate change, sheer population levels, higher impact levels of consumption, and waste by several billion more people.

SELECT BIBLIOGRAPHY

This select bibliography reflects two contrasting challenges presented by the research. First, while water's role in history per se has rarely been a central focus of previous books, many historians and scholars from diverse fields have insightfully treated its influential aspects in their own major works. Part of the bibliography, therefore, reflects my effort to pull these ideas together into a cohesive framework and narrative. Second, today's world water crisis is producing an explosion of wide-ranging and substantive literature on current water issues that is far too extensive to comprehensively list. Regrettably, I have had to exclude all news and most other periodical articles from the bibliography; a few that are the source of cited facts are covered in the notes. I drew heavily from current events reporting from the *New York Times,* the *Economist,* and the *Washington Post,* as well as the BBC, the *Financial Times,* the *Wall Street Journal,* and many magazines. I have omitted separate bibliographic entries for periodical and research articles that are included in listed compendiums; again, some of these works are cited in the notes. There is as well vast informative content—official, academic, reportorial, and eclectic—available on the Internet, which has provided rich background, but which is not referenced either in the select bibliography or the notes.

Achenbach, Joel. "America's River." *Washington Post Magazine,* May 5, 2002.
Aicher, Peter J. *Guide to the Aqueducts of Ancient Rome.* Wauconda, Ill.: Bolchazy-Carducci, 1995.

Allan, J. A. *The Middle East Water Question: Hydropolitics and the Global Economy.* London: I. B. Tauris, 2002.

Alley, Richard B. *The Two-Mile Time Machine: Ice Cores, Abrupt Climate Change, and Our Future.* Princeton, N.J.: Princeton University Press, 2000.

Alwash, Azzam. "Water at War." *Natural History,* November 2007.

Amery, Hussein A., and Aaron T. Wolf. *Water in the Middle East: A Geography of Peace.* Austin: University of Texas Press, 2000.

Appiah, Kwame Anthony. "How Muslims Made Europe." *New York Review of Books* 55, no. 17 (November 6, 2008).

Ball, Philip. *Life's Matrix.* New York: Farrar, Straus & Giroux, 1999.

Barlow, Maude, and Tony Clarke. *Blue Gold: The Fight to Stop the Corporate Theft of the World's Water.* New York: New Press, 2002.

Barnes, Julian. "The Odd Couple." *New York Review of Books* 54, no. 5 (March 29, 2007).

Barry, John M. *Rising Tide: The Great Mississippi Flood of 1927 and How It Changed America.* New York: Touchstone, 1998.

Beasley, W. G. *The Modern History of Japan.* 7th ed. New York: Praeger, 1970.

Belt, Don, ed. "The World of Islam." *National Geographic.* Supplement, 2001.

Bernstein, Peter L. *The Power of Gold: The History of an Obsession.* New York: John Wiley, 2000.

———. *Wedding of the Waters: The Erie Canal and the Making of a Great Nation.* New York: W. W. Norton, 2005.

Biddle, Wayne. *A Field Guide to Germs.* New York: Henry Holt, 1995.

Billington, David P., Donald C. Jackson, and Martin V. Melosi. *The History of Large Federal Dams: Planning, Design, and Construction in the Era of Big Dams.* Denver: U.S. Department of the Interior, Bureau of Reclamation, 2005.

Billington, Ray Allen. *American Frontier Heritage.* Reprint, New York: Holt, Rinehart and Winston, 1968.

Bleier, Ronald. "Will Nile Water Go to Israel?: North Sinai Pipelines and the Politics of Scarcity." *Middle East Policy,* 5, no. 3 (September 1997), 113–124; http://desip.igc.org/willnile1.html.

Boorstin, Daniel J. *The Discoverers: A History of Man's Search to Know His World and Himself.* New York: Random House, 1985.

Boutros-Ghali, Boutros. *Egypt's Road to Jerusalem.* New York: Random House, 1997.

Braudel, Fernand. *Afterthoughts on Material Civilization and Capitalism.* 3rd ed. Translated by Patricia Ranum. Baltimore: Johns Hopkins University Press, 1985.

———. *A History of Civilizations.* Translated by Richard Mayne. New York: Penguin, 1995.

———. *Memory and the Mediterranean.* Translated by Siân Reynolds. New York: Alfred A. Knopf, 2001.

———. *The Perspective of the World.* Vol. 3 of *Civilization and Capitalism, 15th–18th Century.* Translated by Siân Reynolds. New York: Harper & Row, 1984.

———. *The Structures of Everyday Life.* Vol. 1 of *Civilization and Capitalism, 15th–18th Century.* Translated by Siân Reynolds. New York: Harper & Row, 1981.

———. *The Wheels of Commerce.* Vol. 2 of *Civilization and Capitalism, 15th–18th Century.* Translated by Siân Reynolds. New York: Harper & Row, 1982.

Brewer, John. "The Return of the Imperial Hero." *New York Review of Books* 52, no. 17 (November 3, 2005).

Brindley, James. *Power through the Ages.* London: Blackie, 2002.

Bronowski, Jacob. *The Ascent of Man.* Boston: Little, Brown, 1973.

Bronowski, Jacob, and Bruce Mazlish. *The Western Intellectual Tradition: From Leonardo to Hegel.* New York: Harper & Row, 1975.

Brown, Lester. "Aquifer Depletion." *Encyclopedia of Earth.* http://www.eoearth.org/article/Aquifer_depletion (revised February 12, 2007).

———. "Grain Harvest Growth Slowing." Earth Policy Institute. 2002. http://www.earth-policy.org/Indicators/indicator6.htm.

———. "Water Scarcity Spreading." Earth Policy Institute. 2002. http://www.earth-policy.org/Indicator7_print.htm.

Bulloch, John, and Adel Darwish. *Water Wars: Coming Conflicts in the Middle East.* London: Victor Gollancz, 1993.

Butzer, K. W. *Early Hydraulic Civilization in Egypt.* Chicago: University of Chicago Press, 1976.

Byatt, Andrew, Alastair Fothergill, and Martha Homes. *The Blue Planet: Seas of Life.* Foreword by Sir David Attenborough. London: BBC Worldwide Limited, 2001.

Cameron, Rondo. *A Concise Economic History of the World: From Paleolithic Times to the Present.* 2nd ed. New York: Oxford University Press, 1993.

Campbell, Joseph. *The Hero's Journey.* 3rd ed. Novato, Calif.: New World Library, 2003.

Campbell-Green, Tim. "Outline the Nature of Irrigation and Water Management in Southern Mesopotamia in the 3rd Millennium." *Bulletin of Sumerian Agriculture* 5 (1990). Irrigation and Cultivation, pt. 2, Cambridge. www.art.man.ac.uk/ARTHIST/EStates/Campbell.htm.

Cantor, Norman F. *Antiquity: From the Birth of Sumerian Civilization to the Fall of the Roman Empire.* New York: HarperCollins, 2003.

———. *The Civilization of the Middle Ages.* Rev. ed. New York, HarperCollins, 1994.

Carson, Rachel. *Silent Spring.* New York: Houghton Mifflin, 2002.

Cary, M., and E. H. Warmington. *The Ancient Explorers.* Baltimore: Penguin, 1963.

Casson, Lionel. *The Ancient Mariners: Seafarers and Sea Fighters of the Mediterranean in Ancient Times.* 2nd ed. Princeton, N.J.: Princeton University Press, 1991.

Chamberlain, John. *The Enterprising Americans: A Business History of the United States.* Rev. ed. New York: Harper & Row, 1974.

Churchill, Winston S. *A History of the English-Speaking Peoples: The Age of Revolution.* New York: Dodd, Mead, 1957.

Clarke, Robin. *Water: The International Crisis.* Cambridge, Mass.: MIT Press, 1993.

Clarke, Robin, and Jannet King. *The Water Atlas: A Unique Analysis of the World's Most Critical Resource.* New York: New Press, 2004.

Clough, Shepard B. *The Rise and Fall of Civilization: An Inquiry into the Relationship between Economic Development and Civilization.* 2nd ed. New York: Columbia University Press, 1957.

Cockburn, Andrew. "Lines in the Sand: Deadly Time in the West Bank and Gaza." *National Geographic* 202, no. 2 (October 2002).

Collins, Robert O. *The Nile.* New Haven, Conn.: Yale University Press, 2002.

Curtis, John. *Ancient Persia.* 2nd ed. London: British Museum Press, 2000.

Darwish, Adel. "Water Wars." http://www.mideastnews.com/WaterWars.htm. June 1994.

Das, Gurcharan. "The India Model." *Foreign Affairs* 85 (July–August 2006).

Davidson, Basil. *The Lost Cities of Africa.* Rev. ed. Boston: Little, Brown, 1970.

Davies, Norman. *Europe: A History.* New York: HarperPerennial, 1998.

Davis, David Brion. "He Changed the New World." *New York Review of Books* 44, no. 9 (May 31, 2007).

Davis, Paul K. *100 Decisive Battles from Ancient Times to the Present: The World's Major Battles and How They Shaped History.* New York: Oxford University Press, 2001.

De Villiers, Marq. *Water: The Fate of Our Most Precious Resource.* New York: Houghton Mifflin, 2001.

Diamond, Jared. *Collapse: How Societies Choose to Fail or Succeed.* New York: Penguin, 2005.

———. *Guns, Germs, and Steel: The Fates of Human Societies.* New York: W. W. Norton, 1999.

Durant, Will, and Ariel Durant. *The Lessons of History.* New York: Simon & Schuster, 1968.

Economist staff. "An Affair to Remember." Special Report: The Suez Crisis, *Economist,* July 29, 2006.

———. "A Ravenous Dragon." Special report on China's quest for resources, *Economist,* March 5, 2007.

———. "The Story of Wheat." Ears of Plenty: A Special Report, *Economist,* December 24, 2005.

———. "Tapping the Oceans." *Economist Technology Quarterly,* June 7, 2008.

———. *The Economist Pocket World in Figures 2009.* London: Profile Books, 2008.

Edwards, Mike. "Han." *National Geographic* 205, no. 2 (February 2004), 2–29.

Elhance, Arun P. *Hydropolitics in the Third World: Conflict and Cooperation in International River Basins.* Washington, D.C.: United States Institute of Peace Press, 1999.

Elvin, Mark. *The Pattern of the Chinese Past.* Stanford, Calif.: Stanford University Press, 1973.

Erlich, Haggai. *The Cross and the River.* Boulder, Colo.: L. Rienner, 2002.

Evans, Harold. *The American Century.* London: Jonathan Cape/Pimlico, 1998.

Evans, Harry B. *Water Distribution in Ancient Rome: "The Evidence of Frontinus."* Ann Arbor: University of Michigan Press, 1997.

Fairbank, John King, and Merle Goldman. *China: A New History.* 9th ed. Cambridge, Mass.: Belknap Press of Harvard University Press, 2001.

Ferguson, Niall. *Empire: How Britain Made the Modern World.* London: Penguin, 2003.

Fernández-Armesto, Felipe. *Civilizations: Culture, Ambition, and the Transformation of Nature.* New York: Touchstone, 2002.

Fineman, Herman. *Dulles over Suez.* Chicago: Quadrangle, 1964.

Foreman, Laura. *Alexander the Conqueror: The Epic Story of the Warrior King.* Foreword by Professor Eugene N. Borza. Cambridge, Mass.: Da Capo, 2004.

Freely, John. *Istanbul: The Imperial City.* London: Penguin, 1998.

Frontinus, Julius. *De Acqaeductu Urbis Romae* (On the Water-Management of the City of Rome). Translated by R. H. Rodgers. 2003. University of Vermont. http://www.uvm.edu/~rrodgers/Frontinus.html.

———. *De Acqaeductu Urbis Romae.* Edited, introduction, and commentary by R. H. Rodgers. Cambridge, U.K.: Cambridge University Press, 2004.

Fulton, Robert. "Letter from Robert Fulton to President George Washington," London, February 5, 1797. In "History of the Erie Canal," Department of History, University of Rochester. http://www.history.rochester.edu/canal/fulton/feb1797.htm.

———. "Mr. Fulton's Communication." Submitted to Albert Gallatin, Esq., Secretary of the Treasury, Washington, D.C., December 8, 1807. Included in Report of the Secretary of the Treasury, on the Subject of Public Roads and Canals (1808): 100–116. Contributed by Howard B. Winkler. In *Towpath Topics* (Middlesex Canal Association) (September 1994; March 2000). http://www.middlesexcanal.org/towpath/fulton.htm.

Galusha, Diane. *Liquid Assets: A History of New York City's Water System.* Fleischmanns, N.Y.: Purple Mountain Press, 2002.

Ganguly, Sumit. "Will Kashmir Stop India's Rise?" *Foreign Affairs* 85 (July–August 2006).

Gertner, Joe. "The Future Is Drying Up." *New York Times Magazine,* October 21, 2007.

Gibbon, Edward. *The Decline and Fall of the Roman Empire.* Abridgement by D. M. Low. New York: Harcourt, Brace, 1960.

Gies, Frances, and Joseph Gies. *Cathedral, Forge, and Waterwheel: Technology and Invention in the Middle Ages.* New York: HarperCollins Publishers, 1995.

Gifford, Rob. "Yellow River Blues." *Asia Literary Review* 8 (2008).

Gimpel, Jean. *The Medieval Machine.* New York, London: Penguin Group, 1976.

Gleick, Peter H. "Making Every Drop Count." *Scientific American,* February 2001.

———. *The World's Water, 1998–1999: The Biennial Report on Freshwater Resources.* Washington, D.C.: Island Press, 1998.

———. *The World's Water, 2000–2001: The Biennial Report on Freshwater Resources.* Washington, D.C.: Island Press, 2000.

Gleick, Peter H., with William C. G. Burns, Elizabeth L. Chalecki, Michael Cohen, Katherine Kao Cushing, Amar S. Mann, Rachel Reyes, Gary H. Wolff, and Arlene K. Wong. *The World's Water, 2002–2003: The Biennial Report on Freshwater Resources.* Washington, D.C.: Island Press, 2002.

Gleick, Peter H., with Nicholas L. Cain, Dana Haasz, Christine Henges-Jeck, Catherine Hunt, Michael Kiparsky, Marcus Moench, Meena Palaniappan, Veena Srinivasan, and Gary H. Wolff. *The World's Water, 2004–2005: The Biennial Report on Freshwater Resources.* Washington, D.C.: Island Press, 2004.

Gleick, Peter H., with Heather Cooley, David Katz, Emily Lee, Jason Morrison, Meena Palaniappan, Andrea Samulon, and Gary H. Wolff. *The World's Water, 2006–2007: The Biennial Report on Freshwater Resources.* Washington, D.C.: Island Press, 2006.

Glennon, Robert. *Water Follies: Groundwater Pumping and the Fate of America's Fresh Waters.* Washington, D.C.: Island Press, 2002.

Goldschmidt, Arthur, Jr. *A Concise History of the Middle East.* 7th ed. Boulder, Colo.: Westview, 2002.

Gordon, John Steele. *A Thread across the Ocean: The Heroic Story of the Transatlantic Cable.* New York: HarperCollins, Perennial, 2003.

Gore, Rick. "Who Were the Phoenicians? Men of the Sea: A Lost History." *National Geographic* 206, no. 4 (October 2004).

Grann, David. "City of Water." *New Yorker*, September 1, 2003.

Grey, David, and Claudia W. Sadoff. "Sink or Swim? Water Security for Growth and Development." *Water Policy* 9 (2007): 545–571.

Grimal, Nicolas. *A History of Ancient Egypt.* Translated by Ian Shaw. Reprint, Oxford, U.K.: Blackwell, 1992.

Groner, Alex. *The American Heritage History of American Business and Industry.* New York: American Heritage Publishing, 1972.

Guardian (U.K.) staff. "The World's Water." Special section, *Guardian* (U.K.), August 23, 2003.

Gunter, Ann C., ed. *Caravan Kingdoms: Yemen and the Ancient Incense Trade.* Washington, D.C.: Freer Gallery of Art and Arthur M. Sackler Gallery, Smithsonian Institution, 2005.

Halliday, Stephen. *The Great Stink of London: Sir Joseph Bazalgette and the Cleansing of the Victorian Capital.* Foreword by Adam Hart-Davis. Phoenix Mill, U.K.: Sutton Publishing, 2000.

Hammurabi. *The Code of Hammurabi.* Translated by L. W. King (1910). Edited by Richard Hooker (June 6, 1999). In "World Civilizations," Washington State University. http://www.wsu.edu/~dee/MESO/CODE.HTM.

Hansen, Jim. "The Threat to the Planet." *New York Review of Books* 53, no. 12 (July 13, 2006).

Harris, Marvin. *Cannibals and Kings: The Origins of Cultures.* New York: Random House, 1977.

Heilbroner, Robert L. *The Making of Economic Society.* Englewood Cliffs, N.J.: Prentice-Hall, 1980.

———. *The Nature and Logic of Capitalism.* New York: W. W. Norton, 1985.

———. *Visions of the Future: The Distant Past, Yesterday, Today, and Tomorrow.* New York: New York Public Library; Oxford University Press, 1995.

———. *The Worldly Philosophers: The Lives, Times, and Ideas of the Great Economic Thinkers.* 7th ed. New York: Simon & Schuster, 1999.

Heilbroner, Robert L., and Aaron Singer. *The Economic Transformation of America: 1600 to the Present.* 2nd ed. New York: Harcourt Brace Jovanovich, 1984.

Herodotus. *The Histories.* Translated by Aubrey de Selincourt. 1954. Revised translation by John Marincola. Reprint, New York: Penguin, 1972.

———. *The Persian Wars.* Translated by George Rawlinson. Introduction by Francis R. B. Godolphin. New York: Modern Library, 1942.

Hibbert, Christopher. *Rome: The Biography of a City.* London: Penguin, 1985.

Hobsbawm, Eric. *The Age of Extremes: A History of the World, 1914–1991.* New York: Vintage, 1996.

Hollister, C. Warren. *Roots of the Western Tradition: A Short History of the Ancient World.* New York: John Wiley & Sons, 1966.

Hone, Philip, *The Diary of Philip Hone, 1828–1851.* Pt. 2. Edited by Bayard Tuckerman. New York: Dodd, Mead, 1910. Internet Archive. http://www.archive.org/stream/diaryofphiliphon00hone.

Hooke, S. H. *Middle Eastern Mythology: From the Assyrians to the Hebrews.* Reprint, Middlesex, U.K.: Penguin, 1985.

Hosack, David, ed. *Memoir of De Witt Clinton.* Commissioned by New York Literary and Philosophical Society, 1829. Transcribed from original text and html prepared by Bill Carr, updated 7/5/99. In "History of the Erie Canal," Department of History, University of Rochester. http://www.history.rochester.edu/canal/bib/hosack/Contents.html. "Claims of Joshua Forman." http://www.history.rochester.edu/canal/bib/hosack/APP0U.html. "Views of General Washington Relative to the Inland Navigation of the United States." http://www.history.rochester.edu/canal/bib/hosack/APP0P.html.

Hourani, Albert. *A History of the Arab Peoples.* New York: Warner, 1992.

Howarth, David. *Famous Sea Battles.* Boston: Little, Brown, 1981.

Hvistendahl, Mara. "China's Three Gorges Dam: An Environmental Catastrophe?" *Scientific American,* March 25, 2008.

Ibn Battutah. *The Travels of Ibn Battutah.* Edited by Tim Mackitosh-Smith. London: Picador Pan Macmillan, 2002.

Jacobs, Els M. *In Pursuit of Pepper and Tea: The Story of the Dutch East India Company.* 3rd ed. Amsterdam: Netherlands Maritime Museum, 1991.

Johnson, Paul. *A History of the Jews.* New York: Harper & Row, 1987.

Joinville[Jean of], and [Geoffrey of] Villehardouin. *Chronicles of the Crusades.* Translated with an introduction by M. R. B. Shaw. Baltimore: Penguin, 1963.

Jones, W. T. *A History of Western Philosophy.* New York: Harcourt, Brace & World, 1952.

Karlen, Arno. *Man and Microbes: Disease and Plagues in History and Modern Times.* New York: Touchstone, 1996.

Karmon, David. "Restoring the Ancient Water Supply System in Renaissance Rome: The Popes, the Civic Administration, and the Acqua Vergine." *Waters of Rome* 3 (August 2005). http://www.iath.virginia.edu/waters/Journal3KarmonNew.pdf.

Karsh, Efraim, and Inari Karsh. *Empires of the Sand: The Struggle for Mastery in the Middle East, 1789–1923.* Cambridge, Mass.: Harvard University Press, 2001.

Keay, John. *India: A History.* New York: Grove Press, 2000.

Kelly, Bill "Greed Runs through It." *L.A. Weekly,* March 16, 2006. http://www.laweekly.com/2006–03–16/news/greed-runs-through-it/1.

Kennedy, Paul. "The Eagle Has Landed." *Financial Times,* FT Weekend, February 2–3, 2002.

———. "Has the U.S. Lost Its Way?" *Guardian* (U.K.)/*Observer,* March 3, 2002. http://guardian.co.uk/world/2002/mar/03/usa.georgebush/print.

———. *The Rise and Fall of the Great Powers.* New York: Random House, 1989.

Koeppel, Gerard T. *Water for Gotham: A History.* 3rd ed. Princeton, N.J.: Princeton University Press, 2001.

Kolbert, Elizabeth. *Field Notes from a Catastrophe: Man, Nature, and Climate Change.* New York: Bloomsbury, 2006.

Kurlansky, Mark. *Salt: A World History.* New York: Walker, 2002.

Lambert, Andrew. *War at Sea in the Age of Sail, 1650–1850.* London: Cassell, 2000.

Lavelle, Marianne, and Joshua Kurlantzick. "The Coming Water Crisis." *U.S. News & World Report,* August 12, 2002.

Lear, Linda. *Rachel Carson: Witness for Nature.* Boston: Houghton Mifflin Harcourt (Mariner Books), 2009.

Levy, Matthys, and Richard Panchyk. *Engineering the City: How Infrastructure Works.* Chicago: Chicago Review Press, 2000.

Lewis, Bernard. *The Muslim Discovery of Europe.* New York: W. W. Norton, 2001.

———. *What Went Wrong?* Oxford, U.K.: Oxford University Press, 2002.

Lira, Carl T. *Biography of James Watt: A Summary.* 2001. College of Engineering, Michigan State University. http://www.egr.msu.edu/~lira/supp/steam/wattbio.html.

Lopez, Robert S. *The Commercial Revolution of the Middle Ages, 950–1350.* New York: Cambridge University Press, 1976.

Love, Robert W., Jr. *History of the U.S. Navy.* Vol 1, *1775–1941.* Vol. 2, *1945–1991.* Harrisburg, Pa.: Stackpole, 1992.

Luft, Gal. "The Wazzani Water Dispute." *PeaceWatch* (Washington Institute for Near East Policy) 397 (September 20, 2002).

Ma, Jun. *China's Water Crisis.* Translated by Nancy Yang Liu and Lawrence R. Sullivan. Norwalk, Conn.: EastBridge, 2004.

Mahan, A. T. *The Influence of Sea Power upon History, 1660–1783.* 5th ed. Mineola, N.Y.: Dover, 1987.

Markham, Adam. *A Brief History of Pollution.* New York: St. Martin's, 1994.

Matthews, John P. C. "John Foster Dulles and the Suez Crisis of 1956: A Fifty Year Perspective." September 14, 2006. American Diplomacy, University of North Carolina. http://www.unc.edu/depts/diplomat/item/2006/0709/matt/matthews_suez.html.

McAleavy, Henry. *The Modern History of China.* 4th ed. New York: Praeger, 1969.

McCullough, David. *The Path between the Seas: The Creation of the Panama Canal, 1870–1914.* New York: Simon & Schuster, 1977.

McDonald, Bernadette, and Douglas Jehl, eds. *Whose Water Is It? The Unquenchable Thirst of a Water-Hungry World.* Washington, D.C.: National Geographic Society, 2004.

McGuire, V. L. "Water-Level Changes in the High Plains Aquifer, Predevelopment to 2002, 1980 to 2002, and 2001 to 2002." U.S. Geological Survey. http://pubs.usgs.gov/fs/2004/3026/pdf/fs04–3026.pdf.

———. "Water-Level Changes in the High Plains Aquifer, 1980–1999," U.S. Geological Survey http//pubs.usg.sgove/fs/2001–029–01.

McKenzie, A. E. E. *The Major Achievements of Science: The Development of Science from Ancient Times to the Present.* New York: Touchstone, 1973.

McKibben, Bill. *The End of Nature.* New York: Random House, 2006.

———. "Our Thirsty Future." *New York Review of Books* 50, no. 14 (September 25, 2003).

McNeill, J. R. *Something New Under the Sun: An Environmental History of the Twentieth-Century World.* New York: W. W. Norton, 2001.

McNeill, J. R., and William H. McNeill. *The Human Web: A Bird's-Eye View of World History.* New York: W. W. Norton, 2003.

McNeill, William H. *The Global Condition: Conquerors, Catastrophes, and Community.* Princeton, N.J.: Princeton University Press, 1992.

———. *Plagues and Peoples.* New York: Anchor Books, 1989.

———. *The Pursuit of Power: Technology, Armed Force, and Society since A.D. 1000.* Chicago: University of Chicago Press, 1982.

———. *The Rise of the West: A History of Human Community.* Chicago: University of Chicago Press, 1963.

———. *A World History.* 4th ed. New York: Oxford University Press, 1999.

Millennium Ecosystem Assessment. *Ecosystems and Human Well-Being: Synthesis.* Washington, D.C.: Island Press, 2005.

Mirsky, Jonathan. "The China We Don't Know." *New York Review of Books* 56, no. 3 (February 26, 2009).

Mitchell, John G. "Down the Drain? The Incredible Shrinking Great Lakes." *National Geographic* 202, no. 3 (September 2002).

Mohan, C. Raja. "India and the Balance of Power." *Foreign Affairs* 85 (July–August 2006): 17–32.

Montaigne, Fen. "Water Pressure: Challenges for Humanity." *National Geographic* 202, no. 3 (September 2002).

Moorehead, Alan. *The White Nile.* Rev. ed. Middlesex, U.K.: Penguin, 1973.

Morison, Samuel Eliot. *The Oxford History of the American People.* New York: Oxford University Press, 1965.

Mumford, Lewis. *The City in History: Its Origins, Its Transformations, and Its Prospects.* New York: Harcourt, Brace & World, 1961.

Natural History staff. "Water, the Wellspring of Life." Special issue, *Natural History,* November 2007.

Needham, Joseph. *Science and Civilisation in China.* Vol. 4, *Physics and Physical*

Technologies, pt. 3, *Civil Engineering and Nautics.* In collaboration with Wang Ling and Lu Gwei-Djen. Cambridge, U.K.: Cambridge University Press, 1971.

New York Times staff. "Managing Planet Earth." Special issue, "Science Times," *New York Times,* August 20, 2002.

Norwich, John Julius. *A History of Venice.* New York: Vintage, 1989.

———. *The Middle Sea: A History of the Mediterranean.* New York: Doubleday, 2006.

———. *A Short History of Byzantium.* New York: Random House, Vintage, 1999.

"Of Water and Wars: Interview with Dr. Ismail Serageldin." *Frontline* (India) 16, no. 9 (April 24–May 7, 1999), http://www.hindu.com/fline/fl1609/16090890.htm.

Outwater, Alice. *Water: A Natural History.* New York: Basic Books, 1996.

Pacey, Arnold. *Technology in World Civilization.* Cambridge, Mass.: MIT Press, 1991.

Pearce, Fred. *When the Rivers Run Dry: Water—the Defining Crisis of the Twenty-First Century.* Boston: Beacon Press, 2006.

Peet, John. "Priceless: A Survey of Water." *Economist,* July 19, 2003.

Pepys, Samuel. *Diary of Samuel Pepys.* http://www.pepysdiary.com/archive/. Original Source from Project Gutenberg: http://www.gutenberg.org/etext/4125.

Perlin, John. *A Forest Journey: The Role of Wood in the Development of Civilization.* New York: W. W. Norton, 1989.

Pielou, E. C. *Fresh Water.* Chicago: University of Chicago Press, 1998.

Polo, Marco. *The Travels of Marco Polo.* Translated by Ronald Latham. Middlesex, U.K.: Penguin, 1958.

Ponting, Clive. *A Green History of the World: The Environment and the Collapse of Great Civilizations.* New York: Penguin, 1993.

Postel, Sandra. "Growing More Food with Less Water." *Scientific American,* February 2001.

———. "Hydro Dynamics." *Natural History,* May 2003.

———. *Last Oasis: Facing Water Scarcity.* New York: W. W. Norton, 1997.

———. "Sharing the River Out of Eden." *Natural History,* November 2007.

Postel, Sandra, and Aaron Wolf. "Dehydrating Conflict." *Foreign Policy* (September–October 2001): 60–67.

Postel, Sandra, and Brian Richter. *Rivers for Life: Managing Water for People and Nature.* Washington, D.C.: Island Press, 2003.

Potts, Timothy. "Buried between the Rivers." *New York Review* 5, no.14 (September 25, 2003).

Procopius of Caesarea. *The Gothic War.* Bks. 5 and 6, *History of the Wars.* Translated by H. B. Dewey. London: William Heineman, 1919. Project Gutenberg, 2007. http://www.gutenberg.org/files/20298/20298-h/20298-h.htm.

Reade, Julian. *Mesopotamia.* 2nd ed. London: British Museum Press, 2000.

Reinhold, Meyer. *Marcus Agrippa.* Geneva, N.Y.: W. F. Humphrey Press, 1933.

Reisner, Marc. *Cadillac Desert: The American West and Its Disappearing Water.* Rev. ed. New York: Penguin, 1993.

———. "The Age of Dams and Its Legacy." *EARTHmatters* (Earth Institute at Columbia University) (Winter 1999–2000). Columbia Earthscape. http://www.earthscape.org/p2/em/em_win00/win18.html.

Roberts, J. M. *The Penguin History of Europe.* London: Penguin, 1997.

———. *The Penguin History of the World.* 3rd ed. London: Penguin, 1995.

Roesdahl, Else. *The Vikings.* 2nd ed. Translated by Susan M. Margeson and Kirsten Williams. London: Penguin, 1998.

Roosevelt, Theodore. *An Autobiography.* New York: Charles Scribner's Sons, 1913.

———. "Charter Day Address, Berkeley Cal., March 23, 1911." *University of California Chronicle,* April 1911, 139. Cited in "Panama Canal—Roosevelt and." In *Theodore Roosevelt Cyclopedia,* edited by Albert Bushnell Hart and Herbert Ronald Ferleger. Rev. 2nd ed. Theodore Roosevelt Association. http://www.theodoreroosevelt.org/TR%20Web%20Book/Index.html.

———. "State of the Union Message," December 3, 1901." Theodore Roosevelt: Speeches, Quotes, Addresses, and Messages. http://www.theodore-roosevelt.com/sotu1.html. American Presidency Project, Department of Political Science, University of California, Santa Barbara. http://www.polsci.ucsb.edu/projects/presproject/idgrant/site/state.html.

Rothfeder, Jeffrey. *Every Drop for Sale.* New York: Penguin Putnam, 2001.

Sadoff, Claudia W., and David Grey. "Beyond the River: The Benefits of Cooperation on International Rivers." *Water Policy* 4 (2002): 389–403.

———."Cooperation on International Rivers: A Continuum for Securing and Sharing Benefits." *Water International* 30, no. 4 (December 2005): 420–427.

"Secrets of Lost Empires: Roman Bath." *NOVA,* PBS, February 22, 1990. Transcript. PBS. http://www.pbs.org/wgbh/nova/transcripts/27rbroman.html.

Service, Alastair. *Lost Worlds.* New York: Arco, 1981.

Sharon, Ariel with David Chanoff. *Warrior: The Autobiography of Ariel Sharon.* New York: Simon & Schuster, 2001.

Shaw, Ian, ed. *The Oxford History of Ancient Egypt.* Oxford, U.K.: Oxford University Press, 2003.

Sher, Hanan. "Source of Peace." *Jerusalem Report,* March 13, 2000.

Shiklomanov, I. A., and John C. Rodda, eds. *World Water Resources at the Beginning of the Twenty-first Century.* Cambridge, U.K.: Cambridge University Press, 2004.

Shinn, David. "Preventing a Water War in the Nile Basin." *Diplomatic Courier.* http://www.diplomaticcourier.org.

Shipley, Frederick W. "Agrippa's Building Activities in Rome." *Washington University Studies—New Series* (St. Louis) 4 (1933): 20–25.

Shlaim, Avi. *War and Peace in the Middle East: A Concise History.* Rev. ed. New York: Penguin, 1995.

Simmons, I. G. *Changing the Face of the Earth: Culture, Environment, History.* Oxford, U.K.: Basil Blackwell, 1989.

Simon, Paul, Dr. *Tapped Out: The Coming World Crisis in Water and What We Can Do About It.* New York: Welcome Rain, 2001.

Smith, Adam. *The Wealth of Nations.* 1776. In *The Essential Adam Smith,* edited by Robert L. Heilbroner. New York: W. W. Norton, 1986.

Smith, Henry Nash. *Virgin Land: The American West as Symbol and Myth.* Rev. ed. New York: Vintage, 1970.

Smith, Norman. *A History of Dams.* Secaucus, N.J.: Citadel Press, 1972.

———. *Man and Water.* Great Britain: Charles Scribner's Sons, 1975.

Specter, Michael. "The Last Drop." *New Yorker,* October 23, 2006.

Staccioli, Romolo A. *Acquedotti, Fontane e Terme di Roma Antica: I Grandi Monumenti che Celebrarono il 'Trionfo dell'Acqua' nella Città Più Potente dell'Antichità.* Rome: Newton & Compton Editori, 2002.

Sterling, Eleanor. "Blue Planet Blues." Special issue: Water: The Wellspring of Life. *Natural History,* November 2007.

Suetonius. *The Lives of the Twelve Caesars.* Edited by Joseph Gavose. New York: Modern Library, 1931.

Swanson, Peter. *Water: The Drop of Life.* Foreword by Mikhail Gorbachev. Minnetonka, Minn.: NorthWord Press, 2001.

"Talking Point: Ask Boutros Boutros Ghali." Transcript. *BBC News,* June 10, 2003. http://news.bbc.co.uk/2/hi/talking_point/2951028.stm.

Tann, Jennifer, Dr. ed. *The Selected Papers of Boulton and Watt.* Vol 1, *The Engine Partnership, 1775–1825.* Cambridge, Mass.: MIT Press, 1981.

Temple, Robert. *The Genius of China: 3,000 Years of Science, Discovery and Invention.* Introduction by Joseph Needham. New York: Touchstone, 1989.

Thomas, Hugh. *A History of the World.* New York: Harper & Row, 1979.

Tindall, George Brown. *America: A Narrative History.* Vol. 1. 3rd ed. With David E. Shi. New York: W. W. Norton, 1984.

Toynbee, Arnold J. *A Study of History: Abridgement of Volumes I–VI.* Abridgement by D. C. Somervell. London: Oxford University Press, 1974.

———. *Civilization on Trial* and *The World and the West.* New York: World Publishing, 1971.

Trevelyan, George Macaulay. *A Shortened History of England.* New York: Longmans, Green, 1942.

Turner, Frederick Jackson. *The Frontier in American History.* New York: Harry Holt, 1921.

United Nations Millennium Project Task Force on Water and Sanitation. *Health, Dignity, and Development: What Will It Take?* Final report, abr. ed. Coordinated by Roberto Lenton and Wright Albert. New York: United Nations Millennium Project, 2005. http://unmillenniumproject.org/documents/what–will–it–take.pdf.

Urquhart, Brian. "Disaster: From Suez to Iraq." *New York Review of Books* 54, no. 5 (March 29, 2007).

U.S. Army Corps of Engineers. *The History of the U.S. Army Corps of Engineers.* Alexandria, Va.: U.S. Army Corps of Engineers, 1998. http://140.194.76.129/publications/eng-pamphlets/ep870–1–45/entire.pdf.

U.S. Department of Energy. "World Transit Chokepoints." Report, April 2004. Energy Information Administration. www.eia.doe.gov/emeu/cabs/choke.html.

U.S. Department of the Interior, Bureau of Reclamation, Lower Colorado Region. "Hoover Dam." *Reclamation: Managing Water in the West* (January 2006).

U.S. Geological Survey. "Estimated Use of Water in the United States in 2000: Trends in Water Use, 1950–2000." U.S. Geological Survey. http://pubs.usgs.gov/circ/2004/circ1268/htdocs/text-trends.html.

———. "Water Resources of the United States." U.S. Geological Survey. http://water.usgs.gov/.

Usher, Abbott Payson. *A History of Mechanical Inventions.* Boston: Beacon Press, 1959.

Van De Mieroop, Marc. *A History of the Ancient Near East, Ca. 3000–323 BC.* 2nd ed. Malden, Mass.: Blackwell, 2007.

Ward, Diane Raines. *Water Wars: Drought, Flood, Folly, and the Politics of Thirst.* New York: Riverhead, 2003.

Waterbury, John. *Hydropolitics of the Nile Valley.* Syracuse, N.Y.: Syracuse University Press, 1979.

Waterbury, John, and Dale Whittington. "Playing Chicken on the Nile? The Implications of Microdam Development in the Ethiopian Highlands and Egypt's New Valley Project." *Transformations of Middle Eastern Natural Environments: Legacies and Lessons,* Yale School of Forestry and Environmental Studies Bulletin Series, no. 103 (1998): 150–167. http://environment.research.yale.edu/documents/downloads/0–9/103waterbury.pdf.

Webb, Walter Prescott. *The Great Plains.* Lincoln: University of Nebraska Press, 1981.

Weightman, Gavin. *The Frozen-Water Trade.* New York: Hyperion, 2003.

Weiss, Harvey, and Raymond S. Bradley. "What Drives Societal Collapse?" *Science* 291 (January 26, 2001).

Wells, H. G. *The Outline of History.* Revised by Raymond Postgate and G. P. Wells. Garden City, N.Y.: Doubleday, 1971.

Whitaker, Brian. "One River's Journey through Troubled Times." *Guardian* (U.K.), August 23, 2003.

White, Lynn, Jr. *Medieval Technology and Social Change.* London: Oxford University Press, 1964.

White, Richard. *The Organic Machine: The Remaking of the Columbia River.* New York: Hill & Wang, 1996.

Williams, Trevor I. *A History of Invention: From Stone Axes to Silicon Chips.* Rev. ed. London: Little, Brown, 1999.

Wilshire, Howard G., Jane E. Nielson and Richard W. Hazlett. *The Ameican West at Risk.* New York: Oxford University Press, 2008.

Wilson, Edward O. *The Future of Life.* New York: Alfred A. Knopf, 2002.

Wittfogel, Karl A. *Oriental Despotism: A Comparative Study of Total Power.* New York: Vintage, 1981.

Wolf, Aaron T. "Conflict and Cooperation along International Waterways." *Water Policy* 1, no. 2 (1998): 251–265.

Wood, Gordon S. "The Making of a Disaster." *New York Review of Books* 52, no. 7 (April 28, 2005).

World Bank. "Better Management of Indus Basin Waters." January 2006. http://siteresources.worldbank.org/INTPAKISTAN/Data%20and%20Reference/20805819/Brief-Indus-Basin-Water.pdf.

World Economic Forum in partnership with Cambridge Energy Research Associates. *Thirsty Energy: Water and Energy in the 21st Century.* Geneva, Switzerland: World Economic Forum, 2008.

Worster, Donald. *Rivers of Empire: Water, Aridity, and the Growth of the American West.* New York: Oxford University Press, 1992.

Wright, Rupert. *Take Me to the Source: In Search of Water.* London: Harvill Secker, 2008.

Yergin, Daniel. "Ensuring Energy Security." *Foreign Affairs* 85 (March–April 2006): 69–82.

———. *The Prize: The Epic Quest for Oil, Money, and Power.* New York: Simon & Schuster, 1992.

PHOTOGRAPH CREDITS

The New York Public Library, Astor, Lenox and Tilden Foundations: p. 1 top; p. 2 top; p. 10 top and bottom; **Science Museum/Science and Society Picture Library**: p. 1 middle; p. 3 top left and right; p. 4 bottom right; p. 5 middle; p. 6 top and bottom; p. 8 bottom; **Claudine Macé**: p. 1 bottom; p. 2 bottom; p. 6 middle; p. 7 bottom left, top right; p. 14 bottom; p. 16 top and middle; **Library of Congress Prints and Photographs Division**: p. 3 middle and bottom; p. 5 top and bottom; p. 7 top left, bottom right; p. 8 top right; p. 9 top, bottom right, and left; p. 11 top and middle; p. 12 top, middle, bottom; p. 14 middle; **Wikimedia Commons, GNU Free Documentation License**: p. 4 top, bottom left; **Wikimedia Commons, Public Domain**: p. 8 top left; p. 11 bottom, p. 14 top; **Franklin D. Roosevelt Presidential Library**: p. 13 top; **Herbert Hoover Presidential Library**: p. 13 middle; **U.S. National Archives, Ansel Adams series**: p. 13 bottom; **Reproduced from the original held by the Linda Lear Center for Special Collections and Archives, Connecticut College Photographer**: Edwin Gray: p. 15 top; **Reproduction from Cleveland State University Cleveland Press Collection Photographer**: James Thomas: p. 15 middle; **U.S. Geological Survey**: p. 15 bottom; **NASA, Visible Earth**: p. 16 bottom.

INDEX

Page numbers in italics indicate maps.